INTERNATIONAL SERIES OF MONOGRAPHS IN
NATURAL PHILOSOPHY
GENERAL EDITOR: D. TER HAAR

VOLUME 70

QUANTUM ELECTRODYNAMICS

QUANTUM ELECTRODYNAMICS

BY

IWO BIAŁYNICKI-BIRULA

University of Warsaw and University of Pittsburgh

and

ZOFIA BIAŁYNICKA-BIRULA

Institute of Physics, Polish Academy of Sciences

Translated from the Polish by

EUGENE LEPA

PERGAMON PRESS

OXFORD • NEW YORK • TORONTO
SYDNEY

PWN—POLISH SCIENTIFIC PUBLISHERS

WARSZAWA

Pergamon Press Ltd., Headington Hill Hall, Oxford
Pergamon Press Inc., Maxwell House, Fairview Park, Elmsford, New York 10523
Pergamon of Canada Ltd., 207 Queen's Quay West, Toronto 1
Pergamon Press (Aust.) Pty. Ltd., 19a Boundary Street, Rushcutters Bay, N.S.W. 2011, Australia

---

First English edition 1975

This is a translation from the Polish edition *Elektrodynamika kwantowa*, published by Państwowe Wydawnictwo Naukowe

Library of Congress Cataloging in Publication Data

Białynicki-Birula, Iwo.
Quantum electrodynamics.
(International series of monographs in natural philosophy, v. 70).
Translation of Elektrodynamika kwantowa. 1974
1. Quantum electrodynamics. I. Białynicka-Birula, Zofia, joint author. II. Title.
QC680.B513. 1974 537,6 74-4473
ISBN 0-08-017188-5

Printed in Poland

CONTENTS

Preface xi

Introduction . . . 1

Chapter 1: THE GENERAL PRINCIPLES OF QUANTUM THEORY 5

1. The Postulates of Quantum Theory . . . 5
   The Time Evolution of a System . . . 10
   Causality . . . 16
2. Symmetries . . . 19
3. Canonical Quantization . . . 23

Chapter 2: NON-RELATIVISTIC QUANTUM MECHANICS 26

4. The Quantum Mechanics of a Particle . . . 26
   The Particle Propagator . . . 27
   The Scattering Amplitude . . . 29
   The *S* Matrix . . . 35
5. The Many-Particle System . . . 38
   Spin and Statistics . . . 38
   Scattering in an External Potential . . . 40
   The Occupation-Number Representation . . . 42
   Creation and Annihilation Operators . . . 45
   Field Operators . . . 48
   Scattering and the S Operator . . . 61
   The Theory of Propagators and Feynman Diagrams . . . 71

Chapter 3: THE CLASSICAL THEORY OF THE ELECTROMAGNETIC FIELD 87

6. The Tensor Description of the Electromagnetic Field and the Field Equations . . . 87
   The Field Equations . . . 88
   Transformation Laws . . . 90
   Energy-Momentum Tensor of Electromagnetic Field . . . 91
   The Conservation Laws . . . 92
7. Canonical Formalism for the Electromagnetic Field . . . 95
   The Generalized Poisson Brackets . . . 97

Canonical Transformations and Their Generators 100
Poincaré Transformations as Canonical Transformations 103
8. The Electromagnetic Field with Sources 106
Charged Fluid 107
Charged Particles 110
Magnetic Charges 117
Charged Field 119
9. The Maxwellian Field 123
Solution of the Initial-Value Problem and the Poisson Brackets 123
Fourier Analysis of the Field 127
Conformal Transformations 134
The Maxwellian Field with External Sources 138
The Radiation Field 139
Multipole Radiation 144
Radiation of a Point Particle 151

Chapter 4: THE QUANTUM THEORY OF THE ELECTROMAGNETIC FIELD 156

10. Canonical Quantization of the Electromagnetic Field 156
The Poincaré Group as a Symmetry Group 157
11. Quantization of the Free Maxwellian Field 161
Relativistic Invariance 167
Photons 169
Coherent States 173
Coherence of Electromagnetic Radiation 178
12. The Interaction of the Quantum Electromagnetic Field with External Sources 186
13. The Formulation of the Quantum Theory of the Maxwellian Field with the Aid of Potentials 195
The Classical Theory 197
The Quantum Theory of the Free Field 198
The Quantum Theory of the Electromagnetic Field with External Sources 201
The Poincaré Group as the Symmetry Group of Amplitudes . . 205
A Simple Representation of Transition Amplitudes 207
Induced Processes in Intense Photon Beams 212
Many-Photon Propagators 213
14. The Proca Theory 215
The Classical Theory of a Vector Field with Mass 215
The Quantum Theory of a Vector Field with Mass 219
The S Operator in the Presence of External Sources 222
15. The Infrared Catastrophe 224

Chapter 5: RELATIVISTIC QUANTUM MECHANICS OF ELECTRONS 230

16. The Dirac Equation and the Symmetries of Its Solutions 230
    The Properties of Solutions of the Dirac Equation without Potential 241
17. Electron Scattering in the Electromagnetic Field 247
    Electron Wave Functions 260
    The Green's Functions for the Dirac Equation 265
    The Electron in a Static Electromagnetic Field 273
    The Properties of the *S* Matrix for an Electron in an External Field 277
    The Indistinguishability of Particles and the Vacuum-to-Vacuum Transition Amplitude 286
18. Electron Field Operators 293
    Creation and Annihilation Operators 293
    Field Operators 296
    Electrons in an External Electromagnetic Field 300
    The S Operator 302
    Propagators 307

Chapter 6: THE FORMULATION OF QUANTUM ELECTRODYNAMICS 311

19. The General Postulates of Quantum Electrodynamics 311
    The Fundamental Dynamical Postulate 314
20. Perturbation Theory and Feynman Diagrams 316
    The General Principles of Constructing Feynman Diagrams. 318
    Analysis of the Connectedness of Propagators 327
21. Photon Processes 339
    The Relation between Transition Amplitudes and Propagators 340
    The Källén–Lehmann Representation of the Photon Propagator 347
    Renormalization of an External Current 351
22. Electron-Photon Processes 352
    Compensating Current 352
    Electron-Photon Propagators 355
    The Relation between the Compensating Current and the Gauge 356
    Feynman Diagrams 359
    The Relation between Propagators and Transition Amplitudes. 368
    Feynman Diagrams in Momentum Representation 374

Chapter 7: RENORMALIZATION THEORY 382

23. The Necessity for Renormalization 382
    The Electron Propagator 382
    The Photon Propagator 393

The Vertex Function 396
Charge Renormalization 399
24. Equations for Renormalized Propagators 402
Set of Equations for Propagators 403
Elimination of the Field $\mathscr{A}_\mu(z)$ 405
Gauge Invariance and the Ward Identity 410
Symmetric Expressions for the Electron and Photon Propagators and the Vertex Function 413
The Skeleton Structure of Diagrams 419
Renormalization 423
25. Renormalized Perturbation Theory 426
The General Properties of Renormalized Perturbation Theory . 426
Renormalized Transition Probabilities 430
Independence of Transition Probabilities from the Compensating Current 436
The Infrared Catastrophe 437

Chapter 8: APPLICATIONS OF QUANTUM ELECTRODYNAMICS 439

26. Two-Particle Collisions 439
The General Formulae for Two-Particle Processes 439
Negaton-Negaton Scattering 445
Photon-Negaton Scattering 448
Photon-Photon Scattering 451
27. Non-linear Effects in Quantum Electrodynamics 452
28. The Electron in a Static Electromagnetic Field 455
The Effective Field and the Polarization of the Vacuum . . . 455
The Motion of the Electron in a Static Field 459
The Magnetic Moment of the Electron 461
The Lamb–Retherford Shift 464
29. The Limits of Applicability of Quantum Electrodynamics . . . 466
Concluding Remarks 470
Appendices 473
Appendix A: Hilbert Space 473
Linear Operators 473
Appendix B: Chronological and Normal Products 477
The Chronological Product 477
The Normal Product 480
Appendix C: Functional Differentiation 482
Appendix D: The Poincaré Group 485
Appendix E: Green's Functions 491
The Schrödinger Equation 492
The Klein–Gordon and d'Alembert Equations 493
The Dirac and Proca Equations 497

Appendix F: The Symmetric Energy-Momentum Tensor 499
Appendix G: Evaluation of Some Poisson Brackets 502
Appendix H: Some Operator Identities 504
Appendix I: Spinors 508
Appendix J: The Properties of Solutions of the Dirac Equations 516
Appendix K: Regularization 519
Appendix L: Methods of Calculating Integrals over Momentum Space 522
Appendix M: Representation of the *S* Matrix as a Double Limit of the Propagator . . . 525
REFERENCES 527
Articles . . . 527
Textbooks and Monographs . . . 533
INDEX OF SYMBOLS . . . 535
SUBJECT INDEX . . . 541
OTHER TITLES IN THE SERIES IN NATURAL PHILOSOPHY . . . 549

PREFACE

SHOULD we today, in the fashion of the Greek sages, wish to construct the world out of elements, we would name two of them without hesitation—electricity and magnetism. For electromagnetic interactions are a source of forces in a vast number of physical systems, and accordingly no doubt deserve to be so singled out. The quantum theory of these interactions, called *quantum electrodynamics*, underlies the foundations of most modern areas of physics. The whole of optics and electrodynamics, atomic and molecular physics, solid-state physics, the physics of fluids, gases, and plasmas are all, actually speaking, special applications of quantum electrodynamics to selected physical systems. In the physics of the atomic nucleus and elementary particles, strong and weak interactions do play a primary role, it is true, but electromagnetic interactions here, too, occupy a distinguished place as the sole and universal "measuring instrument". Man and all instruments "see and feel" by means of electricity and magnetism. Thus far, there has been no other method of accelerating, sorting, and recording particles, apart from methods based on electromagnetic interactions.

Quantum electrodynamics concerns itself first and foremost with the mutual interaction of electrons and photons and their interaction with given electromagnetic fields and given currents. Since nuclear (strong) interactions do not come within its compass, atomic nuclei in quantum electrodynamics are treated as indivisible carriers of electric charge and magnetic moment. Such replacement of atomic nuclei by given sources of electromagnetic fields is a good approximation in atomic physics.

The description of the electromagnetic interactions of muons in an approximation in which we disregard weak interactions does not differ from the quantum electrodynamics of electrons.

Quantum electrodynamics also concerns itself with the description of the electromagnetic properties of other elementary particles. This is hindered, however, by the lack of a complete theory of the other interactions.

The mathematical structure of quantum electrodynamics, as of every fundamental theory, is highly complex. It would be unreasonable, therefore, to apply the vast apparatus of complete quantum electrodynamics to every physical system. In most cases, this would make it altogether impossible to obtain results. For this reason, most areas of theoretical physics which arise out of quantum electrodynamics in practice develop on the basis of their own, usually phenomenological, models of the physical systems considered. It is our belief that each such model can be justified on the basis of quantum electrodynamics, but showing these connections would go beyond the bounds of our book. In it we shall confine ourselves merely to giving the formulation of quantum electrodynamics in its most general and its most abstract form: relativistic quantum field theory. For pedagogical reasons, we precede this general formulation with several introductory chapters, in which step by step we shall introduce the concepts and methods of this difficult theory. Such a gradual, inductive introduction of quantum electrodynamics is also advisable for another reason. In contrast to classical mechanics, quantum mechanics, thermodynamics, or classical electrodynamics, in quantum electrodynamics we have a theory that is not complete. Owing to the enormous mathematical difficulties, it is not yet known whether the set of postulates adopted is not inconsistent and whether it determines the theory uniquely. All we know is an approximate scheme for carrying out calculations based on perturbation theory. The results of these calculations are in amazing agreement with experimental results but the problem of the convergence of the series obtained has not been solved as yet. Moreover, the results of investigations into the mathematical structure of the theory indicate that perturbation series are divergent asymptotic series.

Quantum electrodynamics thus constitutes a programme rather than a closed theory. As we try to show in the early chapters of this book, this programme rests on two pillars: on the theory of the quantum Maxwellian field interacting with given (external) classical sources of radiation and on the relativistic quantum mechanics of electrons interacting with a given (external) classical electromagnetic field. Both of these theories are closed and relatively straightforward, mathematically speaking. It is not until the complete relativistic theory of the mutual interactions of electrons and photons, which is a synthesis of these two theories, is being formulated that insurmountable difficulties arise. As a consequence of these difficulties, numerous monographs and papers have presented many diverse formulations of quantum electrodynamics. These formulations in the final account lead to the same results, but they are by no means equivalent in the mathematical respect. In our book we give only one formulation of quantum electrodynamics; others may be found in the books of Akhiezer and Berestetskii, Bjorken and Drell, Bogolyubov and Shirkov, Jauch and Rohrlich, Heitler, and others. None of the existing formulations of quantum electrodynamics is satisfactory from the mathematical point of view. In this theory we must on many occasions renounce mathematical rigour and resort to heuristic considerations, which are frequently justified solely by the agreement of their results with experiment.

The very formulation of the theory thus is a controversial matter, one to which we devote much more space in our book than we do to concrete applications of the theory. For it is our conviction that the computational scheme of renormalized perturbation theory is quite simple and upon mastering this scheme, the reader will be able to follow the calculations given in the monographic and original literature. For this reason, we have confined applications to a minimum and little space is devoted to them.

Our book is addressed to readers familiar with quantum mechanics and classical electrodynamics at the level of university courses. The necessary material from the domain of quantum mechanics is given, for instance, in Chapters 1–20 of the textbook by E. Merzbacher, and that from classical electrodynamics in Chapters 1–12 of the textbook by J. D. Jackson.

The scheme on p. xiv illustrates the logical interconnections between the various chapters of the book.

Chapter 1 has been given over to the fundamental principles of quantum theory formulated in a general, abstract fashion. Chapters 2–5 are easier to grasp than the others and may be read independently of the rest of the book since they constitute a self-contained whole made up of two independent parts. The part comprising Chapters 3 and 4 contains primarily the theory of the electromagnetic field interacting with given sources of radiation. This theory is applicable to the study

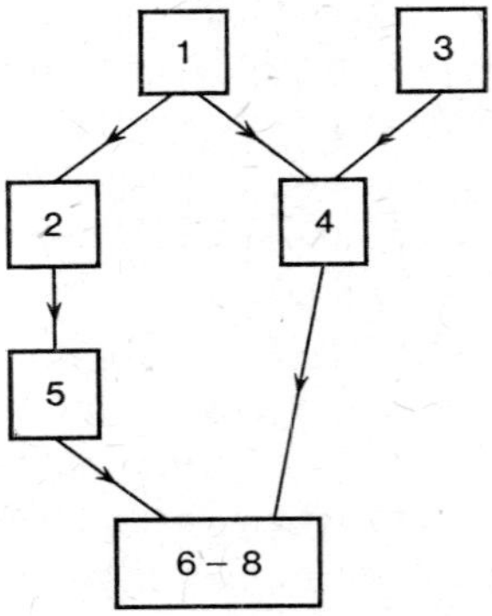

of systems in which the influence of the radiation on the sources may be neglected. An example of such a system is black-body radiation. If the reader wishes to confine himself to only this part of the book, he may omit the final three sections (Sections 13–15).

The part composed of Chapters 2 and 5 concerns the quantum mechanics of particles. Chapter 5 expounds the relativistic theory of mutually non-interacting electrons, moving in a given electromagnetic field. It may be applied, for instance, when one considers the motion of electrons in the field of a nucleus.

Small print has been used in the text for those parts of the material which give the justification for assertions made or contain ancillary material going beyond the logical mainstream of the exposition. The references are given at the end of the book, the listing being made with the authors in alphabetical order. In view of the large number of notions and symbols used, we felt it necessary to give a detailed subject index and an index of the principal symbols. The subject index indicates the page on which the given concept is defined or explained, but the pages where further reference to it is made are not listed. It frequently happens that in various parts of the text, the same term denotes different, though related, concepts. In such cases, several page numbers are given. Most terms listed in the index are italicized in the text. In the index of symbols *bold-face* print is used for the number of the formula which defines the given symbol explicitly or implicitly. Furthermore, we sometimes give the numbers of other formulae in which the given quantity plays a cardinal role.

Footnotes referring to the history of quantum electrodynamics are marked by sharps (#) whereas footnotes explaining the text are indicated by †, ‡, §, etc.

In the course of our work on the first and second editions of this book, our colleagues and co-workers made many suggestions and criticisms. We are indebted to them for pointing out errors and mistakes and for proposing several improvements.

This text has come into being in the creative scientific atmosphere in the Warsaw School of Theoretical Physics, created by the late Professor Leopold Infeld, to whose memory we dedicate this book.

INTRODUCTION

IN THE introduction we shall discuss the system of units and the notation used throughout the text.

Units

We shall use a system of units based on Planck's constant $\hbar$, the velocity of light c, and an arbitrarily chosen unit of length l (most frequently it will be 1 cm, though sometimes it will be 1 fermi, i.e. 10^{-13} cm). Table 1 lists the dimensions of the physical quantities which occur most frequently, in $\hbar cl$ units.

The strengths of electromagnetic fields and charges are defined so that the factor 4π appears in the formula expressing Coulomb's law, but does not figure in Maxwell's equations, just as in the choice of the Heaviside–Lorentz electromagnetic units. To simplify the notation,

TABLE 1

Physical quantity	$\hbar cl$
Length	l
Time	lc^{-1}
Velocity	c
Mass	$\hbar c^{-1}l^{-1}$
Momentum	$\hbar l^{-1}$
Angular momentum	$\hbar$
Energy	$\hbar cl^{-1}$
Action	$\hbar$
Electric charge	$\hbar^{\frac{1}{2}} c^{\frac{1}{2}}$
Strength of electromagnetic field	$\hbar^{\frac{1}{2}} c^{\frac{1}{2}} l^{-2}$
Wave function	$l^{-\frac{3}{2}}$

we shall omit all $\hbar$'s and c 's in the formulae. In order to read the value of a physical quantity given in this abbreviated notation, first complement its dimensions by writing in the missing powers of $\hbar$ and c in accordance with Table 1. For instance, an energy of $(10^{-13}\text{ cm})^{-1}$ should be read as $\hbar c(10^{-13}\text{ cm})^{-1} = 200$ MeV.

TABLE 2

Velocity of light	$c = 2.99792 \times 10^{10}\text{ cm}\cdot\text{sec}^{-1}$
Planck's constant	$\hbar = 6.5822 \times 10^{-22}\text{ MeV}\cdot\text{sec} = 1.0546 \times 10^{-27}\text{ erg}\cdot\text{sec}$
Fine-structure constant	$\alpha = e^2/4\pi\hbar c,\ 1/\alpha = 137.036$
Electron mass	$m = \hbar/\lambda_c c = 0.51100\text{ MeV}/c^2$
Compton wavelength of electron	$\lambda_c = \hbar/mc = 3.8616 \times 10^{-11}\text{ cm}$
Classical radius of electron	$r_0 = e^2/4\pi mc^2 = 2.818 \times 10^{-13}\text{ cm}$

Table 2 lists the relations between $\hbar cl$ units and other units used in quantum electrodynamics.

Vectors, Tensors, and Spinors

The components of various geometrical objects will be labelled with various types of indices. Lower-case Roman letters from the middle of the alphabet—i, j, k, l, m, n—label the components of vectors and tensors in three-dimensional space.

Lower-case Greek letters label the components of vectors and tensors in four-dimensional space-time.

Lower-case Roman letters from the beginning of the alphabet—a, b, c, d, e, f—label the components of bispinors.

Upper-case Roman letters label the components of spinors.

Vectors are denoted by bold-face Roman letters in three-dimensional space, and by ordinary Roman letters in four-dimensional space-time. In Hilbert space, upper-case Greek letters are used for vectors.

The scalar product of three- and four-dimensional vectors is indicated by a dot. For example:

$$\mathbf{k}\cdot\mathbf{x} = k_1 x_1 + k_2 x_2 + k_3 x_3,$$

$$k\cdot x = k_0 x_0 - k_1 x_1 - k_2 x_2 - k_3 x_3.$$

The vector product is denoted by the symbol $\times$. The scalar product of the vectors Ψ and Φ in Hilbert space is written as the symbol $(\Psi|\Phi)$.

The summation convention holds for all components of geometrical objects. This means that the summation is to be performed over repeated indices. For instance:

$$
\begin{aligned}
k_i p_i &= k_1 p_1 + k_2 p_2 + k_3 p_3, \\
k_\mu x^\mu &= k_0 x^0 + k_1 x^1 + k_2 x^2 + k_3 x^3, \\
\bar{\psi}^a \psi_a &= \bar{\psi}^1 \psi_1 + \bar{\psi}^2 \psi_2 + \bar{\psi}^3 \psi_3 + \bar{\psi}^4 \psi_4, \\
\varepsilon^{AB} \psi_B &= \varepsilon^{A1} \psi_1 + \varepsilon^{A2} \psi_2 .
\end{aligned}
$$

Completely antisymmetric tensors, with coordinates equal to 0 and ± 1, in three-dimensional space and four-dimensional space-time are denoted by, respectively, ε_{ijk}, $\varepsilon^{\mu\nu\lambda\varrho}$, and $\varepsilon_{\mu\nu\lambda\varrho}$. We adopt the following convention:

$$
\begin{aligned}
\varepsilon_{123} &= 1, \\
\varepsilon^{0123} &= 1 = -\varepsilon_{0123} .
\end{aligned}
$$

The tensors $\varepsilon^{\mu\nu\lambda\varrho}$ and $\varepsilon_{\mu\nu\lambda\varrho}$ can be used to define $\check{f}^{\mu\nu}$ and $\check{h}_{\mu\nu}$, the duals of arbitrary antisymmetric tensors $f_{\mu\nu}$ and $h^{\mu\nu}$,

$$
\begin{aligned}
\check{f}^{\mu\nu} &\equiv \tfrac{1}{2} \varepsilon^{\mu\nu\lambda\varrho} f_{\lambda\rho}, \\
\check{h}_{\mu\nu} &\equiv \tfrac{1}{2} \varepsilon_{\mu\nu\lambda\varrho} h^{\lambda\rho} .
\end{aligned}
$$

Symmetrization and antisymmetrization are denoted (together with a factor of 2) by brackets:

$$
\begin{aligned}
A_{(\mu\nu)} &\equiv A_{\mu\nu} + A_{\nu\mu}, \\
A_{[\mu\nu]} &\equiv A_{\mu\nu} - A_{\nu\mu} .
\end{aligned}
$$

Fourier Transforms

Fourier transforms of a function will be denoted by the same symbol as the function but with a tilde over it. We shall adopt the following convention concerning the factors 2π and the signs in the exponent:

$$
\begin{aligned}
f(\mathbf{x}) &= \int \frac{d^3p}{(2\pi)^3} e^{i\mathbf{p}\cdot\mathbf{x}} \tilde{f}(\mathbf{p}), \\
\tilde{f}(\mathbf{p}) &= \int d^3x \, e^{-i\mathbf{p}\cdot\mathbf{x}} f(\mathbf{x}),
\end{aligned}
$$

$$f(p) = \int \frac{d^4p}{(2\pi)^4} e^{-ip\cdot x}\tilde{f}(p),$$

$$\tilde{f}(p) = \int d^4x e^{ip\cdot x} f(x).$$

Matrices

A matrix which is the transpose of a given matrix is denoted by the same symbol but with T as a superscript. The Hermitian conjugate is labelled with a dagger, †, and the conjugate (all elements replaced by the conjugate numbers) is indicated by an asterisk, *.

Functionals

Functionals (as opposed to functions) are distinguished by putting the argument of the functional in brackets (whereas that of the function is put in parentheses). To simplify the notation, arguments of functions and arguments of functionals which appear at the same time are written in brackets and separated by a vertical line.

The Functions $\theta(x)$ and $\varepsilon(x)$

The step function $\theta(x)$ and the sign function $\varepsilon(x)$ are treated as distributions which do not possess values at the point $x = 0$. When they are so understood, these functions may be differentiated, yielding the Dirac function $\delta(x)$ and $2\delta(x)$, respectively.

Commutators and Anticommutators

Commutators and anticommutators are denoted in the following manner,

$$[A, B]_- \equiv AB - BA,$$

$$[A, B]_+ \equiv AB + BA.$$

The minus sign subscript at the bracket indicating a commutator will frequently be omitted, wherever this does not result in confusion.

In order to make the notation in this book similar to that in general use in the literature, symbols containing abbreviations are employed to denote many physical quantities. For example, the abbreviations IN and OUT from *incoming* and *outgoing*, RAD from *radiation*, RET and ADV from *retarded* and *advanced*, INT from *interaction*, EXT from *external*, IL from *inhomogeneous Lorentz*, and REN from *renormalized*.

CHAPTER 1

THE GENERAL PRINCIPLES OF QUANTUM THEORY

THE aim of this chapter is to discuss the general principles underlying all quantum theories. We also included in this chapter and in the appendices to it a brief description of mathematical concepts which are used in quantum theories.

1. THE POSTULATES OF QUANTUM THEORY

A set of quantum theory postulates are presented and briefly discussed in this section. This will be a systematic summary, presented in an abstract form, of information given in quantum mechanics courses. This general scheme of quantum theory possesses various concrete realizations.

A particular system is the object under study in quantum theory. The state of the system and elementary questions concerning that state may be adopted as the fundamental concepts. The mathematical model in which all the properties of the system are represented is the Hilbert space over the field of complex numbers and the linear operators acting in that space.† Hilbert space is a vector space in which the scalar product is defined complete with respect to the metric induced by this scalar product. We shall assume that the Hilbert space is separable, that is to say, has a countable basis, since non-separable Hilbert spaces have hitherto found only marginal applications in quantum theories. Let us adopt an operational definition of the state of the system, identifying the notion of the state of the system with the body of information available about the system. Each question pertaining to the system is expressible in terms of elementary questions with only YES or NO

† The properties of Hilbert space are discussed briefly in Appendix A.

for answers. Quantum theory is a probabilistic theory. It enables us to determine only the probability of an affirmative answer to elementary questions.

All the information that we have about physical systems is obtained by using macroscopic instruments. Both the states and the questions thus are always labelled with sets of parameters of a macroscopic nature.

Underlying quantum theory are the following four postulates:

Postulate I

To each elementary question $\mathscr{P}$ in quantum theory there corresponds a projection operator P,

$$P^{\dagger} = P, \tag{1a}$$

$$P^2 = P. \tag{1b}$$

At the same time, the following conditions are satisfied:

1. The operator 1 projecting onto all Hilbert space represents the question to which the answer is always YES.

2. The operator $1-P$ represents a negation of the question $\mathscr{P}$.

3. If an affirmative answer to the question $\mathscr{P}_1$ always leads to an affirmative answer to the question $\mathscr{P}_2$, the subspace onto which P_1 projects is contained in the subspace onto which P_2 projects. The question $\mathscr{P}_1$ is then said to be more restrictive than the question $\mathscr{P}_2$.

4. The question obtained by combining the questions $\mathscr{P}_1$ and $\mathscr{P}_2$ by means of the conjunction AND is represented by the operator $P_1 P_2$, provided that this product is a projection operator.

5. The question obtained by combining the questions $\mathscr{P}_1$ and $\mathscr{P}_2$ by means of the conjunction OR is represented by the operator P_1+P_2, provided that this sum is a projection operator.

Postulate II

To each state of a system there corresponds in quantum theory a self-adjoint, non-negative operator ϱ with trace equal to unity, known as the *density operator*.

$$\varrho^{\dagger} = \varrho, \tag{2a}$$

$$(\Psi|\varrho\Psi) \geqslant 0, \tag{2b}$$

$$\mathrm{Tr}\{\varrho\} = 1. \tag{2c}$$

Postulate III

The number p,

$$p = \text{Tr}\{\varrho P\}, \tag{3}$$

is the probability of getting an affirmative answer to the question $\mathscr{P}$ when the system is in the state described by ϱ.

Postulate IV

To each dynamical variable $\mathscr{A}$ there corresponds a spectral family of projection operators $E_\lambda^{(A)}$. The projection operator $E_\lambda^{(A)}$ represents the question: Is the value of the dynamical variable $\mathscr{A}$ not greater than λ?

Since the set of probabilities p is the only information available in quantum theory about the state of the system, from the operational point of view the concept of state of the system should be identified with the function $p(\mathscr{P})$ defined on the set of all questions. The correspondence between the states of the system and the density operators is one-to-one only when the set of questions is so rich that the equality

$$\text{Tr}\{\varrho_1 P\} = \text{Tr}\{\varrho_2 P\}$$

for all questions $\mathscr{P}$ implies the equality of the density operators.† This assumption excludes from the considerations, among other things, systems with undetermined total electric charge. Because of the formal convenience, physical supersystems consisting of parts possessing a total charge of 0, e, $2e$, $3e$, etc. are considered in quantum field theory. Each subsystem in such a supersystem is subject to quantum laws, but the composition of these systems is classical in character. The only questions allowed concerning the entire system are those which can be formulated independently for each subsystem.

Among projection operators, a distinguished role is played by those which project onto one-dimensional subspaces of Hilbert space, i.e. onto the directions in these spaces. Each such operator can represent not only particular elementary questions, but also a particular state, since it satisfies conditions (2a) to (2c). Such a state is called a *pure*

† This is the essence of Gleason's theorem, which the reader will find in the book by Jauch.

state. One which is not pure is referred to as being *mixed.* The density operator corresponding to this state is not a projection operator. Corresponding to each pure state is a direction in Hilbert space (the so-called ray), that is, the set of all vectors differing only by a numerical factor. A pure state thus also lends itself to representation by an arbitrary vector Ψ belonging to this one-dimensional space. For convenience, a normalized vector is customarily chosen. Then the arbitrariness of the choice of the vector Ψ representing a pure state is limited to arbitrariness in the choice of the phase factor $e^{i\varphi}$. A projection operator representing a pure state is denoted† by the symbol P_Ψ. Briefly, the system is said to be in the state P_Ψ or simply in the state Ψ.

One can deduce from Postulate III which question, $\mathscr{P}_\Psi$, is represented by the operator P_Ψ. The probability of getting an affirmative answer to the question $\mathscr{P}_\Psi$, when the system is in a state described by P_Ψ, is equal to unity,

$$p = \mathrm{Tr}\{P_\Psi P_\Psi\} = \mathrm{Tr}\{P_\Psi\} = 1.$$

Since no questions are more restrictive than those represented by operators projecting onto one-dimensional subspaces (cf. Postulate I, point 3), the question $\mathscr{P}_\Psi$ must be read: "Is the system in the state P_Ψ?"

Now we shall consider two directions in Hilbert space which are specified by two normalized vectors Ψ and Φ. The probability of getting an affirmative answer to the question "Is the system in the state P_Ψ?", when that system is in the state P_Φ, is given by the formula

$$p_{\Psi\Phi} = \mathrm{Tr}\{P_\Phi P_\Psi\} = |(\Psi|\Phi)|^2 = p_{\Phi\Psi}.$$

In view of the interpretation of this formula, the scalar product of the normalized vectors $(\Psi|\Phi)$ is called the *amplitude of the probability* that a system in the state Ψ is in the state Φ.

Every pair of two density operators ϱ_1 and ϱ_2, describing the states of a given system may be used to construct a one-parameter family of density operators $\varrho_{12}(\lambda)$ defined by the formula

$$\varrho_{12}(\lambda) = \lambda\varrho_1 + (1-\lambda)\varrho_2, \quad \text{where} \quad 0 \leqslant \lambda \leqslant 1.$$

† The action of the operator P_Ψ on any vector Φ of Hilbert space is expressed by the relation $P_\Psi\Phi = (\Psi|\Phi)\Psi$, for $||\Psi|| = 1$.

The operation of forming new density operators by adding ϱ_1 and ϱ_2 together is interpreted as the statistical mixing of states described by ϱ_1 and ϱ_2 in the proportions λ and $(1-\lambda)$, respectively. The operation of mixing states has the same meaning in quantum theory as in classical statistical physics.

Mixing pure states in this manner yields all possible mixed states Each density operator ϱ may be written as

$$\varrho = \sum_i p_i P_{\Psi_i}, \tag{4}$$

where P_{Ψ_i} are projection operators onto the orthogonal eigen-directions of the operator ϱ. The sum in this formula may contain a finite or infinite number of terms. The positive coefficients p_i determine the probability of an affirmative answer to the elementary questions $\mathscr{P}_{\Psi_i}$ concerning a system in the state ϱ. If a density operator ϱ is a linear combination of at least two projection operators, it describes a mixed state.

The states of a system constitute a convex set whose interior consists of mixed states and whose boundary consists of pure states.

It follows from Postulate II and the foregoing discussion that the propositional calculus in quantum theory differs from that of classical logic. It is frequently said that quantum logic reigns in quantum theory. For instance, not all questions can be joined by the conjunctions AND and OR. In the first case, this can be done only when the pertinent projection operators commute, whereas in the second case a stronger condition must be met:

$$P_1 \cdot P_2 = 0 \text{ (cf. Postulate I)}.$$

Questions represented by commuting operators are said to be compatible. The measurements which are to yield an answer to these questions are independent of each other.

In expositions of quantum mechanics it is most frequently assumed that to each dynamical quantity $\mathscr{A}$ there corresponds a self-adjoint operator† A. Such procedure is consistent with Postulate IV given

† The term *self-adjoint* will be used in this book too extravagantly from the point of view of mathematical rigour. For we shall apply the term self-adjoint not only to those operators for which an appropriate proof exists, but also to those

above, for exactly one spectral family of projection operators is associated with each self-adjoint operator, and knowing the spectral family of projection operators, we can define the self-adjoint operator by the formula

$$A = \int_{-\infty}^{+\infty} \lambda \mathrm{d}E_\lambda^{(A)}. \tag{5}$$

Inasmuch as discontinuities may occur in $E_\lambda^{(A)}$, this integral must be taken to be a Stieltjes integral (cf. Appendix A). It follows from Postulate III and formula (5) that the average value $\langle \mathscr{A} \rangle$ of the dynamical quantity $\mathscr{A}$ in the state described by ϱ is given by

$$\langle \mathscr{A} \rangle = \mathrm{Tr}\{\varrho A\}. \tag{6}$$

THE TIME EVOLUTION OF A SYSTEM

Now we proceed with a quantum theoretical description of the evolution of a system in time. For this description we shall use one-parameter families of questions $\mathscr{P}(t)$, where $-\infty < t < \infty$. Each such family comprises all those questions which differ only in that they refer to different instants of time. Such a family may be said to be formed by one question being moved through time.

To describe the time evolution of the system we introduce a family of unitary *evolution operators* $\mathsf{U}(t, t_0)$. The evolution operators satisfy the following composition law:

$$\mathsf{U}(t, t_1)\mathsf{U}(t_1, t_0) = \mathsf{U}(t, t_0). \tag{7}$$

The generator of these transformations will be defined by

$$H(t) \equiv -i\hbar \frac{\partial \mathsf{U}(t, t_0)}{\partial t} \mathsf{U}^\dagger(t, t_0). \tag{8}$$

The composition law (7) implies that the operator $H(t)$ does not depend on t_0. In the general case, however, $H(t)$ does depend on t. The defini-

symmetrical (formally self-adjoint) operators which should possess this property on account of their physical interpretation. The same is true of the terms *adjoint* and *unitary* which we shall apply also to formally adjoint and formally unitary operators without determining whether the domains and ranges of these operators satisfy the appropriate conditions.

tion of $H(t)$ may be rewritten as a differential equation known as the *evolution equation*,

$$-i\hbar\frac{\partial \mathsf{U}(t,\, t_0)}{\partial t} = H(t)\mathsf{U}(t,\, t_0). \tag{9}$$

If the one-parameter family of generators $H(t)$ is known, the evolution operator $\mathsf{U}(t, t_0)$ may in principle† be found as the solution of the differential equation satisfying the initial condition

$$\mathsf{U}(t_0, t_0) = 1.$$

From unitarity and from the composition law it follows that $\mathsf{U}^\dagger(t, t_0) = \mathsf{U}(t_0, t)$. From this we obtain another equivalent form of the evolution equation (9):

$$i\hbar\frac{\partial \mathsf{U}(t,\, t_0)}{\partial t_0} = \mathsf{U}(t,\, t_0)H(t_0). \tag{9a}$$

The postulate concerning the time evolution of the system gives the relationship between projection operators representing the same questions but referring to the instants t and t_0.

Postulate V

The evolution of a system in time is described by a family of evolution operators which is generated by the *energy operator* (*Hamiltonian*) $H(t)$. Projection operators representing the same question referred to the instants t and t_0 are related by the formula

$$P(t) = \mathsf{U}(t,\, t_0)P(t_0)\mathsf{U}(t_0,\, t). \tag{10}$$

As the projection operators representing questions change in time, analogous changes take place in the spectral families, and together with them in the operators representing physical quantities

$$A(t) = \mathsf{U}(t,\, t_0)A(t_0)\mathsf{U}(t_0,\, t). \tag{11}$$

Such a description of the time evolution of a system, one in which variations in the results of measurements are associated with changes in the projection operators representing questions and physical quantities, is known as the *Heisenberg picture*.

---

† The formal solution of this equation is given in Appendix B.

In the Heisenberg picture, operators representing physical quantities satisfy the Heisenberg equations of motion

$$\frac{\mathrm{d}A(t)}{\mathrm{d}t} = \frac{1}{i\hbar}[A(t), H(t)]. \tag{12}$$

If the operator $A(t)$ also depends parametrically on time, as is sometimes the case, eq. (12) undergoes the modification

$$\frac{\mathrm{d}A(t)}{\mathrm{d}t} = \frac{\partial A(t)}{\partial t} + \frac{1}{i\hbar}[A(t), H(t)]. \tag{13}$$

Since the trace is invariant under cyclic permutations of the operators, the probability of getting an affirmative answer to the question $\mathscr{P}(t)$ may also be written as

$$p(t) = \mathrm{Tr}\{\varrho_S(t, t_0)P(t_0)\}, \tag{14}$$

where

$$\varrho_S(t, t_0) \equiv \mathsf{U}(t_0, t)\varrho\mathsf{U}(t, t_0). \tag{15}$$

Thus, the probability $p(t)$ may be calculated for any instant by using projection operators $P(t_0)$ referred to a particular fixed instant t_0, on condition that the time-dependent operator $\varrho_S(t, t_0)$ is substituted for the density operator ϱ.

Such a description of time evolution in which variations in the results of measurements in time are associated with changes in the density operator, i.e. are attributed as it were to changes in the state of the system, is called the *Schrödinger picture.*

To simplify the notation, as other authors do, we shall henceforth assume that $t_0 = 0$, that is to say that the Heisenberg and Schrödinger pictures are identical at the instant $t = 0$.

In the Schrödinger picture the density operator satisfies the Liouville–von Neumann equation

$$i\hbar\frac{\partial\varrho_S(t)}{\partial t} = [H_S(t), \varrho_S(t)], \tag{16}$$

where the Hamiltonian $H_S(t)$ in the Schrödinger picture is defined by

$$H_S(t) \equiv \mathsf{U}(0, t)H(t)\mathsf{U}(t, 0). \tag{17}$$

Equation (16) leads to the Schrödinger equation for vectors $\Psi_S(t)$ representing pure states in the Schrödinger picture,

$$i\hbar \frac{\partial \Psi_S(t)}{\partial t} = H_S(t)\Psi_S(t). \tag{18}$$

Time evolution in the Schrödinger picture is conveniently described by using unitary operators $\mathsf{U}_S(t, t_0)$ which are defined as the solution of the differential equation†

$$i\hbar \frac{\partial}{\partial t} \mathsf{U}_S(t, t_0) = H_S(t)\mathsf{U}_S(t, t_0) \tag{19}$$

satisfying the initial condition

$$\mathsf{U}_S(t_0, t_0) = 1.$$

Equations (9), (17) and (19) imply that the evolution operators in the Heisenberg and Schrödinger pictures are related by

$$\mathsf{U}_S(t, t_0) = \mathsf{U}(0, t)\mathsf{U}(t_0, 0). \tag{20}$$

From the fact that the operator $H_S(t)$ is self-adjoint it follows that the operators $\mathsf{U}_S(t, t_0)$ are unitary and subject to the composition law.

A physical system is said to be *conservative* when the Hamiltonian (energy of the system) does not depend on time

$$H(t) = H.$$

The description of the time evolution of a conservative system is especially straightforward and we shall devote further discussion to such systems. The time evolution operator of a system is simply an exponential function of the energy operator. Moreover, the energy operators in the Heisenberg and Schrödinger pictures are equal to each other,

$$H_S = H.$$

† Note the difference in sign between the equations for $\mathsf{U}(t, t_0)$ and for $\mathsf{U}_S(t, t_0)$. Depending on what picture is adopted, the time evolution generator may be the Hamiltonian taken with either a plus sign or a minus sign. In general discussions concerning evolution operators the sign of the generator is customarily taken to be as in eq. (19).

Accordingly, a simple relation holds

$$\mathsf{U}(t, t_0) = \exp\left[\frac{i}{\hbar} H(t-t_0)\right] = \mathsf{U}_S(t_0, t),$$

which implies the following simple forms of formulae (11) and (15),

$$A(t) = \exp\left(\frac{i}{\hbar} Ht\right) A(0) \exp\left(-\frac{i}{h} Ht\right),$$

$$\varrho_S(t) = \exp\left(-\frac{i}{\hbar} Ht\right) \varrho \exp\left(\frac{i}{h} Ht\right).$$

In addition to the Heisenberg and Schrödinger pictures, we also introduce the *Dirac picture* in which projection operators and density operators alike change in time. The change in projection operators is governed in this picture† by the operators $\exp\left(\frac{i}{\hbar} H_0 t\right)$ describing a simple, easily determined time evolution, most frequently the free evolution of the system.

The change in the density operators is governed in the Dirac picture by the products of the unitary operators

$$\exp\left(\frac{i}{\hbar} H_0 t\right) \exp\left(-\frac{i}{\hbar} Ht\right).$$

These operators may be said to describe the evolution relative to free evolution. Of course, the probability $p(t)$ calculated in the Dirac picture has the same value as in the Heisenberg or Schrödinger picture,

$$p(t) = \mathrm{Tr}\{\varrho_D(t) P_D(t)\},$$

where

$$\varrho_D(t) \equiv \exp\left(\frac{i}{\hbar} H_0 t\right) \exp\left(-\frac{i}{\hbar} Ht\right) \varrho \exp\left(\frac{i}{\hbar} Ht\right) \exp\left(-\frac{i}{\hbar} H_0 t\right), \quad (21)$$

$$P_D(t) \equiv \exp\left(\frac{i}{\hbar} H_0 t\right) P \exp\left(-\frac{i}{\hbar} H_0 t\right). \quad (22)$$

† To simplify the notation we once again make the assumption that the Dirac picture coincides with the Heisenberg picture at the time $t = 0$.

In the Dirac picture the density operators and state vectors satisfy the differential equations

$$i\hbar \frac{\partial \varrho_D(t)}{\partial t} = [H_{\text{int}}(t), \varrho_D(t)], \tag{23}$$

$$i\hbar \frac{\partial \Psi_D(t)}{\partial t} = H_{\text{int}}(t)\Psi_D(t) \tag{24}$$

where $H_{\text{int}}(t)$ is the Hamiltonian of interaction in the Dirac picture

$$H_{\text{int}}(t) \equiv \exp\left(\frac{i}{\hbar}H_0 t\right)(H-H_0)\exp\left(-\frac{i}{\hbar}H_0 t\right).$$

It is also convenient to introduce a two-parameter family of evolution operators in the Dirac picture, $\mathsf{V}(t, t_0)$, which has properties characteristic of the aforementioned families of evolution operators in the Heisenberg and Schrödinger pictures, $\mathsf{U}(t, t_0)$ and $\mathsf{U}_S(t, t_0)$,

$$\mathsf{V}(t, t_0) \equiv \exp\left(\frac{i}{\hbar}H_0 t\right)\exp\left[-\frac{i}{\hbar}H(t-t_0)\right]\exp\left(-\frac{i}{\hbar}H_0 t_0\right). \tag{25}$$

These operators are unitary and satisfy the composition law and the differential equation

$$i\hbar\frac{\partial}{\partial t}\mathsf{V}(t, t_0) = H_{\text{int}}(t)\mathsf{V}(t, t_0). \tag{26}$$

In the Dirac picture, when the operator H is appropriately broken up into the parts H_0 and $H-H_0$, there frequently exist limits† of the evolution operator $\mathsf{V}(t, t_0)$ when $t \to \infty$ and $t \to -\infty$. The limiting operator is called the S *operator*‡ and serves to describe scattering processes,

$$\mathsf{S} \equiv \lim_{\substack{t\to\infty \\ t_0\to-\infty}} \mathsf{V}(t, t_0). \tag{27}$$

The S operator describes the time evolution of the states of a system after the free evolution has been eliminated. The canonical transforma-

---

† Such limits do not in general exist for evolution operators $\mathsf{U}(t, t_0)$ unless the evolution of the system "dies out" in the past and in the future.

‡ The notion of the S operator was introduced independently by J. A. Wheeler (1937) and W. Heisenberg (1943).

tions of classical theory correspond to the unitary operators of quantum theory. The S operator also has a corresponding canonical transformation. This is the transformation relating coordinates and momenta $x_{in}(0)$ and $p_{in}(0)$ and $x_{out}(0)$ and $p_{out}(0)$ obtained when the asymptotic, free trajectories of particles in the past and future are extended to the time $t = 0$.

The definition of the S operator may also be formulated by using the Møller operators $\boldsymbol{\Omega}_{\pm}$

$$\boldsymbol{\Omega}_{\pm} \equiv \lim_{t \to \pm\infty} \exp\left(-\frac{i}{\hbar} Ht\right) \exp\left(\frac{i}{\hbar} H_0 t\right), \tag{28}$$

$$\mathsf{S} \equiv \boldsymbol{\Omega}_{-}^{\dagger} \boldsymbol{\Omega}_{+}. \tag{29}$$

Causality

An operator $\mathsf{U}(t, t_0)$ which describes the time evolution of a quantum system from an earlier time t_0 to a later time t may, by the composition law (7), be written as

$$\mathsf{U}(t, t_0) = \mathsf{U}(t, t_n)\mathsf{U}(t_n, t_{n-1}) \dots \mathsf{U}(t_1, t_0), \tag{30}$$

where $t_1, \dots, t_n$ denote a subdivision of the time interval (t_0, t) into subintervals: $(t_0, t_1), (t_1, t_2), \dots, (t_n, t)$. The definition of the generator† $H(t)$ implies that a unitary operator $\mathsf{U}(t_i, t_{i-1})$ may, for a very short interval of time (t_{i-1}, t_i), be approximated by the formula

$$\mathsf{U}(t_i, t_{i-1}) \approx 1 - \frac{i}{\hbar} \int_{t_{i-1}}^{t_i} \mathrm{d}t H(t) \approx \exp\left(-\frac{i}{\hbar} \int_{t_{i-1}}^{t_i} \mathrm{d}t H(t)\right).$$

The operator $\mathsf{U}(t, t_0)$ for the time interval (t_0, t) may thus be presented in the form:‡

† Cf. also the footnote on p. 13.

‡ Here we omit all problems concerning the mathematical propriety of the operations considered. All transformations will have a formal, algebraic character, but attempts to make them rigorous would take us not only far beyond the scope of this book, but also frequently beyond the limits of the present-day state of mathematical knowledge in this field.

$$\lim_{\substack{n\to\infty \\ |t_i-t_{i-1}|\to 0}} \exp\left(-\frac{i}{\hbar}\int_{t_n}^{t} \mathrm{d}t\, H(t)\right)\exp\left(-\frac{i}{\hbar}\int_{t_{n-1}}^{t_n} \mathrm{d}t\, H(t)\right)$$

$$\ldots \exp\left(-\frac{i}{\hbar}\int_{t_0}^{t_1} \mathrm{d}t\, H(t)\right). \tag{31}$$

This limit of the product of exponential functions is called a *chronological exponential operator* defined by the family of self-adjoint operators $H(t)$ and is denoted by

$$T\exp\left(-\frac{i}{\hbar}\int_{t_0}^{t} \mathrm{d}t\, H(t)\right).$$

The name is associated with the following fundamental property of the expression (31): it is a combination of unitary operators

$$\exp\left(-\frac{i}{\hbar}\int_{t_{i-1}}^{t_i} \mathrm{d}t\, H(t)\right)$$

arranged in chronological order (time increases from right to left).

If for any t and t' the relation

$$[H(t), H(t')] = 0$$

holds, the chronological exponential operator is equal to

$$\exp\left(-\frac{i}{\hbar}\int_{t_0}^{t} \mathrm{d}t\, H(t)\right).$$

In the particular case when the Hamiltonian is time-independent the chronological exponential operator reduces to an ordinary exponential function of the operator $H: \exp\left[-\frac{i}{\hbar}H(t-t_0)\right]$.

It is shown in Appendix B that the chronological exponential operator may also be written as the formal series

$$T\exp\left(-\frac{i}{\hbar}\int_{t_0}^{t} \mathrm{d}t\, H(t)\right)$$

$$=\sum_{n=0}^{\infty}\left(-\frac{i}{\hbar}\right)^n \frac{1}{n!}\int_{t_0}^{t} \mathrm{d}t_n \ldots \int_{t_0}^{t} \mathrm{d}t_1\, T\big(H(t_n)\ldots H(t_1)\big). \tag{32}$$

It follows from formulae (31) or (32) that if a one-parameter family of operators $H(t)$ is given, the time evolution of the system can be determined. The composition law (7) furthermore implies that the time evolution of the system as determined by $H(t)$ is causal in the following sense: the operator $\mathsf{U}(t, t_0)$ which describes the time evolution of the system from the time t_0 to the time t does not depend on the form of $H(t)$ at times lying beyond the interval (t_0, t).

We shall illustrate this with the example of a system perturbed by an external agent. The assumption made here is that the Hamiltonian $H(t)$ depends locally in time on the real function $f(t)$ describing the external perturbations, i.e. that the condition†

$$\frac{\delta H(t)}{\delta f(t')} = \delta(t-t')A(t)$$

is met. The operators $\mathsf{U}(t, t_0)$ for any $t > t_0$ then satisfy the condition

$$\frac{\delta}{\delta f(t'')}\left(\mathsf{U}^{\dagger}(t, t_0)\frac{\delta}{\delta f(t')}\mathsf{U}(t, t_0)\right) = 0, \quad \text{when} \quad t'' > t'. \tag{33}$$

When t' lies beyond the interval (t_0, t) the condition (33) is satisfied for every t''. And when t' is contained in the interval (t_0, t), by formula (31) we have

$$\frac{\delta}{\delta f(t'')}\left[\mathsf{U}^{\dagger}(t, t_0)\frac{\delta}{\delta f(t')}\mathsf{U}(t, t_0)\right]$$

$$= \frac{\delta}{\delta f(t'')}\left[-\frac{i}{\hbar}\mathsf{U}^{\dagger}(t, t_0)\lim_{n\to\infty}\exp\left(-\frac{i}{\hbar}\int_{t_n}^{t}\mathrm{d}t\, H(t)\right)\right.$$

$$\left.\ldots\int_{t_{i-1}}^{t_i}\mathrm{d}t\,\frac{\delta H(t)}{\delta f(t')}\ldots\exp\left(-\frac{i}{\hbar}\int_{t_0}^{t_1}\mathrm{d}t\, H(t)\right)\right]$$

† The symbol $\delta/\delta f$ denotes a functional derivative with respect to the function f. The definition and properties of the functional derivative are given in Appendix C.

$$= -\frac{i}{\hbar}\frac{\delta}{\delta f(t'')}[\mathsf{U}^\dagger(t', t_0)A(t')\mathsf{U}(t', t_0)] = 0, \quad \text{when} \quad t'' > t' \quad \text{or} \quad t'' < t_0.$$

The relativistic generalization of this condition

$$\frac{\delta}{\delta f(y)}\left(\mathsf{U}^\dagger \frac{\delta}{\delta f(x)}\mathsf{U}\right) = 0, \quad \text{when} \quad y \gtrsim x, \tag{34}$$

is known as the *Bogolyubov causality condition*.# In this formula x and y denote the coordinates of points in space-time. The external perturbation thus is described by a local function of space-time variables. The symbol $\gtrsim$ denotes the relation: later or spatially separated betweeen points of coordinates y and x.

2. SYMMETRIES

Answers to questions concerning a physical system when it is in various states are not in general correlated with each other. If such correlation does occur, then in accordance with intuitive understanding we are inclined to attribute it to the existence of a symmetry of the system. This notion will now be made precise for relativistic quantum theory.

First of all, we introduce the concept of automorphism in the set of all questions. This is a one-to-one mapping of questions into questions

$$\mathscr{P} \to \mathscr{P}' = f(\mathscr{P}), \tag{1}$$

such that it preserves the structure imposed in the set of questions by Postulate I. Each such automorphism induces an automorphism in the set of projection operators $P \to P'$; and this latter automorphism, by Wigner's theorem,## may be written

$$P \to P' = \mathsf{U}^\dagger P \mathsf{U}, \tag{2}$$

where U is a unitary or anti-unitary operator.†

This condition is given in the book by Bogolyubov and Shirkov as one of the postulates defining the S operator.

The proof of this theorem, along with an extensive discussion can be found in a paper by Bargmann (1964) and in the book by Jauch.

† The definition of anti-unitary operators is given in Appendix A.

In relativistic quantum theory, all measurements refer to bounded regions of space-time because interactions are propagated at finite speed. Questions concerning the properties of a system in the region $\mathcal{O}$ will be denoted by the symbols $\mathcal{P}_{\mathcal{O}}$.

Suppose we consider a set of regions obtained when a given region $\mathcal{O}$ is displaced ($\mathcal{O} \xrightarrow[a,\Lambda]{} \mathcal{O}'$) by using all the Poincaré transformations.†
A whole family of questions $\{\mathcal{P}_{\mathcal{O}}\}$ may be associated with each question $\mathcal{P}_{\mathcal{O}}$, just as a family of questions $\mathcal{P}(t)$ may be associated with each question $\mathcal{P}$. This family comprises all questions which differ solely in that they apply to different regions interrelated by Poincaré transformations. The question $\mathcal{P}_{\mathcal{O}'}$ reads the same with respect to the region $\mathcal{O}'$ as $\mathcal{P}_{\mathcal{O}}$ does with respect to $\mathcal{O}$. Answers to $\mathcal{P}_{\mathcal{O}}$ and $\mathcal{P}_{\mathcal{O}'}$ may be sought by using identical apparatus which is merely displaced in space-time in accordance with the Poincaré transformations.

The homogeneity and isotropy of space-time implies that for an isolated system the transformation $\mathcal{O} \to \mathcal{O}'$ must entail automorphism of the set of questions. Accordingly, there exists a unitary or anti-unitary operator $\mathsf{U}(a, \Lambda)$ such that the projection operators representing the questions $\mathcal{P}_{\mathcal{O}}$ and $\mathcal{P}_{\mathcal{O}'}$ are related by

$$P_{\mathcal{O}'} = \mathsf{U}^{\dagger}(a, \Lambda) P_{\mathcal{O}} \mathsf{U}(a, \Lambda). \tag{3}$$

This formula contains as a special case the postulate about the time evolution of the system. The probability of obtaining an affirmative answer to the question $\mathcal{P}_{\mathcal{O}'}$ is given by

$$p_{\mathcal{O}'} = \mathrm{Tr}\{\varrho \mathsf{U}^{\dagger}(a, \Lambda) P_{\mathcal{O}} \mathsf{U}(a, \Lambda)\},$$

which may be rewritten as

$$p_{\mathcal{O}'} = \mathrm{Tr}\{\varrho' P_{\mathcal{O}}\},$$

where

$$\varrho' = \mathsf{U}(a, \Lambda) \varrho \mathsf{U}^{\dagger}(a, \Lambda). \tag{4}$$

Thus, for every state ϱ there exists a state ϱ', as defined above, which possesses the same properties (this means that all probabilities are the same) with respect to questions from the region $\mathcal{O}'$, as the state ϱ has

† The definition and properties of Poincaré transformations are given in Appendix D.

with respect to questions from the region $\mathcal{O}$. This correlation of questions is attributed to the symmetry of the system. The Poincaré transformations are said to be *symmetry transformations* of the system.

Symmetry transformations may also be transformations of questions which are not related to changes of space-time region. Such symmetries are known as internal symmetries. The symmetry transformations of projection operators

$$P'_{\mathcal{O}} = \mathsf{U}^{\dagger} P_{\mathcal{O}} \mathsf{U} \tag{5}$$

correspond to them by virtue of Wigner's theorem. In relativistic theory such transformations are required not to violate the homogeneity and the isotropy of space-time. This requirement thus means that the projection operators $P'_{\mathcal{O}}$ and $P'_{\mathcal{O}'}$ referring to two different regions related by a Poincaré transformation are also related by formula (3),

$$P'_{\mathcal{O}'} = \mathsf{U}^{\dagger}(a, \Lambda) P'_{\mathcal{O}} \mathsf{U}(a, \Lambda), \tag{6}$$

and that formula (5) is of the same form for all regions related by Poincaré transformations, whereby

$$P'_{\mathcal{O}'} = \mathsf{U}^{\dagger} P_{\mathcal{O}'} \mathsf{U}. \tag{7}$$

Comparison of formulae (3), (5), (6) and (7) yields

$$\mathsf{U}^{\dagger}\mathsf{U}^{\dagger}(a, \Lambda) P_{\mathcal{O}} \mathsf{U}(a, \Lambda) \mathsf{U} = \mathsf{U}^{\dagger}(a, \Lambda) \mathsf{U}^{\dagger} P_{\mathcal{O}} \mathsf{U}\mathsf{U}(a, \Lambda). \tag{8}$$

The operator U represents an internal symmetry transformation only when its action on all projection operators is commutative with that of the operators $\mathsf{U}(a, \Lambda)$.

Formula (2) implies immediately that a combination of symmetry transformations is also a symmetry transformation and that in the set of symmetry transformations there exist a unity (an identity transformation) and inverse transformations (the corresponding operators are $\mathsf{U}^{-1} = \mathsf{U}^{\dagger}$). Symmetry transformations thus form a group. With respect to physical interpretation various subgroups, frequently called symmetry groups, are distinguished in the set of all symmetries. In relativistic quantum theory one of them is the group of Poincaré transformations which plays a fundamental role in that theory. Both continuous and discrete groups appear in quantum theories.

We shall adopt the following general scheme relating abstract groups† to groups of symmetry transformations of quantum theories. If an abstract group is to play a part in physical considerations, we must know its physical interpretation. One and the same abstract group G may have several different physical interpretations. We know most frequently how elements of this group act on observables, e.g. on the average values of physical quantities. Symmetry operators U form a certain unitary‡ representation of the group G. This representation is defined by the mapping

$$G \ni g \to \mathsf{U}(g) \tag{9}$$

satisfying the condition§

$$\mathsf{U}(g_1)\mathsf{U}(g_2) = \mathsf{U}(g_1 g_2). \tag{10}$$

If the elements of the group are labelled by using the parameters $\alpha_1, \ldots, \alpha_n$, the unitary symmetry operators may also be regarded as functions of these parameters

$$g(\alpha_1, \ldots, \alpha_n) \to \mathsf{U}(\alpha_1, \ldots, \alpha_n).$$

For Lie groups the parameters α_i vary continuously. We can then define the concept of the generators of the group. These are the derivatives of the operators U with respect to the parameters α, calculated for values of α corresponding to the identity transformation. The group parametrization customarily adopted is one for which zero values of

† The reader will find the necessary information on group theory in the book by Hamermesh.

‡ Henceforth, we shall confine ourselves to considering only unitary symmetry operators. Anti-unitary operators need be introduced only to describe symmetry with respect to time inversion.

§ In actual fact, by formula (2) describing the manner in which symmetry operators act on projection operators, we can obtain only a weaker condition for the composition law of unitary symmetry operators, viz.

$$\mathsf{U}(g_1)\,\mathsf{U}(g_2) = \omega(g_1, g_2)\mathsf{U}(g_1 g_2), \tag{10a}$$

where $\omega(g_1, g_2)$ is a phase factor. The mapping $g \to \mathsf{U}(g)$ satisfying the condition (10a) is called the projective representation of the group G. Projective representations may be avoided by extending the group to the covering group $\tilde{G}$. For this reason we shall not consider projective representations.

all the parameters correspond to the unit element of the group. The generators G_k are then defined by the formula

$$G_k \equiv i\hbar \left. \frac{\partial U(\alpha_1, \ldots, \alpha_n)}{\partial \alpha_k} \right|_{\alpha_l = 0}. \tag{11}$$

The composition law for symmetry transformations (10) implies the commutation relations for the generators.

$$\frac{1}{i\hbar}[G_i, G_j] = c_{ij}^k G_k. \tag{12}$$

The coefficients c_{ij}^k are called *structure constants.* They depend only on the structure of the group and the parametrization used. The generators of all the representations of a given group satisfy the same commutation relations.

3. CANONICAL QUANTIZATION

A quantum theory is most frequently constructed by using the *method of canonical quantization.* This method is employed first and foremost to quantize mechanical systems, but it may be used when quantizing any system described by classical theory in canonical form.

By the canonical formulation of classical theory we mean a formulation that is modelled after the canonical formalism of classical mechanics. In the canonical formulation of classical theory the state of a system at any instant of time is described by a set of canonical variables $u_A(t)$. Each dynamical quantity F is a function of canonical variables. An operation called the *Poisson bracket,* usually denoted by the symbol $\{F, G\}$, is defined in the set of functions $F(u_A)$. The values of the Poisson brackets of canonical variables are usually numbers, i.e. they do not depend on the canonical variables,

$$\{u_A(t), u_B(t)\} = c_{AB}. \tag{1}$$

The Poisson brackets introduce the structure of Lie algebra into the set of physical quantities, that is to say, the following relations are satisfied:

$$\begin{aligned} \{F, G\} &= -\{G, F\}, \\ \{\lambda_1 F_1 + \lambda_2 F_2, G\} &= \lambda_1\{F_1, G\} + \lambda_2\{F_2, G\}, \\ \{F_1 F_2, G\} &= F_1\{F_2, G\} + \{F_1, G\}F_2, \\ \{F_1, \{F_2, F_3\}\} + \{F_2, \{F_3, F_1\}\} + \{F_3, \{F_1, F_2\}\} &= 0. \end{aligned} \tag{2}$$

It follows from these properties that the Poisson bracket of any two dynamical variables may be determined if the Poisson brackets of the canonical variables are known.

A distinguished role is played among dynamical variables by the energy of the system $H(u_A)$. The Poisson brackets are defined so that the time derivative of any physical quantity $F(u_A, t)$ is expressible in terms of the Poisson bracket of that quantity and the energy H:

$$\frac{\mathrm{d}F}{\mathrm{d}t} = \frac{\partial F}{\partial t} + \{F, H\}. \tag{3}$$

If in particular the canonical variables $u_A(t)$ are substituted for F in this formula the canonical equations of motion for these variables are obtained.

The choice of canonical variables is not unique. Two allowable systems of canonical variables are connected by a canonical transformation. Among canonical transformations a special role is played by symmetry transformations. These are canonical transformations, defined for all values of t, such that they do not alter the form of the canonical equations of motion. If the symmetry transformations constitute a Lie group, the concept of generators of this group may be introduced. These are physical quantities $G_i(u_A)$ whose Poisson brackets with canonical variables give infinitesimal changes of these variables under symmetry transformations,

$$\delta u_A = \{G_i, u_A\}\, \delta\alpha^i.$$

The Poisson brackets of the generators are specified by the structure constants of the symmetry groups:

$$\{G_i, G_j\} = c_{ij}^k G_k. \tag{4}$$

The canonical quantization of classical theory consists in assigning to the canonical variables $u_A(t)$ linear operators $\hat{u}_A(t)$ (usually self-adjoint) acting in the Hilbert spaces of the state vectors, so that the following commutation relations be satisfied:

$$\frac{1}{i\hbar}[\hat{u}_A(t), \hat{u}_B(t)]_- = c_{AB}, \tag{5}$$

where the numbers c_{AB} are equal to the corresponding Poisson brackets in classical theory. Next, we assign linear operators to all dynamical variables, inserting the operators $\hat{u}_A$ in the place of the canonical operators u_A in the relevant functions $F(u_A)$. This prescription does not, in general, lead to a unique result since the operators $\hat{u}_A$ do not commute. Auxiliary conditions restricting this non-uniqueness are obtained from the requirement that the symmetry groups of classical theory also be symmetry groups of quantum theory. The dynamical quantities G_i which are the generators of the symmetry transformations of classical theory must then be represented by the operators $\hat{G}_i$ satisfying the commutation relations (2.12).

Canonical quantization can be carried out in the case of non-relativistic mechanical systems and free fields. In the case of relativistic field theory with interaction, canonical quantization is rather a set of heuristic hints than a well-defined procedure.

CHAPTER 2

NON-RELATIVISTIC QUANTUM MECHANICS

THIS chapter will be devoted to a description of the motion of a particle in a potential field of forces and to a description of many-particle systems in non-relativistic quantum mechanics. Using these examples which are no doubt familiar to the reader, we wish to discuss two methods of describing phenomena in quantum theory since subsequent chapters are based in great measure on these methods. These are the propagator method of describing the time evolution of a system and the method of describing systems of many particles in terms of creation and annihilation operators and field operators.

4. THE QUANTUM MECHANICS OF A PARTICLE

The *Schrödinger coordinate representation* is the representation of state vectors employed most frequently in quantum mechanics. In it, a state vector Ψ is represented by a complex wave function of the variables x, y, and z. In the Schrödinger picture the wave function is also a function of time. Following the notation introduced by Dirac, we denote this representation by the symbol $\langle \mathbf{x} | \Psi \rangle$,

$$\Psi(t) \rightarrow \langle \mathbf{x} | \Psi(t) \rangle = \psi(\mathbf{x}, t). \tag{1}$$

The Hilbert-space scalar product of two vectors Ψ and Φ is equal to the scalar product of the wave functions,

$$(\Psi | \Phi) = \int d^3x \, \psi^*(\mathbf{x}, t) \, \varphi(\mathbf{x}, t). \tag{2}$$

Square-integrable functions† thus correspond to state vectors. Linear operators in Hilbert space are represented by the distributions of two vector variables,

$$A \rightarrow \langle \mathbf{x}|A|\mathbf{y}\rangle = A(\mathbf{x}, \mathbf{y}). \tag{3}$$

The distribution $\delta(\mathbf{x}-\mathbf{y})$ corresponds to the unit operator, $\mathbf{x}\delta(\mathbf{x}-\mathbf{y})$ to the position operator, and the distribution $\frac{\hbar}{i}\nabla_x \delta(\mathbf{x}-\mathbf{y})$ to the momentum operator $\mathbf{p}$. The action of an operator on a wave function is expressed by the integral:

$$(A\Psi)(\mathbf{x}) = \int d^3y\, A(\mathbf{x}, \mathbf{y})\psi(\mathbf{y}). \tag{4}$$

In Dirac's notation, this formula assumes the form:

$$\langle \mathbf{x}|A\Psi\rangle = \int d^3y \langle \mathbf{x}|A|\mathbf{y}\rangle\langle \mathbf{y}|\Psi\rangle, \tag{5}$$

whence it is seen that the relation

$$\int |\mathbf{y}\rangle d^3y \langle \mathbf{y}| = 1$$

holds in this notation.

The Particle Propagator

The distribution representing the time evolution operator in the Schrödinger picture, $\mathsf{U}_S(t, t_0)$, is called a *propagator* and will be denoted by the symbol $\mathscr{K}$. Suppose that the Hamiltonian of a particle in the Schrödinger picture is the sum of the kinetic energy $\mathbf{p}^2/2m$ and the potential function $U(\mathbf{x}, t)$,

$$H(t) = \mathbf{p}^2/2m + U(\mathbf{x}, t). \tag{6}$$

In this case the propagator $\mathscr{K}$ depends functionally on $U(\mathbf{x}, t)$:

$$\mathscr{K}[\mathbf{x}, t; \mathbf{y}, t_0|U] \equiv i\langle \mathbf{x}|\mathsf{U}_S(t, t_0)|\mathbf{y}\rangle. \tag{7}$$

† Strictly speaking, equivalence classes of wave functions, i.e. sets of wave functions differing on sets of measure zero, correspond to vectors in Hilbert space.

The propagator $\mathscr{K}$ propagates wave functions from the time t_0 to t,

$$\psi(\mathbf{x}, t) = -i\int d^3y\mathscr{K}[\mathbf{x}, t; \mathbf{y}, t_0|U]\psi(\mathbf{y}, t_0). \tag{8}$$

With a view to further applications, we break up the propagator $\mathscr{K}$ into two parts,

$$\mathscr{K} = \mathscr{K}_{\mathrm{R}} - \mathscr{K}_{\mathrm{A}},$$

the first of which, called a *retarded propagator*, is zero for $t < t_0$, whereas the second, called an *advanced propagator*, is equal to zero for $t > t_0$:

$$\mathscr{K}_{\mathrm{R}}[\mathbf{x}, t; \mathbf{y}, t_0|U] \equiv \theta(t-t_0)\mathscr{K}[\mathbf{x}, t; \mathbf{y}, t_0|U], \tag{9a}$$

$$\mathscr{K}_{\mathrm{A}}[\mathbf{x}, t; \mathbf{y}, t_0|U] \equiv -\theta(t_0-t)\mathscr{K}[\mathbf{x}, t; \mathbf{y}, t_0|U]. \tag{9b}$$

In keeping with the convention adopted concerning discontinuous functions, we do not define the values of the retarded and advanced propagators for equal times, when $t = t_0$. The limit of the propagator $\mathscr{K}_{\mathrm{R}}$, when $t \to t_0+0$, is equal to $i\delta(\mathbf{x}-\mathbf{y})$, but is zero when $t \to t_0-0$. For the propagator $\mathscr{K}_{\mathrm{A}}$ these limits are 0 and $-i\,\delta(\mathbf{x}-\mathbf{y})$, respectively. Although the propagators $\mathscr{K}_{\mathrm{R}}$ and $\mathscr{K}_{\mathrm{A}}$ themselves are discontinuous, their difference is continuous (as a distribution) with respect to the variables t and t_0. The retarded and advanced propagators obey the *inhomogeneous Schrödinger equations* in both sets of variables:

$$\left[-i\frac{\partial}{\partial t} - \frac{1}{2m}\Delta_x + U(\mathbf{x}, t)\right]\mathscr{K}_{\mathrm{R,A}}[\mathbf{x}, t; \mathbf{y}, t_0|U] = \delta(t-t_0)\,\delta(\mathbf{x}-\mathbf{y}), \tag{10a}$$

$$\mathscr{K}_{\mathrm{R,A}}[\mathbf{x}, t; \mathbf{y}, t_0|U]\left[i\frac{\overleftarrow{\partial}}{\partial t_0} - \frac{1}{2m}\overleftarrow{\Delta}_y + U(\mathbf{y}, t_0)\right] = \delta(t-t_0)\,\delta(\mathbf{x}-\mathbf{y}). \tag{10b}$$

This follows from the equation satisfied by the time evolution operator U_S and from the fact that the differentiation of the step function $\theta(t-t_0)$ yields the function $\delta(t-t_0)$. The operator $\mathsf{U}_S(t, t)$ is a unit operator, and thus is represented by $\delta(\mathbf{x}-\mathbf{y})$.

In the special case, when the particle is free ($U = 0$), these equations can be solved explicitly. The corresponding propagators will be denoted by the symbols $\mathscr{K}_{\mathrm{R,A}}(\mathbf{x}-\mathbf{y}, t-t_0)$,

$$\mathscr{K}_{\mathrm{R,A}}(\mathbf{x}, t) = \int \frac{d^3k\,d\omega}{(2\pi)^4} e^{-i\omega t+i\mathbf{k}\cdot\mathbf{x}} \tilde{\mathscr{K}}_{\mathrm{R,A}}(\mathbf{k}, \omega), \tag{11a}$$

$$\mathscr{K}_{\mathrm{R,A}}(\mathbf{k}, \omega) = \frac{1}{\dfrac{\mathbf{k}^2}{2m} - \omega \mp i\varepsilon}. \tag{11b}$$

Formula (11b) is symbolical in character inasmuch as the parameter ε appears in it. The meaning of formulae of this type is discussed in Appendix E.

The differential equation (10a) and the retardation condition

$$\mathscr{K}_{\mathrm{R}}[\mathbf{x}, t; \mathbf{y}, t_0|U] = 0, \quad \text{when} \quad t < t_0, \tag{12}$$

are equivalent to the integral equation

$$\mathscr{K}_{\mathrm{R}}[\mathbf{x}, t; \mathbf{y}, t'|U] = \mathscr{K}_{\mathrm{R}}(\mathbf{x}-\mathbf{y}, t-t')$$
$$- \int d^3x_1\, dt_1 \mathscr{K}_{\mathrm{R}}(\mathbf{x}-\mathbf{x}_1, t-t_1) U(\mathbf{x}_1, t_1) \mathscr{K}_{\mathrm{R}}[\mathbf{x}_1, t_1; \mathbf{y}, t'|U]. \tag{13}$$

This equation lends itself naturally to solution by the iteration method. In this way we obtain the series

$$\mathscr{K}_{\mathrm{R}}[\mathbf{x}, t; \mathbf{y}, t'|U] = \mathscr{K}_{\mathrm{R}}(\mathbf{x}-\mathbf{y}, t-t')$$
$$- \int d^3x_1\, dt_1 \mathscr{K}_{\mathrm{R}}(\mathbf{x}-\mathbf{x}_1, t-t_1) U(\mathbf{x}_1, t_1) \mathscr{K}_{\mathrm{R}}(\mathbf{x}_1-\mathbf{y}, t_1-t')$$
$$+ \int d^3x_1\, d^3x_2\, dt_1\, dt_2 \mathscr{K}_{\mathrm{R}}(\mathbf{x}-\mathbf{x}_1, t-t_1) U(\mathbf{x}_1, t_1) \mathscr{K}_{\mathrm{R}}(\mathbf{x}_1-\mathbf{x}_2, t_1-t_2)$$
$$\times U(\mathbf{x}_2, t_2) \mathscr{K}_{\mathrm{R}}(\mathbf{x}_2-\mathbf{y}, t_2-t') + \ldots \tag{14}$$

The Scattering Amplitude

Most phenomena associated with the motion of particle in a force field can be described on the assumption that the time-dependent part of the potential is nonzero only in a finite interval of time. Thus, in the remote past ($t < T_1$) and in the far future ($t > T_2$) the potential $U(\mathbf{x}, t)$ goes over into a static potential $U(\mathbf{x})$. We assume that the static potential has a finite range,† that is, that it tends to zero more rapidly than $|\mathbf{x}|^{-1-\varepsilon}$ when $|\mathbf{x}| \to \infty$.

Let us consider the state of a particle which at the time t_0 was described by the state vector Φ_i. The probability amplitude of finding this particle in the state Φ_f at the time t is given by

$$A_{fi} \equiv (\Phi_f(t)|\mathsf{U}_S(t, t_0)\Phi_i(t_0)). \tag{15}$$

In order to take full account of the effect caused by the time-dependent part of the potential, we shall assume that $t > T_2$ and $t_0 < T_1$. Then the time-dependence of the vectors Φ_i and Φ_f is specified by the operator U_S, which has the following energy operator associated with it:

$$H = \frac{\mathbf{p}^2}{2m} + U(\mathbf{x}). \tag{16}$$

An energy operator so defined, with rather mild restrictions on the potential,‡ is a self-adjoint operator. In general it possesses both a continuous spectrum (energy $E > 0$) and a point spectrum ($E < 0$).

The eigenvectors Φ_n belonging to the point spectrum span a subspace which describes the *bound states* of the particle. The time-dependence of the state vectors Ψ_B in this subspace is given by the formula

$$\Psi_B(t) = \sum_n c_n \Phi_n \mathrm{e}^{-iE_n t}. \tag{17}$$

In the subspace corresponding to the continuous spectrum of the energy operator lie vectors Ψ_{SC} which describe the *scattering states* of the particle. It is convenient to expand these vectors into generalized eigen-

† This assumption excludes the Coulomb potential which has an infinite range but no ideal Coulomb potential ever occurs in reality since electric charges are always screened by surrounding charges. Even in the interstellar space, where the density of free electrons and protons is 10 per cm³, the electrostatic field of the charge vanishes at distances of the order of 1 metre. A good approximation of the electrostatic potential in such a situation is given by the Yukawa potential $r^{-1}\,\mathrm{e}^{-\mu r}$ with constant μ equal to $(10^2\ \mathrm{cm})^{-1}$.

‡ The following conditions ensure that the operator H is self-adjoint: there exist numbers M and R such that

$$\int_{|\mathbf{x}| \leqslant R} \mathrm{d}^3x\,|U(\mathbf{x})| \leqslant M \quad \text{and} \quad |U(\mathbf{x})| \leqslant M \quad \text{for} \quad |\mathbf{x}| \geqslant R.$$

vectors of the energy operator (not belonging to Hilbert space). We shall consider two complete sets of such generalized vectors which we denote by $\Phi^+(\mathbf{p})$ and $\Phi^-(\mathbf{p})$. We adopt the normalization condition

$$(\Phi^\pm(\mathbf{p})\,|\,\Phi^\pm(\mathbf{q})) = (2\pi)^3\delta(\mathbf{p}-\mathbf{q}). \tag{18}$$

The generalized eigenvectors Φ^+ and Φ^- may be defined by giving the wave functions $\varphi^\pm(\mathbf{x}, \mathbf{p})$ which represent them. These functions are usually defined as solutions of the integral equations

$$\varphi^+(\mathbf{x}, \mathbf{p}) = e^{i\mathbf{p}\cdot\mathbf{x}} - \frac{m}{2\pi}\int d^3y\,\frac{e^{ip|\mathbf{x}-\mathbf{y}|}}{|\mathbf{x}-\mathbf{y}|}\,U(\mathbf{y})\,\varphi^+(\mathbf{x}, \mathbf{p}), \tag{19a}$$

$$\varphi^-(\mathbf{x}, \mathbf{p}) = e^{i\mathbf{p}\cdot\mathbf{x}} - \frac{m}{2\pi}\int d^3y\,\frac{e^{-ip|\mathbf{x}-\mathbf{y}|}}{|\mathbf{x}-\mathbf{y}|}\,U(\mathbf{y})\,\varphi^-(\mathbf{x}, \mathbf{p}). \tag{19b}$$

It follows from these equations that the functions $\varphi^\pm(\mathbf{x}, \mathbf{p})$ are generalized eigenfunctions of the energy operator corresponding to the eigenvalue $E = \mathbf{p}^2/2m$.

It can also be shown† that the functions φ^+ and φ^- are normalized in the same way as the inhomogeneous terms in the integral equations determining them, i.e. that the normalization conditions (18) are satisfied.

The time evolution of any arbitrary vector Ψ_{SC} from the scattering subspace may be described by each of two formulae:

$$\Psi_{SC}(t) = \int \frac{d^3p}{(2\pi)^3}\,g(\mathbf{p})\,\Phi^+(\mathbf{p})\,e^{-i\mathbf{p}^2t/2m}, \tag{20a}$$

$$\Psi_{SC}(t) = \int \frac{d^3p}{(2\pi)^3}\,f(\mathbf{p})\,\Phi^-(\mathbf{p})\,e^{-i\mathbf{p}^2t/2m}. \tag{20b}$$

In addition to the wave functions $\psi_{SC}(\mathbf{x}, t)$ representing vectors $\Psi_{SC}(t)$, as an auxiliary object we shall also consider the wave functions which satisfy the Schrödinger equation without potential. Each solution of this equation may be represented as a superposition of plane waves, known as a wave packet,

$$\varphi[\mathbf{x}, t\,|\,f] = \int \frac{d^3p}{(2\pi)^3}\,f(\mathbf{p})\,e^{-i\mathbf{p}^2t/2m}\,e^{i\mathbf{p}\cdot\mathbf{x}}. \tag{21}$$

† The proof can be found, for instance, in the book by Merzbacher.

The function $f(\mathbf{p})$ will be called the *profile of the wave packet*. For a potential of finite range it has been proved† that the difference between the wave function $\psi_{SC}(\mathbf{x}, t)$ and the wave packet $\varphi[\mathbf{x}, t|g]$ of profile equal to the function $g(\mathbf{p})$ which appears in the expansion of $\Psi_{SC}(t)$ into the vectors $\Phi^+(\mathbf{p})$, tends (in the norm) to zero when $t \to -\infty$. Similarly, the difference between $\psi_{SC}(\mathbf{x}, t)$ and the packet $\varphi[\mathbf{x}, t|f]$ tends to zero when $t \to \infty$.

The scattering solution of the Schrödinger equation with static potential, or the potential $U(\mathbf{x}, t)$ which goes over asymptotically into a static potential, can be fully characterized either by giving the profile $g(\mathbf{p})$ of the wave packet into which that solution goes over in the remote past, or by giving the profile $f(\mathbf{p})$ of the wave packet into which the solution goes over in the far future. It is not possible to give simultaneously arbitrary shapes of wave packets in both asymptotic time regions. The wave function satisfying the Schrödinger equation of given wave packet profile g in the remote past will be denoted by the symbol $\psi_{\mathrm{R}}[\mathbf{x}, t|g]$, whereas that with a given wave packet profile f in the far future will be denoted by the symbol $\psi_{\mathrm{A}}[\mathbf{x}, t|f]$. These functions satisfy the conditions

$$(\psi_{\mathrm{R}}[\mathbf{x}, t|g]-\varphi[\mathbf{x}, t|g]) \to 0, \quad \text{when} \quad t \to -\infty,$$

$$(\psi_{\mathrm{A}}[\mathbf{x}, t|f]-\varphi[\mathbf{x}, t|f]) \to 0, \quad \text{when} \quad t \to +\infty.$$

The components of the vector $\mathbf{p}$ parametrizing the state vectors $\Phi^+(\mathbf{p})$ and $\Phi^-(\mathbf{p})$ may thus be interpreted as components of the particle momentum in the remote past and far future, respectively.

In keeping with the above classification of the states of a particle in a static potential, the probability amplitude (15) can describe four different processes:

a. Scattering, when both states Φ_i and Φ_f are scattering states;
b. The process of capture into a bound state, when the initial state Φ_i is a scattering state and the final state Φ_f a bound one;
c. A process of the ionization type, the inverse of process b;
d. A transition between two bound states.

---

† These problems are discussed extensively in the monograph by Newton. An exhaustive bibliography on the subject is also given there.

Only the scattering process will be considered here. The other three processes can be described in similar fashion. The probability amplitude of the scattering process is called the *scattering amplitude*. In the Schrödinger representation it has the form

$$A_{if} = S[f, g] = -i\int d^3x d^3y \psi_A^*[\mathbf{x}, t|f]\mathcal{K}_R[\mathbf{x}, t; \mathbf{y}, t_0|U]\psi_R[\mathbf{y}, t_0|g]. \tag{22}$$

If we set $t = t_0$ in the above formula, we obtain a representation of the scattering amplitude as the scalar product of the functions ψ_R and ψ_A:

$$S[f, g] = (\psi_A[f]|\psi_R[g]). \tag{22a}$$

The function $\psi_R[g]$ is determined by the instrument preparing the state of the particle (source), and the function $\psi_A[f]$ by the analysing instrument (detector). In experiments on particle scattering the potential does not, in general, affect the preparation and detection of the particle. Accordingly, the shape of the incident beam may be described by the profile $g(\mathbf{p})$, and the properties of the counter by the profile $f(\mathbf{p})$. The amplitude $S[f, g]$ is the amplitude of the probability that a particle after scattering will be recorded by a counter characterized by the function $f(\mathbf{p})$, when the shape of the incident beam is specified by the function $g(\mathbf{p})$.

The amplitude $S[f, g]$ may be written in the following equivalent form:

$$S[f, g] = (\varphi[f]|\varphi[g])$$
$$+i\int d^3x dt d^3y dt' \varphi^*[\mathbf{x}, t|f]\, S_x \mathcal{K}_R[\mathbf{x}, t; \mathbf{y}, t'|U]\overleftarrow{S}_y \varphi[\mathbf{y}, t'|g], \tag{23}$$

where

$$S_x = -i\frac{\partial}{\partial t} - \frac{1}{2m}\Delta_x,$$

$$\overleftarrow{S}_x = i\frac{\overleftarrow{\partial}}{\partial t} - \frac{1}{2m}\overleftarrow{\Delta}_x.$$

In order to substantiate formula (23), we first note that the amplitude $S[f, g]$ represented by the formula (22) does not depend on the times t and t_0. Thus, the formula

$$S[f, g] = \lim_{\substack{t\to\infty \\ t_0\to-\infty}} \left(-i\int d^3x d^3y \varphi^*[\mathbf{x}, t|f]\mathcal{K}_R[\mathbf{x}, t;\ \mathbf{y}, t_0|U]\varphi[\mathbf{y}, t_0|g]\right) \tag{24}$$

is valid. Applying the identity

$$f(t)-f(t_0) = \int_{t_0}^{t} dt' \frac{df(t')}{dt'}$$

to the function I,

$$I \equiv -i\int d^3x d^3y \varphi^*[\mathbf{x}, t|f]\mathscr{K}_R[\mathbf{x}, t; \mathbf{y}, t_0|U]\varphi[\mathbf{y}, t_0|g],$$

which is treated as a function of t, we obtain

$$I = (f|g) - i\int_{t_0}^{t} dt' \int d^3x d^3y \frac{\partial}{\partial t'}(\varphi^*[\mathbf{x}, t'|f])\mathscr{K}_R[\mathbf{x}, t'; \mathbf{y}, t_0|U]\varphi[\mathbf{y}, t_0|g]$$

$$-i\int_{t_0}^{t} dt' \int d^3x d^3y \varphi^*[\mathbf{x}, t'|f]\frac{\partial}{\partial t'}\mathscr{K}_R[\mathbf{x}, t'; \mathbf{y}, t_0|U]\varphi[\mathbf{y}, t_0|g]$$

$$= (f|g) + \int_{t_0}^{t} dt' \int d^3x d^3y \varphi^*[\mathbf{x}, t'|f]S_x\mathscr{K}_R[\mathbf{x}, t'; \mathbf{y}, t_0|U]\varphi[\mathbf{y}, t_0|g],$$

where

$$(f|g) = (\varphi[f]|\varphi[g]) = \int \frac{d^3p}{(2\pi)^3} f^*(\mathbf{p})g(\mathbf{p}).$$

Applying the same identity to the expression $\mathscr{K}_R[\mathbf{x}, t'; \mathbf{y}, t_0|U]\varphi[\mathbf{y}, t_0|g]$ treated as a function of t_0 and using the retardation condition (12) for the propagator $\mathscr{K}_R$, we arrive at

$$I = (f|g) + i\int_{t_0}^{t} dt' \int_{t_0}^{t} dt'' \int d^3x d^3y \varphi^*[\mathbf{x}, t'|f]S_x\mathscr{K}_R[\mathbf{x}, t'; \mathbf{y}, t''|U]\overleftarrow{S}_y\varphi[\mathbf{y}, t''|g],$$

whence we get formula (23).

When we insert the iterative solution (14) for the propagator $\mathscr{K}_R$ into the formula (23), we obtain the *perturbation solution* for the scattering amplitude:

$$S[f, g] = (f|g) - i\int d^3x dt \varphi^*[\mathbf{x}, t|f]U(\mathbf{x}, t)\varphi[\mathbf{x}, t|g]$$

$$+i\int d^3x d^3y dt_1 dt_2 \varphi^*[\mathbf{x}, t_1|f]U(\mathbf{x}, t_1)$$

$$\times \mathscr{K}_R(\mathbf{x}-\mathbf{y}, t_1-t_2)U(\mathbf{y}, t_2)\varphi[\mathbf{y}, t_2|g] + \ldots$$

$$= S^{(0)} + S^{(1)} + S^{(2)} + \ldots \tag{25}$$

Knowledge of the scattering amplitude merely allows the probability for the scattering process under study to be calculated. The existence of a perturbation expansion, moreover, enables a picture of this process to be formed. We assume that the first term of the expansion, $S^{(0)}$, is the probability amplitude for the particle to undergo a transition without interacting with the potential, while the second term, $S^{(1)}$, is the probability amplitude for a single interaction with the potential. In this process, the free particle described by the wave packet $\varphi[g]$ interacts at the point $\mathbf{x}$ and at the time t with the potential. After the interaction the particle continues to move freely, described by the wave packet $\varphi[f]$. The amplitude $S^{(1)}$ is an integral over three-dimensional space and over time, since the particle may interact with the potential at any point and at any instant of time. The third term, $S^{(2)}$, is the probability amplitude of a two-fold interaction with the potential. This process takes place as follows. The free particle interacts with the potential at the point $\mathbf{y}$ at the instant of time t_2, and interacts again at the point $\mathbf{x}$ at the instant of time t_1. In the interval of time between these two events, the particle is free, and its propagation is described by the free propagator $\mathscr{K}_R(\mathbf{x}-\mathbf{y}, t_1-t_2)$. In some cases, the first few terms in the expansion of the scattering amplitude give a good approximation of this amplitude. If we limit ourselves to the first two terms, we obtain the scattering amplitude in the *Born approximation.*

It should be emphasized that the picture given above for the scattering process on the basis of the perturbation expansion of the transition amplitude is an outcome of the mathematical description we have chosen for the scattering processes and for that reason should not be taken too literally. Nevertheless, for heuristic and didactic purposes, we shall continue to use pictures of this type.

THE *S* MATRIX

We shall now express the scattering amplitude by the *S matrix* which we shall denote by the symbol $S(\mathbf{p}, \mathbf{q})$. The scattering amplitude $S[f, g]$ is a linear functional of the function $g(\mathbf{p})$ and an antilinear functional of the function $f(\mathbf{p})$. If the potential and the functions $f(\mathbf{p})$ and $g(\mathbf{p})$ are assumed to be suitably regular, a distribution $S(\mathbf{p}, \mathbf{q})$ exists such that the relation

$$S[f, g] = \int \frac{d^3p}{(2\pi)^3} \frac{d^3q}{(2\pi)^3} f^*(\mathbf{p}) S(\mathbf{p}, \mathbf{q}) g(\mathbf{q}) \tag{26}$$

is satisfied.

It follows from formula (23) that the S matrix may be represented as a double Fourier transform,†

$$S(\mathbf{p}, \mathbf{q}) = (2\pi)^3 \delta(\mathbf{p}-\mathbf{q})$$
$$+i\int d^3x\, dt\, d^3y\, dt'\, e^{i\mathbf{p}^2 t/2m - i\mathbf{p}\cdot\mathbf{x}} S_x \mathscr{K}_{\mathrm{R}}[\mathbf{x}, t; \mathbf{y}, t'|U] \overleftarrow{S}_y e^{-i\mathbf{q}^2 t'/2m + i\mathbf{q}\cdot\mathbf{y}}. \tag{27}$$

On the other hand, formula (25) leads to the following perturbation expansion of the S matrix:

$$S(\mathbf{p}, \mathbf{q}) = (2\pi)^3 \delta(\mathbf{p}-\mathbf{q}) - i\tilde{U}\left(\mathbf{p}-\mathbf{q}, \frac{\mathbf{p}^2}{2m} - \frac{\mathbf{q}^2}{2m}\right)$$
$$+i\int \frac{d^3k\, d\omega}{(2\pi)^4} \tilde{U}\left(\mathbf{p}-\mathbf{k}, \frac{\mathbf{p}^2}{2m} - \omega\right) \tilde{\mathscr{K}}_{\mathrm{R}}(\mathbf{k}, \omega)\, \tilde{U}\left(\mathbf{k}-\mathbf{q}, \omega - \frac{\mathbf{q}^2}{2m}\right) + \ldots \tag{28}$$

where the symbol $\tilde{U}(\mathbf{k}, \omega)$ denotes the Fourier transform of the potential $U(\mathbf{x}, t)$.

The successive terms in the perturbation expansion of the S matrix $S(\mathbf{p}, \mathbf{q})$ are interpreted, just as in the case of the scattering amplitude expansion, as describing processes without interaction with the potential, with a single interaction, with a two-fold interaction, etc. The arguments of the Fourier transform of the potential in formula (28) are interpreted as the momentum and energy transferred to the particle by an external field in a single act of interaction.

The matrix $S(\mathbf{p}, \mathbf{q})$ is the momentum representation of the S *operator* as defined in Section 1, when the operator H_0 is taken in the form

$$H_0 = \mathbf{p}^2/2m.$$

In the momentum representation, vectors and operators are represented by Fourier transforms of the relevant quantities in the coordinate representation, e.g.

$$\langle \mathbf{p}|\Psi(t)\rangle = \tilde{\psi}(\mathbf{p}, t).$$

† In Appendix M we give a different representation of the S matrix, which is perhaps closer to the intuitive notion of scattering.

In this notation, the relationship between the S matrix and the S operator is of the form

$$S(\mathbf{p}, \mathbf{q}) = \langle \mathbf{p} | \mathsf{S} | \mathbf{q} \rangle . \tag{29}$$

To prove this relation, let us consider the matrix element of the operator $\mathsf{V}(t, t')$ between the vectors Φ_f and Φ_i, whose wave functions in momentum representation are equal to $f(\mathbf{p})$ and $g(\mathbf{p})$, respectively:

$$\langle \mathbf{p} | \Phi_f \rangle = f(\mathbf{p}),$$

$$\langle \mathbf{p} | \Phi_i \rangle = g(\mathbf{p}).$$

Using the relation

$$\langle \mathbf{x} | \exp(-iH_0 t) \Phi_i \rangle = \varphi[\mathbf{x}, t | g]$$

and the analogous relation for the vector Φ_f, we obtain

$$(\Phi_f | \mathsf{V}(t, t') \Phi_i) = -i \int d^3x \, d^3y \, \varphi^*[\mathbf{x}, t | f] \mathscr{K}_{\mathrm{R}}[\mathbf{x}, t; \mathbf{y}, t' | U] \varphi[\mathbf{y}, t' | g].$$

Going to the limit, when $t \to \infty$ and $t' \to -\infty$, we finally arrive at (cf. formulae (1.18) and (24)):

$$(\Phi_f | \mathsf{S} \Phi_i) = S[f, g],$$

whence equation (29) follows directly.

Since the S operator is unitary, the matrix $S(\mathbf{p}, \mathbf{q})$ satisfies two *unitarity conditions*. In the case when processes of the type of ionization and capture into a bound state are not possible, these conditions are of the form:

$$\int \frac{d^3k}{(2\pi)^3} S^*(\mathbf{k}; \mathbf{p}) S(\mathbf{k}; \mathbf{q}) = (2\pi)^3 \delta(\mathbf{p} - \mathbf{q}), \tag{30a}$$

$$\int \frac{d^3k}{(2\pi)^3} S(\mathbf{p}; \mathbf{k}) S^*(\mathbf{q}; \mathbf{k}) = (2\pi)^3 \delta(\mathbf{p} - \mathbf{q}). \tag{30b}$$

The S matrix plays the part of the scattering amplitude for states with well-defined momenta in the simplified description of scattering in which the particles before and after the interaction are represented by plane waves. Such a description is not very precise since the wave functions describing the plane waves cannot be normalized. Unphysical states of particles smeared throughout all space correspond to these functions. In actual experiments we are always dealing with particles localized in a bounded region of space.

5. THE MANY-PARTICLE SYSTEM

This section is devoted to a non-relativistic description of the properties of a system of many identical particles within the framework of quantum mechanics. We shall consider interacting particles moving in an external field of forces. The purpose of this section is to present the specific methods of many-body theory, and particularly the method of *Feynman diagrams* and a method known as the *second quantization*. The latter consists in choosing a special representation of the state vectors of the system such that it permits a convenient, concise description to be introduced for a system of many indistinguishable particles. This method may also be used to describe systems with the number of particles varying in time. Hence, in this way it is a bridge between quantum mechanics and quantum field theory.

Spin and Statistics

We shall consider a system of particles endowed with spin. By the symbol s we shall denote the maximum value of the projection of the spin onto a chosen direction in units of $\hbar$. It takes on the values $0, \frac{1}{2}, 1, \frac{3}{2}, \ldots$ for various kinds of particles. The Schrödinger wave function of one particle with spin s has $2s+1$ components which we shall denote by the symbol $\psi(\mathbf{x}, \sigma, t)$, where the variable σ runs over the values $\sigma = -s, \ -s+1, \ldots, s-1, s$. The components are interpreted as the amplitudes of the probability density that at the time t a particle with a spin projection of $\sigma\hbar$ will be found at the point $\mathbf{x}$.

A system of n identical particles at the time t is described by a wave function of n pairs of variables $(\mathbf{x}_i, \sigma_i)$,

$$\psi(\mathbf{x}_1 \sigma_1, \ldots, \mathbf{x}_i \sigma_i, \ldots, \mathbf{x}_n \sigma_n, t).$$

In view of the assumption that the particles are identical and hence indistinguishable, we shall describe the system by means of only those wave functions which, under exchange of variables of the i-th and j-th particles, change by, at most, a phase factor. This property expresses the fact that the results of measurements made on the state described by the function $\psi(\mathbf{x}_1 \sigma_1, \ldots, \mathbf{x}_n \sigma_n, t)$ do not depend on how the particles are numbered. Hence, the following condition is satisfied,

$$P\psi(\mathbf{x}_1 \sigma_1, \ldots, \mathbf{x}_n \sigma_n, t) = e^{i\,P}\psi(\mathbf{x}_1 \sigma_1, \ldots, \mathbf{x}_n \sigma_n, t), \tag{1}$$

where P is an arbitrary permutation operator changing the order of the arguments $\mathbf{x}_i \sigma_i$ of the wave function, and θ_p is a real number. Condition (1) means that the wave function defines a one-dimensional representation of the permutation group. The group of permutations of n elements has only two one-dimensional representations: symmetric and antisymmetric. Each wave function describing a system of n indistinguishable particles, therefore, must be either symmetric or antisymmetric in the variables of all the particles. The results of experiments and certain theoretical considerations lead to the formulation of the following law, known as the *theorem on the connection between spin and statistics.*#

Particles with integer spin ($s = 0, 1, \ldots$) *are described by symmetric wave functions, and those with half-integer spin* ($s = \frac{1}{2}, \frac{3}{2}, \ldots$) *by antisymmetric wave functions.*

Particles described by symmetric wave functions are called *bosons*; they obey Bose–Einstein statistics. Particles described by antisymmetric wave functions are called *fermions*; they obey Fermi–Dirac statistics.

The Hamiltonian of the system under consideration consists of three parts: the kinetic energy of the particles, the potential energy of the mutual interaction of the particles, and the potential of the external forces. For simplicity, we shall assume that both the interaction between the particles and the interaction with the external field of forces depend not on the spins and velocities of the particles, but only on their positions, and that the particles interact in pairs. The Schrödinger equation in the coordinate representation is thus of the form,

$$i\frac{\partial}{\partial t}\psi(\mathbf{x}_1\sigma_1, \ldots, \mathbf{x}_n\sigma_n, t) = \left[\frac{-1}{2m}\sum_i \Delta_{\mathbf{x}_i} + \sum_i U(\mathbf{x}_i, t) + \frac{1}{2}\sum_{i \neq j} V(\mathbf{x}_i - \mathbf{x}_j)\right]\psi(\mathbf{x}_1\sigma_1, \ldots, \mathbf{x}_n\sigma_n, t). \tag{2}$$

The Hamiltonian of the system commutes with all the permutation operators P. Thus, we may seek the solutions of the Schrödinger equation

The theorem on the connection between the spin and the statistics was first proved by W. Pauli (1940) in the relativistic theory of particles. A modern proof of this theorem can be found in the book by Streater and Wightman, and also in the book by Bogolyubov, Logunov, and Todorov.

with arbitrary symmetry properties, and in particular we may consider symmetric and antisymmetric solutions.

The scalar product of two state vectors Ψ and Φ for a system of n particles is equal to the scalar product of the wave functions,

$$(\Psi|\Phi) = (\psi|\varphi)$$

$$\equiv \sum_{\sigma_1\ldots n} \int d^3x_1 \ldots d^3x_n \psi^*(\mathbf{x}_1\sigma_1, \ldots, \mathbf{x}_n\sigma_n, t)\varphi(\mathbf{x}_1\sigma_1, \ldots, \mathbf{x}_n\sigma_n, t). \quad (3)$$

Scattering in an External Potential

The wave function of a system of n identical mutually non-interacting particles, moving in a given force field of potential $U(\mathbf{x}, t)$, satisfies a Schrödinger equation of the form

$$i\frac{\partial}{\partial t}\psi(\mathbf{x}_1\sigma_1, \ldots, \mathbf{x}_n\sigma_n, t)$$

$$= \left[-\frac{1}{2m}\sum_i \Delta_{x_i} + \sum_i U(\mathbf{x}_i, t)\right]\psi(\mathbf{x}_1\sigma_1, \ldots, \mathbf{x}_n\sigma_n, t). \quad (4)$$

We shall study the time evolution of the function ψ. Let the functions $\psi_i(\mathbf{x}, \sigma, t)$ form a complete set of solutions to the one-particle Schrödinger equation with potential $U(\mathbf{x}, t)$. Out of this system we select n (not necessarily different) functions ψ_{i_k} ($k = 1, \ldots, n$). It may be verified that the function

$$\psi_{\{i_k\}}(\mathbf{x}_1\sigma_1, \ldots, \mathbf{x}_n\sigma_n, t) \equiv N \sum_{\text{perm } j_k} \varepsilon_P \psi_{j_1}(\mathbf{x}_1\sigma_1, t) \ldots \psi_{j_n}(\mathbf{x}_n\sigma_n, t) \quad (5)$$

satisfies the Schrödinger equation (4). In this formula, N is the normalization factor. The summation is taken over all permutations of the set of indices $(i_1, \ldots, i_n)$. The factor ε_P is always unity for bosons, whereas for fermions it is $+1$ for even permutations of the indices j_k and -1 for odd ones. The fundamental permutation of indices, for which $j_k = i_k$, is regarded as even.

The wave functions so constructed have suitable symmetry properties. Antisymmetric wave functions of the form (5) have an important property which is an expression of the Pauli exclusion principle.

These functions are zero if one and the same function ψ_i appears twice in a product. This means that two fermions of the same kind cannot simultaneously be in the same state. The wave functions (5) constitute a complete set in the space of symmetric or antisymmetric wave functions of n pairs of variables $(\mathbf{x}_i, \sigma_i)$.

The wave function (5) describes a system of independently moving particles in states described by one-particle wave functions. Such a straightforward description of the system is possible only because the mutual interaction between particles has been disregarded.

Scattering is described most conveniently by using one-particle functions of the type $\psi_R[\mathbf{x}, \sigma, t|g]$ and $\psi_A[\mathbf{x}, \sigma, t|f]$. These functions are a natural generalization, in the case of particles with spin, of the functions defined in the preceding section. The profile of a wave packet into which solutions ψ_R and ψ_A go over in the remote past and far future, respectively, is a function of $\mathbf{p}$ and σ in the given case. Inasmuch as the interactions are assumed to be spin-independent, the indices σ play a role only under symmetrization or antisymmetrization of the wave functione. To simplify the notation, henceforth in this section we shall frequently omit all the spin indices. If needed, they can be easily added in the final formulae. For denoting n-particle wave functions constructed from the products of the functions ψ_R and ψ_A, we introduce the notation†

$$\psi_{R,A}[\mathbf{x}_1, \ldots, \mathbf{x}_n, t|g_1, \ldots, g_n]$$
$$= \frac{1}{\sqrt{n!}} \sum_{\text{perm } i_k} \varepsilon_P \psi_{R,A}[\mathbf{x}_1, t|g_{i_1}] \ldots \psi_{R,A}[\mathbf{x}_n, t|g_{i_n}]. \tag{6}$$

The wave function $\psi_R[\mathbf{x}_1, \ldots, \mathbf{x}_n, t|g_1, \ldots, g_n]$ describes a state of the system such that in the remote past it contained one particle with wave packet profile $g_1(\mathbf{p}, \sigma)$, one with profile $g_2(\mathbf{p}, \sigma)$, etc. The state described by the function ψ_A may be characterized in a similar manner.

The *scattering amplitude* of n not mutually interacting particles in an external field, which we shall denote by the symbol $S[f_1 \ldots f_n, g_1 \ldots g_n]$, is defined by the formula

$$S[f_1 \ldots f_n, g_1 \ldots g_n] \equiv (\psi_A[f_1 \ldots f_n]|\psi_R[g_1 \ldots g_n]).$$

† The choice of the normalization factor $1/\sqrt{n!}$ in this formula corresponds to the choice of a system of different orthogonal and normalized profiles $g_1, \ldots, g_n$. In the general case, this factor must be modified in an appropriate manner.

The scattering amplitude so defined is the amplitude of the probability of finding a system of n particles with wave packet profiles $f_1, \ldots, f_n$ in the far future if in the remote past this system was characterized by the profiles $g_1, \ldots, g_n$. For a system consisting of one particle, this definition coincides with the definition of one-particle scattering amplitude given in the preceding section (cf. 4.22a). The scattering amplitude of n particles may be expressed in terms of one-particle scattering amplitudes for which we introduce the notation

$$S[f_i, g_j] \equiv S[i, j].$$

From the definition of scattering amplitude and from the definition of $\psi_{\mathrm{R,A}}[g_1 \ldots g_n]$ we have:

$$S[f_1 \ldots f_n, g_1 \ldots g_n] = \frac{1}{n!} \sum_{\text{perm } i_k} \varepsilon_P \sum_{\text{perm } j_k} \varepsilon'_P S[i_1, j_1] \ldots S[i_n, j_n]. \quad (7)$$

The fact that an n-particle scattering amplitude can be expressed in a simple manner in terms of one-particle scattering amplitudes is due to the assumption that no interaction occurs between the particles.

Transition amplitudes with the participation of bound states can be described in similar fashion.

The Occupation-Number Representation

Systems of many identical particles are conveniently described by means of a new representation of wave functions and operators which is particularly suited for the purpose. This is known as the *occupation-number representation* because in it we give direct information about how numerously one-particle states are occupied by the partieles of the system in question.

We shall be dealing with the description of a system of n identical particles possessing spin, bosons or fermions. Let us introduce a complete and orthonormal set of one-particle functions $h_i(\mathbf{x}\,\sigma)$,

$$\sum_{\sigma} \int \mathrm{d}^3x\, h_i^*(\mathbf{x}\,\sigma) h_j(\mathbf{x}\,\sigma) = \delta_{ij}, \quad (8a)$$

$$\sum_{i} h_i(\mathbf{x}\,\sigma) h_i^*(\mathbf{y}\,\sigma') = \delta_{\sigma\sigma'}\, \delta(\mathbf{x}-\mathbf{y}). \quad (8b)$$

In the space of n-particle functions, we choose a basis consisting of the functions $\varphi_{n_1 \ldots n_k \ldots}(\mathbf{x}_1 \sigma_1, \ldots, \mathbf{x}_n \sigma_n)$ constructed according to the prescription

$$\varphi_{n_1 \ldots n_k \ldots}(\mathbf{x}_1 \sigma_1, \ldots, \mathbf{x}_n \sigma_n)$$

$$\equiv \left(\frac{n_1! \ldots n_k! \ldots}{n!} \right)^{\frac{1}{2}} \sum_{\text{perm } i_k} \varepsilon_P h_{i_1}(\mathbf{x}_1 \sigma_1) \ldots h_{i_n}(\mathbf{x}_n \sigma_n), \tag{9}$$

where the summation runs over all nonequivalent permutations of the indices $i_1, \ldots, i_n$ satisfying the condition that the number of indices 1 is n_1, of indices 2 is n_2, etc. The number of such permutations is $n!/(n_1! \ldots n_k! \ldots)$, which justifies the choice of the normalization factor in formula (9). The numbers $n_1, n_2, \ldots$ are called *occupation numbers*. In the many-particle state described by the wave function (9) there are n_1 particles in a state described by the function h_1, n_2 particles in a state described by the function h_2, etc. For bosons, the occupation numbers n_k may take on the values from 0 to n, whereas for fermions n_k may assume only values of 0 or 1 (Pauli exclusion principle).

By virtue of conditions (8), the wave functions (9) form a complete orthonormal set of wave functions in the space of symmetric or antisymmetric wave functions of n variables $\mathbf{x}_i$ and n variables σ_i. The set of coefficients $\psi(n_1, \ldots, n_k, \ldots, t)$ in the expansion of an arbitrary wave function into functions of this system is called a *wave function in the occupation-number representation* in the Schrödinger picture:

$$\psi(\mathbf{x}_1 \sigma_1, \ldots, \mathbf{x}_n \sigma_n, t)$$

$$= \sum_{n_1 \ldots n_k \ldots} \psi(n_1, \ldots, n_k, \ldots, t) \varphi_{n_1 \ldots n_k \ldots}(\mathbf{x}_1 \sigma_1, \ldots, \mathbf{x}_n \sigma_n). \tag{10}$$

The sum of the occupation numbers figuring in formula (10) is equal to the number of particles in the system

$$\sum_{k=1}^{\infty} n_k = n.$$

For bosons, the summation in formula (10) runs over all sets of occupation numbers which satisfy this condition, whereas for fermions, it runs over all sets of occupation numbers consisting of 0's and 1's which also satisfy this condition.

State vectors corresponding to the wave functions $\varphi_{n_1 \dots n_k \dots}$ will be denoted by the symbols $\Phi_{n_1 \dots n_k \dots}$ or $\Phi_{\{n_k\}}$. These vectors should actually be denoted by the symbols $\Phi^t_{\{n_k\}}$ indicating the instant of time t to which the occupation numbers $n_1, \dots, n_\kappa, \dots$ refer. In the discussion concerning the general properties of sets of vectors $\Phi_{\{n_k\}}$, we shall frequently omit the superscript t wherever this causes no confusion.

A wave function in the occupation-number representation which corresponds to the state vector Ψ can be expressed by the scalar product

$$\psi(n_1, \dots, n_k, \dots) = (\Phi_{n_1 \dots n_k \dots} | \Psi).$$

Operators in the occupation-number representation are infinite matrices, whose elements are labelled by two sets of occupation numbers,

$$A(m_1 \dots m_k \dots; n_1 \dots n_k \dots) = (\Phi_{m_1 \dots m_k \dots} | A \Phi_{n_1 \dots n_k \dots}).$$

The vectors $\Phi_{\{n_k\}}$ form an orthonormal basis in the state-vector space of the system, i.e. the following conditions are satisfied:

$$(\Phi_{\{n_k\}} | \Phi_{\{n'_k\}}) = \delta_{n_1 n'_1} \dots \delta_{n_k n'_k} \dots, \tag{11a}$$

$$\sum_{n_1 \dots n_k \dots} P(n_1 \dots n_k \dots) = 1, \tag{11b}$$

where $P(n_1 \dots n_k)$ is a projection operator on the direction $\Phi_{n_1 \dots n_k \dots}$. The summation in formula (11b), for bosons, runs over all sets of numbers n_i satisfying the condition $\sum_{i=1}^{\infty} n_i = n$; for fermions, it runs over sets of numbers n_i consisting of 0's and 1's which also satisfy the condition

$$\sum_{i=1}^{\infty} n_i = n.$$

For every basis $\Phi_{\{n_k\}}$ we can define *occupation-number operators* N_i which are diagonal in the given basis. The eigenvalues of these operators are equal to the occupation numbers:

$$N_i \Phi_{\{n_k\}} = n_i \Phi_{\{n_k\}}. \tag{12}$$

Occupation-number operators are self-adjoint. They form a complete set of commuting operators, i.e. the basis vector is determined by giving the eigenvalues of all the operators N_i. The sum of all occupation-number operators is the operator of the total number of particles. We denote it by the symbol N:

$$N = \sum_{i=1}^{\infty} N_i. \tag{13}$$

Creation and Annihilation Operators

Discarding the condition $\sum_{i=1}^{\infty} n_i = n$, we can in a natural manner extend the space of the state vectors of the system. This extension is necessary if we want to describe systems with a varying number of particles (for instance, systems of photons interacting with sources of radiation). To describe a system of bosons we shall hereafter use a space spanned by all vectors $\Phi_{\{n_k\}}$ for which the occupation numbers n_k assume arbitrary values limited only by the condition† $\sum_{i=1}^{\infty} n_i < \infty$. To describe a system of fermions we shall use a space spanned by all vectors $\Phi_{\{n_k\}}$ for which the occupation numbers are 0 or 1 and satisfy the condition $\sum_{i=1}^{\infty} n_i < \infty$. We shall assume that these vectors in both cases constitute an orthonormal basis in the extended state-vector space, i.e. that relations (11) hold for them, the summation over the sets of numbers n_k in formula (11b) being limited only by the condition $\sum_{i=1}^{\infty} n_i < \infty$. In the case when the particles carry electric charge (or other charges obeying absolute conservation laws), the combination of subspaces with different values of total charge into one large space is purely a convenient mathematical construction. All operators corresponding to physical quantities operate independently in each subspace.

In the extended space, besides the operators of the occupation numbers as defined by formula (12), the *creation operators* $a_k^\dagger$ and *annihilation operators* a_k which, respectively, increase or reduce the occupation

† This condition ensures that the Hilbert space so constructed is separable.

number by one are natural linear operators.† In the given basis we define them in the following manner:#
for bosons:

$$a_k^\dagger \Phi_{n_1 \dots n_k \dots} = \sqrt{n_k+1}\, \Phi_{n_1 \dots n_k+1 \dots}, \tag{14a}$$

$$a_k \Phi_{n_1 \dots n_k \dots} = \begin{cases} \sqrt{n_k}\, \Phi_{n_1 \dots n_k-1 \dots}, & \text{when} \quad n_k > 0, \\ 0 & \text{when} \quad n_k = 0, \end{cases} \tag{14b}$$

for fermions:

$$a_k^\dagger \Phi_{n_1 \dots n_k \dots} = \begin{cases} \nu(n_1 \dots n_{k-1}) \Phi_{n_1 \dots n_k+1 \dots}, & \text{when} \quad n_k = 0, \\ 0, & \text{when} \quad n_k = 1, \end{cases} \tag{15a}$$

$$a_k \Phi_{n_1 \dots n_k \dots} = \begin{cases} 0, & \text{when} \quad n_k = 0, \\ \nu(n_1 \dots n_{k-1}) \Phi_{n_1 \dots n_k-1 \dots}, & \text{when} \quad n_k = 1. \end{cases} \tag{15b}$$

The numbers $\nu(n_1 \dots n_i)$, which bear the name of *Wigner factors*, are defined in the following manner:

$$\nu(n_1 \dots n_i) \equiv (-1)^{\sum_{k=1}^{i} n_k} = \prod_{k=1}^{i} (1-2n_k). \tag{16}$$

It follows from these definitions that the creation and annihilation operators, $a_k^\dagger$ and a_k, are mutually adjoint. The numerical factors in definitions (14) and (15) have been chosen so as to obtain for bosons the simple commutation relations

$$[a_i, a_j^\dagger]_- = \delta_{ij}, \tag{17a}$$

$$[a_i, a_j]_- = 0 = [a_i^\dagger, a_j^\dagger]_-, \tag{17b}$$

and for fermions the simple anticommutation relations

$$[a_i, a_j^\dagger]_+ = \delta_{ij}, \tag{18a}$$

$$[a_i, a_j]_+ = 0 = [a_i^\dagger, a_j^\dagger]_+ . \tag{18b}$$

---

† We recall that because the basis vectors $\Phi_{\{n_k\}}$ are always referred to a certain instant t, the creation and annihilation operators also depend on t. The superscript t will in general be omitted.

Creation and annihilation operators for bosons (to be precise for photons) were introduced by P. A. M. Dirac (1927), and for fermions by P. Jordan and E. Wigner (1928).

In acting on vectors which do not contain particles in the k-th state, the operator a_k, by definition, yields a null vector, in the case of a boson and a fermion system alike. For fermions we also get a null vector from the action of the creation operator $a_k^\dagger$ on state vectors containing one particle in the k-th state. This result is in accordance with the Pauli exclusion principle. The occupation-number operators can be constructed out of the creation and annihilation operators in the following manner:

$$N_k = a_k^\dagger a_k. \tag{19}$$

Among the state vectors a distinguished role is played by the vector $\Phi_{0\ldots 0\ldots}$ describing a state without particles, known as the *vacuum state*. This vector will be denoted by the symbol Ω. The other basis vectors can be obtained by operating on the *vacuum vector* with the creation operators according to the formula

$$\Phi_{n_1\ldots n_k\ldots} = (n_1!)^{-\frac{1}{2}}(a_1^\dagger)^{n_1}\ldots(n_k!)^{-\frac{1}{2}}(a_k^\dagger)^{n_k}\ldots\Omega. \tag{20}$$

To prove this statement, apply the operator N_k to both sides of formula (20), express the operator N_k in its operation on the right-hand side in terms of the creation and annihilation operators in accordance with formula (19), and make use of the commutation or anticommutation rules for the operators $a_i^\dagger$ and a_j. Next, utilize the fact that the operators N_k form a complete set. To complete the proof, verify that vectors constructed according to formula (20) are normalized.

The basis consisting of the vectors $\Phi_{\{n_k\}}$, called the *Fock basis*, can be depicted as an inverted pyramid based on the vacuum vector Ω. The upper levels contain, successively, the vectors of one-particle, two-particle, three-particle states, etc. For bosons this pyramid is of the form

$$\begin{array}{c}
\ldots\ldots\ldots\ldots\ldots\ldots\ldots\ldots\ldots\ldots\ldots\ldots\ldots\ldots\ldots\ldots\\
\frac{1}{\sqrt{3!}}a_1^\dagger a_1^\dagger a_1^\dagger\Omega, \qquad \frac{1}{\sqrt{2!}}a_1^\dagger a_1^\dagger a_2^\dagger\Omega, \ldots\\
\frac{1}{\sqrt{2!}}a_1^\dagger a_1^\dagger\Omega, \qquad a_1^\dagger a_2^\dagger\Omega, \ldots\\
a_1^\dagger\Omega, \qquad a_2^\dagger\Omega, \ldots\\
\Omega
\end{array}$$

The creation and annihilation operators depend on the choice of basis in the state-vector space used in their definition. For each basis we obtain a different set of operators a_k and $a_k^\dagger$. Since the basis is characterized by giving the occupation numbers in relation to a particular instant of time t (or asymptotic region), the symbol denoting an annihilation or creation operator should have the superscript t. We shall introduce this superscript whenever the time-dependence of these operators is essential and in particular when we shall be using several different sets of operators simultaneously.

All operators acting on the vectors of the state of a system of identical particles can be expressed in terms of the creation and annihilation operators. This construction will be given after the introduction of field operators.

Field Operators

A description of many-particle systems by using creation and annihilation operators requires the introduction of a particular basis in the space of one-particle functions. At times this is even an advantage since we can simplify the description of the system by choosing the basis vectors so that they correspond to states of a system distinguished by the physical conditions. In general considerations and in investigations of scattering phenomena, however, it is more convenient to make use of a description of many-particle systems which is based on field operators. This description no longer contains any distinguished system of one-particle states. This may be said to be an abstract treatment of the method of creation and annihilation operators.

The starting point for the construction of field operators is the same extended Hilbert space $\mathscr{H}$ used previously for defining the creation and annihilation operators. This space is the direct sum of a sequence of spaces: zero-particle (vacuum vector), one-particle, two-particle, etc.,

$$\mathscr{H} = \bigoplus_{n=0}^{\infty} \mathscr{H}^{(n)}.$$

Each vector Ψ of this space may have associated with it a sequence of wave functions (symmetric or antisymmetric) taken at a particular instant t,

$$\Psi^{(0)}(t), \quad \Psi^{(1)}(\mathbf{x}\sigma, t), \quad \Psi^{(2)}(\mathbf{x}_1\sigma_1, \mathbf{x}_2\sigma_2, t), \ldots$$

These functions will be called components of the vector and will always be denoted by the same upper-case Greek letters as the vectors to which they correspond. The operations of adding vectors, multiplying them by a number, and taking their scalar product may be expressed in the following manner in terms of the components of the vectors:

$$(\lambda\Psi+\mu\Phi)^{(n)}(\mathbf{x}_1\sigma_1, \ldots, \mathbf{x}_n\sigma_n, t)$$
$$= \lambda\Psi^{(n)}(\mathbf{x}_1\sigma_1, \ldots, \mathbf{x}_n\sigma_n, t)+\mu\Phi^{(n)}(\mathbf{x}_1\sigma_1, \ldots, \mathbf{x}_n\sigma_n, t), \tag{21a}$$

$$(\Psi|\Phi) = \sum_{n=0}^{\infty} \sum_{\sigma_1\ldots\sigma_n} \int d^3x_1 \ldots d^3x_n$$
$$\times \Psi^{(n)*}(\mathbf{x}_1\sigma_1, \ldots, \mathbf{x}_n\sigma_n, t)\Phi^{(n)}(\mathbf{x}_1\sigma_1, \ldots, \mathbf{x}_n\sigma_n, t). \tag{21b}$$

The unique association of sequences of wave functions with state vectors from the abstract Hilbert space $\mathscr{H}$ is, from the mathematical point of view, an isomorphism of two Hilbert spaces. Inasmuch as wave functions have a well-known physical interpretation, this isomorphism plays an important role, determining the physical parametrization† of Hilbert spaces. The parametrization of the Hilbert space by means of sequences of wave functions made up of functions describing zero-particle states, one-particle states, two-particle states, etc., will be called a *Fock parametrization.* Various Fock parametrizations of the space of state vectors will appear in several more places in this section.

In the space $\mathscr{H}$ parametrized by sequences of wave functions taken at the instant of time t we shall define two mutually adjoint operator distributions‡ $\psi(\mathbf{x}\sigma t)$ and $\psi^\dagger(\mathbf{x}\sigma t)$ known as *field operators.* They

---

† Not much concrete information about physical systems could be obtained from quantum theories without introducing the parametrization of Hilbert spaces. It may be said that only Hilbert spaces parametrized in a manner specifying the physical interpretation of state vectors lend themselves to describing physical systems.

‡ An operator (operator-valued) distribution is a linear (antilinear) functional defined on the space of trial functions with values in the set of operators.

associate with each one-particle wave function $h(\mathbf{x}\sigma)$ the operators $\psi^t[h]$ and $\psi^{\dagger t}[h]$:

$$h(\mathbf{x}\sigma)\xrightarrow[\text{antilinear in } h]{\psi(\mathbf{x}\sigma t)}\psi^t[h],$$

$$h(\mathbf{x}\sigma)\xrightarrow[\text{linear in } h]{\psi^\dagger(\mathbf{x}\sigma t)}\psi^{\dagger t}[h].$$

In accordance with the tradition of representing distributions as integrals, the association above may be written as

$$\psi^t[h] = \sum_\sigma \int d^3x h^*(\mathbf{x}\sigma)\psi(\mathbf{x}\sigma t), \tag{22a}$$

$$\psi^{\dagger t}[h] = \sum_\sigma \int d^3x h(\mathbf{x}\sigma)\psi^\dagger(\mathbf{x}\sigma t). \tag{22b}$$

The action of the operators $\psi^t[h]$ and $\psi^{\dagger t}[h]$ in the space $\mathscr{H}$ is defined in the following manner,

$$(\psi^t[h]\Phi)^{(n)}(\mathbf{x}_1\sigma_1, \ldots, \mathbf{x}_n\sigma_n, t)$$
$$= \sqrt{n+1}\sum_\sigma \int d^3x h^*(\mathbf{x}\sigma)\Phi^{(n+1)}(\mathbf{x}\sigma, \mathbf{x}_1\sigma_1, \ldots, \mathbf{x}_n\sigma_n, t), \tag{23a}$$

$$(\psi^{\dagger t}[h]\Phi)^{(n)}(\mathbf{x}_1\sigma_1, \ldots, \mathbf{x}_n\sigma_n, t)$$
$$= \frac{1}{\sqrt{n}}\sum_{i=1}^n (\pm 1)^{i+1} h(\mathbf{x}_i\sigma_i)\Phi^{(n-1)}(\mathbf{x}_1\sigma_1, \ldots, \check{\mathbf{x}}_i\check{\sigma}_i, \ldots, \mathbf{x}_n\sigma_n, t), \tag{23b}$$

$$(\psi^{\dagger t}[h]\Phi)^{(0)} = 0, \tag{23c}$$

where, as usual, the upper sign applies to bosons and the lower one to fermions, whereas the symbol $\check{\mathbf{x}}_i\check{\sigma}_i$ denotes that the argument $\mathbf{x}_i\sigma_i$ does not appear among the arguments of the function $\Phi^{(n-1)}$. The operators $\psi^t[h]$ and $\psi^{\dagger t}[h]$ are mutually adjoint with respect to the scalar product (21b). From the definition of the operators $\psi^t[h]$ and $\psi^{\dagger t}[h]$ it follows that they satisfy the following commutation relations,

$$[\psi^t[h_1], \psi^{\dagger t}[h_2]]_\mp = (h_1|h_2), \tag{24a}$$

$$[\psi^t[h_1], \psi^t[h_2]]_\mp = 0 = [\psi^{\dagger t}[h_1], \psi^{\dagger t}[h_2]]_\mp. \tag{24b}$$

The distribution $\psi(\mathbf{x}\sigma t)$ is simultaneously a linear operator† in the space $\mathscr{H}$, whose operation is defined by the formula

$$(\psi(\mathbf{x}\sigma t)\Phi)^{(n)}(\mathbf{x}_1\sigma_1, \ldots, \mathbf{x}_n\sigma_n, t) = \sqrt{n+1}\,\Phi^{(n+1)}(\mathbf{x}\sigma, \mathbf{x}_1\sigma_1, \ldots, \mathbf{x}_n\sigma_n, t). \tag{25a}$$

The field operator $\psi(\mathbf{x}\sigma t)$ annihilates the vacuum state vector

$$\psi(\mathbf{x}\sigma t)\Omega = 0. \tag{25b}$$

The vector $\psi(\mathbf{x}\sigma t)$ does not, however, have an adjoint among the operators acting in the space $\mathscr{H}$. By analogy to $\psi(\mathbf{x}\sigma t)$, we define the action of the operators $\psi^\dagger(\mathbf{x}\sigma t)$ by means of the formulae

$$(\psi^\dagger(\mathbf{x}\sigma t)\Phi)^{(n)}(\mathbf{x}_1\sigma_1, \ldots, \mathbf{x}_n\sigma_n, t)$$
$$= \frac{1}{\sqrt{n}}\sum_{i=1}^{n}(\pm 1)^{i-1}\delta_{\sigma\sigma_i}\delta(\mathbf{x}-\mathbf{x}_i)\Phi^{(n-1)}(\mathbf{x}_1\sigma_1, \ldots, \check{\mathbf{x}}_i\check{\sigma}_i, \ldots, \mathbf{x}_n\sigma_n, t), \tag{26a}$$

$$(\psi^\dagger(\mathbf{x}\sigma t)\Phi)^{(0)} = 0. \tag{26b}$$

This operator $\psi^\dagger(\mathbf{x}\sigma t)$ belongs to a class of linear operators operating in a vector space broader than the initial Hilbert space, viz. in the space D whose elements are sequences of distributions (symmetric or antisymmetric). In the space D the formulae (26) are not only of symbolic character.

The field operators $\psi(\mathbf{x}\sigma t)$ and $\psi^\dagger(\mathbf{x}\sigma t)$ are merely auxiliary mathematical objects and hence one may accept the fact that they do not operate‡ in the physical Hilbert space of state vectors. All operators representing physical quantities may, however, be constructed out of field operators in a straightforward and instructive manner. The simple rules for operating with the objects $\psi(\mathbf{x}\sigma t)$ and $\psi^\dagger(\mathbf{x}\sigma t)$ make formalism based on field operators a very convenient tool for describing many-particle systems.

---

† Just as distributions with numerical values may simultaneously be ordinary functions.

‡ The action of the operator $\psi(\mathbf{x}\sigma t)$ in the space $\mathscr{H}$ may be defined in non-relativistic theory as we have done above, but this can no longer be done in relativistic theory.

Underlying this operator calculus are the following simple commutation relations:

$$[\psi(\mathbf{x}\sigma t), \psi^{\dagger}(\mathbf{y}\,\sigma'\,t)]_{\mp} = \delta_{\sigma\sigma'}\,\delta(\mathbf{x}-\mathbf{y}), \tag{27a}$$

$$[\psi(\mathbf{x}\sigma t)\,\psi(\mathbf{y}\,\sigma'\,t)]_{\mp} = 0 = [\psi^{\dagger}(\mathbf{x}\sigma t), \psi^{\dagger}(\mathbf{y}\,\sigma'\,t)]_{\mp}. \tag{27b}$$

These relations follow from formulae (25) and (26) which define the action of field operators in the distribution space D.

The basic formula, which we shall henceforth use frequently, specifies the relation between field operators and wave functions:

$$\frac{1}{\sqrt{n!}}(\Omega|\psi(\mathbf{x}_n\sigma_n\,t)\ldots\psi(\mathbf{x}_1\sigma_1\,t)\Phi) = \Phi^{(n)}(\mathbf{x}_1\sigma_1, \ldots, \mathbf{x}_n\sigma_n, t). \tag{28}$$

This formula is proved by rewriting the left-hand side as

$$\frac{1}{\sqrt{n!}}(\psi(\mathbf{x}_n\sigma_n\,t)\ldots\psi(\mathbf{x}_1\sigma_1\,t)\Phi)^{(0)}$$

and applying formula (25a) n times.

Formula (28) means that the expression $\frac{1}{\sqrt{n!}}\psi^{\dagger}(\mathbf{x}_1\sigma_1\,t)\ldots\psi^{\dagger}(\mathbf{x}_n\sigma_n\,t)\Omega$ is a vector-valued distribution associating corresponding wave functions with vectors from Hilbert space. As in the case of a single particle, these distributions are denoted by the Dirac symbol

$$|\mathbf{x}_1\sigma_1, \ldots, \mathbf{x}_n\sigma_n, t\rangle = \frac{1}{\sqrt{n!}}\psi^{\dagger}(\mathbf{x}_1\sigma_1\,t)\ldots\psi^{\dagger}(\mathbf{x}_n\sigma_n\,t)\Omega.$$

Now we shall discuss the relationship between field operators and creation and annihilation operators. For this purpose let us consider a particular orthonormal set of one-particle functions $h_i(\mathbf{x}\sigma)$ described on p. 42 and define the sequences of operators $\psi^t[h_i]$ and $\psi^{\dagger t}[h_i]$ in terms of these functions. With the aid of formulas (23) it is possible to verify that the action of the operators $\psi^t[h_i]$ on the state vectors $\Phi^t_{\{n_k\}}$ with wave functions (5) at the instant of time t is the same as that of the annihilation operators a_i^t, whereas the operation of the operators $\psi^{\dagger t}[h_i]$ is the same as that of the creation operators $a_i^{\dagger t}$.

Since the vectors $\Phi_{\{n_k\}}$ constitute the basis in the Hilbert space $\mathscr{H}$, the following equalities hold:

$$a_i^t = \psi^t[h_i] \equiv \sum_\sigma \int d^3x\, h_i^*(\mathbf{x}\sigma)\psi(\mathbf{x}\sigma t), \tag{29a}$$

$$a_i^{\dagger t} = \psi^{\dagger t}[h_i] \equiv \sum_\sigma \int d^3x\, h_i(\mathbf{x}\sigma)\psi^\dagger(\mathbf{x}\sigma t). \tag{29b}$$

If the completeness of the set of functions h_i is utilized, the relations above can be formally reversed to obtain field operators in the form of infinite series of annihilation or creation operators:

$$\psi(\mathbf{x}\sigma t) = \sum_{i=1}^{\infty} h_i(\mathbf{x}\sigma)a_i^t, \tag{30a}$$

$$\psi^\dagger(\mathbf{x}\sigma t) = \sum_{i=1}^{\infty} h_i^*(\mathbf{x}\sigma)a_i^{\dagger t}. \tag{30b}$$

The convergence of these series should be taken to mean convergence in the sense of a distribution, i.e.

$$\sum_{i=1}^{n}\sum_\sigma \int d^3x\, h^*(\mathbf{x}\sigma)h_i(\mathbf{x}\sigma)a_i^t \xrightarrow[n\to\infty]{} \psi^t[h].$$

Operators representing the fundamental physical quantities for systems of many particles may be presented in easily remembered form by using field operators. We shall confine ourselves to a discussion of one-particle and two-particle operators, $A_{(1)}$ and $A_{(2)}$, i.e. such that their action on the wave function $\psi(\mathbf{x}_1\sigma_1, \ldots, \mathbf{x}_n\sigma_n, t)$ may be expressed by the formulae

$$(A_{(1)}\psi)(\mathbf{x}_1\sigma_1, \ldots, \mathbf{x}_n\sigma_n, t) = \sum_{k=1}^{n} A_{x_k}\psi(\mathbf{x}_1\sigma_1, \ldots, \mathbf{x}_n\sigma_n, t), \tag{31a}$$

$$(A_{(2)}\psi)(\mathbf{x}_1\sigma_1, \ldots, \mathbf{x}_n\sigma_n, t) = \frac{1}{2}\sum_{\substack{k,l=1\\k\neq l}}^{n} A_{x_kx_l}\psi(\mathbf{x}_1\sigma_1, \ldots, \mathbf{x}_n\sigma_n, t), \tag{31b}$$

where A_{x_k} and $A_{x_kx_l}$ are operations which act on the variables $\mathbf{x}_k\sigma_k$ and $\mathbf{x}_l\sigma_l$. The operators for the kinetic energy of a system of particles $\hat{T}$, the potential energy in an external field $\hat{U}$, the total momentum $\hat{\mathbf{P}}$, and the angular momentum $\hat{\mathbf{M}}$ are one-particle operators whereas the operator for the potential energy of mutual interaction $\hat{V}$ is a two-particle operator. The operators $A_{(1)}$ and $A_{(2)}$ expressed in terms of field operators are of the form

$$A_{(1)} = \sum_{\sigma} \int d^3x \psi^{\dagger}(\mathbf{x}\sigma t) A_x \psi(\mathbf{x}\sigma t), \tag{32a}$$

$$A_{(2)} = \frac{1}{2} \sum_{\sigma\sigma'} \int d^3x d^3y \psi^{\dagger}(\mathbf{x}\sigma t) \psi^{\dagger}(\mathbf{y}\sigma' t) A_{xy} \psi(\mathbf{y}\sigma' t) \psi(\mathbf{x}\sigma t), \tag{32b}$$

where the operations A_x and A_{xy} operate on the variables $\mathbf{x}\sigma$ or $\mathbf{x}\sigma$ and $\mathbf{y}\sigma'$ in the field operators.

These formulae may be proved by substituting the vectors $A_{(1)}\Phi$ and $A_{(2)}\Phi$, respectively, for the vector Φ in formula (28). The formulae thus obtained for $(A_{(1,2)}\Phi)^{(n)}(\mathbf{x}_1\sigma_1, \ldots, \mathbf{x}_n\sigma_n, t)$ can then be transformed, by repeated application of the commutation rules, to the form of the definition formulae (31). In this process the operators $\psi^{\dagger}$ should be moved to the left until an expression of the type $(\Omega|\psi^{\dagger}(\mathbf{x}\sigma t)\ldots)$ is obtained; by the condition (25b) this expression is zero.

In particular, for the operators given above we get:

$$\hat{T} = -\frac{1}{2m} \sum_{\sigma} \int d^3x \psi^{\dagger}(\mathbf{x}\sigma t)\, \Delta \psi(\mathbf{x}\sigma t), \tag{33a}$$

$$\hat{U} = \sum_{\sigma} \int d^3x \psi^{\dagger}(\mathbf{x}\sigma t)\, U(\mathbf{x}, t) \psi(\mathbf{x}\sigma t), \tag{33b}$$

$$\hat{\mathbf{P}} = \sum_{\sigma} \int d^3x \psi^{\dagger}(\mathbf{x}\sigma t) \frac{1}{i} \nabla \psi(\mathbf{x}\sigma t), \tag{33c}$$

$$\hat{\mathbf{M}} = \sum_{\sigma} \int d^3x \psi^{\dagger}(\mathbf{x}\sigma t) \left[\left(\frac{1}{i} \mathbf{x} \times \nabla \right) \delta_{\sigma\sigma'} + \mathbf{s}_{\sigma\sigma'} \right] \psi(\mathbf{x}\sigma' t), \tag{33d}$$

$$\hat{V} = \frac{1}{2} \sum_{\sigma\sigma'} \int d^3x d^3y \psi^{\dagger}(\mathbf{x}\sigma t) \psi^{\dagger}(\mathbf{y}\sigma' t) V(\mathbf{x}-\mathbf{y}) \psi(\mathbf{y}\sigma' t) \psi(\mathbf{x}\sigma t). \tag{33e}$$

All of these expressions are constructed in exactly the same way as the average values of the corresponding physical quantities, except that field operators appear in the place of wave functions.

If into these formulae we substitute field operators expressed in terms of the creation and annihilation operators, in the general case we arrive at:

$$A_{(1)} = \sum_{i,j=1}^{\infty} a_i^{\dagger} A(i,j) a_j, \tag{34a}$$

$$A_{(2)} = \frac{1}{2} \sum_{i,j,k,l=1}^{\infty} a_i^{\dagger} a_j^{\dagger} A(ij,kl) a_k a_l, \tag{34b}$$

where

$$A(ij) = \sum_{\sigma} \int d^3x\, h_i^*(\mathbf{x}\sigma) A_x h_j(\mathbf{x}\sigma), \tag{35a}$$

$$A(ij,kl) = \sum_{\sigma\sigma'} \int d^3x\, d^3y\, h_i^*(\mathbf{x}\sigma) h_j^*(\mathbf{y}\sigma') A_{xy} h_k(\mathbf{y}\sigma') h_l(\mathbf{x}\sigma). \tag{35b}$$

These formulae (especially formula (34b)) are quite tedious to derive without using field operators.

Hitherto we have dealt with field operators constructed by using wave functions at a given instant of time t. Now we shall examine the dependence of the field operators on the variable t. For this purpose, let us assume that the wave functions describing n-particle states are subject to the Schrödinger equation (2). Having defined the time-dependence of the wave functions in this way, we can then find the time-dependence of the field operators. The differential equation determining this relationship can be obtained from formulae (25) and (26) which define the action of the field operators; this is done by differentiating these formulae with respect to t and using the Schrödinger equation for the wave functions $\Phi^{(n)}$ and $\Phi^{(n+1)}$ and again using the definition equations. In this way, we get

$$\left(i\frac{\partial}{\partial t}\psi(\mathbf{x}\sigma t)\Phi\right)^{(n)}(\mathbf{x}_1\sigma_1, \ldots, \mathbf{x}_n\sigma_n, t)$$

$$= \sqrt{n+1}\left(-\frac{1}{2m}\Delta_x + U(\mathbf{x}, t)\right)\Phi^{(n+1)}(\mathbf{x}\sigma, \mathbf{x}_1\sigma_1, \ldots, \mathbf{x}_n\sigma_n, t)$$

$$+\sqrt{n+1}\sum_{i=1}^{n} V(\mathbf{x}-\mathbf{x}_i)\Phi^{(n+1)}(\mathbf{x}\sigma, \mathbf{x}_1\sigma_1, \ldots, \mathbf{x}_n\sigma_n, t).$$

Repeated application of the definition equation for the operator ψ enables us to reduce the right-hand side of the equality above to the form

$$\left(\left[-\frac{1}{2m}\Delta_x + U(\mathbf{x}, t)\right]\psi(\mathbf{x}\sigma t)\Phi\right)^{(n)}(\mathbf{x}_1\sigma_1, \ldots, \mathbf{x}_n\sigma_n, t)$$

$$+\left(\sum_{\sigma'}\int d^3y\,\psi^\dagger(\mathbf{y}\sigma' t)\psi(\mathbf{y}\sigma' t)V(\mathbf{x}-\mathbf{y})\psi(\mathbf{x}\sigma t)\Phi\right)^{(n)}(\mathbf{x}_1\sigma_1, \ldots, \mathbf{x}_n\sigma_n, t).$$

Since the vector Φ is arbitrary, the result obtained may be written as an operator equation, known as a *field equation*:

$$i\frac{\partial}{\partial t}\psi(\mathbf{x}\sigma t) = \left(-\frac{1}{2m}\Delta + U(\mathbf{x}, t)\right)\psi(\mathbf{x}\sigma t)$$

$$+\sum_{\sigma'}\int d^3y\,\psi^\dagger(\mathbf{y}\sigma' t)\psi(\mathbf{y}\sigma' t)V(\mathbf{x}-\mathbf{y})\psi(\mathbf{x}\sigma t). \tag{36a}$$

In similar fashion, for the operator $\psi^\dagger(\mathbf{x}\sigma t)$ we obtain a field equation which, as was to be expected, is adjoint to the equation for ψ:

$$-i\frac{\partial}{\partial t}\psi^\dagger(\mathbf{x}\sigma t) = \left(-\frac{1}{2m}\Delta + U(\mathbf{x}, t)\right)\psi^\dagger(\mathbf{x}\sigma t)$$

$$+\sum_{\sigma'}\int d^3y\,\psi^\dagger(\mathbf{x}\sigma t)V(\mathbf{x}-\mathbf{y})\psi^\dagger(\mathbf{y}\sigma' t)\psi(\mathbf{y}\sigma' t). \tag{36b}$$

Field equations may also be presented in the form of Heisenberg equations of motion,

$$i\frac{\partial}{\partial t}\psi(\mathbf{x}\sigma t) = [\psi(\mathbf{x}\sigma t), H(t)], \tag{37a}$$

$$i\frac{\partial}{\partial t}\psi^\dagger(\mathbf{x}\sigma t) = [\psi^\dagger(\mathbf{x}\sigma t), H(t)], \tag{37b}$$

if the Hamiltonian is written as

$$H(t) = -\frac{1}{2m}\sum_{\sigma}\int d^3x\psi^{\dagger}(\mathbf{x}\sigma t)\Delta\psi(\mathbf{x}\sigma t)$$

$$+\sum_{\sigma}\int d^3x\psi^{\dagger}(\mathbf{x}\sigma t)\,U(\mathbf{x}, t)\psi(\mathbf{x}\sigma t)$$

$$+\frac{1}{2}\sum_{\sigma\sigma'}\int d^3x d^3y\psi^{\dagger}(\mathbf{x}\sigma t)\psi^{\dagger}(\mathbf{y}\sigma' t)\,V(\mathbf{x}-\mathbf{y})\psi(\mathbf{y}\sigma' t)\psi(\mathbf{x}\sigma t), \tag{38}$$

and use is made of the commutation relations for the field operators.

By formulae (33), the three terms comprising the Hamiltonian are identified, respectively, with the operators of the kinetic energy, of the potential energy in an external field, and of the potential energy of the mutual interaction of the particles. Their sum is the operator of the total energy of the system. The time-dependence postulated above for the field operators thus is also their time-dependence in the Heisenberg picture. The field operators at different instants of time t and t_0 are related by the unitary evolution operator $\mathsf{U}(t, t_0)$ whose generator is $H(t)$:

$$\psi(\mathbf{x}\sigma t) = \mathsf{U}(t, t_0)\psi(\mathbf{x}\sigma t_0)\mathsf{U}^{\dagger}(t, t_0), \tag{39a}$$

$$\psi^{\dagger}(\mathbf{x}\sigma t) = \mathsf{U}(t, t_0)\psi^{\dagger}(\mathbf{x}\sigma t_0)\mathsf{U}^{\dagger}(t, t_0). \tag{39b}$$

The discussion of this section could easily be generalized to the momentum representation. To this end it is necessary first to parametrize (in Fock fashion) the space $\mathscr{H}$ with a sequence of wave functions in the momentum representation and then introduce further elements of the field-operator formalism in the manner presented above.

The momentum representation is particularly convenient for describing free particles (without mutual interaction and without a field of external forces) and for describing scattering phenomena. The association of these two problems is no accident since, as we have already shown, the free motion of particles before and after scattering is a source of convenient parametrization of scattering states. As an introduction to the description of scattering phenomena with the aid of field operators, let us now enter into a more detailed discussion of the description of free particles in this formalism.

The starting point of our considerations will once again be the concept of the Fock parametrization of the space of state vectors. Each wave function φ describing the free motion of particles may be expressed in terms of the (symmetric or antisymmetric) wave-packet profile $g(\mathbf{p}_1\sigma_1, \ldots, \mathbf{p}_n\sigma_n)$ by the formula

$$\varphi[\mathbf{x}_1\sigma_1, \ldots, \mathbf{x}_n\sigma_n, t|g]$$
$$= \int \mathrm{d}\Gamma_1 \ldots \mathrm{d}\Gamma_n \mathrm{e}^{-i(E_{p1}+\ldots+E_{pn})t+i(\mathbf{p}_1\cdot\mathbf{x}_1+\ldots+\mathbf{p}_n\cdot\mathbf{x}_n)} g(\mathbf{p}_1\sigma_1, \ldots, \mathbf{p}_n\sigma_n) \quad (40)$$

where†

$$\int \mathrm{d}\Gamma_i = \int \frac{\mathrm{d}^3 p_i}{(2\pi)^3}.$$

The vectors of Hilbert space which describe the states of free particles can thus be Fock parametrized by means of sequences of wave-packet profiles‡

$$g^{(0)},\ g^{(1)}(\mathbf{p}\sigma),\ g^{(2)}(\mathbf{p}_1\sigma_1, \mathbf{p}_2\sigma_2), \ldots$$

The sequences of wave-packet profiles $\{g\}$ form a Hilbert space with the scalar product

$$(\{g_1\}|\{g_2\})$$
$$\equiv \sum_{n=0}^{\infty} \sum_{\sigma_1\ldots\sigma_n} \int \mathrm{d}\Gamma_1 \ldots \mathrm{d}\Gamma_n g_1^{(n)*}(\mathbf{p}_1\sigma_1, \ldots, \mathbf{p}_n\sigma_n)$$
$$\times g_2^{(n)}(\mathbf{p}_1\sigma_1, \ldots, \mathbf{p}_n\sigma_n). \quad (41)$$

The vector Ψ corresponding to the sequence $\{g\}$ of wave-packet profiles will be denoted by the symbol $\Psi\{g\}$. In the Hilbert space so parametrized we introduce two mutually adjoint distributions with the operator values $a(\mathbf{p}\sigma)$ and $a^\dagger(\mathbf{p}\sigma)$. These distributions associate with each one-particle profile $f(\mathbf{p}\sigma)$ the annihilation and creation operators $a[f]$ and $a^\dagger[f]$:

$$f(\mathbf{p}\sigma) \xrightarrow[\text{antilinear in } f]{a(\mathbf{p}\sigma)} a[f],$$

$$f(\mathbf{p}\sigma) \xrightarrow[\text{linear in } f]{a^\dagger(\mathbf{p}\sigma)} a^\dagger[f].$$

† At this stage we are already preparing the notation for later generalization to relativistic theory.

‡ Note that this parametrization is independent of time.

The action of the operators $a[f]$ and $a^\dagger[f]$ will be defined by the formulae

$$a[f]\Psi\{g\} = \Psi\{\tilde{g}\}, \tag{42a}$$

$$a^\dagger[f]\Psi\{g\} = \Psi\{\hat{g}\}, \tag{42b}$$

where the sequences $\{\tilde{g}\}$ and $\{\hat{g}\}$ are composed of the functions $\hat{g}^{(n)}$ and $\tilde{g}^{(n)}$, defined as

$$\tilde{g}^{(n)}(\mathbf{p}_1\sigma_1, \ldots, \mathbf{p}_n\sigma_n) = \sum_\sigma \int \mathrm{d}\Gamma f^*(\mathbf{p}\sigma) g^{(n+1)}(\mathbf{p}\sigma, \mathbf{p}_1\sigma_1, \ldots, \mathbf{p}_n\sigma_n),$$

$$\hat{g}^{(n)}(\mathbf{p}_1\sigma_1, \ldots, \mathbf{p}_n\sigma_n)$$

$$= \frac{1}{n}\sum_{i=1}^{n} (\pm 1)^{i+1} f(\mathbf{p}_i\sigma_i) g^{(n-1)}(\mathbf{p}_1\sigma_1, \ldots, \check{\mathbf{p}}_i\check{\sigma}_i, \ldots, \mathbf{p}_n\sigma_n).$$

From these definitions we have the following commutation relations for $a[f]$ and $a^\dagger[f]$:

$$[a[f_1], a^\dagger[f_2]]_\mp = (f_1|f_2), \tag{43a}$$

$$[a[f_1], a[f_2]]_\mp = 0 = [a^\dagger[f_1], a^\dagger[f_2]]_\mp. \tag{43b}$$

The formulae most frequently presented in the physical literature are in a form "stripped" of wave-packet profiles. To get such formulae, in addition to the operator distributions $a(\mathbf{p}\sigma)$ and $a^\dagger(\mathbf{p}\sigma)$ one must also define vector-valued distributions $|\mathbf{p}_1\sigma_1, \ldots, \mathbf{p}_n\sigma_n\rangle$. Such a distribution determines the following association,

$$g^{(n)}(\mathbf{p}_1\sigma_1, \ldots, \mathbf{p}_n\sigma_n) \xrightarrow[\text{linear in } g^{(n)}]{|\mathbf{p}_1\sigma_1, \ldots, \mathbf{p}_n\sigma_n\rangle} \Psi\{g^{(n)}\},$$

where $\{g^{(n)}\}$ denotes the sequence $(0, 0, \ldots, g^{(n)}(\mathbf{p}_1\sigma_1, \ldots, \mathbf{p}_n\sigma_n), 0, \ldots)$. This association is represented in the following symbolic form after the fashion of other relations of this type,

$$\Psi\{g\} = \sum_{n=0}^{\infty} \sum_{\sigma_1 \ldots \sigma_n} \int \mathrm{d}\Gamma_1 \ldots \mathrm{d}\Gamma_n g^{(n)}(\mathbf{p}_1\sigma_1, \ldots, \mathbf{p}_n\sigma_n)|\mathbf{p}_1\sigma_1, \ldots, \mathbf{p}_n\sigma_n\rangle. \tag{44}$$

The objects $|\mathbf{p}_1\sigma_1, \ldots, \mathbf{p}_n\sigma_n\rangle$ are also frequently called *generalized state vectors.* For uniformity of notation, the vacuum vector, as one of the vectors $|\mathbf{p}_1\sigma_1, \ldots, \mathbf{p}_n\sigma_n\rangle$, will be denoted by the symbol $|0\rangle$.

Below we give the set of the fundamental properties of the distributions introduced above:

$$a(\mathbf{p}\sigma)|\mathbf{p}_1\sigma_1, \ldots, \mathbf{p}_n\sigma_n\rangle$$

$$= \frac{1}{\sqrt{n}} \sum_{i=1}^{n} (\pm 1)^{i+1} \delta_{\sigma\sigma_i} \delta_\Gamma(\mathbf{p}, \mathbf{p}_i)|\mathbf{p}_1\sigma_1, \ldots, \check{\mathbf{p}}_i\check{\sigma}_i, \ldots, \mathbf{p}_n\sigma_n\rangle, \tag{45a}$$

$$a^\dagger(\mathbf{p}\sigma)|\mathbf{p}_1\sigma_1, \ldots, \mathbf{p}_n\sigma_n\rangle = \sqrt{n+1}\,|\mathbf{p}\sigma, \mathbf{p}_1\sigma_1, \ldots, \mathbf{p}_n\sigma_n\rangle, \tag{45b}$$

$$[a(\mathbf{p}\sigma), a^\dagger(\mathbf{p}'\sigma')]_\mp = \delta_{\sigma\sigma'}\delta_\Gamma(\mathbf{p}, \mathbf{p}'), \tag{45c}$$

$$[a(\mathbf{p}\sigma), a(\mathbf{p}'\sigma')]_\mp = 0 = [a^\dagger(\mathbf{p}\sigma), a^\dagger(\mathbf{p}'\sigma')]_\mp, \tag{45d}$$

$$|\mathbf{p}_1\sigma_1, \ldots, \mathbf{p}_n\sigma_n\rangle = \frac{1}{\sqrt{n!}} a^\dagger(\mathbf{p}_n\sigma_n) \ldots a^\dagger(\mathbf{p}_1\sigma_1)|0\rangle, \tag{45e}$$

$$\frac{1}{\sqrt{n!}} \langle \mathbf{p}_1\sigma_1, \ldots, \mathbf{p}_n\sigma_n|\Psi\{g\}\rangle = g^{(n)}(\mathbf{p}_1\sigma_1, \ldots, \mathbf{p}_n\sigma_n), \tag{45f}$$

$$\langle \mathbf{p}_1\sigma_1, \ldots, \mathbf{p}_n\sigma_n|\mathbf{p}'_1\sigma'_1, \ldots, \mathbf{p}'_m\sigma'_m\rangle$$

$$= \delta_{nm}\frac{1}{n!}\sum_{\text{perm } i} \delta_{\sigma_1\sigma'_{i_1}} \ldots \delta_{\sigma_n\sigma'_{i_n}} \delta_\Gamma(\mathbf{p}_1, \mathbf{p}'_{i_1}) \ldots \delta_\Gamma(\mathbf{p}_n, \mathbf{p}'_{i_n}), \tag{45g}$$

where

$$\delta_\Gamma(\mathbf{p}, \mathbf{p}') = (2\pi)^3\delta(\mathbf{p}-\mathbf{p}').$$

Comparison of the formulae defining field operators and the operators $a(\mathbf{p}\sigma)$ and $a^\dagger(\mathbf{p}\sigma)$ shows that in the case of non-interacting particles the field operators $\psi_0(\mathbf{x}\sigma t)$ and $\psi_0^\dagger(\mathbf{x}\sigma t)$ may be written as

$$\psi_0(\mathbf{x}\sigma t) = \int \mathrm{d}\Gamma a(\mathbf{p}\sigma)\mathrm{e}^{-iEt+i\mathbf{p}\cdot\mathbf{x}},$$

$$\psi_0^\dagger(\mathbf{x}\sigma t) = \int \mathrm{d}\Gamma a^\dagger(\mathbf{p}\sigma)\mathrm{e}^{iEt-i\mathbf{p}\cdot\mathbf{x}}.$$

The commutation relations for these field operators at any arbitrary instants (not necessarily the same) are

$$[\psi_0(\mathbf{x}\sigma t), \psi_0^\dagger(\mathbf{x}'\sigma' t')]_\mp = \frac{1}{i}\delta_{\sigma\sigma'}\mathscr{K}(\mathbf{x}-\mathbf{x}', t-t'), \tag{46}$$

where

$$\mathscr{K}(\mathbf{x}, t) = i\int \mathrm{d}\Gamma \mathrm{e}^{-iE_p t + i\mathbf{p}\cdot\mathbf{x}}.$$

Scattering and the S Operator

Let us now go about describing scattering; in connection with this we shall confine ourselves to the scattering subspace of the Hilbert space of the state vectors. To begin with let us base ourselves on the properties of the wave function for a single particle and let us assume that the wave function describing the scattering state of a many-particle system also goes over in the remote past and far future into wave packets which satisfy the free Schrödinger equation. This property may be regarded in a sense as a definition of scattering states.

For the vectors of scattering states, therefore, two Fock parametrizations may be introduced by means of sequences of wave-packet profiles. The vector Ψ may have associated with it either a set of profiles $\{g\}$ of wave packets into which the wave functions $\Psi^{(n)}$ corresponding to that vector go over in the past,

$$(\Psi^{(n)}(\mathbf{x}_1\sigma_1, \ldots, \mathbf{x}_n\sigma_n, t) - \varphi[\mathbf{x}_1\sigma_1, \ldots, \mathbf{x}_n\sigma_n, t|g^{(n)}]) \to 0, \qquad t \to -\infty,$$

or a set of profiles $\{g'\}$ of wave packets into which the wave functions $\Psi^{(n)}$ go over in the future,

$$(\Psi^{(n)}(\mathbf{x}_1\sigma_1, \ldots, \mathbf{x}_n\sigma_n, t) - \varphi[\mathbf{x}_1\sigma_1, \ldots, \mathbf{x}_n\sigma_n, t|g'^{(n)}]) \to 0, \qquad t \to +\infty.$$

State vectors parametrized by a sequence $\{g\}$ in the past will be denoted by the symbol $\Psi^{\text{in}}\{g\}$, and those in the future by $\Psi^{\text{out}}\{g\}$. Each vector may, of course, be written in two forms:

$$\Psi^{\text{out}}\{g'\} = \Psi = \Psi^{\text{in}}\{g\}.$$

Just as in the case of a single particle, the two aforementioned parametrizations can be described in terms of wave functions, the functions ψ_{R} and ψ_{A}—which are dependent on the profile g—being defined by

$$\psi_{\mathrm{R}}[\mathbf{x}_1\sigma_1, \ldots, \mathbf{x}_n\sigma_n, t|g^{(n)}] \leftrightarrow \Psi^{\text{in}}\{g\},$$

and

$$\psi_{\mathrm{A}}[\mathbf{x}_1\sigma_1, \ldots, \mathbf{x}_n\sigma_n, t|g^{(n)}] \leftrightarrow \Psi^{\text{out}}\{g\}.$$

The construction discussed earlier for the case of a system of free particles can also be carried out both for the IN and OUT parametrizations. When the parametrization $\Psi^{\text{in}}\{g\}$ is used, it is possible to introduce operator-valued distributions $a^{\text{in}}(\mathbf{p}\sigma)$ and $a^{\dagger\text{in}}(\mathbf{p}\sigma)$ and field operators $\psi^{\text{in}}(\mathbf{x}\sigma t)$ and $\psi^{\dagger\text{in}}(\mathbf{x}\sigma t)$, as well as vector-valued distributions $|^{\text{in}}\mathbf{p}_1\sigma_1, \ldots, \mathbf{p}_n\sigma_n\rangle$. Using the parametrization $\Psi^{\text{out}}\{g\}$, we can introduce analogous operators with the superscript OUT.

All the formulae derived earlier for noninteracting particles also hold for the IN and OUT operators. In all these formulae a^{in}, $a^{\dagger\text{in}}$, and $\Psi^{\text{in}}\{g\}$ or a^{out}, $a^{\dagger\text{out}}$ and $\Psi^{\text{out}}\{g\}$ may be introduced in place of the operators a, $a^{\dagger}$, and the vectors $\Psi\{g\}$.

The scalar product of the vectors $\Psi^{\text{out}}\{g_1\}$ and $\Psi^{\text{in}}\{g_2\}$ is the *scattering amplitude A*. This is the amplitude of finding in a state, which in the remote past is described by the set of profiles $\{g_2\}$, such a state which in the far future is described by the set of profiles $\{g_1\}$:

$$A(\{g_1\}, \{g_2\}) \equiv (\Psi^{\text{out}}\{g_1\}|\Psi^{\text{in}}\{g_2\}). \tag{47}$$

In terms of wave functions the scattering amplitude is defined as the scalar product $(\psi_{\text{A}}|\psi_{\text{R}})$.

Scattering amplitudes are matrix elements of the S operator, calculated between the vectors Ψ^{in},

$$(\Psi^{\text{in}}\{g_1\}|\mathsf{S}\Psi^{\text{in}}\{g_2\}) = (\Psi^{\text{out}}\{g_1\}|\Psi^{\text{in}}\{g_2\}). \tag{48}$$

The S operator is unitary in the subspace of scattering states. For equation (48) implies the relation

$$\Psi^{\text{out}}\{g\} = \mathsf{S}^{\dagger}\Psi^{\text{in}}\{g\}$$

whence

$$(\mathsf{S}^{\dagger}\Psi^{\text{in}}\{g_1\}|\mathsf{S}^{\dagger}\Psi^{\text{in}}\{g_2\}) = (\Psi^{\text{out}}\{g_1\}|\Psi^{\text{out}}\{g_2\})$$
$$= (\Psi^{\text{in}}\{g_1\}|\Psi^{\text{in}}\{g_2\}),$$

that is,

$$\mathsf{S}\mathsf{S}^{\dagger} = 1.$$

From the fact that the mapping $\Psi^{\text{in}}\{g\} \to \Psi^{\text{out}}\{g\}$ is one-to-one it follows that also the equality

$$\mathsf{S}^{\dagger}\mathsf{S} = 1$$

is satisfied.

From the fact that the operation of the IN operators on the vectors Ψ^{in} is the same as that of the OUT operators on the vectors Ψ^{out}, it follows that the S operator takes the IN operators over into the OUT operators,

$$\mathsf{S}^{\dagger}a^{\text{in}}(\mathbf{p}\sigma)\mathsf{S} = a^{\text{out}}(\mathbf{p}\sigma), \tag{49a}$$

$$\mathsf{S}^{\dagger}a^{\dagger\text{in}}(\mathbf{p}\sigma)\mathsf{S} = a^{\dagger\text{out}}(\mathbf{p}\sigma). \tag{49b}$$

The scattering amplitude stripped of the wave-packet profile is the S matrix. The components $S^{(n)}$ of this matrix are defined by the relations

$$(\Psi^{\text{in}}\{g_1\}|\mathsf{S}\Psi^{\text{in}}\{g_2\}) = \sum_{n=0}^{\infty}\sum_{\substack{\sigma_1\ldots\sigma_n\\ \sigma_1'\ldots\sigma_n'}}\int \mathrm{d}\Gamma_1 \ldots \mathrm{d}\Gamma_n \mathrm{d}\Gamma_1' \ldots \mathrm{d}\Gamma_n'$$
$$\times\, g_1^{(n)*}(\mathbf{p}_1\sigma_1, \ldots, \mathbf{p}_n\sigma_n)\, S^{(n)}(\mathbf{p}_1\sigma_1, \ldots, \mathbf{p}_n\sigma_n; \mathbf{p}_n'\sigma_n', \ldots, \mathbf{p}_1'\sigma_1')$$
$$\times\, g_2^{(n)}(\mathbf{p}_1'\sigma_1', \ldots, \mathbf{p}_n'\sigma_n') \tag{50}$$

and may be represented by the formulae

$$S^{(n)}(\mathbf{p}_1\sigma_1, \ldots, \mathbf{p}_n\sigma_n; \mathbf{p}_n'\sigma_n', \ldots, \mathbf{p}_1'\sigma_1')$$
$$= \langle \mathbf{p}_1\sigma_1, \ldots, \mathbf{p}_n\sigma_n^{\text{out}}|^{\text{in}}\mathbf{p}_1'\sigma_1', \ldots, \mathbf{p}_n'\sigma_n'\rangle$$
$$= \frac{1}{n!}(\Omega|a^{\text{in}}(\mathbf{p}_1\sigma_1) \ldots a^{\text{in}}(\mathbf{p}_n\sigma_n)\mathsf{S}a^{\dagger\text{in}}(\mathbf{p}_n'\sigma_n) \ldots a^{\dagger\text{in}}(\mathbf{p}_1'\sigma_1')\Omega). \tag{50a}$$

The IN and OUT operators are, in a sense, the limits of the field operators. For they may be related to the field operators by means of so-called *asymptotic conditions*. These conditions are of the form

$$\lim_{t\to\mp\infty}(\Phi|a^t[f]\Psi) = (\Phi|a^{\substack{\text{in}\\ \text{out}}}[f]\Psi), \tag{51}$$

where

$$a^t[f] \equiv \sum_{\sigma}\int \mathrm{d}^3x\varphi^*[\mathbf{x}\sigma t|f]\psi(\mathbf{x}\sigma t),$$

and the vectors Φ and Ψ describe the scattering states. By taking the Hermitian conjugate of the asymptotic conditions for the annihilation operators, we obtain analogous conditions for the creation operators.

The proof will be given for the asymptotic condition when $t \to -\infty$. The state vectors Φ and Ψ will be parametrized by using the wave-packet profiles $\{g_1\}$ and $\{g_2\}$

in the past. The matrix element on the left-hand side of eq. (51) can then be rewritten as

$$(\Phi^{\text{in}}\{g_1\}|a^t[f]\Phi^{\text{in}}\{g_2\})$$

$$= \sum_{n=1}^{\infty} \sqrt{n+1} \sum_{\sigma,\sigma_1\ldots\sigma_n} \int d^3x d^3x_1 \ldots d^3x_n \varphi^*[\mathbf{x}\sigma t|f]$$

$$\times \psi_R^*[\mathbf{x}_1\sigma_1, \ldots, \mathbf{x}_n\sigma_n, t|g_1^{(n)}]\psi_R[\mathbf{x}\sigma, \mathbf{x}_1\sigma_1, \ldots, \mathbf{x}_n\sigma_n, t|g_2^{(n+1)}].$$

The difference between the functions ψ_R and the wave packets $\varphi[\mathbf{x}_1\sigma_1, \ldots, \mathbf{x}_n\sigma_n, t|g_2^{(n)}]$ tends in the norm to zero when $t \to -\infty$. Thus, in the limit this expression becomes

$$\lim_{t\to\infty} \sum_{n=1}^{\infty} \sqrt{n+1} \sum_{\sigma,\sigma_1\ldots\sigma_n} \int d^3x d^3x_1 \ldots d^3x_n$$

$$\times \varphi^*[\mathbf{x}\sigma t|f]\varphi^*[\mathbf{x}_1\sigma_1, \ldots, \mathbf{x}_n\sigma_n, t|g_1^{(n)}]\varphi[\mathbf{x}\sigma, \mathbf{x}_1\sigma_1, \ldots, \mathbf{x}_n\sigma_n, t|g_2^{(n+1)}]$$

$$= \sum_{n=1}^{\infty} \sqrt{n+1} \sum_{\sigma,\sigma_1\ldots\sigma_n} \int d\Gamma d\Gamma_1 \ldots d\Gamma_n$$

$$\times f^*(\mathbf{p}\sigma) g_1^{(n)}(\mathbf{p}_1\sigma_1, \ldots, \mathbf{p}_n\sigma_n) g_2^{(n+1)}(\mathbf{p}\sigma, \mathbf{p}_1\sigma_1, \ldots, \mathbf{p}_n\sigma_n).$$

This expression may be rewritten in the form $(\Phi^{\text{in}}\{g_1\}|a^{\text{in}}[f]\Phi^{\text{in}}\{g_2\})$, which completes the proof of the asymptotic condition when $t \to -\infty$. The proof of the asymptotic condition when $t \to \infty$ proceeds in similar fashion.

After these general considerations, let us go on to derive formulae for the S operator and its matrix elements. The S operator will be written as an exponential chronological operator, whereas its matrix elements will be expressed in terms of the average values of the products of the field operators in the vacuum state.

Exponential chronological operator $\mathsf{V}(t, t_0)$ is the name given to a formal series of operators of the form

$$\mathsf{V}(t, t_0) = T\exp\left(-i\int_{t_0}^{t} dt' H(t')\right)$$

$$= \sum_{n=0}^{\infty} \frac{(-i)^n}{n!} \int_{t_0}^{t} dt_1 \ldots \int_{t_0}^{t} dt_n T(H(t_1) \ldots H(t_n)), \tag{52}$$

where $H(t)$ is a one-parameter family of self-adjoint operators, whereas the symbol T denotes the operation of chronologically ordering the operators. This operation should be applied to each term in the expansion of the exponential function before integration with respect to time is performed. The product of the operators ordered chronologically (arranged in order of time arguments increasing from right to left) is called the chronological product.†

Exponential chronological operators are unitary and they satisfy the *composition law*

$$\mathsf{V}(t, t_1)\mathsf{V}(t_1, t_0) = \mathsf{V}(t, t_0),$$

and the differential *evolution equations* with respect to both arguments

$$i\frac{\partial}{\partial t}\mathsf{V}(t, t_0) = H(t)\mathsf{V}(t, t_0),$$

$$-i\frac{\partial}{\partial t}\mathsf{V}(t, t_0) = \mathsf{V}(t, t_0)H(t_0).$$

In our further considerations we shall use the operators $H(t)$ constructed out of the IN or OUT operators. These will be denoted by $H^{\text{in}}(t)$ and $H^{\text{out}}(t)$, respectively, and they will be used to define‡ two field operators $\mathbf{\Phi}_{\text{R}}(\mathbf{x}\sigma t)$ and $\mathbf{\Phi}_{\text{A}}(\mathbf{x}\sigma t)$,

$$\mathbf{\Phi}_{\text{R}}(\mathbf{x}\sigma t) \equiv \mathsf{V}^{\dagger\text{in}}(t, -\infty)\psi^{\text{in}}(\mathbf{x}\sigma t)\mathsf{V}^{\text{in}}(t, -\infty), \tag{53a}$$

$$\mathbf{\Phi}_{\text{A}}(\mathbf{x}\sigma t) \equiv \mathsf{V}^{\text{out}}(\infty, t)\psi^{\text{out}}(\mathbf{x}\sigma t)\mathsf{V}^{\dagger\text{out}}(\infty, t), \tag{53b}$$

where

$$\mathsf{V}^{\overset{\text{out}}{\text{in}}}(t, t_0) = T\exp\left(-\int_{t_0}^{t} \mathrm{d}t' H^{\overset{\text{out}}{\text{in}}}(t')\right).$$

The operators $H^{\overset{\text{out}}{\text{in}}}(t)$ will be taken in the form of potential energy operators, with the operators ψ^{in} or ψ^{out} substituted for the field operators ψ:

$$H^{\overset{\text{out}}{\text{in}}}(t) = \sum_{\sigma}\int \mathrm{d}^3x\psi^{\overset{\dagger\text{out}}{\text{in}}}(\mathbf{x}\sigma t)U(\mathbf{x}, t)\psi^{\overset{\text{out}}{\text{in}}}(\mathbf{x}\sigma t)$$

$$+\frac{1}{2}\sum_{\sigma\sigma'}\int \mathrm{d}^3x\,\mathrm{d}^3y\psi^{\overset{\dagger\text{out}}{\text{in}}}(\mathbf{x}\sigma t)\psi^{\overset{\dagger\text{out}}{\text{in}}}(\mathbf{y}\sigma' t)V(\mathbf{x}-\mathbf{y})\psi^{\overset{\text{out}}{\text{in}}}(\mathbf{y}\sigma' t)\psi^{\overset{\text{out}}{\text{in}}}(\mathbf{x}\sigma t). \tag{54}$$

† The properties of chronological products and exponential chronological operators are discussed in Appendix B.

‡ Limits of the operators $\mathsf{V}(t, t_0)$, when t and $t_0 \to \pm\infty$, do not, of course, exist for all operators $H(t)$.

With such a choice of Hamiltonian $\overset{\text{out}}{H^{\text{in}}}(t)$, the operators $\mathbf{\Phi}_{\text{R}}$ and $\mathbf{\Phi}_{\text{A}}$ satisfy field equations of the same form as do the field operators ψ.

The proof will be given for the operator $\mathbf{\Phi}_{\text{R}}$. Differentiating both sides of the definition (53a) with respect to t and using the evolution equation, we obtain

$$i\frac{\partial}{\partial t}\mathbf{\Phi}_{\text{R}}(\mathbf{x}\sigma t) = \mathsf{V}^{\dagger\text{in}}(t,-\infty)\, i\frac{\partial}{\partial t}\psi^{\text{in}}(\mathbf{x}\sigma t)\mathsf{V}^{\text{in}}(t,-\infty)$$
$$+\mathsf{V}^{\dagger\text{in}}(t,-\infty)[\psi^{\text{in}}(\mathbf{x}\sigma t), H^{\text{in}}(t)]\mathsf{V}^{\text{in}}(t,-\infty).$$

The right-hand side of this equation can be rearranged into the form of the right-hand side of the field equation (36a) by means of the field equation and commutation relation for IN operators and the unitarity of the operators V^{in}.

The operator $\mathbf{\Phi}_{\text{R}}$ satisfies in the past the same asymptotic condition as does the field operator ψ, whereas the operator $\mathbf{\Phi}_{\text{A}}$ does the same in the future.

This follows immediately from the definition of the operators $\mathbf{\Phi}_{\text{R, A}}$, if it is assumed that the passages to the limit in the operators $\mathsf{V}(t, t_0)$, when $t \to \pm\infty$ and $t_0 \to \pm\infty$, are sufficiently regular so as to be performable simultaneously. For example,

$$\lim_{t\to-\infty}\mathsf{V}(t,-\infty) = \lim_{\substack{t\to-\infty\\ t_0\to-\infty}}\mathsf{V}(t,t_0) = \lim_{t\to-\infty}\mathsf{V}(t,t) = \lim_{t\to-\infty}1 = 1.$$

The theory of integro-differential equations satisfied by operator-valued distributions has not yet been worked out. Accordingly, we may only rely on our intuition when we discuss the uniqueness of the solutions of the field equations. It seems to us that at this point reference may be made to analogy with equations satisfied by ordinary functions and one can assume that the asymptotic conditions in the past or in the future determine the solutions. We then obtain the important result that the field operators $\mathbf{\Phi}_{\text{R}}$ and $\mathbf{\Phi}_{\text{A}}$ are identical with the original field operator ψ,

$$\mathbf{\Phi}_{\text{R}}(\mathbf{x}\sigma t) = \psi(\mathbf{x}\sigma t) = \mathbf{\Phi}_{\text{A}}(\mathbf{x}\sigma t).$$

The equality $\mathbf{\Phi}_{\text{R}} = \mathbf{\Phi}_{\text{A}}$ leads to a very important representation of the S operator as an exponential chronological operator. When we use the asymptotic conditions in the past and in the future, then by virtue of relations (49) we obtain

$$\mathsf{S} = T\exp\Big(-i\int_{-\infty}^{+\infty} \mathrm{d}t\, \overset{\text{out}}{H^{\text{in}}}(t)\Big). \tag{55}$$

In this way, we have arrived at an explicit (though formal) representation of the S operator. It is used very frequently in relativistic quantum electrodynamics.

The operator S given by formula (55) satisfies the causality condition in the form

$$\frac{\delta}{\delta U(\mathbf{y}, t')}\left(\mathsf{S}^{\dagger}\frac{\delta \mathsf{S}}{\delta U(\mathbf{x}, t)}\right) = 0, \quad \text{when} \quad t' > t. \tag{56}$$

This condition is derived just like the causality condition (1.33), except that this time the potential $U(\mathbf{x}, t)$ is treated as an arbitrary function characterizing the external perturbation.

Since the exponential chronological operator has only the sense of a formal series of operators, the S operator presented in this form may be employed only for perturbation calculations.

Though less convenient a procedure in perturbation calculations, it is more useful in general considerations to represent the S operator as the sum of the normal products of IN and OUT operators. In the *normal product*, all the creation operators $\psi^{\dagger}$ are to the left of all the annihilation operators ψ. In the case of fermions we introduce an additional factor of -1 when the normal order differs from the initial one by an odd number of operator transpositions.†

Every operator $\hat{A}$ acting in Hilbert space Fock-parametrized by sequences of wave-packet profiles $\{g\}$ may be written as

$$\hat{A} = \sum_{n,m}\frac{1}{n!\,m!}\sum_{\sigma\sigma'}\int \mathrm{d}\Gamma_1 \ldots \mathrm{d}\Gamma_n \mathrm{d}\Gamma'_1 \ldots \mathrm{d}\Gamma'_m a^{\dagger}(\mathbf{p}_1\,\sigma_1) \ldots a^{\dagger}(\mathbf{p}_n\sigma_n)$$
$$\times\, \alpha^{(n,m)}(\mathbf{p}_1\sigma_1, \ldots, \mathbf{p}_n\sigma_n; \mathbf{q}_m\sigma'_m, \ldots, \mathbf{q}_1\,\sigma'_1)\, a(\mathbf{q}_m\,\sigma'_m) \ldots a(\mathbf{q}_1\,\sigma'_1). \tag{57}$$

To prove this, let us consider an arbitrary matrix element of the operator $\hat{A}$ between (generalized) state vectors $|\mathbf{p}_1\,\sigma_1, \ldots, \mathbf{p}_n\sigma_n\rangle$ and $|\mathbf{q}_1\,\sigma'_1, \ldots, \mathbf{q}_m\sigma'_m\rangle$. If eq. (57) holds, this matrix element may be presented as

$$\sqrt{n!\,m!}\,\langle \mathbf{p}_1\,\sigma_1, \ldots, \mathbf{p}_n\sigma_n|\hat{A}|\mathbf{q}_1\,\sigma'_1, \ldots, \mathbf{q}_m\sigma'_m\rangle$$
$$= \alpha^{(n,m)}(\mathbf{p}_1\,\sigma_1, \ldots, \mathbf{p}_n\sigma_n; \mathbf{q}_m\sigma'_m, \ldots, \mathbf{q}_1\sigma'_1)$$
$$+ \sum_{i=1}^{n}\sum_{j=1}^{m}(\pm 1)^{i+j}\delta_{\sigma_i\sigma'_j}\delta_{\Gamma}(\mathbf{p}_i, \mathbf{q}_j)\,\alpha^{(n-1,m-1)}(\mathbf{p}_1\,\sigma_1, \ldots, \check{\mathbf{p}}_i\check{\sigma}_i,$$
$$\ldots, \mathbf{p}_n\sigma_n; \mathbf{q}_m\sigma'_m, \ldots, \check{\mathbf{q}}_j\check{\sigma}'_j, \ldots, \mathbf{q}_1\,\sigma'_1) + \ldots \tag{58}$$

† Further properties of normal products are discussed in Appendix B.

These equations may be regarded as a system of equations for the functions $\alpha^{(n,m)}$. These equations can always be solved for arbitrarily chosen matrix elements of the operator $\hat{A}$ by successively solving equations with increasing m's and n's. In this way we obtain:

$$\alpha^{(0,0)} = \langle 0|\hat{A}|0\rangle,$$

$$\alpha^{(1,0)}(\mathbf{p}\sigma; -) = \langle \mathbf{p}\sigma|\hat{A}|0\rangle,$$

$$\alpha^{(0,1)}(-; \mathbf{q}\sigma) = \langle 0|\hat{A}|\mathbf{q}\sigma\rangle,$$

$$\alpha^{(1,1)}(\mathbf{p}\sigma; \mathbf{q}\sigma') = \langle \mathbf{p}\sigma|\hat{A}|\mathbf{q}\sigma'\rangle - \delta_{\sigma\sigma'}\delta_\Gamma(\mathbf{p}, \mathbf{q})\langle 0|\hat{A}|0\rangle,$$

$$\alpha^{(2,2)}(\mathbf{p}_1\sigma_1, \mathbf{p}_2\sigma_2; \mathbf{q}_2\sigma_2', \mathbf{q}_1\sigma_1')$$
$$= 2\langle \mathbf{p}_1\sigma_1, \mathbf{p}_2\sigma_2|\hat{A}|\mathbf{q}_1\sigma_1', \mathbf{q}_2\sigma_2'\rangle - \delta_{\sigma_1\sigma_1'}\delta_\Gamma(\mathbf{p}_1, \mathbf{q}_1)\langle \mathbf{p}_2\sigma_2|\hat{A}|\mathbf{q}_2\sigma_2'\rangle$$
$$\pm \delta_{\sigma_1\sigma_2'}\delta_\Gamma(\mathbf{p}_1, \mathbf{q}_2)\langle \mathbf{p}_2\sigma_2|\hat{A}|\mathbf{q}_1\sigma_1'\rangle \pm \delta_{\sigma_2\sigma_1'}\delta_\Gamma(\mathbf{p}_2, \mathbf{q}_1)\langle \mathbf{p}_1\sigma_1|\hat{A}|\mathbf{q}_2\sigma_2'\rangle$$
$$- \delta_{\sigma_2\sigma_2'}\delta_\Gamma(\mathbf{p}_2, \mathbf{q}_2)\langle \mathbf{p}_1\sigma_1|\hat{A}|\mathbf{q}_1\sigma_1'\rangle + [\delta_{\sigma_1\sigma_1'}\delta_\Gamma(\mathbf{p}_1, \mathbf{q}_1)\delta_{\sigma_2\sigma_2'}\delta_\Gamma(\mathbf{p}_2, \mathbf{q}_2)$$
$$\pm \delta_{\sigma_1\sigma_2'}\delta_\Gamma(\mathbf{p}_1, \mathbf{q}_2)\delta_{\sigma_2\sigma_1'}\delta_\Gamma(\mathbf{p}_2, \mathbf{q}_1)]\langle 0|\hat{A}|0\rangle$$

and so forth.

Thus, for every operator $\hat{A}$ and, even more, for every bilinear form with known matrix elements, the functions $\alpha^{(m,n)}$ in formula (57) can be chosen so that both sides have the same matrix elements.

Formula (57) may also be presented in a form that is solved for the functions $\alpha^{(n,m)}$. We obtain this form by calculating the multiple commutators (or, alternatively, anticommutators and commutators when we have fermions) and taking the expectation value in the vacuum state. For bosons, this formula is of the form

$$\alpha^{(m,n)}(\mathbf{p}_1\sigma_1, \ldots, \mathbf{p}_n\sigma_n; \mathbf{q}_m\sigma_m', \ldots, \mathbf{q}_1, \sigma_1')$$
$$= (\Omega|[\ldots[a(\mathbf{p}_n\sigma_n), \ldots [a(\mathbf{p}_2\sigma_2), [a(\mathbf{p}_1\sigma_1), \hat{A}]] \ldots,$$
$$a^\dagger(\mathbf{q}_1\sigma_1')] \ldots, a^\dagger(\mathbf{q}_m\sigma_m')]\Omega). \tag{59}$$

For fermions the innermost brackets $[a(\mathbf{p}_1\sigma_1), \hat{A}]$ denote a commutator or anticommutator, depending on whether $n+m$ is even or odd. The other brackets are alternately anticommutators and commutators.

Now let us apply the general formalism to the S operator. Since in the nonrelativistic theory under consideration this operator acts invariantly in every n-particle subspace, the functions $\sigma^{(n,m)}$ appearing in the expansion of the S operator in products of the creation and annihilation operators are zero when $n \neq m$:

$$\mathsf{S} = \sum_{n=1}^{\infty} \frac{1}{(n!)^2} \sum_{\sigma\sigma'} \int \mathrm{d}\Gamma_1 \ldots \mathrm{d}\Gamma_n \mathrm{d}\Gamma'_1 \ldots \mathrm{d}\Gamma'_n a^\dagger(\mathbf{p}_1\sigma_1) \ldots a^\dagger(\mathbf{p}_n\sigma_n)$$

$$\times \sigma^{(n)}(\mathbf{p}_1\sigma_1, \ldots, \mathbf{p}_n\sigma_n; \mathbf{q}_n\sigma'_n, \ldots, \mathbf{q}_1\sigma'_1) a(\mathbf{q}_n\sigma'_n) \ldots a(\mathbf{q}_1\sigma'_1). \quad (60)$$

The foregoing series expansion of the S operator allows the proper scattering amplitude of n particles to be separated from the total scattering amplitude of n particles. The proper amplitude is described by the n-th term in the series (60). The terms $S^{(k)}$, when $k < n$, also make contributions to the scattering amplitude of n particles. The role of these terms may be described in the following manner by resorting to the concept of proper amplitudes. The k-th term describes scattering in which only k particles undergo actual scattering (this is described by the proper k-particle amplitude). The other n–k particles do not interact at all. These particles should be chosen in every possible way and the contributions from all configurations added together. We have already observed a similar situation in the case of a single particle. The second term in formula (4.27) represents the proper one-particle amplitude, whereas the first describes transition without interaction with the potential.

The creation and annihilation operators were hitherto associated with arbitrary Fock parametrization of the vector-state space. For a scattering subspace we have two physically distinguished parametrizations: IN and OUT. To focus attention let us take the IN parametrization. It turns out that the multiple commutators of the operators a^{in} and $a^{\dagger\mathrm{in}}$ (as well as the operators a^{out} and $a^{\dagger\mathrm{out}}$) with the S operator may be expressed in terms of the field operators by means of asymptotic conditions. This purpose is served by the so-called *reduction formulae*

$$[a^{\mathrm{in}}[f], \mathsf{S}T(\mathcal{O}_1 \ldots \mathcal{O}_n)]_\mp$$

$$= i\mathsf{S} \sum_\sigma \int \mathrm{d}t\mathrm{d}^3x \varphi^*[\mathbf{x}\sigma t|f] S_x T(\psi(\mathbf{x}\sigma t)\mathcal{O}_1 \ldots \mathcal{O}_n), \quad (61\mathrm{a})$$

$$[\mathsf{S}T(\mathcal{O}_1 \ldots \mathcal{O}_n), a^{\dagger\mathrm{in}}[f]]_\mp$$

$$= i\mathsf{S} \sum_\sigma \int \mathrm{d}t\mathrm{d}^3y T(\mathcal{O}_1 \ldots \mathcal{O}_n\psi^\dagger(\mathbf{y}\sigma t)) \overleftarrow{S}_y \varphi[\mathbf{y}\sigma t|f]. \quad (61\mathrm{b})$$

In these formulae $\mathcal{O}_1 \ldots \mathcal{O}_n$ denotes an arbitrary product of the field operators ψ and $\psi^\dagger$, whereas the commutators refer to bosons and, when n is even, to fermions.

We shall give the proof of the first of the reduction formulae; the second proof is analogous. By virtue of the asymptotic conditions and relations (49), the left-hand side of eq. (61a) is written as

$$\sum_\sigma \int d^3x \varphi^*[\mathbf{x}\sigma t|f] \mathsf{S} T(\psi(\mathbf{x}\sigma t)\mathcal{O}_1 \ldots \mathcal{O}_n)\Big|_{t\to-\infty}^{t\to+\infty}.$$

Next, we use the identity

$$f(t)\Big|_{t\to-\infty}^{t\to+\infty} = \int_{-\infty}^{+\infty} dt \frac{df(t)}{dt}$$

and the Schrödinger equations satisfied by the function $\varphi[\mathbf{x}\sigma t|f]$. Integration by parts yields the expression on the right-hand side of eq. (61a), in which all the differentiations act to the right.

The reduction formulae will now be used to determine the coefficients $\sigma^{(n)}$ in the expansion of the S operator in terms of the operators $a^{\dagger\text{in}}$ and a^{in}. When we apply formula (61a) n times and formula (61b) n times, after stripping away the wave-packet profiles on both sides we obtain

$$\begin{aligned}
&\Big[\ldots\Big[[a^{\text{in}}(\mathbf{p}_1\sigma_1), \ldots, [a^{\text{in}}(\mathbf{p}_n\sigma_n), \mathsf{S}]\ldots], a^{\dagger\text{in}}(\mathbf{q}_1\sigma_1')\Big], \ldots, a^{\dagger\text{in}}(\mathbf{q}_n\sigma_n')\Big] \\
&= i^{2n}\mathsf{S}\int dx_1 \ldots dx_n dy_1 \ldots dy_n e^{ip_1\cdot x_1+\ldots+ip_n\cdot x_n} e^{-iq_1\cdot y_1-\ldots-iq_n\cdot y_n} \\
&\qquad \times[S_{x_1} \ldots S_{x_n} T(\psi(\mathbf{x}_1\sigma_1 t_1) \ldots \psi(\mathbf{x}_n\sigma_n t_n) \\
&\qquad \times\psi^\dagger(\mathbf{y}_n\sigma_n' t_n') \ldots \psi^\dagger(\mathbf{y}_1\sigma_1' t_1'))\overleftarrow{S}_{y_1} \ldots \overleftarrow{S}_{y_n}],
\end{aligned} \tag{62}$$

where we have introduced abbreviated notation suitable for writing formulae in relativistic theory:

$$\int dx_i = \int d^3x_i dt_i, \quad \int dy_i = \int d^3y_i dt_i',$$

$$p_i\cdot x_i = E_{p_i}t_i - \mathbf{p}_i\cdot\mathbf{x}_i, \quad q_i\cdot y_i = E_{q_i}t_i' - \mathbf{q}_i\cdot\mathbf{y}_i.$$

Brackets stand for commutators or anticommutators, just as in formula (59), which we now use in order to write the final result:

$$\sigma^{(n)}(\mathbf{p}_1\sigma_1, \ldots, \mathbf{p}_n\sigma_n; \mathbf{q}_n\sigma'_n, \ldots, \mathbf{q}_1\sigma'_1)$$
$$= i^{2n}\int \mathrm{d}x_1 \ldots \mathrm{d}x_n \mathrm{d}y_1 \ldots \mathrm{d}y_n \mathrm{e}^{ip_i\cdot x_i + \ldots + ip_n\cdot x_n} \mathrm{e}^{-iq_1\cdot y_1 - \ldots - iq_n\cdot y_n}$$
$$\times [S_{x_1} \ldots S_{x_n} T(x_1 \ldots x_n; y_n \ldots y_1)\overleftarrow{S}_{y_1} \ldots \overleftarrow{S}_{y_n}], \tag{63}$$

where

$$T(x_1 \ldots x_n; y_n \ldots y_1) = T(\mathbf{x}_1\sigma_1 t_1, \ldots, \mathbf{x}_n\sigma_n t_n; \mathbf{y}_n\sigma'_n t'_n, \ldots, \mathbf{y}_1\sigma'_1 t'_1)$$
$$\equiv (\Omega|T(\psi(\mathbf{x}_1\sigma_1 t_1) \ldots \psi(\mathbf{x}_n\sigma_n t_n)\psi^\dagger(\mathbf{y}_n\sigma'_n t'_n) \ldots \psi^\dagger(\mathbf{y}_1\sigma'_1 t'_1))\Omega), \tag{64}$$

whereas x_i and y_j denote the sets of variables $(\mathbf{x}_i\sigma_i t_i)$ and $(\mathbf{y}_j\sigma'_j t'_j)$.

As promised, therefore, we have expressed the S operator in terms of the vacuum-state expectation value of the chronological products of the field operators. The expressions $T(x_1 \ldots x_n; y_n \ldots y_1)$ are called *propagators*; these are distributions in many variables.

It follows from the commutation relations for field operators that propagators are discontinuous functions of the variables t_i and t'_j. The location of the discontinuities is determined by n^2 equations $t_i = t'_j$, whereas the jumps of the propagators for $t_i = t'_j$ are proportional to the functions $\delta(\mathbf{x}_i - \mathbf{y}_j)$.

In the simplest case we have

$$\lim_{0<\varepsilon\to 0} \big(T(\mathbf{x}\sigma t, \mathbf{y}\sigma' t+\varepsilon) - T(\mathbf{x}\sigma t, \mathbf{y}\sigma' t-\varepsilon)\big) = -\delta_{\sigma\sigma'}\,\delta(\mathbf{x}-\mathbf{y}).$$

In accordance with the general principle adopted here (cf. the introduction) we shall not assign any value to the propagators at the points of discontinuity. Accordingly, wherever the values of propagators at the points of discontinuity appear in formulae it will be necessary to indicate whether we have the right-hand or left-hand limit in mind.

The Theory of Propagators and Feynman Diagrams

The propagators T are a natural generalization of the retarded one-particle propagator $\mathscr{K}_{\mathrm{R}}$. In the special case when all the times t_i and all the times t'_j are equal,

$$t_1 = t_2 = \ldots = t_n = t,$$
$$t'_1 = t'_2 = \ldots = t'_n = t',$$

the propagators T are simply matrix elements in the coordinate representation of the evolution operator in the Schrödinger picture. Using the relation (1.11) between operators in the Heisenberg picture at various instants of time and the relation (1.20) between U and U_s, when $t > t'$ we get

$$T(x_1 \dots x_n; y_n \dots y_1)\Big|_{\substack{t_i=t \\ t_j=t'}}$$

$$= (\Omega|\psi(\mathbf{x}_1\sigma_1 0) \dots \psi(\mathbf{x}_n\sigma_n 0)\mathsf{U}(0, t)\mathsf{U}(t', 0)\psi^\dagger(\mathbf{y}_n\sigma_n' 0) \dots \psi^\dagger(\mathbf{y}_1\sigma_1' 0)\Omega)$$

$$= n!\,\langle \mathbf{x}_1\sigma_1, \dots, \mathbf{x}_n\sigma_n|\mathsf{U}_s(t, t')|\mathbf{y}_1\sigma_1', \dots, \mathbf{y}_n\sigma_n'\rangle .$$

At the outset of the general theory of propagators, we shall use formulae which express the field operators ψ and $\psi^\dagger$ in terms of IN operators. To begin with, we shall assume that the succession of times is

$$t_1 > t_2 > \dots > t_n > t_n' > \dots > t_1' .$$

The chronological product of field operators can then be written as

$$T(\psi(x_1) \dots \psi(x_n)\psi^\dagger(y_n) \dots \psi^\dagger(y_1))$$
$$= \mathsf{V}^{\dagger\text{in}}(t_1, -\infty)\psi^{\text{in}}(x_1)\mathsf{V}^{\text{in}}(t_1, t_2)\psi^{\text{in}}(x_2) \dots \mathsf{V}^{\text{in}}(t_n, t_n')\psi^{\dagger\text{in}}(y_n)$$
$$\dots \psi^{\dagger\text{in}}(y_1)\mathsf{V}^{\text{in}}(t_1', -\infty),$$

or as

$$T(\psi(x_1) \dots \psi(x_n)\psi^\dagger(y_n) \dots \psi^\dagger(y_1))$$
$$= \mathsf{S}^\dagger T(\mathsf{S}\psi^{\text{in}}(x_1) \dots \psi^{\text{in}}(x_n)\psi^{\dagger\text{in}}(y_n) \dots \psi^{\dagger\text{in}}(y_1))$$
$$\equiv \mathsf{S}^\dagger \sum_{k=0}^{\infty} \frac{(-\imath)^k}{k!} \int d\tau_1 \dots d\tau_k$$
$$\times T(H^{\text{in}}(\tau_1) \dots H^{\text{in}}(\tau_k)\psi^{\text{in}}(x_1) \dots \psi^{\text{in}}(x_n)\psi^{\dagger\text{in}}(y_n) \dots \psi^{\dagger\text{in}}(y_1)).$$

The latter representation of the chronological product holds for any succession of times. Calculating the vacuum-state expectation values by sides and taking account of the stability of the vacuum

$$\mathsf{S}\Omega = \Omega,$$

we arrive at

$$T(x_1 \dots x_n; y_n \dots y_1)$$
$$= (\Omega|T(\mathsf{S}\psi^{\text{in}}(x_1) \dots \psi^{\text{in}}(x_n)\psi^{\dagger\text{in}}(y_n) \dots \psi^{\dagger\text{in}}(y_1))\Omega). \qquad (65)$$

In our discussion thus far, the potential U described a particular field of external forces. This was, therefore, a given function of the variables $(\mathbf{x}, t)$. Further on, the function $U(\mathbf{x}, t)$ will appear in another role as well. The function $U(\mathbf{x}, t)$ (from a certain set of regular functions defined in space-time) will be regarded as a new variable of the theory which will not take on a definite form, dictated by physical conditions, until the final formulae. To emphasize this new role of the function U we introduce new notations for the propagators T, similar to those used for the one-particle propagators $\mathscr{K}$. Henceforth, we shall write

$$T[x_1 \dots x_n; y_n \dots y_1 | U],\ \mathsf{S}[U], \text{etc.}$$

With the meaning of the external potential so extended, functional derivatives† with respect to U can be considered. The method of functional derivatives will play an important role in the formulation of relativistic quantum electrodynamics. In this section we shall apply it to non-relativistic quantum mechanics so as to explain, with the simplest example, how it operates. It will be our purpose first and foremost to derive concise, though only formal, formulae for propagators.#

The Hamiltonian $H^{\text{in}}(t)$ specifying the time evolution of a system of particles will be expressed in terms of the particle density operator $\varrho^{\text{in}}(\mathbf{x}, t)$ by the formula

$$H^{\text{in}}(t) = \frac{\lambda^2}{2} \int d^3x d^3y \varrho^{\text{in}}(\mathbf{x}, t) V(\mathbf{x}-\mathbf{y}) \varrho^{\text{in}}(\mathbf{y}, t)$$

$$-\lambda^2 \int d^3x V_0 \varrho^{\text{in}}(\mathbf{x}, t) + \lambda \int d^3x\, U(\mathbf{x}, t) \varrho^{\text{in}}(\mathbf{x}, t), \tag{66}$$

where

$$\varrho^{\text{in}}(\mathbf{x}, t) \equiv \sum_{\sigma} \psi^{\dagger\text{in}}(\mathbf{x}\sigma t) \psi^{\text{in}}(\mathbf{x}\sigma t),$$

---

† The definition and fundamental properties of functional derivatives are given in Appendix C.

The functional formalism based on the concept of the propagator, as given in our book, is patterned after the papers of Feynman (1949a, b) and (1950) and Schwinger (1951), (1953a, b) and (1954a, b). The second volume of the monograph by Rzewuski contains an extensive exposition of functional methods and a recent book by Fried describes various applications of this method to quantum field theory.

whereas

$$V_0 \equiv V(\mathbf{x}-\mathbf{y})|_{\mathbf{x}=\mathbf{y}}.$$

The parameter λ has been introduced here for the sake of a better analogy with quantum electrodynamics.

Since all the operators ϱ^{in} under the chronological product sign may be interchanged, the operator $\mathsf{S}[U]$ may be expressed as follows,†

$$\mathsf{S}[U] = T\left\{\exp\left(-\frac{i\lambda^2}{2}\int \mathrm{d}x\,\mathrm{d}y\,\varrho(x)\,V(x-y)\,\varrho(y)\right)\right.$$
$$\left.\times \exp\left(i\lambda^2\int \mathrm{d}x\,V_0\,\varrho(x)\right)\exp\left(-i\lambda\int \mathrm{d}x\,U(x)\,\varrho(x)\right)\right\}, \qquad (67)$$

where

$$V(x-y) \equiv \delta(t-t')\,V(\mathbf{x}-\mathbf{y}).$$

First of all, let us consider the operator $\mathsf{S}_0[U]$ which describes a system of non-interacting particles moving in a field of potential U:

$$\mathsf{S}_0[U] = T\exp\left(-i\lambda\int \mathrm{d}x\,U(x)\,\varrho(x)\right)$$
$$\equiv \sum_{n=0}^{\infty}\frac{(-i\lambda)^n}{n!}\int \mathrm{d}x_1 \ldots \mathrm{d}x_n\,U(x_1)\ldots U(x_n)\,T\big(\varrho(x_1)\ldots\varrho(x_n)\big).$$

The operator S_0 satisfies the following relation which derives from the fundamental properties of functional derivatives:

$$\frac{\delta \mathsf{S}_0[U]}{\delta U(x)} = -i\lambda T\big(\mathsf{S}_0[U]\varrho(x)\big).$$

Applying this formula repeatedly, we obtain

$$\frac{\delta \mathsf{S}_0[U]}{\delta U(x_1)\ldots\delta U(x_n)} = (-i\lambda)^n\,T\big(\mathsf{S}_0[U]\varrho(x_1)\ldots\varrho(x_n)\big).$$

Thus, the operators ϱ which appear together with $\mathsf{S}_0[U]$ under the chronological product sign may be replaced by the functional derivatives $i\lambda^{-1}\delta/\delta U$. Hence, we have the relationship

$$\mathsf{S}[U] = \exp\left(-\frac{\lambda}{2}\int V_0\frac{\delta}{\delta U}\right)\exp\left(\frac{i}{2}\int\frac{\delta}{\delta U}\,V\,\frac{\delta}{\delta U}\right)\mathsf{S}_0[U], \qquad (68)$$

† For simplicity, we henceforth omit the superscript IN.

where

$$\int V_0 \frac{\delta}{\delta U} \equiv \int \mathrm{d}x\, V_0 \frac{\delta}{\delta U(x)},$$

$$\int \frac{\delta}{\delta U} V \frac{\delta}{\delta U} \equiv \int \mathrm{d}x\,\mathrm{d}y \frac{\delta}{\delta U(x)} V(x-y) \frac{\delta}{\delta U(y)}.$$

By this relation and the expression (65) for the propagator, we have

$$T[x_1 \ldots x_n; y_n \ldots y_1|U] = \exp\left(-\frac{\lambda}{2}\int V_0 \frac{\delta}{\delta U}\right)$$

$$\times \exp\left(\frac{i}{2}\int \frac{\delta}{\delta U} V \frac{\delta}{\delta U}\right) T_0[x_1 \ldots x_n; y_n \ldots y_1|U], \tag{69}$$

where

$$T_0[x_1 \ldots x_n; y_n \ldots y_1|U]$$

$$\equiv \left(\Omega|T(\mathsf{S}_0[U]\psi^{\mathrm{in}}(x_1) \ldots \psi^{\mathrm{in}}(x_n)\psi^{\dagger}(y_n) \ldots \psi^{\dagger\mathrm{in}}(y_1))\Omega\right).$$

For the moment, we have put aside the question of how to calculate effectively the result of the exponential operations on the propagators T_0. We have reduced the propagators T—which describe a system of particles with interaction—to an expression constructed out of the propagators of non-interacting particles which, as will later turn out, are simple in construction.

The propagators T_0 are most easily defined as the solution of a system of differential equations satisfied by those objects. It is worthwhile deriving these equations in the more general case, for the propagators T, since they have many applications in the theory of many particles.

The equations for propagators constitute an infinite system of coupled integro-differential equations,

$$[S_{x_i} + \lambda U(x_i)]\, T[x_1 \ldots x_n; y_n \ldots y_1|U]$$

$$+ i\lambda^2 \sum_{\sigma(y)} \int \mathrm{d}y\, V(x_i - y)\, T[x_1 \ldots x_n y; y^+ y_n \ldots y_1|U]$$

$$= -i \sum_{j=1}^{n} (\pm 1)^{j-i} \delta(x_i - y_j)\, T[x_1 \ldots \check{x}_i \ldots x_n; y_n \ldots \check{y}_j \ldots y_1|U], \tag{70}$$

where

$$\delta(x_i-y_j) \equiv \delta_{\sigma_i\sigma'_j}\delta(\mathbf{x}_i-\mathbf{y}_j)\delta(t_i-t'_j).$$

In order to derive these equations, it is sufficient to differentiate the propagator T with respect to t_i by using the relation

$$\begin{aligned}
&\partial_{t_i} T(\psi(x_1)\ldots\psi(x_n)\psi^\dagger(y_n)\ldots\psi^\dagger(y_1)) \\
&= T(\psi(x_1)\ldots\partial_{t_i}\psi(x_i)\ldots\psi(x_n)\psi^\dagger(y_n)\ldots\psi^\dagger(y_1)) \\
&+\sum_{k=1}^{n}(\pm 1)^{i-k+1}T(\psi(x_1)\ldots\check{\psi}(x_i)\ldots\psi(x_{k-1})[\psi(x_i),\psi(x_k)]_{\mp} \\
&\quad\times\delta(t_i-t_k)\psi(x_{k+1})\ldots\psi(x_n)\psi^\dagger(y_n)\ldots\psi^\dagger(y_1)) \\
&+\sum_{j=1}^{n}(\pm 1)^{i-j}T(\psi(x_1)\ldots\check{\psi}(x_i)\ldots\psi(x_n)\psi^\dagger(y_n)\ldots\psi^\dagger(y_{j+1}) \\
&\quad\times[\psi(x_i),\psi^\dagger(y_j)]_{\mp}\delta(t_i-t'_j)\psi^\dagger(y_{j-1})\ldots\psi^\dagger(y_1)),
\end{aligned}$$

which has been proved in Appendix B, as well as the field equations and the commutation relations for the field operators.

A similar system of equations, differing only in that S_{x_i} has been replaced by $\overleftarrow{S}_{y_j}$, is satisfied by the propagators T with respect to the variables y_j (cf. also eq. (4.10b)). The propagator $T[\ldots y, y^+\ldots|U]$ is defined as the limit

$$\lim_{0<\varepsilon\to 0} T[\ldots, \mathbf{y}t; \mathbf{y}t+\varepsilon, \ldots|U].$$

The equations for the propagators T_0 are obtained by setting $V = 0$:

$$\begin{aligned}
&[S_{x_i}+\lambda U(x_i)]\,T_0[x_1\ldots x_n; y_n\ldots y_1|U] \\
&= -i\sum_{j=1}^{n}(\pm 1)^{j-i}\delta(x_i-y_j)\,T_0[x_1\ldots\check{x}_i\ldots x_n; y_n\ldots\check{y}_j\ldots y_1|U]. \qquad (71)
\end{aligned}$$

By the definition of the chronological product, both the propagators T and T_0 satisfy the (anti)-symmetry conditions and the *retardation conditions*

$$T[x_1\ldots x_n; y_n\ldots y_1|U] = 0, \qquad T_0[x_1\ldots x_n; y_n\ldots y_1|U] = 0,$$

whenever any one of the time arguments t_i is earlier than all the times t'_j. The foregoing conditions are of service in choosing the proper solutions out of the many possible for the equations for the propagators. In the case of the propagators T_0, these solutions are symmetrized or anti-symmetrized products of the propagators $\mathscr{K}_{\mathrm{R}}$, describing the propagation of a single particle†

$$T_0[x_1 \ldots x_n; y_n \ldots y_1|U] = (-i)^n \mathscr{K}_{\mathrm{R}}[x_1 \ldots x_n; y_n \ldots y_1|U]$$
$$\equiv (-i)^n \sum_{\mathrm{perm}\, i} \varepsilon_p \mathscr{K}_{\mathrm{R}}[x_1; y_{i_1}|U] \ldots \mathscr{K}_{\mathrm{R}}[x_n; y_{i_n}|U]. \tag{72}$$

The one-particle propagators are now also matrices in the spin variables

$$\mathscr{K}_{\mathrm{R}}[x; y|U] = \delta_{\sigma\sigma'} \mathscr{K}_{\mathrm{R}}[\mathbf{x}t; \mathbf{y}t'|U].$$

The distributions defined by eq. (72) satisfy both the system of equations for propagators and the retardation conditions.

Now we can proceed to discuss the complete propagators T. How the exponential operators appearing in formula (69) operate is best elucidated by the Feynman diagram method. Our description of this method will be based on the following fundamental relation which is satisfied by the one-particle propagator:

$$\frac{\delta \mathscr{K}_{\mathrm{R}}[x; y|U]}{\delta U(z)} = -\lambda \mathscr{K}_{\mathrm{R}}[x; z|U] \mathscr{K}_{\mathrm{R}}[z; y|U]. \tag{73}$$

This relation is derived from the differential equation (4.10a) satisfied by the propagator $\mathscr{K}_{\mathrm{R}}$. To this end, we functionally differentiate both sides of these equations with respect to $U(z)$, multiply by $\mathscr{K}_{\mathrm{R}}[w; x|U]$, and integrate with respect to x over all space-time. On integrating by parts (the boundary terms are zero because of the retardation conditions satisfied by $\mathscr{K}_{\mathrm{R}}$) and on utilizing eq. (4.10b), we obtain relation (73).

For particles with spin, the product of propagators is taken to mean the matrix product

$$\mathscr{K}_{\mathrm{R}}[x; z|U] \mathscr{K}_{\mathrm{R}}[z; y|U]$$
$$= \sum_{\sigma''} \mathscr{K}_{\mathrm{R}}[\mathbf{x}\sigma t; \mathbf{z}\sigma'' t''|U] \mathscr{K}_{\mathrm{R}}[\mathbf{z}\sigma'' t''; \mathbf{y}\sigma' t'|U].$$

† By substituting the propagator T_0 into formulae (60) and (63) we obtain the S operator for a system of particles not interacting mutually. Scattering amplitudes calculated then by means of the S operator split up in the familiar way (cf. formula (7)) into products of one-particle amplitudes.

Repeated differentiation of the propagator $\mathscr{K}_{\mathrm{R}}$ yields

$$\frac{\delta^n}{\delta U(z_1)\,\dots\,\delta U(z_n)}\mathscr{K}_{\mathrm{R}}[x;y|U]$$

$$= (-\lambda)^n \sum_{\mathrm{perm}\,i} \mathscr{K}_{\mathrm{R}}[x;z_{i_1}|U]\mathscr{K}_{\mathrm{R}}[z_{i_1};z_{i_2}|U]\dots\mathscr{K}_{\mathrm{R}}[z_{i_n};y|U].$$

When we multiply both sides of these equations by $U_1(z_1)\dots U_1(z_n)$ and integrate with respect to $z_1,\dots,z_n$, we arrive at

$$\frac{1}{n!}\int \mathrm{d}z_1\dots\mathrm{d}z_n\,U_1(z_1)\dots U_1(z_n)\frac{\delta^n}{\delta U(z_1)\,\dots\,\delta U(z_n)}\mathscr{K}_{\mathrm{R}}[x;y|U]$$

$$= (-\lambda)^n \int \mathrm{d}z_1\dots\mathrm{d}z_n \mathscr{K}_{\mathrm{R}}[x;z_1|U]\,U_1(z_1)\mathscr{K}_{\mathrm{R}}[z_1;z_2|U]$$

$$\dots U_1(z_n)\mathscr{K}_{\mathrm{R}}[z_n;y|U].$$

On summation over n, we can identify the series on the right-hand side as an iterative solution of the integral equation for $\mathscr{K}_{\mathrm{R}}[x;y|U+U_1]$ (cf. eq. (4.14)). In this way we obtain an expansion of the propagator $\mathscr{K}_{\mathrm{R}}$ in terms of the Volterra series,

$$\mathscr{K}_{\mathrm{R}}[x;y|U+U_1] = \exp\left(\int U_1\frac{\delta}{\delta U}\right)\mathscr{K}_{\mathrm{R}}[x;y|U]$$

$$\equiv \sum_{n=0}^{\infty}\frac{1}{n!}\int \mathrm{d}z_1\dots\mathrm{d}z_n\,U_1(z_1)\dots U_1(z_n)\frac{\delta^n}{\delta U(z_1)\,\dots\,\delta U(z_n)}\mathscr{K}_{\mathrm{R}}[x;y|U]. \tag{74}$$

This is the functional counterpart of Taylor's formula. It holds not only for $\mathscr{K}_{\mathrm{R}}$, but for all sufficiently regular (analytical) functionals.

After these introductory remarks, let us return to our discussion of the properties of the propagators T on the basis of the representation (69). Unfortunately, we cannot determine exactly the result of the action of the operation $\exp\left(\frac{i}{2}\int\frac{\delta}{\delta U}V\frac{\delta}{\delta U}\right)$ on the propagator T_0, but we can instead determine the successive terms in the series formed by expanding the exponential function. To determine this *perturbation expansion* of the propagator as a series in λ^2 (or V) it is sufficient to use the formula for the n-th derivative of the propagator and the rules for the differentiation of a product. The only difficulties encountered in developing such a perturbation theory stem from the occurrence of a large number of terms, which grows rapidly with the order of the perturbation expansions.

Feynman diagrams are used for classifying the terms which appear in an arbitrary order of the perturbation calculations. In addition to advantages of a formal nature, Feynman diagrams have a picturesque and direct physical interpretation.

The following series, representing the result of the exponential operation on the propagator T_0, is the object of our considerations:

$$(-i)^n \sum_{m=0}^{\infty} \left(\frac{i}{2}\right)^m \frac{1}{m!} \int dz_1 \dots dz_m dz_1' \dots dz_m' V(z_1 - z_1') \dots V(z_m - z_m')$$

$$\times \frac{\delta}{\delta U(z_1)} \cdots \frac{\delta}{\delta U(z_m)} \frac{\delta}{\delta U(z_1')} \cdots \frac{\delta}{\delta U(z_m')} \mathscr{K}_{\mathrm{R}}[x_1 \dots x_n; y_n \dots y_1 | U]. \quad (75)$$

The Feynman diagram method consists in introducing a one-to-one correspondence between all terms of the perturbation expansion of T and their graphical representations, known as Feynman diagrams.

On substituting the propagator $\mathscr{K}_{\mathrm{R}}[x_1 \dots x_n; y_n \dots y_1 | U]$ as the sum of the products of one-particle propagators into formula (75) and on performing all differentiations with respect to U, we obtain an infinite series, each term of which will be constructed out of the products of the functions $\mathscr{K}_{\mathrm{R}}$ integrated with the product of the functions V. All the terms of this series are represented graphically by associating particular graphical elements constituting the *diagram* with the functions $\mathscr{K}_{\mathrm{R}}$ and V and with the other components of the integral expressions.

With each function $-i\mathscr{K}_{\mathrm{R}}[x; y | U]$ we associate a segment of a continuous line. The ends of this line are labelled with the coordinates x and y; a line is oriented by providing it with an arrow directed from the point y to x. If the same argument appears twice, the corresponding ends of the line meet at this point.

With each function $iV(z - z')$ we associate the dashed-line segment joining points with coordinates z and z'.

The so-called vertex of the diagram is associated with each variable of integration z_i or z_j', together with summation over the corresponding spin variables and the factor $-i\lambda$. This is the meeting point of two continuous lines representing two functions $\mathscr{K}_R$ containing the repeated argument and a dashed line representing the function V.

The rules above are illustrated by the table below.

Analytical expression	Graphical element
$-i\mathscr{K}_R[x;y\|U]$	
$iV(z-z')$	
$-i\lambda\sum_{\sigma(z)}$	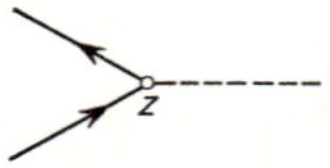

By these rules it is possible to draw a diagram representing every integrand function in formula (75) as well as conversely, to reproduce the function corresponding to a given diagram.

The Feynman diagram method, however, also gives us something more. By this method we can, without performing separate calculations in each individual case, immediately write out all the terms constituting an n-particle propagator in any order of the perturbation calculations. For as we shall explain below, it is easy to draw all the diagrams which are appropriate in the case under consideration, whence by applying the rules given in the table, we obtain all the terms constituting the successive approximations to the propagator.

The problem of classifying all the diagrams will be resolved by studying the operation of functional derivatives acting on the function $\mathscr{K}_R[x_1 \ldots x_n; y_n \ldots y_1|U]$. To begin with, we shall consider the simplest case of the function $\mathscr{K}_R[x; y|U]$.

In graphic language, the action of the functional derivative $\delta/\delta U(z)$ on the propagator $\mathscr{K}_R[x; y|U]$ (cf. formula (73)) reduces to that of breaking the line and inserting a vertex in the break (see Fig. 5.1).

Two-fold action of the functional derivative on the propagator $\mathscr{K}_R[x; y|U]$ is depicted in Fig. 5.2. The result of k-fold operation of the functional derivative on the propagator $\mathscr{K}_R[x; y|U]$ will be indicated by $k!$ broken lines, each with k vertices. These broken lines differ as to the permutations of those vertices, each permutation appearing exactly once.

The diagrammatic representation of the action of functional derivatives can be extended in a natural manner to many-particle propagators. Such a propagator is a sum of $n!$ products of the propagators $\mathscr{K}_{\mathrm{R}}[x; y|U]$. The operation of the functional derivatives on each term

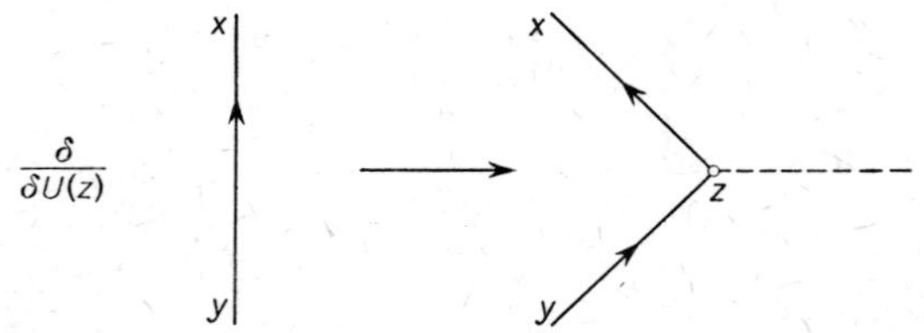

FIG. 5.1. The "splitting" property of the propagator $\mathscr{K}_{\mathrm{R}}[x, y|U]$.

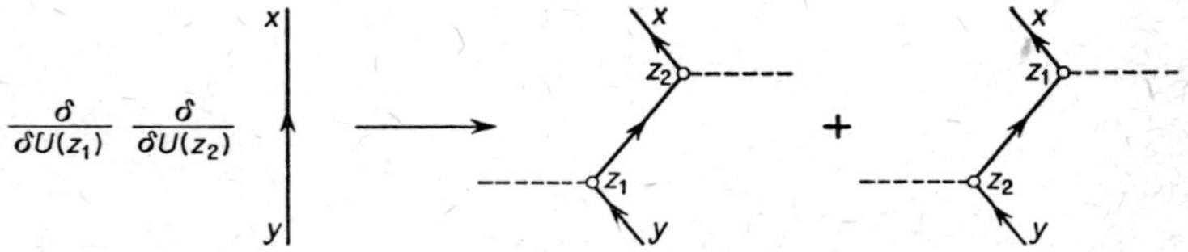

FIG. 5.2. The result of two-fold functional differentiation of the propagator $\mathscr{K}_{\mathrm{R}}[x, y|U]$.

in this sum will be determined by the laws of differentiation of products. By k-fold differentiation of the product of n one-particle propagators we obtain $(n+k-1)!n!/(n-1)!$ terms. When depicted graphically these terms differ as to the distribution of the vertices on n broken lines and the permutations of the vertices on each broken line. Every possible arrangement of the vertices will appear exactly once. The result of two-fold differentiation of a two-particle propagator is represented graphically in Fig. 5.3.

In the m-th order of perturbation calculation, in addition to the $2m$-fold functional derivative of the n-particle propagator we also have the product of m functions V. These functions are represented by a set of m dashed-line segments joining the vertices z_i and z_i'. Since the variables labelling the vertices are integration variables, the integrand expressions represented by diagrams differing only as to the labelling of the vertices will become equal after integration. For this reason we frequently omit the labels of the vertices and draw only diagrams which differ from each other in the structure of lines and their connections, and not in the labelling of vertices. Such diagrams with unla-

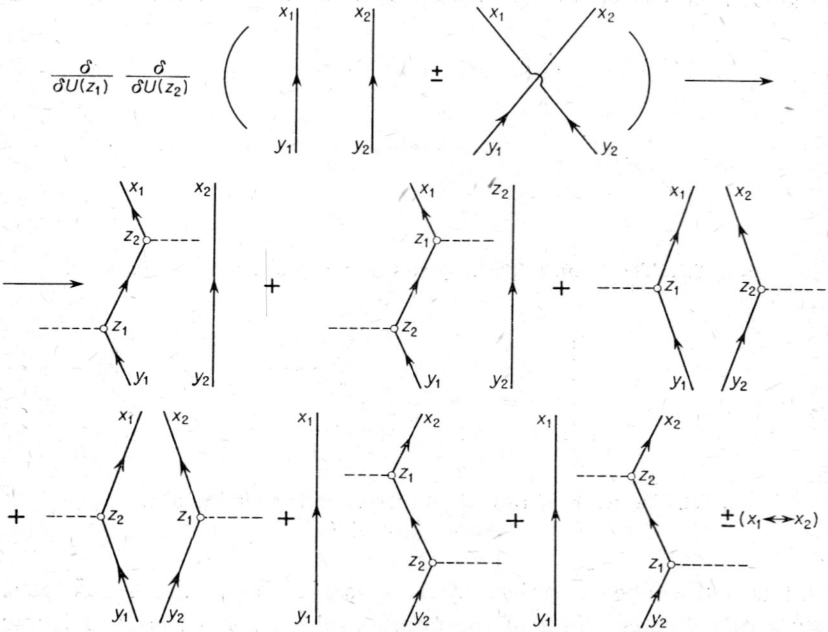

FIG. 5.3. The result of two-fold functional differentiation of a two-particle propagator.

belled vertices will be referred to as *Feynman diagrams*. Each Feynman diagram represents a set of all diagrams differing only as to numeration of vertices.† To one Feynman diagram with m dashed lines there correspond $2^m m!$ diagrams with labelled vertices which differ from each other by the position of the variables z_i and z_i' and the permutations of pairs of variables $(z_1, z_1') \dots (z_m, z_m')$. To put it figuratively, it might be said that labelled diagrams corresponding to a given Feynman diagram differ as to the directions and permutations of the dashed lines.

† Examples of Feynman diagrams and the corresponding diagrams with labelled vertices are shown in Fig. 5.4.

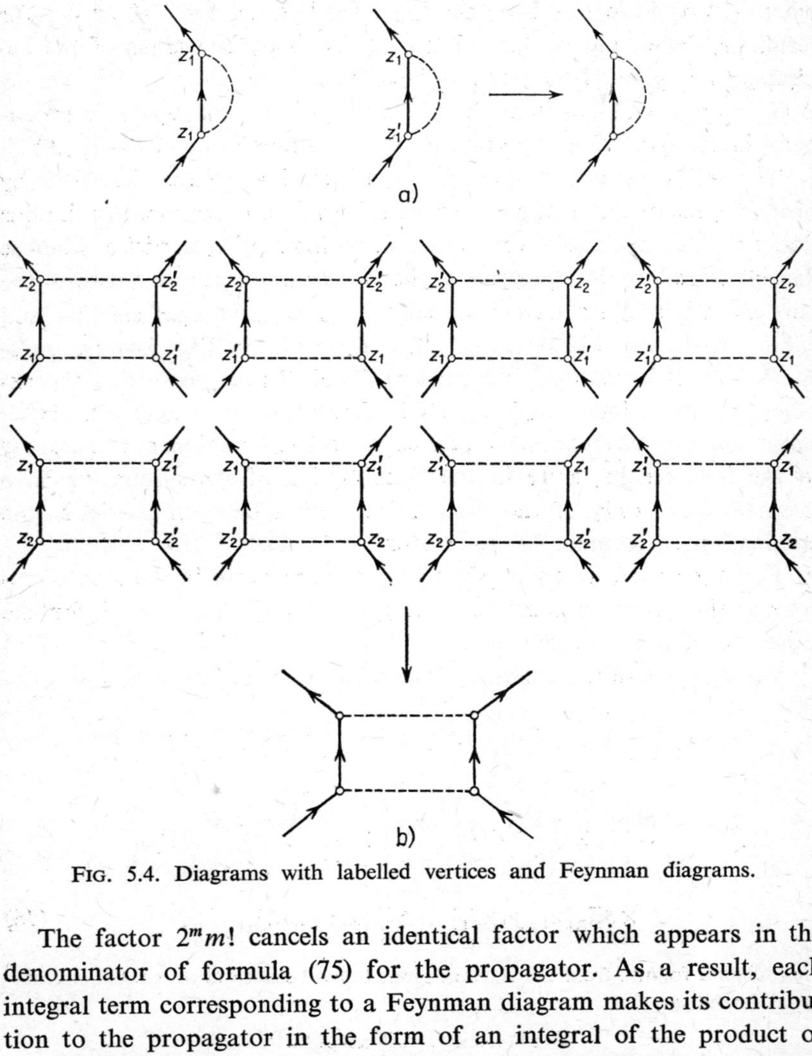

FIG. 5.4. Diagrams with labelled vertices and Feynman diagrams.

The factor $2^m m!$ cancels an identical factor which appears in the denominator of formula (75) for the propagator. As a result, each integral term corresponding to a Feynman diagram makes its contribution to the propagator in the form of an integral of the product of $n+2m$ functions $\mathscr{K}_R$ and m functions V, with the following numerical factor,

$$\varepsilon_p \underbrace{(-i)^{n+2m}}_{\text{continuous lines}} \underbrace{(i)^m}_{\text{dashed lines}} \underbrace{(-i\lambda)^{2m}}_{\text{vertices}} = \varepsilon_p(-i)^n(i\lambda^2)^m,$$

where the coefficient ε_p is equal to $+1$ for bosons, and $+1$ or -1 for fermions, depending on the relative parity of permutations of the beginnings and ends of the fermion lines.

The number of Feynman diagrams representing an n-particle propagator in the m-th order of perturbation calculations is $n(n+2m-1)!/2^m m!$.

The following mental pictures of physical processes described by propagators are often associated with Feynman diagrams. Continuous lines are interpreted as the motions of non-interacting particles, whereas dashed lines are interpreted as elementary interactions between the particles whose lines have been so joined. Dashed lines starting and ending at the same broken line (cf. diagram (a) in Fig. 5.4) are interpreted as self-interaction of particles. We shall now prove that the exponential operation containing V_0 in formula (69) exactly cancels all self-interaction effects. One would expect self-interaction not to appear in the final results, since in the Hamiltonian of the system we have taken account only of the mutual interaction between particles (the summation in formula (2) runs over all $i \neq j$).

For our proof, let us consider only the first term in the sum representing the propagator $\mathscr{K}_{\mathrm{R}}[x_1 \dots x_n; y_n \dots y_1|U]$. The proof for the other terms runs the same way.

We shall prove the validity of the relation

$$\exp\left(\frac{i}{2}\int\frac{\delta}{\delta U}V\frac{\delta}{\delta U}\right)\mathscr{K}_{\mathrm{R}}[x_1; y_1|U]\dots\mathscr{K}_{\mathrm{R}}[x_n; y_n|U]$$

$$= \exp\left(\frac{\lambda}{2}\int V_0\frac{\delta}{\delta U}\right)\left\{\exp\left(\frac{i}{2}\sum_{i\neq j}\int\frac{\delta}{\delta U_i}V\frac{\delta}{\delta U_j}\right)\right.$$

$$\left.\times\mathscr{K}_{\mathrm{R}}[x_1; y_1|U_1]\dots\mathscr{K}_{\mathrm{R}}[x_n; y_n|U_n]\right\}\Big|_{U_i=U}. \tag{76}$$

This relation follows from the relation given below between terms containing functional derivatives,

$$\mathscr{K}_{\mathrm{R}}[x_1; y_1|U]\dots\mathscr{K}_{\mathrm{R}}[x_n; y_n|U]$$

$$= \left\{\exp\left(\int U\frac{\delta}{\delta U_1}\right)\dots\exp\left(\int U\frac{\delta}{\delta U_n}\right)\mathscr{K}_{\mathrm{R}}[x_1; y_1|U_1]\dots\mathscr{K}_{\mathrm{R}}[x_n; y_n|U_n]\right\}\Big|_{U_i=U},$$

$$\frac{\delta}{\delta U(z)}\exp\left(\int U\frac{\delta}{\delta U_i}\right) = \frac{\delta}{\delta U_i(z)}\exp\left(\int U\frac{\delta}{\delta U_i}\right),$$

whence

$$\exp\left(\frac{i}{2}\int\frac{\delta}{\delta U}V\frac{\delta}{\delta U}\right)\mathscr{K}_{\mathrm{R}}[x_1;y_1|U]\ldots\mathscr{K}_{\mathrm{R}}[x_n;y_n|U]$$

$$=\left\{\exp\left(\frac{i}{2}\sum_{i\neq j}\int\frac{\delta}{\delta U_i}V\frac{\delta}{\delta U_j}\right)\exp\left(\frac{i}{2}\int\frac{\delta}{\delta U_1}V\frac{\delta}{\delta U_1}\right)\mathscr{K}_{\mathrm{R}}[x_1;y_1|U_1]\right.$$

$$\left.\ldots\exp\left(\frac{i}{2}\int\frac{\delta}{\delta U_n}V\frac{\delta}{\delta U_n}\right)\mathscr{K}_{\mathrm{R}}[x_n;y_n|U_n]\right\}\Bigg|_{U_i=0}.$$

The assumption that the potential V acts instantaneously enables us to calculate the result of the action of the exponential operation $\exp\left(\frac{i}{2}\int\frac{\delta}{\delta U}V\frac{\delta}{\delta U}\right)$ on the one-particle propagator, which takes place in the formula above. For this purpose, as an auxiliary measure let us calculate the second functional derivative of the propagator,

$$\frac{\delta}{\delta U(z_1,t_1)}\frac{\delta}{\delta U(z_2,t_2)}\mathscr{K}_{\mathrm{R}}[x;y|U],$$

when $t_1=t_2$. This derivative is continuous (as a distribution) when $t_1=t_2$ and it can be calculated as both a right-hand and a left-hand limit, the result being the same in both cases (the spin indices are omitted):

$$\frac{\delta}{\delta U(z_1,t_1)}\frac{\delta}{\delta U(z_1,t_2)}\mathscr{K}_{\mathrm{R}}[x;y|U]|_{t_1=t_2}$$
$$=\lambda^2\{\mathscr{K}_{\mathrm{R}}[x;z_1|U]\mathscr{K}_{\mathrm{R}}[z_1;z_2|U]\mathscr{K}_{\mathrm{R}}[z_2;y|U]$$
$$+\mathscr{K}_{\mathrm{R}}[x;z_2|U]\mathscr{K}_{\mathrm{R}}[z_2;z_1|U]\mathscr{K}_{\mathrm{R}}[z_1;y|U]\}|_{t_1=t_2}$$
$$=i\lambda^2\delta(z_1-z_2)\mathscr{K}_{\mathrm{R}}[\mathbf{x}t;\mathbf{z}_1t_1|U]\mathscr{K}_{\mathrm{R}}[\mathbf{z}_1t_1;\mathbf{y}t'|U].$$

Making use of this equality, we obtain

$$\frac{i}{2}\int\frac{\delta}{\delta U}V\frac{\delta}{\delta U}\mathscr{K}_{\mathrm{R}}[x;y|U]$$
$$=\frac{\lambda}{2}\int d^3z_1\,dt_1V_0\frac{\delta}{\delta U(\mathbf{z}_1,t_1)}\mathscr{K}_{\mathrm{R}}[x;y|U],$$

which, together with the result arrived at above, yields formula (76).

On eliminating the self-interaction, therefore, we get the following representation of the propagator,

$$T[x_1,\ldots,x_n;y_n\ldots y_1|U]=(-i)^n\left\{\exp\left(\frac{i}{2}\sum_{i\neq j}\int\frac{\delta}{\delta U_i}V\frac{\delta}{\delta U_j}\right)\right.$$
$$\left.\times\sum_{\mathrm{perm}\ i}\varepsilon_p\mathscr{K}_{\mathrm{R}}[x_1;y_{i_1}|U_1]\ldots\mathscr{K}_{\mathrm{R}}[x_n;y_{i_n}|U_n]\right\}_{U_i=U}$$

$$\equiv \exp\left(\frac{i}{2}\int \frac{\delta}{\delta U} V \frac{\delta}{\delta U}\right)_{\text{ex}} T_0[x_1 \ldots x_n; y_n \ldots y_1|U], \tag{77}$$

where the symbol ex means that all the expressions containing self-interaction are to be discarded. In other words, the two functional differentiations in the expression $\int \frac{\delta}{\delta U} V \frac{\delta}{\delta U}$ do not act on the same one-particle propagator. In the language of diagrams, on the other hand, this means the non-occurrence of diagrams in which any dashed line begins and ends at the same broken line.

In the non-relativistic quantum mechanics of two or three particles, use is generally made of the Schrödinger equation. Propagators and the diagram method would be an unnecessary encumbrance in such cases. The methods discussed above are very useful in describing many-particle systems, especially in quantum-statistical physics, even in non-relativistic theory. The role of propagators and diagrams grows immeasurably in relativistic theory. Much attention to these subjects will be devoted in subsequent chapters.

CHAPTER 3

THE CLASSICAL THEORY OF THE ELECTROMAGNETIC FIELD

6. THE TENSOR DESCRIPTION OF THE ELECTROMAGNETIC FIELD AND THE FIELD EQUATIONS

THE subject of this chapter is the classical (non-quantum) description of the electromagnetic field, both when free and when interacting with charges and currents generated by classical systems. We shall not confine ourselves to the simplest case, i.e. to the Maxwell theory. Our discussion will also apply to more complicated, non-linear theories of the electromagnetic field.# Non-linear theories describe fields interacting with themselves. Such self-interactions evoke effects analogous to certain phenomena familiar from the electrodynamics of continuous media.

The electromagnetic field is described with the aid of six real-valued functions of position and time, transforming under Poincaré transformations as the components of an antisymmetric tensor $f_{\mu\nu}(x)$. The components of $f_{\mu\nu}(x)$ satisfy a set of first-order partial differential equations. Some of the equations follow from the assumption that the tensor $f_{\mu\nu}$ can be represented as a four-dimensional rotation; the others may be derived from the principle of least action. The density of the Lagrangian is the sum of two terms of which the first, $\mathscr{L}$, describes the electromagnetic field and the second, $\mathscr{L}_{\text{int}}$, the interaction of the field

A non-linear theory of the electromagnetic field which satisfies all the postulates of relativity theory was formulated by M. Born (1934). The best-known version of this theory is that given by M. Born and L. Infeld (1934a). It is discussed in Section 8.

with sources. To ensure relativistic invariance of the theory, we assume that the density of the Lagrangian is a function of two invariants† constructed out of the tensor $f_{\mu\nu}$, the scalar S and the pseudoscalar P,

$$S \equiv -\frac{1}{4} f_{\mu\nu} f^{\mu\nu}, \tag{1a}$$

$$P \equiv -\frac{1}{4} f_{\mu\nu} \check{f}^{\mu\nu}. \tag{1b}$$

The electric field intensity and the magnetic induction are described by the vectors **E** and **B**, respectively. These vectors are constructed in a particular coordinate system out of the components of the tensor $f_{\mu\nu}$ in the following manner:

$$\mathbf{E} \equiv (f_{01}, f_{02}, f_{03}) = (\check{f}^{23}, \check{f}^{31}, \check{f}^{12}), \tag{2a}$$

$$\mathbf{B} \equiv -(f_{23}, f_{31}, f_{12}) = -(\check{f}^{01}, \check{f}^{02}, \check{f}^{03}). \tag{2b}$$

These quantities are determined by measuring the Lorentz force acting on a test charge.

The Field Equations

The assumption that the tensor $f_{\mu\nu}$ can be represented as a four-dimensional curl means that there is a *potential four-vector* $A_\mu(x)$ such that

$$f_{\mu\nu}(x) = \partial_{[\mu} A_{\nu]}(x) \equiv \partial_\mu A_\nu(x) - \partial_\nu A_\mu(x). \tag{3}$$

This gives us the *first set of field equations* which may be written in two equivalent forms,

$$\partial_\mu \check{f}^{\mu\nu} = 0 \tag{4a}$$

or

$$\partial_\lambda f_{\mu\nu} + \partial_\mu f_{\nu\lambda} + \partial_\nu f_{\lambda\mu} = 0. \tag{4b}$$

† The derivatives of $f_{\mu\nu}$ are excluded from appearing in the Lagrangian, for, if they did, the state of the electromagnetic field at a given instant would not be determined by the components of the tensor $f_{\mu\nu}$ alone. In our considerations we also exclude so-called non-local theories in which the Lagrangian density $\mathscr{L}(x)$ of the field depends functionally on the field, hence not only through the values of the fields at the point x. An extensive discussion of non-local theories can be found in the book by Rzewuski.

The *second set of field equations* consists of Euler–Lagrange equations obtained by varying the action with respect to the potential vector A_μ. These equations also have two equivalent forms

$$\partial_\mu h^{\mu\nu} = j^\nu \tag{5a}$$

or

$$\partial_\lambda \check{h}_{\mu\nu} + \partial_\nu \check{h}_{\lambda\mu} + \partial_\mu \check{h}_{\nu\lambda} = \varepsilon_{\lambda\mu\nu\varkappa} j^\varkappa, \tag{5b}$$

where†

$$h^{\mu\nu}(x) \equiv -\frac{\partial \mathscr{L}}{\partial f_{\mu\nu}(x)} = \frac{\partial \mathscr{L}}{\partial S} f^{\mu\nu}(x) + \frac{\partial \mathscr{L}}{\partial P} \check{f}^{\mu\nu}(x), \tag{6}$$

and the *current four-vector* $j^\mu(x)$ is defined as

$$j^\mu(x) \equiv -\frac{\partial \mathscr{L}_{\text{int}}(x)}{\partial A_\mu(x)}. \tag{7}$$

A necessary condition for solutions to exist for the field equations (4) and (5) is that the equation of continuity,

$$\partial_\mu j^\mu(x) = 0, \tag{8}$$

be satisfied by the current four-vector. Out of the components of the antisymmetric tensor $h^{\mu\nu}$ we form the vectors **D** and **H** which we shall call, respectively, the *electric displacement* and the *magnetic field intensity*:

$$\mathbf{D} \equiv -(h^{01}, h^{02}, h^{03}) = (\check{h}_{23}, \check{h}_{31}, \check{h}_{12}), \tag{9a}$$

$$\mathbf{H} \equiv -(h^{23}, h^{31}, h^{12}) = (\check{h}_{01}, \check{h}_{02}, \check{h}_{03}). \tag{9b}$$

Knowing the Lagrangian density $\mathscr{L}$ of the electromagnetic field, we can construct the tensors ε_{ik} and μ_{ik} which define the relations between the vectors **E** and **B** and the vectors **D** and **H**,

$$D_i = \varepsilon_{ik} E_k, \tag{10a}$$

$$H_i = (\mu^{-1})_{ik} B_k. \tag{10b}$$

† To differentiate with respect to the components of an antisymmetric tensor means to differentiate with respect to its independent components. For example, we have

$$\frac{\partial}{\partial f_{\mu\nu}} (f_{\lambda\varrho} f^{\lambda\varrho}) = 4 f^{\mu\nu}.$$

The quantities ε_{ik} and μ_{ik} are the tensors of *dielectric* and *magnetic permeabilities*. In non-linear theory they are functions of the field intensities

$$\varepsilon_{ik} \equiv \frac{\partial \mathcal{L}}{\partial S}\,\delta_{ik} + \frac{\partial \mathcal{L}}{\partial P}\,\frac{B_i B_k}{\mathbf{E}\cdot\mathbf{B}}, \tag{11a}$$

$$\mu_{ik}^{-1} \equiv \frac{\partial \mathcal{L}}{\partial S}\,\delta_{ik} - \frac{\partial \mathcal{L}}{\partial P}\,\frac{E_i E_k}{\mathbf{E}\cdot\mathbf{B}}. \tag{11b}$$

Transformation Laws

The tensor transformation laws for the electromagnetic field and current guarantee that the theory of the electromagnetic field as formulated above is relativistically invariant. The proper Poincaré group is said to be the symmetry group of this theory. This means that in a set of geometric objects $f_{\mu\nu}(x)$, $h^{\mu\nu}(x)$ and $j^{\mu}(x)$ satisfying the field equations (4) and (5) there operates the realization of this group, which is defined by the following transformations of field and current:†

$$f_{\mu\nu}(x) \xrightarrow[(a,\,\Lambda)]{} {}^{\mathrm{IL}}f_{\mu\nu}(x) = \Lambda_{\mu}{}^{\lambda}\Lambda_{\nu}{}^{\varrho} f_{\lambda\varrho}(\Lambda^{-1}(x-a)), \tag{12a}$$

$$h^{\mu\nu}(x) \xrightarrow[(a,\,\Lambda)]{} {}^{\mathrm{IL}}h^{\mu\nu}(x) = \Lambda^{\mu}{}_{\lambda}\Lambda^{\nu}{}_{\varrho} h^{\lambda\varrho}(\Lambda^{-1}(x-a)), \tag{12b}$$

$$j^{\mu}(x) \xrightarrow[(a,\,\Lambda)]{} {}^{\mathrm{IL}}j^{\mu}(x) = \Lambda^{\mu}{}_{\nu} j^{\nu}(\Lambda^{-1}(x-a)). \tag{12c}$$

These transformations are interpreted as translations and rotations of the geometric objects $f_{\mu\nu}$, $h^{\mu\nu}$ and j^{μ} in Minkowski space. If $f_{\mu\nu}$, $h^{\mu\nu}$, and j^{μ} were solutions of equations (4) and (5), the transformed objects will also be solutions of the same equations.

The full Poincaré group is a symmetry group of the theory under consideration if the Lagrangian density is an even function of the invariant P. This group is obtained by adding to the proper Poincaré group two transformations: space inversion and time reversal of processes. The following transformations of the fields $f_{\mu\nu}$ and $h^{\mu\nu}$ and the current j^{μ} constitute the realization of these transformations.

Space inversion:

$$f_{\mu\nu}(x) \to {}^{P}f_{\mu\nu}(x) = \mathcal{P}_{\mu}^{\lambda}\mathcal{P}_{\nu}^{\varrho} f_{\lambda\varrho}(\mathcal{P}x), \tag{13a}$$

$$h^{\mu\nu}(x) \to {}^{P}h^{\mu\nu}(x) = \mathcal{P}^{\mu}_{\lambda}\mathcal{P}^{\nu}_{\varrho} h^{\lambda\varrho}(\mathcal{P}x), \tag{13b}$$

$$j^{\mu}(x) \to {}^{P}j^{\mu}(x) = \mathcal{P}^{\mu}_{\nu} j^{\nu}(\mathcal{P}x). \tag{13c}$$

† The notation associated with the Poincaré group is explained in Appendix D.

Time reversal:

$$f_{\mu\nu}(x) \to {}^{T}f_{\mu\nu}(x) = -\mathscr{P}^{\lambda}_{\mu}\mathscr{P}^{\varrho}_{\nu}f_{\lambda\varrho}(-\mathscr{P}x), \tag{14a}$$

$$h^{\mu\nu}(x) \to {}^{T}h^{\mu\nu}(x) = -\mathscr{P}^{\mu}_{\lambda}\mathscr{P}^{\nu}_{\varrho}h^{\lambda\varrho}(-\mathscr{P}x), \tag{14b}$$

$$j^{\mu}(x) \to {}^{T}j^{\mu}(x) = \mathscr{P}^{\mu}_{\nu}j^{\nu}(-\mathscr{P}x), \tag{14c}$$

where

$$(\mathscr{P}^{\mu}_{\lambda}) \equiv \begin{pmatrix} 1 & 0 & 0 & 0 \\ 0 & -1 & 0 & 0 \\ 0 & 0 & -1 & 0 \\ 0 & 0 & 0 & -1 \end{pmatrix}.$$

Energy-Momentum Tensor of Electromagnetic Field

Let us now consider the properties of the energy-momentum tensor of the electromagnetic field. This tensor is used to construct the energy-momentum four-vector and the four-dimensional angular momentum tensor of the field; in the case of an isolated field these tensors satisfy the conservation laws.

The energy-momentum tensor of the electromagnetic field is of the form†

$$T^{\mu\nu}(x) = f^{\mu\lambda}(x)h_{\lambda}^{\ \nu}(x) - g^{\mu\nu}\mathscr{L}(x). \tag{15}$$

The definition of $h^{\mu\nu}(x)$ implies that the energy-momentum density tensor is symmetric.‡ For, if we use eq. (6) and the identity

$$f_{\mu\lambda}\check{f}^{\lambda}_{\ \nu} \equiv g_{\mu\nu}P$$

we obtain the equivalent form of the tensor $T^{\mu\nu}$,

$$T^{\mu\nu} = \frac{\partial\mathscr{L}}{\partial S} f^{\mu\lambda}f_{\lambda}^{\ \nu} + g^{\mu\nu}\left(\frac{\partial\mathscr{L}}{\partial P}P - \mathscr{L}\right). \tag{16}$$

---

† It is shown in Appendix F that this tensor is obtained by varying the Lagrangian density of the field with respect to the components of the metric tensor.

‡ Often, the canonical energy-momentum tensor of the electromagnetic field is first given and then built up to a symmetric tensor. The canonical tensor does not, however, describe any physical quantity and is not even defined in a manner independent of the choice of potential gauge. For this reason, we shall consider only the symmetric energy-momentum tensor. This tensor is physically meaningful inasmuch as it is a source of gravitational field. An extensive discussion of these aspects can be found in the books by Pauli or Landau and Lifshitz.

The energy-momentum tensor of a source-free electromagnetic field satisfies the continuity equation. For, by the field equations, the equation

$$\partial_\nu T^{\mu\nu} = (\partial_\nu f^{\mu\lambda}) h_\lambda^{\ \nu} + f^{\mu\lambda}\partial_\nu h_\lambda^{\ \nu} - \frac{1}{2}\frac{\partial \mathscr{L}}{\partial f_{\lambda\nu}}\partial^\mu f_{\lambda\nu}$$

$$= \frac{1}{2}(\partial^\nu f^{\mu\lambda} + \partial^\lambda f^{\nu\mu} + \partial^\mu f^{\lambda\nu}) h_{\lambda\nu} + f^{\mu\lambda}\partial_\nu h_\lambda^{\ \nu} = -f^{\mu\lambda} j_\lambda$$

is satisfied. When $j^\lambda = 0$, this yields the continuity equation:

$$\partial_\nu T^{\mu\nu} = 0. \tag{17}$$

Out of the energy–momentum tensor we build the four-dimensional angular momentum density tensor $M^{\mu\nu\lambda}$:

$$M^{\mu\nu\lambda}(x) \equiv x^\mu T^{\nu\lambda}(x) - x^\nu T^{\mu\lambda}(x). \tag{18}$$

The properties of the tensor $T^{\mu\nu}$ imply the following equation for the tensor $M^{\mu\nu\lambda}$:

$$\partial_\lambda M^{\mu\nu\lambda}(x) = -x^{[\mu} f^{\nu]\lambda}(x) j_\lambda(x).$$

In the case of a source-free field this leads to the continuity equation for the four-dimensional angular momentum density tensor

$$\partial_\lambda M^{\mu\nu\lambda}(x) = 0. \tag{19}$$

The Conservation Laws

The tensors $T^{\mu\nu}$ and $M^{\mu\nu\lambda}$ will serve to construct the energy-momentum four-vector of the field and the four-dimensional angular momentum tensor which, as we shall demonstrate in Section 7, constitute ten generators of Poincaré transformations. They are obtained by integrating the tensors $T^{\mu\nu}(x)$ and $M^{\mu\nu\lambda}(x)$ over the three-dimensional spacelike†

† The hypersurface σ is defined as a set of points which satisfy the equation $\sigma(x) = 0$ in a fixed coordinate system. The field of non-normalized vectors normal to the surface, $n_\mu(x)$, is defined by the formula $n_\mu(x) = \partial_\mu \sigma(x)$, where $x \in \sigma$. A hypersurface is said to be spacelike if the condition $n_\mu(x) n^\mu(x) > 0$ is satisfied for all its points. The surface element is a vector with the same direction as the normal vector. In the special case, when the normal vector has the direction of time, $n_\mu = (n, 0, 0, 0)$, the element $d\sigma_\nu$ is of the form $d\sigma_\nu = (dx dy dz, 0, 0, 0)$. A hypersurface for which the condition $n_\mu(x) n^\mu(x) < 0$ is satisfied is called a timelike hypersurface.

hypersurface σ:

$$P^{\mu}[\sigma] \equiv \int_{\sigma} \mathrm{d}\sigma_{\nu} T^{\mu\nu}, \tag{20a}$$

$$M^{\mu\nu}[\sigma] \equiv \int_{\sigma} \mathrm{d}\sigma_{\lambda} M^{\mu\nu\lambda}. \tag{20b}$$

The energy-momentum four-vector P^{μ} and the four-dimensional angular momentum tensor $M^{\mu\nu}$ thus are, in general, functionals of the hypersurface σ. In the particular case, if we restrict our attention to the family of hyperplanes given by the equation t = const, dependence on (independence of) the hyperplanes denotes a dependence on (independence of) time. We shall show that the energy, momentum, and angular momentum of an isolated electromagnetic field satisfy the conservation laws, i.e. do not depend on σ. To this end we shall apply Gauss' theorem to

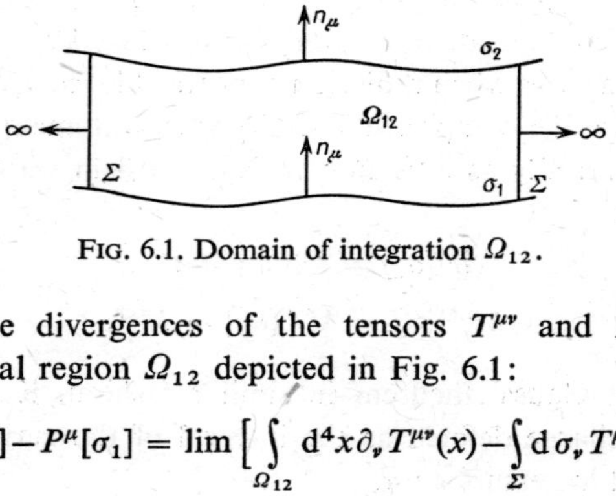

FIG. 6.1. Domain of integration Ω_{12}.

integrals of the divergences of the tensors $T^{\mu\nu}$ and $M^{\mu\nu\lambda}$ over the four-dimensional region Ω_{12} depicted in Fig. 6.1:

$$P^{\mu}[\sigma_2]-P^{\mu}[\sigma_1] = \lim \Big[\int_{\Omega_{12}} \mathrm{d}^4x \partial_{\nu} T^{\mu\nu}(x) - \int_{\Sigma} \mathrm{d}\sigma_{\nu} T^{\mu\nu}(x) \Big]$$

$$= \lim \Big[-\int_{\Omega_{12}} \mathrm{d}^4x f^{\mu}{}_{\lambda}(x) j^{\lambda}(x) - \int_{\Sigma} \mathrm{d}\sigma_{\nu} T^{\mu\nu}(x) \Big],$$

$$M^{\mu\nu}[\sigma_2]-M^{\mu\nu}[\sigma_1] = \lim \Big[\int_{\Omega_{12}} \mathrm{d}^4x \partial_{\lambda} M^{\mu\nu\lambda}(x) - \int_{\Sigma} \mathrm{d}\sigma_{\lambda} M^{\mu\nu\lambda}(x) \Big]$$

$$= \lim \Big[-\int_{\Omega_{12}} \mathrm{d}^4x x^{[\mu} f^{\nu]\lambda}(x) j_{\lambda}(x) - \int_{\Sigma} \mathrm{d}\sigma_{\lambda} M^{\mu\nu\lambda}(x) \Big].$$

The limit symbol denotes the removal of the three-dimensional timelike hypersurface Σ to infinity. The formulae above indicate that there are two reasons why the energy, momentum, and four-dimensional angular momentum of the field may change under passage from the hypersurface σ_1 to the hypersurface σ_2. The first is the occurrence of external sources of field. Four-dimensional integrals describe the exchange of energy, momentum, and four-dimensional angular momentum between the field and the sources. That these integrals appear is due to the fact that the continuity equation is not satisfied for the energy-momentum tensor in the presence of sources. The second reason why there may be a change of energy, momentum, and four-dimensional angular momentum is the existence, at spatial infinity, of radiation which may be regarded as the interaction of the field with sources located at infinity. The integrals over three-dimensional hypersurfaces express the energy, momentum and four-dimensional angular momentum radiated to infinity or from infinity. Now suppose that the field is isolated, i.e. that it is a source-free field, for which the components of the tensor $T^{\mu\nu}$ tend to zero so quickly, as we move away to infinity in spatial directions, that the limits of the integrals over the timelike hypersurfaces $\sum$ are equal to zero. In cases when radiation appears at infinity, this assumption is not valid since the components of the energy-momentum tensor then behave asymptotically as r^{-2}. For an isolated field we obtain the conservation laws:

$$P^\mu[\sigma_2] = P^\mu[\sigma_1] = P^\mu,$$

$$M^{\mu\nu}[\sigma_2] = M^{\mu\nu}[\sigma_1] = M^{\mu\nu}.$$

By applying Gauss' theorem in similar fashion, it may be shown that the total charge defined as the integral of the current four-vector over the spatial hypersurface σ,

$$Q \equiv \int_\sigma d\sigma_\mu j^\mu(x), \tag{21}$$

does not depend on σ by virtue of the continuity equations for current.

It is frequently convenient to choose the hypersurface σ in the form of the hyperplane t = const. The representation of energy, momentum, angular momentum, and moment of energy is then obtained in the form of integrals over three-dimensional space:

$$P^0 = \int d^3x T^{00}(\mathbf{x}, t), \tag{22a}$$

$$P^k = \int d^3x T^{k0}(\mathbf{x}, t), \tag{22b}$$

$$M^{ij} = \int d^3x [x^i T^{j0}(\mathbf{x}, t) - x^j T^{i0}(\mathbf{x}, t)], \tag{22c}$$

$$M^{k0} = \int d^3x [x^k T^{00}(\mathbf{x}, t) - t T^{k0}(\mathbf{x}, t)]. \tag{22d}$$

These integrals are independent of time, by virtue of the conservation laws. If in particular we set $t = 0$, we obtain the following expression for the components M^{k0}:

$$M^{k0} = \int d^3x x^k T^{00}(\mathbf{x}, t). \tag{23}$$

This formula justifies the term *moment of energy* which we have used.

Now, let us express, in three-vector notation, the components of the tensor $T^{\mu\nu}$ and the quantities conserved by the vectors **E**, **B**, **D**, and **H**:

$$T^{00} = \mathbf{E} \cdot \mathbf{D} - \mathscr{L}, \tag{24a}$$

$$T^{k0} = (\mathbf{D} \times \mathbf{B})_k, \tag{24b}$$

$$T^{0k} = (\mathbf{E} \times \mathbf{H})_k, \tag{24c}$$

$$T^{kl} = -E_k D_l - H_k B_l + \delta_{kl}(\mathscr{L} + \mathbf{H} \cdot \mathbf{B}). \tag{24d}$$

The *energy density* T^{00} will frequently be denoted by the symbol $\mathscr{H}$. Taking the hyperplane $t = \text{const}$ instead of the hypersurface σ in eqs. (20), we obtain the *energy* E, *momentum* **P**, *angular momentum* **M**, and *moment of energy* **N** in the form:

$$E = P^0 = \int d^3x \mathscr{H}, \tag{25a}$$

$$\mathbf{P} = (P^1, P^2, P^3) = \int d^3x \mathbf{D} \times \mathbf{B} = \int d^3x \mathbf{E} \times \mathbf{H}, \tag{25b}$$

$$\mathbf{M} = (M^{23}, M^{31}, M^{12}) = \int d^3x \mathbf{x} \times (\mathbf{D} \times \mathbf{B}), \tag{25c}$$

$$\mathbf{N} = (M^{10}, M^{20}, M^{30}) = \int d^3x [-t(\mathbf{D} \times \mathbf{B}) + \mathbf{x}\mathscr{H}]. \tag{25d}$$

7. CANONICAL FORMALISM FOR THE ELECTROMAGNETIC FIELD

With future quantization in mind, we shall now give the canonical formulation of the classical electromagnetic field theory, that is, we shall introduce canonical variables, canonical equations, and the generalized

Poisson brackets.# All the equations in this section will be relations between the field vectors at a particular instant. For that reason we shall not indicate the time-dependence of these vectors. We shall confine ourselves to the case of the electromagnetic field without sources. Three-vector notation will be used. In this notation, the system of field equations introduced in Section 6 breaks down into three groups of equations. The first group contains the time derivatives of the field vectors:

$$\dot{\mathbf{D}}(\mathbf{x}) = \nabla\times\mathbf{H}(\mathbf{x}), \tag{1a}$$

$$\dot{\mathbf{B}}(\mathbf{x}) = -\nabla\times\mathbf{E}(\mathbf{x}). \tag{1b}$$

The second group of equations consists of relations between the vectors **D**, **H**, **E**, and **B** which follow from the definition (6.6) of the tensor $h^{\mu\nu}$:

$$\mathbf{D}(\mathbf{x}) = \frac{\partial\mathscr{L}(\mathbf{E}(\mathbf{x}), \mathbf{B}(\mathbf{x}))}{\partial\mathbf{E}(\mathbf{x})}, \tag{2a}$$

$$\mathbf{H}(\mathbf{x}) = -\frac{\partial\mathscr{L}(\mathbf{E}(\mathbf{x}), \mathbf{B}(\mathbf{x}))}{\partial\mathbf{B}(\mathbf{x})}. \tag{2b}$$

Equations (1) and (2) constitute a complete system of equations, that is to say, they enable the values of the field vectors in all space-time to be found, given the initial conditions for the fields **D** and **B**.

The third group of conditions comprises auxiliary conditions which must be obeyed by the vectors **D** and **B**:

$$\nabla\cdot\mathbf{D} = 0, \tag{3a}$$

$$\nabla\cdot\mathbf{B} = 0. \tag{3b}$$

If these conditions are satisfied at a particular instant, they are met at any instant by virtue of eqs. (1) and (2).

Equations (1) resemble Hamilton's canonical equations in classical mechanics. To make full use of this analogy when building the canonical formalism for the electromagnetic field, we shall treat the fields **D** and **B** as independent variables, being the equivalents of the canonical variables p and q. The field strengths **E** and **H** can be expressed with the

---

The canonical formulation of non-linear electromagnetic field theory was given by M. Born and L. Infeld (1934b, 1935).

aid of eqs. (2) as functions of the variables **D** and **B**. The transition from the variables **E** and **B**, which occur in the Lagrangian, to the canonical variables **D** and **B** corresponds to the Legendre transformation. Making use of the correspondence

$$p \to \mathbf{D}, \qquad q \to \mathbf{B}, \qquad \dot{q} \to \mathbf{E},$$

we introduce the analogue of Hamilton's function:

$$p\dot{q} - L \to \mathbf{D} \cdot \mathbf{E}(\mathbf{D}, \mathbf{B}) - \mathscr{L}\big(\mathbf{E}(\mathbf{D}, \mathbf{B}), \mathbf{B}\big).$$

The function so built is equal to the energy density of the electromagnetic field discussed in the preceding section:

$$\mathscr{H}(\mathbf{x}) = \mathbf{D}(\mathbf{x}) \cdot \mathbf{E}\big(\mathbf{D}(\mathbf{x}), \mathbf{B}(\mathbf{x})\big) - \mathscr{L}\Big(\mathbf{E}\big(\mathbf{D}(\mathbf{x}), \mathbf{B}(\mathbf{x})\big), \mathbf{B}(\mathbf{x})\Big). \tag{4}$$

The integral of the function $\mathscr{H}(\mathbf{x})$ over three-dimensional space will be called the *Hamiltonian H*:

$$H \equiv \int d^3x \mathscr{H}(\mathbf{x}). \tag{5}$$

Equations (2) can be used to express the field strengths **E** and **H** in terms of the derivatives of the function $\mathscr{H}(\mathbf{x})$ with respect to the canonical variables:

$$\mathbf{E}\big(\mathbf{D}(\mathbf{x}), \mathbf{B}(\mathbf{x})\big) = \frac{\partial \mathscr{H}(\mathbf{x})}{\partial \mathbf{D}(\mathbf{x})}, \tag{6a}$$

$$\mathbf{H}\big(\mathbf{D}(\mathbf{x}), \mathbf{B}(\mathbf{x})\big) = \frac{\partial \mathscr{H}(\mathbf{x})}{\partial \mathbf{B}(\mathbf{x})}. \tag{6b}$$

These relations enable the field equations (1) to be rewritten so as to contain the functional derivatives of the Hamiltonian:

$$\dot{\mathbf{D}}(\mathbf{x}) = \nabla \times \frac{\delta H}{\delta \mathbf{B}(\mathbf{x})}, \tag{7a}$$

$$\dot{\mathbf{B}}(\mathbf{x}) = -\nabla \times \frac{\delta H}{\delta \mathbf{D}(\mathbf{x})}. \tag{7b}$$

The Generalized Poisson Brackets

In the next step we define the generalized Poisson brackets. We shall introduce them so that symmetry transformations associated with the Poincaré group be also canonical transformations. In particular, so that the time derivative of any functional of the field inductions **D** and **B** be expressed by the Poisson bracket of this functional with the Hamiltonian:

$$\dot{\mathscr{F}}[\mathbf{D}, \mathbf{B}] = -\{H, \mathscr{F}\}, \tag{8}$$

when **D** and **B** satisfy the field equations (1) and (2). The *generalized Poisson bracket* $\{\mathscr{F}, \mathscr{G}\}$ of two analytic functionals $\mathscr{F}$ and $\mathscr{G}$ in the variables $\mathbf{D}(\mathbf{x})$ and $\mathbf{B}(\mathbf{x})$ is defined by the expression

$$\{\mathscr{F}, \mathscr{G}\} \equiv \int d^3x \left[\frac{\delta\mathscr{F}}{\delta\mathbf{D}(\mathbf{x})}\cdot\nabla\times\frac{\delta\mathscr{G}}{\delta\mathbf{B}(\mathbf{x})} - \frac{\delta\mathscr{G}}{\delta\mathbf{D}(\mathbf{x})}\cdot\nabla\times\frac{\delta\mathscr{F}}{\delta\mathbf{B}(\mathbf{x})}\right]. \tag{9a}$$

Integration by parts allows the Poisson bracket to be presented in a somewhat different, equivalent form:

$$\{\mathscr{F}, \mathscr{G}\} = \int d^3x \left[\frac{\delta\mathscr{G}}{\delta\mathbf{B}(\mathbf{x})}\cdot\nabla\times\frac{\delta\mathscr{F}}{\delta\mathbf{D}(\mathbf{x})} - \frac{\delta\mathscr{F}}{\delta\mathbf{B}(\mathbf{x})}\cdot\nabla\times\frac{\delta\mathscr{G}}{\delta\mathbf{D}(\mathbf{x})}\right]. \tag{9b}$$

The generalized Poisson brackets possess the following properties:†

$$\{\mathscr{F}, \mathscr{G}\} = -\{\mathscr{G}, \mathscr{F}\},$$

$$\{\lambda_1\mathscr{F}_1+\lambda_2\mathscr{F}_2, \mathscr{G}\} = \lambda_1\{\mathscr{F}_1, \mathscr{G}\}+\lambda_2\{\mathscr{F}_2, \mathscr{G}\},$$

$$\{\mathscr{F}_1, \{\mathscr{F}_2, \mathscr{F}_3\}\}+\{\mathscr{F}_3, \{\mathscr{F}_1, \mathscr{F}_2\}\}+\{\mathscr{F}_2, \{\mathscr{F}_3, \mathscr{F}_1\}\} = 0,$$

$$\{\mathscr{F}_1\mathscr{F}_2, \mathscr{G}\} = \mathscr{F}_1\{\mathscr{F}_2, \mathscr{G}\}+\{\mathscr{F}_1, \mathscr{G}\}\mathscr{F}_2.$$

With the aid of the generalized Poisson brackets we write the first group of field equations in a form that is fully analogous to the corresponding equations of classical mechanics:

$$\dot{\mathbf{D}}(\mathbf{x}) = -\{H, \mathbf{D}\}, \tag{10a}$$

$$\dot{\mathbf{B}}(\mathbf{x}) = -\{H, \mathbf{B}\}. \tag{10b}$$

In the special case when $\mathscr{F}$ and $\mathscr{G}$ depend on the fields **D** and **B** at one point, that is, are functions of the variables **D** and **B**, the Poisson bracket is of the form

$$\{F(\mathbf{D}(\mathbf{x}), \mathbf{B}(\mathbf{x})), G(\mathbf{D}(\mathbf{y}), \mathbf{B}(\mathbf{y}))\}$$

$$= \left[\frac{\partial F(\mathbf{x})}{\partial D_i(\mathbf{x})}\frac{\partial G(\mathbf{y})}{\partial B_j(\mathbf{y})} - \frac{\partial F(\mathbf{x})}{\partial B_i(\mathbf{x})}\frac{\partial G(\mathbf{y})}{\partial D_j(\mathbf{y})}\right]\varepsilon_{ikj}\partial_k\delta(\mathbf{x}-\mathbf{y}). \tag{11}$$

---

† A set of functionals in which addition and multiplication by a number are defined and the commutator of two functionals is defined as their Poisson bracket has the structure of a Lie algebra.

From this we obtain the Poisson brackets for the fields **D** and **B**:

$$\{B_i(\mathbf{x}), D_j(\mathbf{y})\} = \varepsilon_{ijk}\,\partial_k\,\delta(\mathbf{x}-\mathbf{y}) = -\{D_i(\mathbf{x}), B_j(\mathbf{y})\}, \tag{12a}$$

$$\{D_i(\mathbf{x}), D_j(\mathbf{y})\} = 0 = \{B_i(\mathbf{x}), B_j(\mathbf{y})\}. \tag{12b}$$

These equalities imply immediately that the Poisson brackets of $\nabla \cdot \mathbf{B}$ and any component of the field, **B** or **D**, are equal to zero. The same is true of $\nabla \cdot \mathbf{D}$:

$$\{\nabla \cdot \mathbf{D}, B_i\} = 0 = \{\nabla \cdot \mathbf{D}, D_i\},$$

$$\{\nabla \cdot \mathbf{B}, B_i\} = 0 = \{\nabla \cdot \mathbf{B}, D_i\}.$$

By virtue of these equalities the Poisson brackets of $\nabla \cdot \mathbf{D}$ (or $\nabla \cdot \mathbf{B}$) and any functional of the fields **D** or **B** are equal to zero. This fact plays an essential role in the quantization of the electromagnetic field.

The auxiliary conditions (3), constituting part of the field equations, will also be expressed with the aid of Poisson brackets just as eqs. (1) were. For this purpose let us calculate the Poisson brackets of the components of the momentum vector **P** and the components of the vectors **D** and **B**:

$$\{P^i, D_k(\mathbf{x})\} = \partial_i D_k(\mathbf{x}) + \delta_{ik}\,\partial_j D_j(\mathbf{x}),$$

$$\{P^i, B_k(\mathbf{x})\} = \partial_i B_k(\mathbf{x}) + \delta_{ik}\,\partial_j B_j(\mathbf{x}).$$

The field equations (3) thus are equivalent to the equations

$$\{P^i, D_k(\mathbf{x})\} = \partial_i D_k(\mathbf{x}), \tag{13a}$$

$$\{P^i, B_k(\mathbf{x})\} = \partial_i B_k(\mathbf{x}). \tag{13b}$$

Similar relations can also be obtained for **E** and **H** if use is made of the fact that they are functions of the variables **D** and **B**. To summarize, we find that the system of field equations (1) and (3) is equivalent to the equations

$$\{P^\mu, f_{\lambda\varrho}(\mathbf{x}, t)\} = -\partial^\mu f_{\lambda\varrho}(\mathbf{x}, t), \tag{14a}$$

$$\{P^\mu, h^{\lambda\varrho}(\mathbf{x}, t)\} = -\partial^\mu h^{\lambda\varrho}(\mathbf{x}, t). \tag{14b}$$

We can also calculate the Poisson brackets of the components of the tensor $M^{\mu\nu}$ and the components of the field tensors $f_{\lambda\varrho}$ and $h^{\lambda\varrho}$ satisfying the field equations. In consequence, we arrive at

$$\{M^{\mu\nu}, f_{\lambda\varrho}(\mathbf{x}, t)\} = -x^{[\mu}\partial^{\nu]} f_{\lambda\varrho}(\mathbf{x}, t) - \delta^{\sigma}_{[\lambda}\delta_{\varrho]}{}^{[\mu} g^{\nu]\tau} f_{\sigma\tau}(\mathbf{x}, t), \tag{15a}$$

$$\{M^{\mu\nu}, h^{\lambda\varrho}(\mathbf{x}, t)\} = -x^{[\mu}\partial^{\nu]} h^{\lambda\varrho}(\mathbf{x}, t) - \delta^{[\lambda}_{\sigma} g^{\varrho][\mu} \delta^{\nu]}_{\tau} h^{\sigma\tau}(\mathbf{x}, t). \tag{15b}$$

Equations (15), just as eqs. (14), are equivalent to the field equations (1) and (3).

To conclude this section, let us calculate the Poisson brackets of the components of the four-vector P^{μ} and the tensor $M^{\mu\nu}$. A convenient point of departure for determining these brackets is given by the following relations which are satisfied by the Poisson brackets of the components of the energy-momentum tensor:#

$$\{T^{00}(\mathbf{x}, t), T^{00}(\mathbf{y}, t)\} = -[T^{0k}(\mathbf{x}, t) + T^{0k}(\mathbf{y}, t)]\partial_k \delta(\mathbf{x}-\mathbf{y}), \tag{16a}$$

$$\{T^{00}(\mathbf{x}, t), T^{0k}(\mathbf{y}, t)\} = -[T^{ki}(\mathbf{x}, t) + T^{00}(\mathbf{y}, t)\,\delta_{ki}]\partial_i \delta(\mathbf{x}-\mathbf{y}), \tag{16b}$$

$$\{T^{0k}(\mathbf{x}, t), T^{0l}(\mathbf{y}, t)\} = -[T^{0l}(\mathbf{x}, t)\partial_k + T^{0k}(\mathbf{y}, t)\,\partial_l]\delta(\mathbf{x}-\mathbf{y}). \tag{16c}$$

The proof of these relations involves quite tedious calculations and is given in Appendix F. To calculate the Poisson brackets of the components of the momentum four-vector and the four-dimensional angular momentum tensor, we express P^{μ} and $M^{\mu\nu}$ in terms of the components of the tensor $T^{\mu\nu}$ and then apply formula (16). The result is written in the form of three equations

$$\{P^{\mu}, P^{\nu}\} = 0, \tag{17a}$$

$$\{M^{\mu\nu}, P^{\lambda}\} = -g^{\mu\lambda} P^{\nu} + g^{\nu\lambda} P^{\mu}, \tag{17b}$$

$$\{M^{\mu\nu}, M^{\lambda\varrho}\} = -g^{\mu\lambda} M^{\nu\varrho} + g^{\mu\varrho} M^{\nu\lambda} - g^{\nu\varrho} M^{\mu\lambda} + g^{\nu\lambda} M^{\mu\varrho}. \tag{17c}$$

Canonical Transformations and Their Generators

Canonical formalism was introduced in a manner dependent on the reference frame. It was done by separating the field tensors $f_{\mu\nu}$ and $h^{\mu\nu}$ into pairs of three-vectors, and in this way we distinguished in space-

---

The commutation relations for the components of the energy-momentum tensor and their connection with the relativistic invariance of the theory was first investigated by J. Schwinger (1962, 1963) and P. A. M. Dirac (1962). The complete set of commutation relations discussed here has been given by one of the present authors (I. Białynicki-Birula (1965a)).

time a three-dimensional hyperplane σ described by the equation $t = \text{const}$. The same hyperplane was also used to define the generalized Poisson brackets. It will now be shown that the Poisson bracket does not depend on that hyperplane, i.e. that the Poisson bracket of two geometric objects—for instance, fields $f_{\mu\nu}(x)$ and $h^{\mu\nu}(x)$—depends only on those objects. To prove that the Poisson brackets are independent of the choice of hyperplane we shall employ the concept of canonical transformation, defined as in classical mechanics.

Canonical transformation of the variables **D** and **B** is such a functional mapping,

$$\mathbf{D}(\mathbf{x}) \to \mathbf{D}^*(\mathbf{x}) = \mathscr{D}^*[\mathbf{x}|\mathbf{D}, \mathbf{B}],$$

$$\mathbf{B}(\mathbf{x}) \to \mathbf{B}^*(\mathbf{x}) = \mathscr{B}^*[\mathbf{x}|\mathbf{D}, \mathbf{B}],$$

which does not change the Poisson brackets:

$$\{D_i^*, B_k^*\} = \{D_i, B_k\},$$

$$\{D_i^*, D_k^*\} = 0 = \{B_i^*, B_k^*\}.$$

As in classical mechanics, it can be shown that if $\mathbf{D}^*$, $\mathbf{B}^*$ are connected with **D**, **B** by a canonical transformation, then

$$\{\mathscr{F}[\mathscr{D}[\mathbf{D}^*, \mathbf{B}^*], \mathscr{B}[\mathbf{D}^*, \mathbf{B}^*]], \mathscr{G}[\mathscr{D}[\mathbf{D}^*, \mathbf{B}^*], \mathscr{B}[\mathbf{D}^*, \mathbf{B}^*]]\}_{D^*, B^*}$$
$$= \{\mathscr{F}[\mathbf{D}, \mathbf{B}], \mathscr{G}[\mathbf{D}, \mathbf{B}]\}_{D, B}, \tag{18}$$

where $\{,\}_{D^*, B^*}$ denotes the Poisson bracket calculated with respect to the variables $\mathbf{D}^*$ and $\mathbf{B}^*$.

The canonical transformations constitute a group of transformations. Further on we shall be interested in canonical transformations determined by the *generating functionals* $\mathscr{K}[\mathbf{D}, \mathbf{B}]$, called the *generators* of these transformations. Namely, we shall consider the transformation of any functional $\mathscr{F}[\mathbf{D}, \mathbf{B}]$ given by the formula

$$\mathscr{F}[\mathbf{D}, \mathbf{B}] \to \mathscr{F}^*[\mathbf{D}, \mathbf{B}]$$

$$= \mathscr{F}[\mathbf{D}, \mathbf{B}] + \frac{1}{1!}\{\mathscr{K}, \mathscr{F}\} + \frac{1}{2!}\{\mathscr{K}, \{\mathscr{K}, \mathscr{F}\}\} + \dots \tag{19}$$

It will be shown that, when applied to the fields **D** and **B**, this is a canonical transformation. For this purpose let us consider a one-parameter group of transformations that is generated by the functional $\mathscr{K}$:

$$\mathscr{F}[\mathbf{D}, \mathbf{B}] \to \mathscr{F}^{(\lambda)}[\mathbf{D}, \mathbf{B}]$$

$$= \mathscr{F}[\mathbf{D}, \mathbf{B}] + \frac{\lambda}{1!}\{\mathscr{K}, \mathscr{F}\} + \frac{\lambda^2}{2!}\big\{\mathscr{K}, \{\mathscr{K}, \mathscr{F}\}\big\} + \ldots \tag{20}$$

Differentiation of both sides with respect to the parameter λ yields

$$\frac{\mathrm{d}\mathscr{F}^{(\lambda)}}{\mathrm{d}\lambda} = \{\mathscr{K}, \mathscr{F}^{(\lambda)}\}. \tag{21}$$

On using the Jacobi identity, we get the following relation for any two functionals $\mathscr{F}$ and $\mathscr{G}$:

$$\frac{\mathrm{d}}{\mathrm{d}\lambda}\{\mathscr{F}^{(\lambda)}, \mathscr{G}^{(\lambda)}\} = \big\{\mathscr{K}, \{\mathscr{F}^{(\lambda)}, \mathscr{G}^{(\lambda)}\}\big\}.$$

A power-series solution of this differential equation is given by

$$\{\mathscr{F}^{(\lambda)}, \mathscr{G}^{(\lambda)}\} = \{\mathscr{F}, \mathscr{G}\} + \frac{\lambda}{1!}\{\mathscr{K}, \{\mathscr{F}, \mathscr{G}\}\}$$

$$+ \frac{\lambda^2}{2!}\big\{\mathscr{K}, \{\mathscr{K}, \{\mathscr{F}, \mathscr{G}\}\}\big\} + \ldots = \{\mathscr{F}, \mathscr{G}\}^{(\lambda)}.$$

It thus follows that when applied to the fields **D** and **B**, this transformation is canonical, for, if D_i and B_k are substituted for $\mathscr{F}$ and $\mathscr{G}$ in the formula above, then all the Poisson brackets, except the first, on the right-hand side are zero since the Poisson brackets of D_i and B_k are functions independent of the fields **D** and **B**. For each value of λ we thus have:

$$\big\{D_i^{(\lambda)}, B_k^{(\lambda)}\big\} = \{D_i, B_k\},$$

$$\big\{D_i^{(\lambda)}, D_k^{(\lambda)}\big\} = 0 = \big\{B_i^{(\lambda)}, B_k^{(\lambda)}\big\}.$$

The transformations generated by $\lambda\mathscr{K}$ convert any functional $\mathscr{F}[\mathbf{D}, \mathbf{B}]$ into a functional $\mathscr{F}^{(\lambda)}[\mathbf{D}, \mathbf{B}]$ which is equal to the functional obtained when the variables transformed by means of the same generator $\lambda\mathscr{K}$ are substituted in $\mathscr{F}[\mathbf{D}, \mathbf{B}]$:

$$\mathscr{F}^{(\lambda)}[\mathbf{D}, \mathbf{B}] = \mathscr{F}[\mathbf{D}^{(\lambda)}, \mathbf{B}^{(\lambda)}]. \tag{22}$$

This theorem is proved by demonstrating that the two sides of eq. (22) satisfy the same differential equation in the variable λ with the same initial condition. For we have

$$\frac{d}{d\lambda}\mathscr{F}[\mathbf{D}^{(\lambda)},\mathbf{B}^{(\lambda)}] = \int d^3x \left[\frac{\delta\mathscr{F}}{\delta D_i^{(\lambda)}(\mathbf{x})}\{\mathscr{K}, D_i^{(\lambda)}(\mathbf{x})\} + \frac{\delta\mathscr{F}}{\delta B_i^{(\lambda)}(\mathbf{x})}\{\mathscr{K}, B_i^{(\lambda)}(\mathbf{x})\}\right]$$

$$= \int d^3x d^3y \left[\frac{\delta\mathscr{F}}{\delta D_i^{(\lambda)}(\mathbf{x})}\frac{\delta\mathscr{K}}{\delta D_k(\mathbf{y})}\varepsilon_{kln}\frac{\partial}{\partial y_l}\frac{\delta D_i^{(\lambda)}(\mathbf{x})}{\delta B_n(\mathbf{y})}\right.$$

$$-\frac{\delta\mathscr{F}}{\delta D_i^{(\lambda)}(\mathbf{x})}\frac{\delta\mathscr{K}}{\delta B_k(\mathbf{y})}\varepsilon_{kln}\frac{\partial}{\partial y_l}\frac{\delta D_i^{(\lambda)}(\mathbf{x})}{\delta D_n(\mathbf{y})}$$

$$+\frac{\delta\mathscr{F}}{\delta B_i^{(\lambda)}(\mathbf{x})}\frac{\delta\mathscr{K}}{\delta D_k(\mathbf{y})}\varepsilon_{kln}\frac{\partial}{\partial y_l}\frac{\delta B_i^{(\lambda)}(\mathbf{x})}{\delta B_n(\mathbf{y})}$$

$$\left.-\frac{\delta\mathscr{F}}{\delta B_i^{(\lambda)}(\mathbf{x})}\frac{\delta\mathscr{K}}{\delta B_k(\mathbf{y})}\varepsilon_{kln}\frac{\partial}{\partial y_l}\frac{\delta B_i^{(\lambda)}(\mathbf{x})}{\delta D_n(\mathbf{y})}\right] = \{\mathscr{K}, \mathscr{F}[\mathbf{D}^{(\lambda)},\mathbf{B}^{(\lambda)}]\}.$$

The functional $\mathscr{F}[\mathbf{D}^{(\lambda)},\mathbf{B}^{(\lambda)}]$ thus satisfies the same equation as does $\mathscr{F}^{(\lambda)}[\mathbf{D},\mathbf{B}]$ (cf. eqs. (2)). These functionals are equal to each other for $\lambda = 0$. This implies their equality for any value of λ.

Poincaré Transformations as Canonical Transformations

The foregoing considerations will now be applied to functionals of a special type, viz. to the field tensors $f_{\mu\nu}$ and $h^{\mu\nu}$ which satisfy the field equations. These tensors are functionals of the initial conditions, i.e. of the vectors **D** and **B** given on the hyperplane t = const:

$$f_{\mu\nu}(x) = \mathscr{F}_{\mu\nu}[x|\mathbf{D},\mathbf{B}], \tag{23a}$$

$$h^{\mu\nu}(x) = \mathscr{H}^{\mu\nu}[x|\mathbf{D},\mathbf{B}]. \tag{23b}$$

Among all the canonical transformations of the initial conditions, a distinguished role is played by transformations:

$$\mathbf{D} \to {}'\mathbf{D} = {}'\mathscr{D}[\mathbf{D},\mathbf{B}],$$

$$\mathbf{B} \to {}'\mathbf{B} = {}'\mathscr{B}[\mathbf{D},\mathbf{B}],$$

which lead to Poincaré transformations of the tensors $f_{\mu\nu}$ and $h^{\mu\nu}$:

$$\mathscr{F}_{\mu\nu}[x|{}'\mathbf{D},{}'\mathbf{B}] = {}^{\mathrm{IL}}f_{\mu\nu}(x), \tag{24a}$$

$$\mathscr{H}^{\mu\nu}[x|{}'\mathbf{D},{}'\mathbf{B}] = {}^{\mathrm{IL}}h^{\mu\nu}(x). \tag{24b}$$

The Poincaré transformation (a, Λ) of the tensors $f_{\mu\nu}$ and $h^{\mu\nu}$ is generated by the functional $\mathscr{K}^{\text{IL}}$ which is a linear combination of the vector P^μ and the tensor $M^{\mu\nu}$:

$$\mathscr{K}^{\text{IL}} = \alpha^\mu P_\mu + \tfrac{1}{2}\alpha^{\mu\nu} M_{\mu\nu}. \tag{25}$$

The parameters $\alpha^{\mu\nu}$ are related to the matrix elements of the Lorentz transformation Λ in the following manner:

$$\Lambda^\mu{}_\nu = (\exp(-\hat{\alpha}))^\mu{}_\nu, \tag{26a}$$

$$(\hat{\alpha})^\mu{}_\nu = \alpha^{\mu\lambda} g_{\lambda\nu}. \tag{26b}$$

Here, the parameters α^μ are related to the components of the translation vector a by the formula

$$a^\mu = \int_0^1 \mathrm{d}\tau\, \Lambda^\mu{}_\nu(\tau)\, \alpha^\nu, \tag{27a}$$

where

$$\Lambda^\mu{}_\nu(\tau) = (\exp(-\tau\hat{\alpha}))^\mu{}_\nu. \tag{27b}$$

To obtain the proof, let us examine the one-parameter family of transformations generated by $\mathscr{K}^{\text{IL}}$. They take the tensor $f_{\mu\nu}$ into $f^{(\lambda)}_{\mu\nu}$. When we use eq. (21) and the relations (14a) and (15a), we arrive at the equation

$$\frac{\mathrm{d}f^{(\lambda)}_{\lambda\varrho}(x)}{\mathrm{d}\lambda} = -\left[\alpha^\mu \partial_\mu \frac{1}{2}\delta^\sigma_{[\lambda}\delta^\tau_{\varrho]} + \frac{1}{2}\alpha^{\mu\nu}\left((x_\mu\partial_\nu - x_\nu\partial_\mu)\frac{1}{2}\delta^\sigma_{[\lambda}\delta^\tau_{\varrho]} + \delta^\sigma_{[\lambda} g_{\varrho][\mu}\delta^\tau_{\nu]}\right)\right] f^{(\lambda)}_{\sigma\tau}(x). \tag{28}$$

For simplicity of notation, the differential operator figuring on the right-hand side of this equation will be written as

$$\alpha^\mu \mathscr{T}_\mu + \frac{1}{2}\alpha^{\mu\nu}\mathscr{T}_{\mu\nu},$$

where $\mathscr{T}_\mu$ and $\mathscr{T}_{\mu\nu}$ are matrices with indices $\lambda\varrho$ and $\sigma\tau$ as defined by the formulae

$$(\mathscr{T}_\mu)_{\lambda\varrho}{}^{\sigma\tau} \equiv \frac{1}{2}\delta^\sigma_{[\lambda}\delta^\tau_{\varrho]}\partial_\mu, \tag{29a}$$

$$(\mathscr{T}_{\mu\nu})_{\lambda\varrho}{}^{\sigma\tau} \equiv \frac{1}{2}\delta^\sigma_{[\lambda}\delta^\tau_{\varrho]}(x_\mu\partial_\nu - x_\nu\partial_\mu) + \delta^\sigma_{[\lambda} g_{\varrho][\mu}\delta^\tau_{\nu]}. \tag{29b}$$

The matrices $\mathscr{T}_\mu$ and $\mathscr{T}_{\mu\nu}$ satisfy the commutation relations specific to the generators of the Poincaré transformations (cf. Appendix D):

$$[\mathscr{T}_\mu, \mathscr{T}_\nu] = 0, \tag{30a}$$

$$[\mathscr{T}_{\mu\nu}, \mathscr{T}_\lambda] = -g_{\mu\lambda}\mathscr{T}_\nu + g_{\nu\lambda}\mathscr{T}_\mu, \tag{30b}$$

$$[\mathscr{T}_{\mu\nu}, \mathscr{T}_{\lambda\varrho}] = -g_{\mu\lambda}\mathscr{T}_{\nu\varrho} + g_{\mu\varrho}\mathscr{T}_{\nu\lambda} - g_{\nu\varrho}\mathscr{T}_{\mu\lambda} + g_{\nu\lambda}\mathscr{T}_{\mu\varrho}. \tag{30c}$$

In matrix notation, eq. (28) becomes

$$\frac{\mathrm{d}f^{(\lambda)}}{\mathrm{d}\lambda} = -\left(\alpha^{\mu}\mathcal{T}_{\mu}+\frac{1}{2}\,\alpha^{\mu\nu}\mathcal{T}_{\mu\nu}\right)f^{(\lambda)}.$$

The formal solution of this equation can be written as an exponential operator acting on f:

$$f^{(\lambda)} = \exp\left[-\lambda\left(\alpha^{\mu}\mathcal{T}_{\mu}+\frac{1}{2}\,\alpha^{\mu\nu}\mathcal{T}_{\mu\nu}\right)\right]f.$$

In Appendix D we have shown that this operator can be represented as the product of two exponential operators:

$$\exp\left[-\lambda\left(\alpha^{\mu}\mathcal{T}_{\mu}+\frac{1}{2}\,\alpha^{\mu\nu}\mathcal{T}_{\mu\nu}\right)\right] = \exp\left[-a^{\mu}(\lambda)\mathcal{T}_{\mu}\right]\exp\left[-\frac{1}{2}\lambda\alpha^{\mu\nu}\mathcal{T}_{\mu\nu}\right] \quad (31)$$

where

$$a^{\mu}(\lambda) = \int_{0}^{\lambda}\mathrm{d}\tau\,\Lambda^{\mu}{}_{\nu}(\tau)\alpha^{\nu}.$$

The operator $\exp[-a^{\mu}\mathcal{T}_{\mu}]$ is an operator of translation by vector a, that is

$$\exp\,[-a^{\mu}\mathcal{T}_{\mu}]f = f(x-a). \quad (32)$$

On expanding the exponential function in a series, we obtain the Taylor-series expansion of the right-hand side of the equation. The second exponential operator on the right-hand side of identity (31) acts on the field $f_{\mu\nu}(x)$ in the following manner:

$$\left(\exp\left(-\frac{1}{2}\lambda\alpha^{\mu\nu}\mathcal{T}_{\mu\nu}\right)\right)_{\lambda\varrho}^{\sigma\tau} f_{\sigma\tau}(x) = \Lambda_{\lambda}{}^{\sigma}(\lambda)\Lambda_{\varrho}{}^{\tau}(\lambda)f_{\sigma\tau}(\Lambda^{-1}(\lambda)x). \quad (33)$$

In order to prove this equality, it is necessary to verify that, by virtue of eqs. (27b) and (30c), both sides satisfy the same differential equation and the same initial conditions for $\lambda = 0$. Combining the results obtained in eqs. (32) and (33), we finally arrive at

$$f_{\mu\nu}^{(\lambda)}(x) = \Lambda_{\mu}{}^{\sigma}(\lambda)\Lambda_{\nu}{}^{\tau}(\lambda)f_{\sigma\tau}\big(\Lambda^{-1}(x-a(\lambda))\big). \quad (34)$$

If we set $\lambda = 1$, we get the transformation law for the tensor $f_{\mu\nu}(x)$ under the transformation (a, Λ).

Thus, we have shown that the functional $\mathcal{K}^{\mathrm{IL}}$ generates the Poincaré transformation of the field $f_{\mu\nu}(x)$. In similar fashion $\mathcal{K}^{\mathrm{IL}}$ can be shown also to generate the Poincaré transformation of the tensor $h^{\lambda\varrho}(x)$. Therefore, we have shown that there exist transformations of the initial conditions **D** and **B** such that they lead to Poincaré transformations of the fields $f_{\mu\nu}$ and $h^{\mu\nu}$ which satisfy the field equations.

It will now be shown that the Poisson brackets of the fields $f_{\mu\nu}$ and $h^{\lambda\varrho}$ do not depend on the choice of the hyperplane σ used in the definition (9) of those brackets. Let us consider two spacelike hyperplanes σ and σ', related by the Poincaré transformation. The fields $f_{\mu\nu}(x)$ and $h^{\lambda\varrho}(x)$ can be defined by giving the initial conditions **D** and **B** in the hyperplane σ or the equivalent initial conditions $'\mathbf{D}$ and $'\mathbf{B}$ in the hyperplane $'\sigma$. (In both cases we choose the coordinate system so that the hyperplane is described by the equation t = const). The fields **D** and **B** in the hyperplane σ are associated with $'\mathbf{D}$ and $'\mathbf{B}$ in $'\sigma$ by means of the functional relationships

$$'\mathbf{D}(\mathbf{x}) = '\mathscr{D}[\mathbf{x}|\mathbf{D}, B],$$
$$'\mathbf{B}(\mathbf{x}) = '\mathscr{B}[\mathbf{x}|\mathbf{D}, \mathbf{B}].$$

It follows from geometric considerations that the transformation of the inductions

$$\mathbf{D}(\mathbf{x}) \to '\mathscr{D}[\mathbf{x}|\mathbf{D}, \mathbf{B}],$$
$$\mathbf{B}(\mathbf{x}) \to '\mathscr{B}[\mathbf{x}|\mathbf{D}, \mathbf{B}]$$

on a particular hyperplane σ leads to the Poincaré transformation of the fields $f_{\mu\nu}(x)$ and $h^{\lambda\varrho}(x)$, and hence is a canonical transformation. Thus, if we utilize the property (18) of canonical transformations, we obtain the following equality for the Poisson brackets of the fields $f_{\mu\nu}$ and $h^{\lambda\varrho}$ which satisfy the field equations:

$$\big\{\mathscr{F}_{\mu\nu}[x|\mathscr{D}['\mathbf{D}, '\mathbf{B}], \mathscr{B}['\mathbf{D}, '\mathbf{B}]], \mathscr{H}^{\lambda\varrho}[y|\mathscr{D}['\mathbf{D}, '\mathbf{B}], \mathscr{B}['\mathbf{D}, '\mathbf{B}]]\big\}_{'\mathbf{D}, '\mathbf{B}}$$
$$= \{\mathscr{F}_{\mu\nu}[x|\mathbf{D}, \mathbf{B}], \mathscr{H}^{\lambda\varrho}[y|\mathbf{D}, \mathbf{B}]\}_{\mathbf{D},\mathbf{B}}. \qquad (35)$$

This equality means that the Poisson bracket of two components of the fields $f_{\mu\nu}$ and $h^{\lambda\varrho}$ depends only on these fields.

8. THE ELECTROMAGNETIC FIELD WITH SOURCES

The distinction between the fields **D** and **H** and **E** and **B** has a profound physical sense. It is evident from the field equations (6.5) that the vectors **D** and **H** describe the original fields produced by the source. On the other hand, **E** and **B** are the effective fields, modified in comparison with **D** and **H** by interaction. This interaction may have its source in the non-linear structure of the field. In relativistic theory, such self-interaction of the field with itself is most easily specified by giving the Lagrangian density. The interaction may also be due to the influence of the material medium. In that event, it is described by giving the relevant constitutive relation due to the averaging of the results of microscopic theory which describes the interaction of the field with charged particles.

In this section we shall discuss simple examples of interaction between the electromagnetic field and its sources.

Charged Fluid

The simplest model of a physical system interacting with an electromagnetic field is that of a charged ideal fluid with a constant ratio e/m of charge density to mass density. Let us denote the field of the velocity four-vector of this fluid by $u^\mu(x)$, $\left(u_\mu(x)u^\mu(x) = 1\right)$, and the scalar density of the mass of the fluid by $m\varrho(x)$. The four-vector j^μ of the electric current density and the energy-momentum tensor $T^{\mu\nu}_{(m)}$ for the fluid are of the form

$$j^\mu = e\varrho u^\mu, \tag{1a}$$

$$T^{\mu\nu}_{(m)} = m\varrho u^\mu u^\nu. \tag{1b}$$

The electromagnetic field equations and the equations of motion for the fluid are postulated to be

$$\partial_\mu h^{\mu\nu} = e\varrho u^\nu, \tag{2a}$$

$$\partial_\mu \check{f}^{\mu\nu} = 0, \tag{2b}$$

$$u^\mu \partial_\mu u^\nu = \frac{e}{m} f^{\nu\lambda} u_\lambda. \tag{2c}$$

These equations imply the continuity equations for current and for the total energy-momentum tensor $T^{\mu\nu}$ of the system, which is the sum of the energy-momentum tensor of the fluid and the energy-momentum tensor of the field, defined earlier by formula (6.15):

$$T^{\mu\nu} = T^{\mu\nu}_{(f)} + T^{\mu\nu}_{(m)}.$$

The system field–fluid thus is an isolated system.† The laws of conservation of charge, energy, momentum, and four-dimensional angular momentum hold for this system. From the equations of motion (2c) for the fluid we obtain an interpretation of the vectors **E** and **B**. These are the effective fields acting on the fluid. The structure of the theory of a system composed of a fluid and a field and the role of the field equations and the equations of motion are illustrated by the diagram in Fig. 8.1.

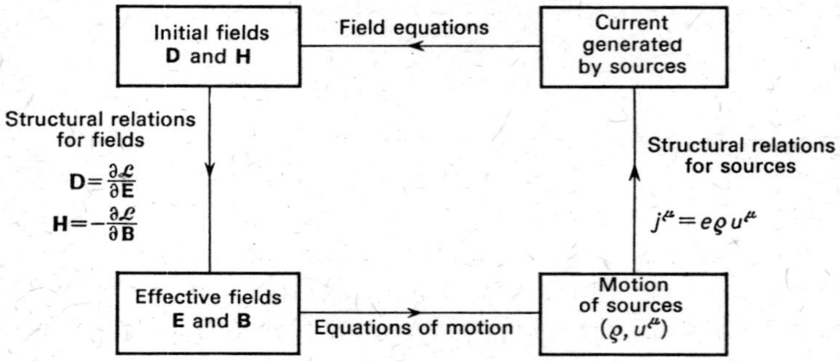

FIG. 8.1. The structure of the electrodynamics of a charged fluid.

Note that this diagram is closed only when we move in the directions indicated by arrows. In general it is not possible to determine the effective fields from the motion of the sources. Nor is it possible in the general case‡ to determine the motion of sources when only the current j^{μ} is known. The closure of the diagram in the direction indicated means that the theory is self-consistent and does not contain external elements.

The theory discussed here can also be described by using canonical formalism. As canonical variables we choose the quantities

$$\phi \equiv \varrho u^0,$$

$$\boldsymbol{\pi} \equiv m\varrho u^0 \mathbf{u},$$

† The field and the fluid are here assumed to be localized.

‡ For example, when the current is generated by the motion of two types of fluid with different charges.

for the fluid and **D** and **B** for the electromagnetic field. The field equations and the equations of motion for the fluid in three-dimensional notation are of the form

$$\frac{\partial}{\partial t}\mathbf{D} = \nabla\times\mathbf{H} - e\phi\mathbf{v},$$

$$\nabla\cdot\mathbf{D} = e\phi,$$

$$\frac{\partial}{\partial t}\mathbf{B} = -\nabla\times\mathbf{E},$$

$$\nabla\cdot\mathbf{B} = 0,$$

$$\frac{\partial}{\partial t}\boldsymbol{\pi} = -\partial_i(v_i\boldsymbol{\pi}) + e\phi\mathbf{E} + e\phi\mathbf{v}\times\mathbf{B},$$

$$\frac{\partial}{\partial t}\phi = -\nabla\cdot(\phi\mathbf{v}),$$

where

$$\mathbf{v} \equiv \mathbf{u}/u^0 = \boldsymbol{\pi}/\sqrt{m^2\phi^2+\boldsymbol{\pi}^2}\,.$$

These equations can be expressed in terms of the Poisson brackets of the canonical variables and the Hamiltonian. To this end, it is necessary to postulate the following values of the Poisson brackets for the canonical variables:

$$\{\phi(\mathbf{x}), \phi(\mathbf{y})\} = 0,$$

$$\{\phi(\mathbf{x}), \boldsymbol{\pi}(\mathbf{y})\} = -\phi(\mathbf{y})\nabla\delta(\mathbf{x}-\mathbf{y}),$$

$$\{\pi_i(\mathbf{x}), \pi_j(\mathbf{y})\} = -\big(\pi_j(\mathbf{x})\,\partial_i + \pi_i(\mathbf{y})\,\partial_j\big)\,\delta(\mathbf{x}-\mathbf{y})$$
$$+e\phi(\mathbf{x})\,\varepsilon_{ijk}B_k(\mathbf{x})\,\delta(\mathbf{x}-\mathbf{y}),$$

$$\{B_i(\mathbf{x}), \boldsymbol{\pi}(\mathbf{y})\} = 0 = \{B_i(\mathbf{x}), \phi(\mathbf{y})\},$$

$$\{D_i(\mathbf{x}), \pi_j(\mathbf{y})\} = -\delta_{ij}e\phi(\mathbf{x})\,\delta(\mathbf{x}-\mathbf{y}),$$

$$\{D_i(\mathbf{x}), \phi(\mathbf{y})\} = 0,$$

$$\{B_i(\mathbf{x}), B_j(\mathbf{y})\} = 0 = \{D_i(\mathbf{x}), D_j(\mathbf{y})\},$$

$$\{B_i(\mathbf{x}), D_j(\mathbf{y})\} = \varepsilon_{ijk}\partial_k\delta(\mathbf{x}-\mathbf{y}).$$

The Hamiltonian, defined as the integral, over three-dimensional space, of the component T^{00} of the energy-momentum density tensor of the system, is the sum of the Hamiltonian $H_{(m)}$ of the fluid,

$$H_{(m)} = \int d^3x \sqrt{m^2\phi(x)+\boldsymbol{\pi}^2(x)}\,,$$

and the Hamiltonian $H_{(f)}$ of the field, defined by the formulae (7.4) and (7.5):

$$H = H_{(m)}+H_{(f)}.$$

The Hamiltonian of the fluid can be constructed in a fashion similar to that used in the case of the electromagnetic field, taking the Lagrangian

$$L = -m\int d^3x\varrho(x) = -m\int d^3x\,\phi(x)\sqrt{1-\mathbf{v}^2}$$

as the starting point, and employing the analogy with classical mechanics. In this case we use the correspondence

$$p\to\boldsymbol{\pi},\quad q\to\varrho,\quad \dot{q}\to\mathbf{v},$$

$$p\dot{q}-L\to\int d^3x[\boldsymbol{\pi}\cdot\mathbf{v}(\phi,\boldsymbol{\pi})+\phi\sqrt{1-\mathbf{v}^2(\phi,\boldsymbol{\pi})}\,,$$

and the Legendre transformation

$$\boldsymbol{\pi} = \frac{\partial\mathscr{L}}{\partial\mathbf{v}}.$$

It may also be shown that in the theory of a charged fluid under discussion, the relations (7.16) are satisfied for the components of the energy-momentum tensor for the entire system. From this it follows that the Poincaré transformations are canonical transformations and are generated by the ten constants of motion built out of the energy-momentum tensor.

Charged Particles

It is much more difficult to provide a description of charged particles interacting with an electromagnetic field than it is for a charged fluid. Hitherto it has not been possible to describe extended particles in agreement with the special theory of relativity. Thus, it remains to consider point particles. Major difficulties are encountered in this case, however, owing to the singularities of the field produced by point particles. We shall discuss these difficulties briefly and sketch out methods of removing them within the framework of classical theory.

The equations of the field produced by the current which is generated by the motion of N point charges and the equations of motion of particles under the influence of the Lorentz force are of the form

$$\partial_\mu h^{\mu\nu}(x) = \sum_{A=1}^{N} e_A \int \mathrm{d}\xi_A^\nu \delta^{(4)}(x-\xi_A), \tag{3a}$$

$$\partial_\mu \check{f}^{\mu\nu}(x) = 0, \tag{3b}$$

$$\frac{\mathrm{d}^2\xi_A^\mu(\tau)}{\mathrm{d}\tau_A^2} = \frac{e_A}{m_A} f^{\mu\nu}(\xi_A) \frac{\mathrm{d}\xi_{A\nu}(\tau)}{\mathrm{d}\tau_A}, \tag{3c}$$

where the four-vector $\xi_A^\mu(\tau)$ describes the trajectory of the A-th particle, τ_A is the proper time of the particle, $\mathrm{d}\tau_A = \sqrt{1-v_A^2}\,\mathrm{d}t$, e_A is the charge, and m_A the mass. These equations are usually derived from the variational principle, with the action W having the form

$$W = \int \mathrm{d}^4x \mathscr{L}_f(x) - \sum_{A=1}^{N} m_A \int \mathrm{d}\tau_A - \sum_{A=1}^{N} e_A \int \mathrm{d}\xi_A^\mu A_\mu(\xi_A). \tag{4}$$

Close inspection of the system of eqs. (3) reveals serious shortcomings. In contradistinction to the eqs. (2) for a charged fluid, the equations for charged particles in the form written above cannot have solutions.# We shall elucidate this with the simplest example of a single particle.

First of all, let us find the electrostatic solutions in the case of a single particle. Setting $\mathbf{B} = 0 = \mathbf{H}$ and assuming that the particle is at rest, we obtain† the following simplified equations:

$$\nabla \cdot \mathbf{D}(\mathbf{x}) = e\delta(\mathbf{x}-\boldsymbol{\xi}), \tag{5a}$$

$$\nabla \times \mathbf{E}(\mathbf{x}) = 0. \tag{5b}$$

Incidentally, there exists a solution of a simplified non-physical model of the electrodynamics of particles in a two-dimensional Minkowski space (cf. I. Białynicki-Birula (1971)).

† The Lagrangian is assumed to be an even function of the invariant P (this means that the theory is invariant under space inversion). The condition $\mathbf{B} = 0$ then implies $\mathbf{H} = 0$ (cf. eqs. (6.11)).

The spherically symmetric solution of eq. (5a) is of the form

$$\mathbf{D} = \frac{e}{4\pi} \frac{(\mathbf{x}-\boldsymbol{\xi})}{|\mathbf{x}-\boldsymbol{\xi}|^3}. \tag{6a}$$

Since the fields **D** and **E** are parallel in the case in question,

$$\mathbf{E} = \frac{(\mathbf{x}-\boldsymbol{\xi})}{|\mathbf{x}-\boldsymbol{\xi}|} f(|\mathbf{x}-\boldsymbol{\xi}|), \tag{6b}$$

eq. (5b) is satisfied regardless of the form of the Lagrangian. In linear electrodynamics **D** and **E** are equal. Both thus increase without bound on approaching the point at which the charge is located. As a result of this growth of **E**, in linear theory the energy of the electrostatic field produced by a point particle, given by the formula

$$E_0 = \frac{e^2}{(4\pi)^2} \int d^3x \frac{1}{|\mathbf{x}-\boldsymbol{\xi}|^4},$$

is infinite. The divergence of this integral is the source of fundamental difficulties. With some additional assumptions it is possible[#] to derive from linear theory the equations of motion for a point charge that take into account the interaction of the charge with its own field. However, efforts to build up a complete theory of fields and particles have not been successful.

The construction of such a theory might be regarded as an academic exercise since quantum effects introduce essential modifications in any event. On the other hand, however, as we shall discuss at length in subsequent chapters, the quantum theory is also not without difficulties similar to those which occur in the classical theory. The elimination of all the difficulties in classical electrodynamics could, therefore, facilitate the construction of a complete quantum electrodynamics. Experiment gives no hint as to what lines the modification of the classical linear theory should take. All attempts in this field thus are of a speculative nature. In our opinion, the concept of a non-linear electrodynamics[†] is the most fortunate. It has been discussed in general terms in

This was proved by P. A. M. Dirac (1938).

† Attempts have also been made to introduce non-local interactions (discussed in the book by Rzewuski), theories with higher derivatives, and other hypotheses.

earlier sections. Now we shall go into a more detailed discussion of a concrete version of that theory—the *Born–Infeld theory*, constructed specifically in order to eliminate the infinite energy of the field of a point particle.

The Lagrangian of the electromagnetic field in the Born–Infeld theory is of the form

$$\mathscr{L}_{\text{BI}} = b^2 [1 - \sqrt{1 - 2b^{-2}S - b^{-4}P^2}], \tag{7}$$

where b is a constant with the same dimension as the field strength. For fields that are weak compared to the critical strength b, the Born–Infeld Lagrangian goes over into the Lagrangian of Maxwellian theory. Thus, linear electrodynamics is valid for weak fields. Departures from the linear theory occur in the vicinity of the field source, at distances comparable with the elementary length l defined by the formula

$$l = \sqrt{\frac{|e|}{4\pi b}}.$$

In particular, the strength of the electric field due to a point particle,

$$\mathbf{E} = \frac{\mathbf{D}}{\sqrt{1 + b^{-2}\mathbf{D}^2}} = \frac{e}{4\pi} \frac{(\mathbf{x} - \boldsymbol{\xi})}{|\mathbf{x} - \boldsymbol{\xi}|} (|\mathbf{x} - \boldsymbol{\xi}|^4 + l^4)^{-\frac{1}{2}}, \tag{8}$$

does not increase without bound as the point where the charge is located is approached, but instead tends to a finite limit

$$|\mathbf{E}|_{\max} = \frac{|e|}{4\pi l^2}.$$

In consequence, the energy E_0 of a field produced by a point particle is finite:

$$E_0 = \int d^3x b^2 (\sqrt{1 + b^{-2}\mathbf{D}^2} - 1) = \frac{e^2}{4\pi l} \int_0^\infty d\lambda (\sqrt{\lambda^4 + 1} - \lambda^2).$$

It is to be expected that for a charge in motion and for a system of point charges, the energy density near a charge also tends to infinity as r^{-2} and, consequently, the energy of the field created by any system of charges is finite.

The structure of the electrodynamics of point charges is illustrated by the diagram in Fig. 8.2. As in the case of a charged fluid, in order to close the diagram, i.e. to build a self-consistent theory, one must give the equations of motion for the charges. This is impossible in linear theory. The equations of motion for point particles are meaningless since the field intensity **E** is infinite at the point where a particle is located.

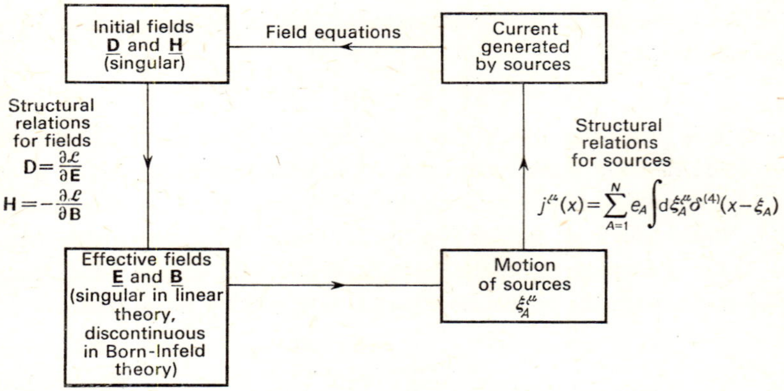

FIG. 8.2. The structure of the electrodynamics of charged point particles.

In the Born–Infeld theory, the intensity **E** is finite, it is true, but is also discontinuous. For there is no distinguished direction at the point where a particle is located. Accordingly, we cannot write the equations of particle motion in the form (3c) without giving an additional prescription for determining the direction of the field at the point ξ_A. This difficulty can, however, be avoided in a natural manner in the Born–Infeld theory by deriving the equations of motion for the particles from the continuity equation for the energy-momentum tensor.

In the electrodynamics of point particles we shall postulate the continuity equation for the energy-momentum tensor of the entire system since we regard it as one of the fundamental equations of the theory. In the charged fluid theory this equation emerged from the field equations and the equations of motion.

The energy-momentum tensor for point particles obtained from the variational principle (cf. Appendix F) is of the form

$$T_{(m)}^{\mu\nu}(x) = \sum_{A=1}^{N} m_A \int d\xi_A^{\mu} \frac{d\xi_A^{\nu}}{d\tau_A} \delta^{(4)}(x - \xi_A). \tag{9}$$

The components of this tensor in three-dimensional notation can be written as

$$T^{00}(\mathbf{x}, t) = \sum_{A=1}^{N} \sqrt{m_A^2 + \boldsymbol{\pi}_A^2(t)}\, \delta\big(\mathbf{x} - \boldsymbol{\xi}_A(t)\big),$$

$$T^{0k}(\mathbf{x}, t) = T^{k0}(\mathbf{x}, t) = \sum_{A=1}^{N} \pi_{Ak}(t)\, \delta\big(\mathbf{x} - \boldsymbol{\xi}_A(t)\big),$$

$$T^{kl}(\mathbf{x}, t) = T^{lk}(\mathbf{x}, t) = \sum_{A=1}^{N} \pi_{Ak}(t) v_{Al}(t)\, \delta\big(\mathbf{x} - \boldsymbol{\xi}_A(t)\big),$$

where

$$\mathbf{v}_A(t) = \frac{\mathrm{d}\boldsymbol{\xi}_A}{\mathrm{d}t}, \qquad \boldsymbol{\pi}_A(t) = \frac{m_A \mathbf{v}_A(t)}{\sqrt{1 - \mathbf{v}_A^2(t)}}.$$

At this point we resort to formal manipulations in which we use the field eqs. (3a,b) and the equations of motion for particles (3c) and for the moment we leave aside the aforementioned difficulties in determining the field. It is then possible to verify that the total energy-momentum tensor

$$T^{\mu\nu}(x) = T^{\mu\nu}_{(f)}(x) + T^{\mu\nu}_{(m)}(x)$$

satisfies the continuity equation.

Now let us take the continuity equation for $T^{\mu\nu}(x)$ as the starting point for deriving the equations of motion for the particles. For this purpose, let us rewrite this equation for the spatial components in three-vector notation (cf. formulae (6.24)):

$$\frac{\partial}{\partial t}\Big[\sum_{A=1}^{N} \pi_{Ak}\, \delta(\mathbf{x} - \boldsymbol{\xi}_A) + (\mathbf{D} \times \mathbf{B})_k\Big]$$

$$= \partial_l \Big[-\sum_{A=1}^{N} \pi_{Ak} v_{Al}\, \delta(\mathbf{x} - \boldsymbol{\xi}_A) + E_k D_l + H_k B_l - \delta_{kl}(\mathscr{L} + \mathbf{H} \cdot \mathbf{B})\Big].$$

Both sides of this equation are then integrated with respect to $\mathbf{x}$ over a small† three-dimensional sphere S_A of radius r and centre at the

† The radius of this sphere must be small enough for all other particles to lie outside the sphere.

point ξ_A. By Gauss' theorem, the integral of divergence on the right-hand side of the equation is changed into a surface integral over the surface of the sphere,

$$\frac{d}{dt}\pi_A+\frac{d}{dt}\int\limits_{S_A} d^3x\,(\mathbf{D}\times\mathbf{B})$$

$$= r^2\int d\Omega[(\hat{\mathbf{r}}\cdot\mathbf{D})\mathbf{E}+(\hat{\mathbf{r}}\cdot\mathbf{B})\mathbf{H}-\hat{\mathbf{r}}(\mathscr{L}+\mathbf{H}\cdot\mathbf{B})],$$

where $\int d\Omega$ denotes integration with respect to the angular variables, and $\hat{\mathbf{r}}$ is a unit vector normal to the surface of the sphere. In accordance with the results obtained for a single particle at rest, we assume that the original fields **D** and **H** produced by the charges tend to infinity as r^{-2} as the point at which the charged particle is located is approached, whereas the fields **E** and **B** remain bounded. The fields on the surface of the small sphere surrounding the point ξ_A can, therefore, be written as

$$\mathbf{D}(\mathbf{x}) = r^{-2}\mathbf{d}_A(\hat{\mathbf{r}}),$$

$$\mathbf{H}(\mathbf{x}) = r^{-2}\mathbf{h}_A(\hat{\mathbf{r}}),$$

$$\mathbf{E}(\mathbf{x}) = \mathbf{E}_A(\hat{\mathbf{r}}),$$

$$\mathbf{B}(\mathbf{x}) = \mathbf{B}_A(\hat{\mathbf{r}}).$$

The volume integral and the surface integral with the Lagrangian can thus be neglected and, finally, we have

$$\frac{d}{dt}\pi_A = \int d\Omega[(\hat{\mathbf{r}}\cdot\mathbf{d}_A(\hat{\mathbf{r}}))\mathbf{E}_A(\hat{\mathbf{r}})+(\hat{\mathbf{r}}\times\mathbf{h}_A(\hat{\mathbf{r}}))\times\mathbf{B}_A(\hat{\mathbf{r}})]. \qquad (10)$$

The expression on the right-hand side of this equation represents the force acting on the particle. It is well defined, but it can be calculated only when the intensities of the fields around the particles are known. In order to determine these fields, one would have to solve the equations for the fields created by a point source, which is an extremely difficult task in the general case. In the particular case of a single particle at rest (cf. eqs. (6a) and (8)) the force is zero, as was to be expected.

Magnetic Charges

Maxwellian theory without sources is invariant under so-called *dual rotations*

$$f_{\mu\nu} \to {}'f_{\mu\nu} = \cos\varphi f_{\mu\nu} + \sin\varphi \check{f}_{\mu\nu},$$
$$\check{f}_{\mu\nu} \to {}'\check{f}_{\mu\nu} = -\sin\varphi f_{\mu\nu} + \cos\varphi \check{f}_{\mu\nu}.$$

This invariance encompasses the energy-momentum tensor as well as the field equations. A dual rotation through an angle of $\varphi = \pi/2$ is simply a replacement of the electric field by the magnetic field and the magnetic field by the electric field, taken with the minus sign:

$$\mathbf{E} \to {}'\mathbf{E} = \mathbf{B},$$
$$\mathbf{B} \to {}'\mathbf{B} = -\mathbf{E}.$$

The free Maxwellian theory thus displays full symmetry with respect to an interchange of "electricity" and "magnetism". The presence of sources of field always violates this symmetry, but the asymmetry observed in nature is, in a sense, minimal. The known sources of electromagnetic field can always be described by a single vector function (current density) $j^\mu(x)$ and consequently the field equations can be broken up into the inhomogeneous equations

$$\partial_\mu f^{\mu\nu} = j^\nu \tag{11a}$$

and the homogeneous equations

$$\partial_\mu \check{f}^{\mu\nu} = 0. \tag{11b}$$

Note than even if the inhomogeneity described by the same current j^ν were to occur in both equations,

$$\partial_\mu f^{\mu\nu} = \alpha j^\nu,$$
$$\partial_\mu f^{\mu\nu} = \beta j^\nu,$$

the equations could be reduced to the form (11) by dual rotation through the angle $\varphi = \arctan\dfrac{\beta}{\alpha}$.

The situation would change fundamentally if the description of the sources of the field required two linearly independent currents: the electric charge current $j^\mu(x)$ and the magnetic charge current $\check{j}^\mu(x)$.

In that case homogeneous equations could not be isolated from the field equations

$$\partial_\mu f^{\mu\nu} = j^\nu,$$
$$\partial_\mu \check{f}^{\mu\nu} = \check{j}^\nu.$$

We shall rewrite these field equations in non-relativistic notation for the general case of the non-linear theory,

$$-\frac{\partial}{\partial t}\mathbf{D}+\nabla\times\mathbf{H} = \mathbf{j}_e,$$

$$\nabla\cdot\mathbf{D} = \varrho_e,$$

$$-\frac{\partial}{\partial t}\mathbf{B}-\nabla\times\mathbf{E} = \mathbf{j}_m,$$

$$\nabla\cdot\mathbf{B} = \varrho_m,$$

where $(\varrho_e, \mathbf{j}_e)$ and $(\varrho_m, \mathbf{j}_m)$ are the components of the four-vectors j^μ and $\check{j}^\mu$.

Point particles endowed with charges provide a simple model of sources. When the particles carry both electric and magnetic charges (e_A and g_A), two currents—electric and magnetic—occur. Point magnetic charges are frequently called *magnetic monopoles*.

The classical theory of electromagnetism indicates no reason for excluding the possibility that magnetic monopoles do exist.† Nor have explicit contradictions been found on the basis of the quantum theory when electric and magnetic charges coexist. An interesting result obtained in the quantum theory of magnetic monopoles is the relation# between the elementary electric charge e and magnetic charge g,

$$\frac{eg}{4\pi\hbar c} = \frac{1}{2}. \tag{12}$$

† In non-linear theories the existence of magnetic monopoles would, however, violate the scheme described on p. 114.

This relation was discovered by P. A. M. Dirac (1931). A simple derivation of this relation has been given by the present authors (I. Białynicki-Birula and Z. Białynicka-Birula (1971)). An extensive discussion of the theory of magnetic monopoles can be found in the monograph by Schwinger.

It follows from this relation that the equivalent of the fine-structure constant for the magnetic charge has a value of 137/4.

Despite an intensive search, no magnetic monopole has hitherto been discovered in nature.

Charged Field

To conclude this section we shall discuss the theory of the electromagnetic field coupled with a scalar complex field. This is the simplest relativistic model of field theory with gauge affecting both electromagnetic potentials and the phase of the field describing the source. We shall return to gauge invariance in Chapter 6 where we shall use the conclusions reached with the aid of this simple model.

The scalar complex field $\phi(x)$ satisfies the equation

$$[(\partial_\mu + ieA_\mu)(\partial^\mu + ieA^\mu) + m^2]\phi = 0, \tag{13}$$

which is obtained from the Klein–Gordon equation (14.4a) when the ordinary derivatives ∂_μ are replaced by the derivatives D_μ which include the field

$$D_\mu \equiv \partial_\mu + ieA_\mu. \tag{14}$$

This method of introducing coupling with the electromagnetic field, called the *minimal coupling method*, is patterned after quantum mechanics (cf. Section 16).

The minimal electromagnetic coupling establishes the relation between the *gauge of the potential* and the *gauge of the fields describing charged particles*. Only if both of these quantities are gauged simultaneously by means of the same function $\Lambda(x)$ is invariance ensured:

$$A_\mu(x) \to {}'A_\mu(x) = A_\mu(x) + \partial_\mu \Lambda(x), \tag{15a}$$

$$\phi(x) \to {}'\phi(x) = e^{-ie\Lambda(x)}\phi(x). \tag{15b}$$

The interrelationship between the unobservable phase and the unobservable longitudinal part of the potential is a profound theoretical idea, which perhaps is not yet fully understood in all its ramifications.

Minimal coupling can be justified if the unobservability of the phase of the complex function ϕ and the locality of the field description are taken as the point of departure. No observable consequences should ensue when the phase of the function ϕ is changed differently at different points (cf. formula (15b)) since the values of the field ϕ at various points are independent physical quantities. If the function satisfies the differential equation determining the dynamic development of the system, the function obtained by changing the phase must also satisfy the same equation. If the phase changes, however, the derivatives of the function ϕ suffer changes and only in the combination (14) of the derivatives and the potential do the changes in derivatives and changes in potential cancel each other.

Sets of variables ϕ, ϕ^*, and A_μ, differing by gauge transformation (15), describe the same physical situation. An equivalence class of fields ϕ, ϕ^* and A_μ therefore corresponds to each state of a system of interaction fields.

The Klein–Gordon equation with electromagnetic field (13) is the Euler–Lagrange equation for the action functional W_ϕ defined by

$$W_\phi = \int d^4x \mathscr{L}_\phi(x) = \int d^4x[(D_\mu\phi)^*(D^\mu\phi) - m^2\phi^*\phi]. \tag{16}$$

The full action functional for a system consisting of a field ϕ and the electromagnetic field is obtained by adding to W_ϕ the action functional W_{em} for the electromagnetic field. We shall consider only the linear variant of the theory, assuming

$$W_{em} = \int d^4x \mathscr{L}_{em}(x) = -\frac{1}{4}\int d^4x f_{\mu\nu}(x) f^{\mu\nu}(x). \tag{17}$$

The complete set of field equations is obtained by varying the action functional $W = W_\phi + W_{em}$ with respect to ϕ^*, ϕ, and A_μ:

$$(D_\mu D^\mu + m^2)\phi(x) = 0, \tag{18a}$$

$$(D_\mu^* D^{\mu *} + m^2)\phi^*(x) = 0, \tag{18b}$$

$$\partial_\mu f^{\mu\nu}(x) = j^\nu(x), \tag{18c}$$

where

$$j^\mu(x) \equiv ie\big(\phi^*(x) D^\mu\phi(x) - (D^\mu\phi)^*(x)\phi(x)\big). \tag{19}$$

By equations (18a) and (18b), the current vector j^μ satisfies the continuity equation

$$\partial_\mu j^\mu(x) = 0.$$

This equation is the condition for the consistency of Maxwell's equations (18c). The electromagnetic field cannot be coupled to a current which does not satisfy the continuity equation. In the hierarchy of conservation laws, the *principle of conservation of charge*, $Q \equiv \int d\sigma_\mu j^\mu = \text{const}$, thus has the highest status.

The symmetric energy-momentum tensor for the system in question is given by the formula

$$T^{\nu\mu}(x) \equiv (D^\mu \phi)^* (D^\nu \phi) + (D^\nu \phi)^* (D^\mu \phi) + f^\mu{}_\lambda f^{\lambda\nu} - g^{\mu\nu}(\mathscr{L}_\phi + \mathscr{L}_{em}). \tag{20}$$

This expression can be obtained by varying the action functional W written in arbitrary coordinates with respect to the coordinates of the metric tensor $g_{\mu\nu}$ (cf. Appendix F).

Using the field equations, one may show by direct calculation that the tensor $T^{\mu\nu}$ satisfies the continuity equation

$$\partial_\nu T^{\mu\nu}(x) = 0.$$

It follows from this equation that the energy, momentum, angular momentum, and moment of the energy are constants of motion.

The current vector and the energy-momentum tensor are invariant under gauge transformations (15).

The ambiguity in the choice of variables ϕ, ϕ^*, and A_μ for describing a given physical situation is a source of major difficulties in the canonical formulation of the theory under consideration, difficulties which are further carried over into quantum theory. For, also inherent in the quantities ϕ, ϕ^*, and A_μ are non-dynamical parts, on which the generators of the Poincaré transformations "do not catch". This is already clearly evident in the case of translations. The relation

$$\partial_\mu \phi = -\{P_\mu, \phi\}$$

cannot hold, since it is not invariant under gauge transformations.†
Similar difficulties also crop up in the case of potentials.

† Nor can the gauge-invariant modification of this formula

$$D_\mu \phi = -\{P_\mu, \phi\}$$

be valid since the operations D_μ and D_ν do not commute,

$$[D_\mu, D_\nu] = ief_{\mu\nu},$$

which violates the Abelian structure of the translation subgroup.

The requirement of relativistic invariance thus cannot be made compatible with gauge invariance even in classical theory if it is insisted that the gauge-dependent quantities ϕ, ϕ^*, and A_μ be chosen as the dynamical variables of the system. The generators P_μ may be said figuratively to "displace" only the dynamical part of the gauge-dependent variables. The non-dynamical dependence should be translated separately, just as is done in classical mechanics with respect to translation in time in the case when there is an explicit dependence on time. The correct formulae for the derivatives and Poisson brackets can thus be written in symbolical form:

$$\partial_\mu \phi = (\partial_\mu \phi)_{\text{nd}} - \{P_\mu, \phi\}.$$

The non-dynamical derivative with respect to the coordinates as written above can be determined only by a choice of a particular gauge, i.e. by violation of the gauge invariance.

The aforementioned difficulties do not occur for any gauge-independent dynamical quantities. For the gauge-independent physical quantities F and G constructed from the fields ϕ, ϕ^*, and $f_{\mu\nu}$ and from the derivatives $D_\mu\phi$ and $(D_\mu\phi)^*$ the Poisson bracket is defined# by the formula:†

$$\{F, G\} \equiv \int d^3z \Bigg\{\left(\frac{\delta F}{\delta\phi(\mathbf{z})} - \mathbf{D}^* \cdot \frac{\delta F}{\delta(\mathbf{D}\phi)(\mathbf{z})}\right) \frac{\delta G}{\delta(D_0\phi)^*(\mathbf{z})}$$
$$+ \left(\frac{\delta F}{\delta^*\phi(\mathbf{z})} - \mathbf{D} \cdot \frac{\delta F}{\delta(\mathbf{D}\phi)^*(\mathbf{z})}\right) \frac{\delta G}{\delta(D_0\phi)(\mathbf{z})} + \frac{\delta F}{\delta \mathbf{E}(\mathbf{z})} \Bigg[\nabla \times \frac{\delta G}{\delta \mathbf{B}(\mathbf{z})}$$
$$+ ie\left(\phi(\mathbf{z}) \frac{\delta G}{\delta(\mathbf{D}\phi)(\mathbf{z})} - \phi^*(\mathbf{z}) \frac{\delta G}{\delta(\mathbf{D}\phi)^*(\mathbf{z})}\right)\Bigg] - (F \leftrightarrow G)\Bigg\}. \tag{21}$$

It may be verified that the Poisson bracket so defined guarantees relativistic invariance of the theory since the components of the energy-momentum tensor then satisfy the relations (7.16); in turn, relations (7.17) for generators characteristic of the Poincaré group follow from (7.16). The Poisson bracket of the charge Q with any physical quantity is equal to zero.

#A canonical formalism invariant under gauge transformations has been given by I. Białynicki-Birula (1970).

† In this formula **D** stands for the spatial part of the operation D^μ.

Thus, we have succeeded in giving the canonical formulation only for physical quantities which are gauge-independent. A relativistically-invariant canonical structure exists only for truly dynamical quantities and the extension of this structure to gauge-dependent objects violates the invariance with respect to gauge transformations. To put it jocularly, the theory has defended itself against all attempts to give a physical meaning to gauge-dependent quantities.

These conclusions emerging from analysis of classical theory with gauge will be utilized in quantum theory.

9. THE MAXWELLIAN FIELD

Solution of the Initial-Value Problem and the Poisson Brackets

We shall now occupy ourselves with *Maxwell's theory*, the simplest—and yet most important—example of relativistic electrodynamics. To begin with, we shall discuss the theory of the electromagnetic field in vacuum. The Lagrangian density in this case is of the simple form:

$$\mathscr{L} = -\frac{1}{4} f_{\mu\nu} f^{\mu\nu} . \tag{1}$$

The relations between **E** and **D** and **B** and **H** are linear.† The field equations then take on the form of *Maxwell's equations*:

$$\partial^\mu f_{\mu\nu} = 0, \tag{2a}$$

$$\partial_\mu \check{f}^{\mu\nu} = 0. \tag{2b}$$

To facilitate analysis of this set of equations, we introduce an auxiliary complex antisymmetric tensor $F_{\mu\nu}$ constructed out of the tensors $f_{\mu\nu}$ and $\check{f}_{\mu\nu}$:

$$F_{\mu\nu}(x) \equiv \frac{1}{2} \left(f_{\mu\nu}(x) + i \check{f}_{\mu\nu}(x) \right) . \tag{3}$$

The tensor $F_{\mu\nu}$ is *self-dual*, which is to say that it satisfies the condition

$$\check{F}_{\mu\nu}(x) = -iF_{\mu\nu}(x). \tag{4}$$

† In the system of units adopted here, we simply have **E** = **D** and **B** = **H**.

The field equations (2) lead to the following equations for the tensor $F_{\mu\nu}(x)$:

$$\partial^{\mu} F_{\mu\nu}(x) = 0. \tag{5}$$

Equations (4) and (5) imply that all the components of the tensor $F_{\mu\nu}(x)$ satisfy the *d'Alembert equation*

$$\Box F_{\mu\nu}(x) = 0. \tag{6}$$

Because $F_{\mu\nu}$ is an antisymmetric self-dual tensor, only three of its components are independent. For the independent components we take the coordinates F_{0k} which we shall denote by the symbols F_k since they constitute a three-vector $\mathbf{F}$. This vector # is a linear combination of the vectors $\mathbf{D}$ and $\mathbf{B}$:

$$F_k(x) \equiv F_{k0}(x) = -i\varepsilon_{klm} F_{lm}(x), \tag{7a}$$

$$\mathbf{F}(x) = \frac{1}{2}\left(\mathbf{D}(x) + i\mathbf{B}(x)\right). \tag{7b}$$

The components $F_k(x)$ satisfy the equation

$$\frac{\partial}{\partial t}\mathbf{F} = -i\nabla \times \mathbf{F} \tag{8}$$

by virtue of eq. (5).

Now we shall proceed to solve the initial-value problem for Maxwell's equations or the equivalent equations (8). To do this we shall use the function† $D(x)$ defined below:

$$D(x) \equiv i\int d\Gamma (e^{-ik\cdot x} - e^{ik\cdot x}), \tag{9}$$

where the symbol $k \cdot x$ denotes the scalar product of the four-vectors k_μ and x_μ. The symbol $d\Gamma$ stands for a suitably normalized, invariant

The classical and quantum electromagnetic fields have been described in a similar fashion, with the aid of a complex vector $\mathbf{F}$, by H. A. Kramers (cf. the book by Kramers).

† Strictly speaking, this is a distribution. The properties of the function $D(x)$ and its relationship with Green's functions for the d'Alembert equation are given in Appendix E.

measure on the surface of the cone:

$$\mathrm{d}\Gamma \equiv \frac{\mathrm{d}^3 k}{2\omega(\mathbf{k})\,(2\pi)^3}\,, \tag{10a}$$

$$\omega(\mathbf{k}) \equiv |\mathbf{k}|\,. \tag{10b}$$

The symbol $\int \mathrm{d}\Gamma$ will henceforth always designate an integral over the upper half of the cone. Accordingly, the vector k_μ which appears as an argument of the integrand satisfies the conditions

$$k^2 = 0,$$

$$k_0 = \omega(\mathbf{k})\,.$$

In subsequent considerations we shall make use of the following properties of the function $D(x)$:

$$\Box D(x) = 0, \tag{11a}$$

$$D(\mathbf{x}, 0) = 0, \tag{11b}$$

$$\frac{\partial}{\partial t} D(\mathbf{x}, t)\Big|_{t=0} = \delta(\mathbf{x})\,. \tag{11c}$$

Since all of the components of the tensor $F_{\mu\nu}(x)$ satisfy the d'Alembert equation, the identity†

$$F_{\mu\nu}(x) = \int \mathrm{d}\sigma^\lambda D(x-y)\,\overleftrightarrow{\partial_\lambda}\,F_{\mu\nu}(y) \tag{12}$$

is satisfied; here, σ is an arbitrary spacelike hypersurface.

To prove the identity (12) we shall first show that the integral appearing in it does not depend on the hypersurface σ. By d'Alembert's equations (6) and (11a), the functional derivative of this integral with respect to the hypersurface σ vanishes at any arbitrary point:

$$\frac{\delta}{\delta\sigma(y)} \int_\sigma \mathrm{d}\sigma^\lambda D(x-y)\,\partial_\lambda F_{\mu\nu}(y) = \frac{\partial}{\partial y_\lambda}\left[D(x-y)\frac{\overleftrightarrow{\partial}}{\partial y^\lambda} F_{\mu\nu}(y)\right] = 0.$$

This means that this integral is independent of the hypersurface σ. If σ is a hyperplane, which is determined by the equation $x^0 = \mathrm{const}$ in a particular coordinate

† The symbol $\overleftrightarrow{\partial_\mu}$ is defined by the formula

$$A(x)\overleftrightarrow{\partial_\mu}B(x) = A(x)\,\partial_\mu B(x) - (\partial_\mu A(x))\,B(x).$$

system, then the properties (11b) and (11c) of the function $D(x)$ can be used to show that the identity (12) is satisfied. Thus, this identity holds for every spacelike hypersurface σ.

Using Maxwell's equations (or the eq. (8) derived from them for the components of the vector $\mathbf{F}(x)$) we can present identity (12) in the form of the solution of the initial-value problem for independent components of the tensor $F_{\mu\nu}$,

$$\mathbf{F}(\mathbf{x}, t) = \int \mathrm{d}^3 y \left[\frac{\partial}{\partial t} D(\mathbf{x}-\mathbf{y}, t-t_0) \mathbf{F}(\mathbf{y}, t_0) \right.$$
$$\left. - iD(\mathbf{x}-\mathbf{y}, t-t_0) \nabla \times \mathbf{F}(\mathbf{y}, t_0) \right]. \tag{13}$$

Comparison of the corresponding real and imaginary parts of both sides of the equation yields the solution of the initial-value problem for the vectors $\mathbf{D}$ and $\mathbf{B}$ (the function $D(x)$ is real):

$$\mathbf{D}(\mathbf{x}, t) = \int \mathrm{d}^3 y \left[\frac{\partial}{\partial t} D(\mathbf{x}-\mathbf{y}, t-t_0) \mathbf{D}(\mathbf{y}, t_0) \right.$$
$$\left. + D(\mathbf{x}-\mathbf{y}, t-t_0) \nabla \times \mathbf{B}(\mathbf{y}, t_0) \right],$$

$$\mathbf{B}(\mathbf{x}, t) = \int \mathrm{d}^3 y \left[\frac{\partial}{\partial t} D(\mathbf{x}-\mathbf{y}, t-t_0) \mathbf{B}(\mathbf{y}, t_0) \right.$$
$$\left. - D(\mathbf{x}-\mathbf{y}, t-t_0) \nabla \times \mathbf{D}(\mathbf{y}, t_0) \right].$$

The Poisson brackets for the components of the field $\mathbf{F}$, taken at the same time, are calculated by using the Poisson brackets of the components of $\mathbf{D}$ and $\mathbf{B}$,

$$\{F_i(\mathbf{x}, t), F_j^*(\mathbf{y}, t)\} = \frac{i}{2} \varepsilon_{ijk} \partial_k \delta(\mathbf{x}-\mathbf{y}), \tag{14a}$$

$$\{F_i(\mathbf{x}, t), F_j(\mathbf{y}, t)\} = 0. \tag{14b}$$

To evaluate the Poisson bracket of two components of the field $\mathbf{F}$, taken at two different instants, e.g. $F_i(\mathbf{x}, t)$ and $F_k^*(\mathbf{y}, t_0)$, we express the vector $\mathbf{F}(\mathbf{x}, t)$ in terms of the initial value $\mathbf{F}(\mathbf{x}, t_0)$ with the aid of the solution (13) and then use the values of the equal-time Poisson brackets of the components of the vector $\mathbf{F}$,

$$\{F_i(\mathbf{x}, t), F_k^*(\mathbf{y}, t_0)\}$$

$$= \frac{1}{2}(\delta_{ik}\,\partial_l\,\partial_l - \partial_i\,\partial_k + i\varepsilon_{ikj}\,\partial_j\,\partial_t)\,D(\mathbf{x}-\mathbf{y}, t-t_0), \tag{15a}$$

$$\{F_i(\mathbf{x}, t), F_k(\mathbf{y}, t_0)\} = 0. \tag{15b}$$

The foregoing relations can be rewritten in the equivalent tensor form

$$\{f_{\mu\nu}(x), f_{\lambda\varrho}(y)\} = \partial_{[\mu}g_{\nu][\lambda}\partial_{\varrho]}D(x-y). \tag{16}$$

Knowledge of the Poisson brackets for the field intensities plays an important role in the canonical quantization in classical theory. The covariant form of the relations (16) is essential for the relativistic invariance of quantum theory of the electromagnetic field.

Fourier Analysis of the Field

Since all the components of the tensor $F_{\mu\nu}(x)$ satisfy the d'Alembert equation, they can be written as superpositions of plane waves. This is so because each function $f(x)$ which satisfies the d'Alembert equation may be written as the sum of two integrals,

$$f(x) = \int d\Gamma f_+(k)e^{-ik\cdot x} + \int d\Gamma f_-(k)e^{ik\cdot x}.$$

The first of these integrals is a superposition of plane waves containing only *positive frequencies*, which means that its dependence on the time x^0 is determined by the factor $\exp(-i\omega x^0)$, where $\omega > 0$. The second integral contains only *negative frequencies*, which means that its time dependence is determined by the factor $\exp(i\omega x^0)$, where $\omega > 0$. The parts of the function $f(x)$ which contain only positive and only negative frequencies will be denoted, respectively, by the symbols $f^{(+)}(x)$ and $f^{(-)}(x)$,

$$f(x) = f^{(+)}(x) + f^{(-)}(x).$$

Thus, for the field $F_{\mu\nu}(x)$ we have

$$F_{\mu\nu}(x) = \int \mathrm{d}\Gamma[a^{+}_{\mu\nu}(\mathbf{k})\mathrm{e}^{-ik\cdot x}+a^{-}_{\mu\nu}(\mathbf{k})\mathrm{e}^{ik\cdot x}]. \tag{17}$$

The antisymmetric tensors $a^{+}_{\mu\nu}$ and $a^{-}_{\mu\nu}$ are coefficients in the decomposition of the field $F_{\mu\nu}$ into plane waves. They satisfy the following relations which follow from the field equations (5) and the self-duality condition (4) for the tensor $F_{\mu\nu}$,

$$\check{a}^{\pm}_{\mu\nu}(\mathbf{k}) = -ia^{\pm}_{\mu\nu}(\mathbf{k}), \tag{18a}$$

$$k^{\mu}a^{\pm}_{\mu\nu}(\mathbf{k}) = 0. \tag{18b}$$

These relations determine the tensors $a^{\pm}_{\mu\nu}$ to within a factor. The tensors $a^{\pm}_{\mu\nu}$ can thus be written in the form

$$a^{+}_{\mu\nu}(\mathbf{k}) = e_{\mu\nu}(\mathbf{k})f(\mathbf{k}, +1), \tag{19a}$$

$$a^{-}_{\mu\nu}(\mathbf{k}) = e_{\mu\nu}(\mathbf{k})f^{*}(\mathbf{k}, -1), \tag{19b}$$

where the antisymmetric tensor $e_{\mu\nu}(\mathbf{k})$ satisfies the algebraic relations

$$k^{\mu}e_{\mu\nu}(\mathbf{k}) = 0, \tag{20a}$$

$$\check{e}_{\mu\nu}(\mathbf{k}) = -ie_{\mu\nu}(\mathbf{k}), \tag{20b}$$

and the normalization condition

$$e_{\mu\nu}(\mathbf{k})e^{*\nu}{}_{\lambda}(\mathbf{k}) = k_{\mu}k_{\lambda}. \tag{21}$$

The following relation for the tensor $e_{\mu\nu}(\mathbf{k})$ follows from eqs. (20) and (21):

$$e_{\mu\nu}(\mathbf{k})e^{*}_{\lambda\varrho}(\mathbf{k}) = \frac{1}{2}k_{[\mu}g_{\nu][\lambda}k_{\varrho]}+\frac{1}{4}ik^{\alpha}\varepsilon_{\alpha\mu\nu[\lambda}k_{\varrho]}-\frac{1}{4}ik^{\alpha}\varepsilon_{\alpha\lambda\varrho[\mu}k_{\nu]}. \tag{22}$$

This condition determines the tensor $e_{\mu\nu}(\mathbf{k})$ up to a phase factor. Let us examine this arbitrariness and we shall then find its geometric interpretation. To this end we introduce a family of three-vectors $\mathbf{e}(\mathbf{k})$ whose components are related to the components of the tensor $e_{\mu\nu}(\mathbf{k})$ by the formulae

$$e_{0i}(\mathbf{k}) = i\omega(\mathbf{k})e_{i}(\mathbf{k}) = -\frac{1}{2}i\varepsilon_{ijk}e_{jk}(\mathbf{k}). \tag{23}$$

Conditions (20), (21), and (22) imply the following system of equations for the components of the vector $\mathbf{e}(\mathbf{k})$:

$$\mathbf{k}\cdot\mathbf{e}(\mathbf{k}) = 0, \tag{24a}$$

$$\mathbf{k}\times\mathbf{e}(\mathbf{k}) = -i\omega(\mathbf{k})\mathbf{e}(\mathbf{k}), \tag{24b}$$

$$\mathbf{e}^*(\mathbf{k})\cdot\mathbf{e}(\mathbf{k}) = 1, \tag{24c}$$

$$\mathbf{e}(\mathbf{k})\cdot\mathbf{e}(\mathbf{k}) = 0 = \mathbf{e}^*(-\mathbf{k})\cdot\mathbf{e}(\mathbf{k}), \tag{24d}$$

$$\mathbf{e}^*(\mathbf{k})\times\mathbf{e}(\mathbf{k}) = i\mathbf{n}, \tag{24e}$$

$$\mathbf{e}^*(-\mathbf{k})\times\mathbf{e}(\mathbf{k}) = 0, \tag{24f}$$

where $\mathbf{n}$ is the unit vector in the $\mathbf{k}$ direction,

$$\mathbf{n} \equiv \frac{\mathbf{k}}{\omega(\mathbf{k})}.$$

Let us decompose the complex vector $\mathbf{e}(\mathbf{k})$ into real and imaginary parts,

$$\mathbf{e}(\mathbf{k}) = \frac{1}{\sqrt{2}}(\mathbf{l}_1(\mathbf{k})+i\mathbf{l}_2(\mathbf{k})). \tag{25}$$

The real vectors $\mathbf{l}_1(\mathbf{k})$ and $\mathbf{l}_2(\mathbf{k})$ are unit vectors. Together with the unit vector $\mathbf{n}$ they form for each $\mathbf{k}$ an orthonormal system of vectors by virtue of eqs. (24) and (25),

$$\mathbf{l}_1\times\mathbf{l}_2 = \mathbf{n},$$
$$\mathbf{l}_2\times\mathbf{n} = \mathbf{l}_1,$$
$$\mathbf{n}\times\mathbf{l}_1 = \mathbf{l}_2.$$

For a given vector $\mathbf{k}$, the vectors $\mathbf{l}_1$ and $\mathbf{l}_2$ are defined up to a rotation in the plane perpendicular to the vector $\mathbf{k}$. The arbitrary phase factor, to within which the tensor $e_{\mu\nu}(\mathbf{k})$ was determined, is related to the angle of that rotation by the following formula:

$$\sqrt{2}\, e^{i\varphi}\mathbf{e} = \mathbf{l}_1\cos\varphi-\mathbf{l}_2\sin\varphi+i(\mathbf{l}_1\sin\varphi+\mathbf{l}_2\cos\varphi).$$

The phase φ may in general depend on the wave vector $\mathbf{k}$.

Now let us calculate the derivatives of the vector $\mathbf{e}(\mathbf{k})$ with respect to the components of the vector $\mathbf{k}$, as these derivatives will be needed in subsequent considerations. The orthonormality conditions of the vectors $\mathbf{l}_1$, $\mathbf{l}_2$ and $\mathbf{n}$ imply that

$$\partial_i e_j(\mathbf{k}) = -\omega^{-1}e_i(\mathbf{k})n_j(\mathbf{k})-i\alpha_i(\mathbf{k})e_j(\mathbf{k}), \tag{26}$$

where $\boldsymbol{\alpha}(\mathbf{k})$ is a vector function which is not fixed by the orthonormality conditions. It follows from eq. (26), however, that

$$\boldsymbol{\alpha}(\mathbf{k}) = ie_i^*(\mathbf{k})\,\nabla e_i(\mathbf{k}),$$

whence, after using eq. (26) again, we obtain

$$\nabla\times\boldsymbol{\alpha}(\mathbf{k}) = -\frac{\mathbf{n}(\mathbf{k})}{\mathbf{k}^2}.$$

The right-hand side of this equation is the gradient of the function $|\mathbf{k}|^{-1}$, while the left-hand side is a curl. The vector field $\boldsymbol{\alpha}(\mathbf{k})$ thus is singular (at least along the line passing through the point $\mathbf{k} = 0$).

The vectors $\mathbf{l}_1$ and $\mathbf{l}_2$ can, for instance, be taken in the form

$$\mathbf{l}_1(\mathbf{k}) = \frac{1}{\sqrt{n_1^2+n_2^2}}(n_1 n_3, n_2 n_3, -n_1^2-n_2^2),$$

$$\mathbf{l}_2(\mathbf{k}) = \frac{1}{\sqrt{n_1^2+n_2^2}}(-n_2, n_1, 0).$$

The function $\boldsymbol{\alpha}(\mathbf{k})$ then is of the form

$$\boldsymbol{\alpha}(\mathbf{k}) = \frac{\omega^{-1}n_3}{n_1^2+n_2^2}(-n_2, n_1, 0).$$

The decomposition of the field $F_{\mu\nu}(x)$ into plane waves can finally be represented in the form

$$F_{\mu\nu}(x) = \int d\Gamma e_{\mu\nu}(\mathbf{k})[f(\mathbf{k}, +1)\mathrm{e}^{-ik\cdot x}+f^*(\mathbf{k}, -1)\mathrm{e}^{ik\cdot x}]. \tag{27}$$

From this we obtain the following decomposition of the field $f_{\mu\nu}(x)$ into plane waves:

$$f_{\mu\nu}(x) = \int d\Gamma[e_{\mu\nu}(\mathbf{k})f(\mathbf{k}, +1)+e_{\mu\nu}^*(\mathbf{k})f(\mathbf{k}, -1)]\mathrm{e}^{-ik\cdot x}+\text{c.c.}, \tag{28}$$

where the symbol c.c. denotes the expression obtained by the complex conjugation of the preceding expression. The coefficients $f(\mathbf{k}, \pm 1)$ will be denoted by the symbol $f(\mathbf{k}, \lambda)$, where λ takes on two values, $+1$ and -1.

The tensor $e_{\mu\nu}(\mathbf{k})$ is determined to within a phase factor, whereas each change in its phase

$$e_{\mu\nu}(\mathbf{k}) \to \mathrm{e}^{-i\varphi(k)}e_{\mu\nu}(\mathbf{k}),$$

entails the following change in the phase of the coefficients $f(\mathbf{k}, \lambda)$:

$$f(\mathbf{k}, \lambda) \to \mathrm{e}^{i\lambda\varphi(k)}f(\mathbf{k}, \lambda).$$

Under the influence of such changes in the phase of the tensor $e_{\mu\nu}$, the vector field $\boldsymbol{\alpha}(\mathbf{k})$ transforms like an electromagnetic potential under the gauge transformation:

$$\boldsymbol{\alpha}(\mathbf{k}) \to \boldsymbol{\alpha}(\mathbf{k}) + \nabla\varphi(\mathbf{k}). \tag{29}$$

This analogy can be built up[#] by introducing the phase-invariant derivatives D_j of the function $f(\mathbf{k}, \lambda)$:

$$D_j f(\mathbf{k}, \lambda) \equiv \nabla_j f(\mathbf{k}, \lambda) - i\lambda\alpha_j(\mathbf{k}) f(\mathbf{k}, \lambda). \tag{30}$$

Changes of phase induce the transformations

$$D_j f(\mathbf{k}, \lambda) \to e^{i\lambda\varphi} D_j f(\mathbf{k}, \lambda),$$

the forms of which imply that expressions of the type $f^* D_j f$ and $(D_i f)^*(D_j f)$ do not depend on the choice of phase.

The commutator of the derivatives D_i and D_j can be expressed by the formula

$$[D_i, D_j] = i\lambda\omega^{-2}\varepsilon_{ijk} n_k. \tag{31}$$

The notion of the derivative D_i will be used to express the generators P_μ and $M_{\mu\nu}$ in terms of the coefficients $f(\mathbf{k}, \lambda)$. For this purpose we shall first express the components of the energy-momentum tensor for the Maxwellian field with the aid of the components of the vectors $\mathbf{F}$ and $\mathbf{F}^*$:

$$T^{00}(x) = 2F_k^*(x) F_k(x), \tag{32a}$$

$$T^{k0}(x) = -2i\varepsilon_{klm} F_l^*(x) F_m(x), \tag{32b}$$

$$T^{kl}(x) = -2F_k^*(x) F_l(x) - 2F_l^*(x) F_k(x) + 2\delta_{kl} F_m^*(x) F_m(x). \tag{32c}$$

Now let us insert the components of the tensor $T^{\mu\nu}$ given by these formulae into eqs. (6.22) which represent the components P^μ and $M^{\mu\nu}$ in the form of three-dimensional integrals. These integrals are most conveniently calculated at the instant $t = 0$ by means of the following representation of the vector $\mathbf{F}(\mathbf{x})$:

$$\mathbf{F}(\mathbf{x}, 0) = \frac{i}{2(2\pi)^3} \int d^3k [\mathbf{e}(\mathbf{k}) f(\mathbf{k}, +1) + \mathbf{e}(-\mathbf{k}) f^*(-\mathbf{k}, -1)] e^{i\mathbf{k}\cdot\mathbf{x}}.$$

[#] Recently, A. Staruszkiewicz (1973) has given a geometrical theory for such covariant derivatives on a cone.

Integration with respect to **x** and one integration with respect to **k** yield†

$$H = \sum_{\lambda} \int d\Gamma \omega(\mathbf{k}) f^*(\mathbf{k}, \lambda) f(\mathbf{k}, \lambda), \tag{33a}$$

$$\mathbf{P} = \sum_{\lambda} \int d\Gamma \mathbf{k} f^*(\mathbf{k}, \lambda) f(\mathbf{k}, \lambda), \tag{33b}$$

$$\mathbf{M} = \sum_{\lambda} \int d\Gamma \left[f^*(\mathbf{k}, \lambda) \left(\mathbf{k} \times \frac{1}{i} \mathbf{D} \right) f(\mathbf{k}, \lambda) + \lambda \mathbf{n}(\mathbf{k}) f^*(\mathbf{k}, \lambda) f(\mathbf{k}, \lambda) \right], \tag{33c}$$

$$\mathbf{N} = \frac{i}{2} \sum_{\lambda} \int d\Gamma \omega(\mathbf{k}) [f^*(\mathbf{k}, \lambda) \mathbf{D} f(\mathbf{k}, \lambda) - (\mathbf{D} f(\mathbf{k}, \lambda))^* f(\mathbf{k}, \lambda)]. \tag{33d}$$

The generators H, **P**, **M**, and **N**, expressed by these formulae, no longer contain the (spurious) time dependence which appeared in the definition of these generators through the integrals over three-dimensional space.

Now let us find the Poisson brackets of the coefficients $f(\mathbf{k}, \lambda)$. When the representations of the fields as Fourier integrals are substituted into the formulae for the Poisson brackets of the fields **F** and **F***, we obtain the relations

$$i\{f(\mathbf{k}, \lambda), f^*(\mathbf{k}', \lambda')\} = \delta_{\lambda\lambda'} \delta_\Gamma(\mathbf{k}, \mathbf{k}'), \tag{34a}$$

$$\{f(\mathbf{k}, \lambda), f(\mathbf{k}', \lambda')\} = 0 = \{f^*(\mathbf{k}, \lambda), f^*(\mathbf{k}', \lambda')\}, \tag{34b}$$

where

$$\delta_\Gamma(\mathbf{k}, \mathbf{k}') \equiv (2\pi)^3 2\omega(\mathbf{k}) \delta(\mathbf{k} - \mathbf{k}'). \tag{35}$$

We shall also write out the Poisson brackets of the coefficients $f(\mathbf{k}, \lambda)$ and the energy, momentum, and the four-dimensional angular momentum:

$$i\{H, f(\mathbf{k}, \lambda)\} = -\omega(\mathbf{k}) f(\mathbf{k}, \lambda), \tag{36a}$$

$$i\{\mathbf{P}, f(\mathbf{k}, \lambda)\} = -\mathbf{k} f(\mathbf{k}, \lambda), \tag{36b}$$

$$i\{\mathbf{M}, f(\mathbf{k}, \lambda)\} = -\left[\left(\mathbf{k} \times \frac{1}{i} \mathbf{D} \right) + \lambda \mathbf{n}(\mathbf{k}) \right] f(\mathbf{k}, \lambda), \tag{36c}$$

$$i\{\mathbf{N}, f(\mathbf{k}, \lambda)\} = -\omega(\mathbf{k}) i \mathbf{D} f(\mathbf{k}, \lambda). \tag{36d}$$

† The relation (24d) should be used when calculating these integrals.

These relations may also be arrived at either by inserting into eqs. (7.14a) and (7.15a) the expressions for the fields $f_{\mu\nu}$ written as superpositions of plane waves, or by performing a direct calculation with the aid of relations (33) and (34).

Knowing the transformation laws of the tensor field $F_{\mu\nu}(x)$ under the Poincaré transformations, we can find the transformation laws for the function $f(\mathbf{k}, \lambda)$. Note first of all that the tensor $e_{\mu\nu}(\mathbf{k})$ must satisfy the identity

$$\Lambda_\mu^{\ \lambda}\Lambda_\nu^{\ \varrho}e_{\lambda\varrho}(\Lambda^{-1}\mathbf{k}) = \mathrm{e}^{i\theta(\mathbf{k},\Lambda)}e_{\mu\nu}(\mathbf{k}), \tag{37}$$

where $\Lambda^{-1}\mathbf{k}$ denotes the spatial part of the vector $\Lambda^{-1\mu}{}_\nu k^\nu$ and $\theta(\mathbf{k}, \Lambda)$ is a real function of the vector $\mathbf{k}$ and the matrix Λ. The expression on the left-hand side obeys all the relations (20) and (21); therefore, it can differ from $e_{\mu\nu}(\mathbf{k})$ only by a phase factor. Next, if into the formula

$$^{IL}F_{\mu\nu}(x) = \Lambda_\mu^{\ \lambda}\Lambda_\nu^{\ \varrho}F_{\lambda\varrho}\big(\Lambda^{-1}(x-a)\big),$$

we substitute the expression for the field $^{IL}F_{\mu\nu}$ in the form

$$^{IL}F_{\mu\nu}(\mathbf{x}) = \int d\Gamma e_{\mu\nu}(\mathbf{k})[^{IL}f(\mathbf{k}, +1)\mathrm{e}^{-ik\cdot x} + {}^{IL}f^*(\mathbf{k}, -1)\mathrm{e}^{ik\cdot x}],$$

then, on making use of the identity (37), we arrive at the transformation laws for the function $f(\mathbf{k}, \lambda)$:

$$^{IL}f(\mathbf{k}, \lambda) = \mathrm{e}^{i\lambda\theta(\mathbf{k},\Lambda)}\mathrm{e}^{ik\cdot a}f(\Lambda^{-1}\mathbf{k}, \lambda). \tag{38}$$

A similar method can be used to obtain the following transformation laws for the functions $f(\mathbf{k}, \lambda)$ under improper Poincaré transformations:

$$^{P}f(\mathbf{k}, \lambda) = f(-\mathbf{k}, -\lambda), \tag{39a}$$

$$^{T}f(\mathbf{k}, \lambda) = -f^*(-\mathbf{k}, \lambda). \tag{39b}$$

It is very convenient to describe fields by using the coefficients of the decomposition into plane waves $f(\mathbf{k}, \lambda)$ since this allows the field equations to be taken into account automatically. The field equations are satisfied for any choice of these coefficients. The coefficients $f(\mathbf{k}, \lambda)$ thus describe the independent degrees of freedom of the field.

In many cases, the fact that the wave vector $\mathbf{k}$ has a continuous set of values constitutes a serious mathematical difficulty. In order to limit the set of values of this vector to a countable set, a three-dimensional space of finite volume V is introduced as an auxiliary measure. If the volume V is sufficiently large, then far from the boundaries of this space physical phenomena occur in almost the same way as in an infinite three-dimensional space. To obviate difficulties associated with the possible interaction of the physical system with the walls confining the space, we introduce the finite volume V without walls by enclosing the space "within itself". This space will have the shape of a cube with edges parallel to the x, y, z axes; we identify the opposite walls of the cube with each other in pairs. In the enclosed space so constructed, the conditions for the continuity of the function $f(x)$ at the identified walls have the form of the periodicity conditions

$$f(x, y, z) = f(x+L, y, z) = f(x, y+L, z) = f(x, y, z+L),$$

where L is the length of the cube edge.

The introduction of such a space enables the integral formulae expressing the decomposition of the functions into plane waves to be replaced by formulae containing series. Each function satisfying the periodicity conditions can be written as a series

$$f(\mathbf{k}) = \frac{1}{\sqrt{V}} \sum_{\mathbf{k}} e^{i\mathbf{k}\cdot\mathbf{x}} f_{\mathbf{k}}.$$

The summation is over all vectors $\mathbf{k}$ of the form

$$\mathbf{k} = \frac{2\pi}{L}(n_x, n_y, n_z),$$

where n_x, n_y, and n_z are integers.

The decomposition of the field $F_{\mu\nu}(x)$ into plane waves in the enclosed space of finite volume is in the form of the series

$$F_{\mu\nu}(x) = \frac{1}{\sqrt{V}} \sum_{\mathbf{k}} e_{\mu\nu}(\mathbf{k})[f_{\mathbf{k},+} e^{-i\mathbf{k}\cdot\mathbf{x}} + f_{\mathbf{k},-} e^{i\mathbf{k}\cdot\mathbf{x}}]. \tag{40}$$

Conformal Transformations

Maxwell's equations in the vacuum are invariant not only under the ten-parameter group of Poincaré transformations but also under a larger, 15-parameter group known as the *group of conformal transformations*.# The term *conformal transformations* means *not altering the shape*, for the angles between vectors do not change under these transformations.

Conformal transformations are those transformations of the points of Minkowski space

$$x^\mu \to {}'x^\mu = {}'x^\mu(x),$$

which satisfy the set of equations

$$\frac{\partial' x^\mu}{\partial x^\lambda} \frac{\partial' x^\nu}{\partial x^\varrho} g_{\mu\nu} = \left|\frac{\partial' x}{\partial x}\right|^{\frac{1}{2}} g_{\lambda\varrho}, \tag{41}$$

where $\left|\frac{\partial' x}{\partial x}\right|$ is the Jacobian of the transformations.

This invariance was discovered by H. Bateman (1909).

Among the conformal transformations we distinguish *dilatations*

$$\mathrm{D}: {}'x^\mu = \alpha x^\mu,$$

and *accelerations*

$$\mathrm{A}: {}'x^\mu = \frac{x^\mu + x^2 a^\mu}{1+2a\cdot x + a^2x^2}.$$

The Jacobians of these transformations are α^4 and $(1+2a\cdot x+a^2x^2)^2$, respectively. Each conformal transformation can be obtained by composition of the dilatations and accelerations with the Poincaré transformations. The interval between two events does change under conformal transformations, except when these events can be connected by a light signal. In the special case of dilatation and acceleration we obtain:

$$\mathrm{D}: ({}'x - {}'y)^2 = \alpha^2(x-y)^2,$$

$$\mathrm{A}: ({}'x - {}'y)^2 = (1+2a\cdot x+a^2x^2)^{-1}(1+2a\cdot y+a^2y^2)^{-1}(x-y)^2.$$

It is natural to assume that under conformal transformations the electromagnetic field intensities transform like a covariant tensor field,

$$'f_{\mu\nu}({}'x) = \frac{\partial x^\lambda}{\partial' x^\mu}\frac{\partial x^\varrho}{\partial' x^\nu} f_{\lambda\varrho}(x). \tag{42}$$

Proof of the invariance of Maxwell's equations consists in demonstrating that the field $'f_{\mu\nu}(x)$ satisfies Maxwell's equations if they are satisfied by the field $f_{\mu\nu}(x)$.

The equations

$$\partial_\lambda f_{\mu\nu}(x) + \partial_\nu f_{\lambda\mu}(x) + \partial_\mu f_{\nu\lambda}(x) = 0$$

are invariant under arbitrary transformations of Minkowski space onto itself. After straightforward calculations we arrive at

$$'\partial_\lambda {}'f_{\mu\nu}({}'x) + {}'\partial_\nu {}'f_{\lambda\mu}({}'x) + {}'\partial_\mu {}'f_{\nu\lambda}({}'x)$$

$$= \frac{\partial x^\alpha}{\partial' x^\mu}\frac{\partial x^\beta}{\partial' x^\nu}\frac{\partial x^\gamma}{\partial' x^\lambda}\left(\partial_\gamma f_{\alpha\beta}(x) + \partial_\beta f_{\gamma\alpha}(x) + \partial_\alpha f_{\beta\gamma}(x)\right).$$

To prove the invariance of the equations

$$\partial_\mu f^{\mu\nu}(x) = 0$$

we first determine the transformation laws of the contravariant components of the field tensor. With the aid of the relationship

$$\frac{\partial x_\mu}{\partial' x_\nu} \equiv g_{\mu\lambda}\frac{\partial x^\lambda}{\partial' x^\varrho} g^{\varrho\nu} = \frac{\partial' x^\nu}{\partial x^\mu}\left|\frac{\partial' x}{\partial x}\right|^{-\frac{1}{2}}, \tag{43}$$

which derives from the definition of conformal transformations, we find

$$'f^{\mu\nu}({}'x) = \left|\frac{\partial' x}{\partial x}\right|^{-1}\frac{\partial' x^\mu}{\partial x^\lambda}\frac{\partial' x^\nu}{\partial x^\varrho} f^{\lambda\varrho}(x). \tag{44}$$

Now let us calculate the divergence of this tensor,

$$'\partial_\mu{}'f^{\mu\nu}('x) = {}'\partial_\mu\left(\left|\frac{\partial' x}{\partial x}\right|^{-1}\frac{\partial' x^\mu}{\partial x^\lambda}\right)\frac{\partial' x^\nu}{\partial x^\varrho}f^{\lambda\varrho}(x)$$

$$+\left|\frac{\partial' x}{\partial x}\right|^{-1}\frac{\partial^2{}' x^\nu}{\partial x^\lambda \partial x^\varrho}f^{\lambda\varrho}(x)+\left|\frac{\partial' x}{\partial x}\right|^{-1}\frac{\partial' x^\nu}{\partial x^\varrho}\partial_\lambda f^{\lambda\varrho}(x). \tag{45}$$

From formula (41), which defines conformal transformations, and eq. (43) it follows that

$$\frac{\partial}{\partial x^\lambda}\left|\frac{\partial' x}{\partial x}\right|^{\frac{1}{2}} = \frac{1}{2}\left(\frac{\partial}{\partial x^\lambda}\frac{\partial' x^\mu}{\partial x^\nu}\right)\frac{\partial' x_\mu}{\partial x_\nu}$$

$$= \frac{1}{2}\left|\frac{\partial' x}{\partial x}\right|^{\frac{1}{2}}\frac{\partial^2{}' x^\mu}{\partial x^\nu \partial x^\lambda}\frac{\partial x^\nu}{\partial' x^\mu} = \frac{1}{2}\left|\frac{\partial' x}{\partial x}\right|^{\frac{1}{2}}\frac{\partial}{\partial' x^\mu}\frac{\partial' x^\mu}{\partial x^\lambda}.$$

As a result, the first term on the right-hand side of eq. (45) vanishes:

$$\frac{\partial}{\partial' x^\mu}\left(\left|\frac{\partial' x}{\partial x}\right|^{-1}\frac{\partial' x^\mu}{\partial x^\lambda}\right) = \frac{\partial}{\partial x^\lambda}\left|\frac{\partial' x}{\partial x}\right|^{-1} + \left|\frac{\partial' x}{\partial x}\right|^{-1}\frac{\partial}{\partial' x^\mu}\frac{\partial' x^u}{\partial x^\lambda} = 0.$$

The second term in that equation vanishes because of the antisymmetry of the tensor $f^{\mu\nu}$. Thus, we finally arrive at

$$'\partial_\mu{}'f^{\mu\nu}('x) = \left|\frac{\partial' x}{\partial x}\right|^{-1}\frac{\partial' x^\nu}{\partial x^\varrho}\partial_\lambda f^{\lambda\varrho}(x),$$

which completes the proof that Maxwell's equations are invariant with respect to conformal transformations.

Under transformations of dilatation and acceleration, the transformed fields ${}^D f_{\mu\nu}(x)$ and ${}^A f_{\mu\nu}(x)$ are given by the formulae

$${}^D f_{\mu\nu}(\alpha x) = \alpha^{-2} f_{\mu\nu}(x),$$

$${}^A f_{\mu\nu}\left(\frac{x+x^2 a}{1+2a\cdot x+a^2x^2}\right) = A_\mu{}^\lambda(x)A_\nu{}^\varrho(x)f_{\lambda\varrho}(x),$$

where

$$A_\mu{}^\lambda(x) \equiv \frac{\partial x^\lambda}{\partial' x^\mu}$$

$$= (1+2a\cdot x+a^2x^2)(\delta^\lambda_\mu+2a_\mu x^\lambda)-2(x_\mu+x^2a_\mu)(a^\lambda+a^2x^\lambda).$$

For infinitesimal transformations, we have

D:
$$\delta x^\mu = \delta\alpha x^\mu,$$
$$\delta f_{\mu\nu}(x) = -\delta\alpha x^\lambda \partial_\lambda f_{\mu\nu}(x),$$

A:
$$\delta x^\mu = x^2\,\delta a^\mu - 2\delta a\cdot x x^\mu,$$
$$\delta f_{\mu\nu}(x) = -(x^2\delta a^\lambda - 2\delta a\cdot x x^\lambda)\partial_\lambda f_{\mu\nu}(x)$$
$$+2\delta a^\sigma x^\varkappa[(g_{\mu\sigma}\delta^\lambda_\varkappa - g_{\mu\varkappa}\delta^\lambda_\sigma)\delta^\varrho_\nu + (g_{\nu\sigma}\delta^\varrho_\varkappa - g_{\nu\varkappa}\delta^\varrho_\sigma)\delta^\lambda_\mu + 2g_{\varkappa\sigma}\delta^\lambda_\mu\delta^\varrho_\nu]f_{\lambda\varrho}(x).$$

The generators of these transformations are the following quantities constructed out of the components of the energy-momentum tensor:

$$D[\sigma] \equiv -\int_{\sigma} d\sigma_\nu x_\mu T^{\mu\nu}(x), \tag{46}$$

$$K^\lambda[\sigma] \equiv \int_{\sigma} d\sigma_\nu (2x^\lambda x_\mu - x^2 \delta^\lambda_\mu) T^{\mu\nu}(x). \tag{47}$$

The change of these quantities under deformations of the hypersurface σ is determined by the expressions

$$\frac{\delta D[\sigma]}{\delta\sigma(x)} = -T^{\mu}_{\mu}(x) - x_\mu \partial_\nu T^{\mu\nu}(x),$$

$$\frac{\delta K^\lambda[\sigma]}{\delta\sigma(x)} = 2x^\lambda T^{\mu}_{\mu}(x) + (2x^\lambda x_\mu - x^2\delta^\lambda_\mu)\partial_\nu T^{\mu\nu}(x).$$

A characteristic feature of theories which are invariant under conformal transformations is that the trace of the energy-momentum tensor vanishes:

$$T^{\mu}_{\mu}(x) = 0.$$

For an isolated system the other terms on the right-hand side also vanish and the generators D and K^λ are constants of motion.

In non-relativistic notation the generators of conformal transformations are of the form

$$D = \int d^3x [x^k T^{k0}(\mathbf{x}, t) - t T^{00}(\mathbf{x}, t)],$$

$$K^0 = \int d^3x [(\mathbf{x}^2 + t^2) T^{00}(\mathbf{x}, t) - 2t x^k T^{k0}(\mathbf{x}, t)],$$

$$K^i = \int d^3x [(\mathbf{x}^2 - t^2) T^{i0}(\mathbf{x}, t) - 2x^i x^j T^{j0}(\mathbf{x}, t) + 2t x^i T^{00}(\mathbf{x}, t)].$$

When use is made of the condition that the trace of the energy-momentum tensor vanishes, the Poisson brackets for the components of this tensor give us the following Poisson brackets for the fifteen generators of the conformal group:

$$\begin{aligned}
\{P_\mu, P_\nu\} &= 0,\\
\{K_\mu, K_\nu\} &= 0,\\
\{P_\mu, K_\nu\} &= -2M_{\mu\nu} - 2g_{\mu\nu} D,\\
\{D, P_\mu\} &= P_\mu,\\
\{D, K_\mu\} &= -K_\mu,\\
\{D, M_{\mu\nu}\} &= 0,\\
\{M_{\mu\nu}, P_\lambda\} &= -g_{\mu\lambda} P_\nu + g_{\nu\lambda} P_\mu,\\
\{M_{\mu\nu}, K_\lambda\} &= -g_{\mu\lambda} K_\nu + g_{\nu\lambda} K_\mu,\\
\{M_{\mu\nu}, M_{\lambda\varrho}\} &= -g_{\mu\lambda} M_{\nu\varrho} + g_{\mu\varrho} M_{\nu\lambda} - g_{\nu\varrho} M_{\mu\lambda} + g_{\nu\lambda} M_{\mu\varrho}.
\end{aligned} \tag{48}$$

In conclusion, let us express the generators of dilatation and acceleration in terms of the amplitudes $f(\mathbf{k}, \lambda)$ and calculate the Poisson brackets of these amplitudes and generators. Putting $t = 0$ considerably facilitates calculation of the integrals figuring in the formulae for D, K^0, and $\mathbf{K}$:

$$D = \frac{1}{2}\sum_{\lambda}\int d\Gamma \mathbf{k}\cdot(f^*(\mathbf{k}, \lambda) i\overleftrightarrow{\mathbf{D}} f(\mathbf{k}, \lambda)),$$

$$K_0 = \sum_{\lambda}\int d\Gamma[\omega(\mathbf{D}f(\mathbf{k}, \lambda))^*\cdot\mathbf{D}f(\mathbf{k}, \lambda)+\omega^{-1}f^*(\mathbf{k}, \lambda)f(\mathbf{k}, \lambda)],$$

$$\mathbf{K} = \sum_{\lambda}\int d\Gamma\lfloor\mathbf{k}(\mathbf{D}f(\mathbf{k}, \lambda))^*\cdot\mathbf{D}f(\mathbf{k}, \lambda)$$

$$-(\mathbf{k}\cdot\mathbf{D}f(\mathbf{k}, \lambda))^*\mathbf{D}f(\mathbf{k}, \lambda)-(\mathbf{D}f(\mathbf{k}, \lambda))^*\mathbf{k}\cdot\mathbf{D}f(\mathbf{k}, \lambda)$$

$$+\lambda\omega^{-1}\mathbf{k}\times((f^*(\mathbf{k}, \lambda) i\overleftrightarrow{\mathbf{D}} f(\mathbf{k}, \lambda))-\omega^{-2}\mathbf{k}f^*(\mathbf{k}, \lambda)f(\mathbf{k}, \lambda)],$$

$$i\{D, f(\mathbf{k}, \lambda)\} = i\mathbf{k}\cdot\mathbf{D}f(\mathbf{k}, \lambda)+if(\mathbf{k}, \lambda),$$

$$i\{K_0, f(\mathbf{k}, \lambda)\} = (\omega\mathbf{D}\cdot\mathbf{D}-\omega^{-1})f(\mathbf{k}, \lambda),$$

$$i\{\mathbf{K}, f(\mathbf{k}, \lambda)\} = (\mathbf{k}(\mathbf{D}\cdot\mathbf{D}) - 2(\mathbf{k}\cdot\mathbf{D})\mathbf{D}-2i\lambda\omega^{-1}\mathbf{k}\times\mathbf{D}-\omega^{-2}\mathbf{k}-2\mathbf{D})f(\mathbf{k}, \lambda).$$

The Maxwellian Field with External Sources

Now we shall generalize the considerations above so that they apply to the case of an electromagnetic field in the presence of *external sources*† $\mathscr{J}^{\mu}(x)$. In doing so, we shall assume that the sources occur only in a bounded (in time and space) region of space-time. Accordingly, the electromagnetic field is free outside that region. This assumption can be reconciled with the principle of conservation of charge if the creation of charge distribution ϱ at a certain instant is taken to mean the separation of charges of opposite sign, whereas the disappearance of charge density is understood as their recombination. In the presence of external sources, the field equations in Maxwell's theory are inhomogeneous equations:

$$\frac{\partial}{\partial t}\mathbf{D}(\mathbf{x}, t) = \nabla\times\mathbf{B}(\mathbf{x}, t)-\mathscr{J}(\mathbf{x}, t), \tag{49a}$$

$$\frac{\partial}{\partial t}\mathbf{B}(\mathbf{x}, t) = -\nabla\times\mathbf{D}(x, t), \tag{49b}$$

$$\nabla\cdot\mathbf{D}(\mathbf{x}, t) = \mathscr{J}^{0}(\mathbf{x}, t), \tag{49c}$$

$$\nabla\cdot\mathbf{B}(\mathbf{x}, t) = 0. \tag{49d}$$

† The term *external sources* is used when we disregard the influence of the field on the motion of the sources.

The general solution of this system of equations is the sum of the general solution of the homogeneous equations (2) and one special solution of the inhomogeneous equations. Of all the special solutions of inhomogeneous equations the retarded and advanced solutions will continue to be of greatest use to us. They are customarily expressed in terms of the *retarded* and *advanced potentials*, A_μ^{ret} and A_μ^{adv},

$$A_\mu^{\text{ret}}(\mathbf{x}, t) = \int d^4y D_R(x-y)\mathscr{I}_\mu(y) = \frac{1}{4\pi}\int d^3y \frac{\mathscr{I}_\mu(\mathbf{y}, t-|\mathbf{x}-\mathbf{y}|)}{|\mathbf{x}-\mathbf{y}|}, \tag{50a}$$

$$A_\mu^{\text{adv}}(\mathbf{x}, t) = \int d^4y D_A(x-y)\mathscr{I}_\mu(y) = \frac{1}{4\pi}\int d^3y \frac{\mathscr{I}_\mu(\mathbf{y}, t+|\mathbf{x}-\mathbf{y}|)}{|\mathbf{x}-\mathbf{y}|}, \tag{50b}$$

where the *Green's functions* $D_R(x)$ and $D_A(x)$ *for the d'Alembert equation* are defined by

$$D_R(x) \equiv \theta(x^0)D(x), \qquad D_A(x) \equiv -\theta(-x^0)D(x). \tag{51}$$

The general solution of the system of equations (49), written in terms of retarded potentials, is of the form

$$\mathbf{D}(\mathbf{x}, t) = \mathbf{D}^{\text{in}}(\mathbf{x}, t) - \frac{\partial}{\partial t}\mathbf{A}^{\text{ret}}(\mathbf{x}, t) - \nabla A_0^{\text{ret}}(\mathbf{x}, t), \tag{52a}$$

$$\mathbf{B}(\mathbf{x}, t) = \mathbf{B}^{\text{in}}(\mathbf{x}, t) + \nabla \times \mathbf{A}^{\text{ret}}(\mathbf{x}, t). \tag{52b}$$

The fields $\mathbf{D}^{\text{in}}$ and $\mathbf{B}^{\text{in}}$ constitute the general solution of the homogeneous equations. They are determined by giving the initial conditions in the remote past (before the sources were switched on). The general solution of eqs. (49) can also be expressed in terms of the advanced potentials,

$$\mathbf{D}(\mathbf{x}, t) = \mathbf{D}^{\text{out}}(\mathbf{x}, t) - \frac{\partial}{\partial t}\mathbf{A}^{\text{adv}}(\mathbf{x}, t) - \nabla A_0^{\text{adv}}(\mathbf{x}, t), \tag{53a}$$

$$\mathbf{B}(\mathbf{x}, t) = \mathbf{B}^{\text{out}}(\mathbf{x}, t) + \nabla \times \mathbf{A}^{\text{adv}}(\mathbf{x}, t), \tag{53b}$$

where $\mathbf{D}^{\text{out}}$ and $\mathbf{B}^{\text{out}}$ constitute the solution of the homogeneous equations and are characterized by giving the values of these vectors at an arbitrary instant in the future after the sources are switched off.

The Radiation Field

In the remote past the field $f_{\mu\nu}$ is equal to the field $f_{\mu\nu}^{\text{in}}$ while in the far future it is equal to $f_{\mu\nu}^{\text{out}}$. The difference between the fields $f_{\mu\nu}^{\text{out}}$ and $f_{\mu\nu}^{\text{in}}$ will be referred to as the *radiation field*, which we shall denote by

$$f_{\mu\nu}^{\text{rad}} \equiv f_{\mu\nu}^{\text{out}} - f_{\mu\nu}^{\text{in}}. \tag{54}$$

The radiation field can also be expressed as the difference between the retarded and advanced solutions:

$$f_{\mu\nu}^{\mathrm{rad}} = f_{\mu\nu}^{\mathrm{ret}} - f_{\mu\nu}^{\mathrm{adv}}. \tag{55}$$

Thus, it can be expressed in terms of the invariant solution of the d'Alembert equation $D(x)$,

$$f_{\mu\nu}^{\mathrm{rad}}(x) = \partial_{[\mu}\int \mathrm{d}^4y D(x-y)\mathscr{I}_{\nu]}(y). \tag{56}$$

The radiation field $f_{\mu\nu}^{\mathrm{rad}}$ satisfies the free field equations in all space-time and is a convenient auxiliary construction. For we can subject this field to Fourier analysis in the manner described in this section. Moreover, as the radiation field in the far future is equal to the field $f_{\mu\nu}^{\mathrm{ret}}$, it can be used to construct important physical quantities which characterize the radiating system. In any region of space-time, the radiation field contains the field $f_{\mu\nu}^{\mathrm{adv}}(x)$, absorbed by the source, in addition to the field $f_{\mu\nu}^{\mathrm{ret}}(x)$, which is radiated by the source.

In the presence of sources, the energy, momentum, and angular momentum of the field undergo changes. This is because the field $f_{\mu\nu}$ performs work in the interaction with the source and may also carry off momentum and energy to infinity in the form of radiation. Now we shall calculate the energy and momentum of the radiation from the source. They are given by

$$P_{\mu}^{\mathrm{rad}} = \int_{\sigma\to+\infty} \mathrm{d}\sigma^{\nu}\, T_{\mu\nu}^{\mathrm{ret}}(x) = \int \mathrm{d}^4x\, \partial^{\nu} T_{\mu\nu}^{\mathrm{ret}}(x) = -\int \mathrm{d}^4x f_{\mu\nu}^{\mathrm{ret}}(x)\mathscr{I}^{\nu}(x). \tag{57}$$

In eq. (57) we substitute the field $f_{\mu\nu}^{\mathrm{ret}}$, expressed in terms of the function $D_{\mathrm{R}}(x-y)$,

$$f_{\mu\nu}^{\mathrm{ret}}(x) = \partial_{[\mu}\int \mathrm{d}^4y D_{\mathrm{R}}(x-y)\mathscr{I}_{\nu]}(y),$$

and make use of the continuity equation for the current. In consequence, we obtain the following expression to describe the radiated energy and momentum:

$$\begin{aligned} P_{\mu}^{\mathrm{rad}} &= -\frac{1}{2}\int \mathrm{d}^4x\,\mathrm{d}^4y\,\mathscr{I}_{\nu}(x)D(x-y)\,\partial_{\mu}\mathscr{I}^{\nu}(y) \\ &= -\int \mathrm{d}\Gamma k_{\mu}\tilde{\mathscr{I}}_{\nu}^{*}(k)\tilde{\mathscr{I}}^{\nu}(k), \end{aligned} \tag{58}$$

where $\tilde{\mathscr{I}}_\mu(k)$ is the Fourier transform of the current,

$$\tilde{\mathscr{I}}_\mu(k) \equiv \int d^4x e^{ik\cdot x} \mathscr{I}_\mu(x),$$

$$\tilde{\mathscr{I}}_\mu^*(k) = \tilde{\mathscr{I}}_\mu(-k).$$

In similar fashion we can arrive at an integral expression for the four-dimensional angular momentum radiated by the sources,

$$M_{\mu\nu}^{\mathrm{rad}} = -\frac{1}{2}\int d^4x d^4y \mathscr{I}_\lambda(x) D(x-y) (y_{[\mu}\partial_{\nu]}\delta_\varrho^\lambda + \delta_{[\mu}^\lambda g_{\nu]\varrho}) \mathscr{I}^\varrho(y). \quad (59)$$

If to the expression (57) for the radiated energy and momentum we add an expression for the work done by the Lorentz force and the momentum transferred to the sources as a result of the action of this force (this work and momentum come from the field $f_{\mu\nu}^{\mathrm{in}}$),

$$P_\mu^w = -\int d^4x f_{\mu\nu}^{\mathrm{in}}(x) \mathscr{I}^\nu(x),$$

we obtain the total change of energy and momentum of the field,

$$P_\mu[\sigma \to +\infty] - P_\mu[\sigma \to -\infty] = -\int d^4x f_{\mu\lambda}(x) \mathscr{I}^\lambda(x)$$

$$= -\int d^4x [f_{\mu\lambda}^{\mathrm{in}}(x) + f_{\mu\lambda}^{\mathrm{ret}}(x)] \mathscr{I}^\lambda(x).$$

It has been assumed here that the field $f_{\mu\nu}^{\mathrm{in}}$ does not carry off energy and momentum to infinity in a finite interval of time, i.e. is bounded in space.

The energy, momentum and four-dimensional angular momentum radiated by the sources can also be determined if the radiation field $f_{\mu\nu}^{\mathrm{rad}}$ is known. They are given by the formulae

$$P_\mu^{\mathrm{rad}} = \int_\sigma d\sigma^\nu T_{\mu\nu}^{\mathrm{rad}},$$

$$M_{\mu\nu}^{\mathrm{rad}} = \int_\sigma d\sigma^\lambda M_{\mu\nu\lambda}^{\mathrm{rad}}.$$

Since the field $f_{\mu\nu}^{\mathrm{rad}}$ satisfies the free Maxwell equations, these integrals do not depend on the choice of hypersurface. We can choose this hypersurface in the future, beyond the region in which the sources occur. In the far future, on the other hand, $f_{\mu\nu}^{\mathrm{rad}} = f_{\mu\nu}^{\mathrm{ret}}$.

The fields $f_{\mu\nu}^{\mathrm{in}}$, $f_{\mu\nu}^{\mathrm{out}}$ and $f_{\mu\nu}^{\mathrm{rad}}$ satisfy the free Maxwell's equations.

Thus, Fourier analysis can be performed on these fields in the manner described earlier. We construct the complex tensors $F^{\text{in}}_{\mu\nu}$, $F^{\text{out}}_{\mu\nu}$, and $F^{\text{rad}}_{\mu\nu}$ which we then decompose into plane waves,

$$F^{\substack{\text{rad}\\ \text{in}\\ \text{out}}}_{\mu\nu}(x) = \int \mathrm{d}\Gamma e_{\mu\nu}(\mathbf{k})[f^{\substack{\text{rad}\\ \text{in}\\ \text{out}}}(\mathbf{k}, +1)\mathrm{e}^{-ik\cdot x} + (f^{\substack{\text{rad}\\ \text{in}\\ \text{out}}}(\mathbf{k}, -1))^* \mathrm{e}^{ik\cdot x}]. \tag{60}$$

By this formula we have defined the coefficients $f^{\text{rad}}(\mathbf{k}, \lambda)$, $f^{\text{in}}(\mathbf{k}, \lambda)$ and $f^{\text{out}}(\mathbf{k}, \lambda)$. Knowing the dependence of the field $f^{\text{rad}}_{\mu\nu}$ on the external current (cf. eq. (56)), we shall calculate the coefficients $f^{\text{rad}}(\mathbf{k}, \lambda)$:

$$e_{\mu\nu}(\mathbf{k}) f^{\text{rad}}(\mathbf{k}, +1) = \frac{1}{2} k_{[\mu}\tilde{\mathcal{J}}_{\nu]}(\mathbf{k}, \omega) + \frac{1}{2} i\varepsilon_{\mu\nu}{}^{\lambda\varrho} k_\lambda \tilde{\mathcal{J}}_\varrho(\mathbf{k}, \omega), \tag{61a}$$

$$e_{\mu\nu}(\mathbf{k}) (f^{\text{rad}}(\mathbf{k}, -1))^* = \frac{1}{2} k_{[\mu}\tilde{\mathcal{J}}^*_{\nu]}(\mathbf{k}, \omega) + \frac{1}{2} i\varepsilon_{\mu\nu}{}^{\lambda\varrho} k_\lambda \tilde{\mathcal{J}}^*_\varrho(\mathbf{k}, \omega). \tag{61b}$$

In non-relativistic notation these relations are of the form

$$i\mathbf{e}(\mathbf{k}) f^{\text{rad}}(\mathbf{k}, +1) = \frac{1}{2}[\mathbf{n}\times(\mathbf{n}\times\tilde{\mathcal{J}}(\mathbf{k}, \omega)) - i\mathbf{n}\times\tilde{\mathcal{J}}(\mathbf{k}, \omega)], \tag{61c}$$

$$-i\mathbf{e}^*(\mathbf{k}) f^{\text{rad}}(\mathbf{k}, -1) = \frac{1}{2}[\mathbf{n}\times(\mathbf{n}\times\tilde{\mathcal{J}}(\mathbf{k}, \omega)) + i\mathbf{n}\times\tilde{\mathcal{J}}(\mathbf{k}, \omega)]. \tag{61d}$$

The operations $\frac{1}{2}[\mathbf{n}\times(\mathbf{n}\times \mp i\mathbf{n}\times)]$ on the right-hand side of these equations produce projections of the vector onto the "directions" of two circular polarizations.

It follows from eqs. (61) that a system of currents and charges radiates if and only if the Fourier transform of the transverse part of the current does not vanish on the light cone [cf. also eq. (58)].

To solve eqs. (61) for the coefficients $f^{\text{rad}}(\mathbf{k}, \lambda)$, we introduce classes of vectors $\varepsilon_\mu(\mathbf{k})$ which satisfy the relationships

$$e_{\mu\nu}(\mathbf{k}) = -ik_{[\mu}\varepsilon_{\nu]}(\mathbf{k}). \tag{62}$$

The vectors $\varepsilon_\mu(\mathbf{k})$ for a given $\mathbf{k}$ are determined by these relations to within the transformation

$$\varepsilon_\mu(\mathbf{k}) \to \varepsilon_\mu(\mathbf{k}) + i\alpha(\mathbf{k}) k_\mu, \tag{63}$$

which we shall call the *gauge transformation of the vectors* ε^μ. The properties (20) of the tensor $e_{\mu\nu}(\mathbf{k})$ can be used to show that in any coordinate system one of the vectors $\varepsilon^\mu(\mathbf{k})$ is of the form

$$\varepsilon^\mu(\mathbf{k}) = \big(0, \mathbf{e}(\mathbf{k})\big).$$

The vector ε^μ in this form will be referred to as the vector ε^μ in the *radiation gauge.*

Each of the vectors $\varepsilon_\mu(\mathbf{k})$ satisfies the relations

$$k^\mu \varepsilon_\mu(\mathbf{k}) = 0, \tag{64a}$$

$$\varepsilon_\mu^*(\mathbf{k})\varepsilon^\mu(\mathbf{k}) = -1, \tag{64b}$$

$$\varepsilon_\mu(\mathbf{k})\varepsilon^\mu(\mathbf{k}) = 0 = \varepsilon_\mu^*(\mathbf{k})\varepsilon^{\mu *}(\mathbf{k}). \tag{64c}$$

These relations are satisfied by the vector $\varepsilon_\mu(\mathbf{k})$ in the radiation gauge and are invariant under gauge transformation. Vectors related by a gauge transformation will be said to be *equivalent vectors.* The identity (37) for the tensor $e_{\mu\nu}(\mathbf{k})$ merely implies for the four-vector $\varepsilon_\mu(\mathbf{k})$ that

$$\Lambda_\mu{}^\nu \varepsilon_\nu(\Lambda^{-1}\mathbf{k}) = \mathrm{e}^{i\theta(\mathbf{k},\Lambda)}\varepsilon_\mu(\mathbf{k}) + \beta(\Lambda, \mathbf{k})k_\mu, \tag{65}$$

where $\beta(\Lambda, \mathbf{k})$ is an unknown coefficient.

Now let us go on to solve eqs. (61). For this purpose we multiply through by $e^{\lambda\mu *}(\mathbf{k})$ and sum over μ. Making use of the normalization condition (21) for the tensor $e_{\mu\nu}$, the continuity equation for the current, and the relation (62) which defines the class of vectors $\varepsilon_\mu(\mathbf{k})$, we obtain the solution

$$f^{\mathrm{rad}}(\mathbf{k}, +1) = -i\varepsilon^{\mu *}(\mathbf{k})\tilde{\mathscr{I}}_\mu(\mathbf{k}, \omega), \tag{66a}$$

$$f^{\mathrm{rad}}(\mathbf{k}, -1) = i\varepsilon^{\mu}(\mathbf{k})\tilde{\mathscr{I}}_\mu(\mathbf{k}, \omega). \tag{66b}$$

Because of the continuity equation for the current, the expressions on the right-hand side do not depend on the choice of the vector ε^μ out of the class of equivalent vectors. The projections of the Fourier transform of the current onto the vectors $\varepsilon^{\mu *}(\mathbf{k})$ and $\varepsilon^\mu(\mathbf{k})$ will henceforth be denoted by the symbols $\iota(\mathbf{k}, \lambda)$,

$$\iota(\mathbf{k}, +1) \equiv -\varepsilon^{\mu *}(\mathbf{k})\tilde{\mathscr{I}}_\mu(\mathbf{k}, \omega), \tag{67a}$$

$$\iota(\mathbf{k}, -1) \equiv \varepsilon^{\mu}(\mathbf{k})\tilde{\mathscr{I}}_\mu(\mathbf{k}, \omega). \tag{67b}$$

The field $f_{\mu\nu}^{\mathrm{rad}}$ is the difference between the fields $f_{\mu\nu}^{\mathrm{out}}$ and $f_{\mu\nu}^{\mathrm{in}}$. This relation is expressible as a relation between the functions $f^{\mathrm{out}}(\mathbf{k}, \lambda)$ and $f^{\mathrm{in}}(\mathbf{k}, \lambda)$,

$$f^{\mathrm{out}}(\mathbf{k}, \lambda) = f^{\mathrm{in}}(\mathbf{k}, \lambda) + i\iota(\mathbf{k}, \lambda). \tag{68}$$

In what follows we shall discuss the radiation field in detail and give its decomposition into the field radiated and the field absorbed by the sources.

Let us now calculate the expression $\sum_{\lambda} |\iota(\mathbf{k}, \lambda)|^2$, which will be used in subsequent sections. Relations (22) and (62) imply the following identity for the vectors $\varepsilon_\mu(\mathbf{k})$,

$$k_{[\mu}\varepsilon_{\nu]}k_{[\lambda}\varepsilon^*_{\varrho]} + k_{[\mu}\varepsilon^*_{\nu]}k_{[\lambda}\varepsilon_{\varrho]} = k_{[\mu}g_{\nu][\lambda}k_{\varrho]}. \tag{69}$$

Multiplying this identity through by $\tilde{\mathcal{J}}^{\nu *}$ and $\tilde{\mathcal{J}}^{\varrho}$ and taking the sum over ν and over ϱ, we obtain the desired results

$$\sum_{\lambda} \iota^*(\mathbf{k}, \lambda)\iota(\mathbf{k}, \lambda) = -\tilde{\mathcal{J}}^*_\mu(\mathbf{k}, \omega)\tilde{\mathcal{J}}^\mu(\mathbf{k}, \omega). \tag{70}$$

Multipole Radiation

The dimensions of radiating systems are frequently much smaller than the wavelength of the radiation emitted. The product kl of the length of the wave vector $\mathbf{k}$ and the characteristic length l specifying the linear dimensions of the region in which charges and currents occur is then a small dimensionless parameter. There is reason to believe that it is sufficient to confine oneself to the first few terms of the expansion in this parameter. Such an expansion is known as a *multipole expansion.* It plays an important part in investigations of the radiation emitted by real physical systems, both in classical theory and in quantum theory. If the higher terms are discarded, the end results are simplified significantly and a detailed study of the properties of the field is made possible. In addition, the classification of various types of radiation obtained by the multipole expansions is useful even when the condition $kl \ll 1$ is not met. It does also happen that radiation from some physical systems contains only some of the terms in the multipole expansion.

The multipole expansion will be discussed on the example of the classical theory of radiation originating from a given system of charges and currents. No difficulty is encountered in extending these considerations to the quantum theory of radiation.

To obtain the multipole expansion of the current and the field, we make use of a system of vector spherical functions $\mathbf{Y}_{JM}(\hat{\mathbf{r}})$. These functions are also important in quantum theory. For these are the simultaneous eigenfunctions of the operators of the square of the total angular momentum and of the projection of the total angular momentum on the z-axis for spin 1 particles:

$$(\mathbf{L}+\mathbf{s})^2\mathbf{Y}_{JM}(\hat{\mathbf{r}}) = J(J+1)\mathbf{Y}_{JM}(\hat{\mathbf{r}}), \tag{71a}$$

$$(L_z+s_z)\mathbf{Y}_{JM}(\hat{\mathbf{r}}) = M\mathbf{Y}_{JM}(\hat{\mathbf{r}}), \tag{71b}$$

where $\hat{\mathbf{r}}$ denotes the unit vector in the $\mathbf{r}$-direction,

$$\mathbf{L} = -i(\mathbf{r}\times\nabla),$$

whereas the action of the spin projection operators s_x, s_y, and s_z on the cartesian components of the vectors $\mathbf{Y}_{JM}(\hat{\mathbf{r}})$ is defined as multiplication by the matrices

$$s_x = \begin{pmatrix} 0 & 0 & 0 \\ 0 & 0 & -i \\ 0 & i & 0 \end{pmatrix}, \quad s_y = \begin{pmatrix} 0 & 0 & i \\ 0 & 0 & 0 \\ -i & 0 & 0 \end{pmatrix}, \quad s_z = \begin{pmatrix} 0 & -i & 0 \\ i & 0 & 0 \\ 0 & 0 & 0 \end{pmatrix}.$$

The number J takes on all positive integer values $0, 1, \ldots$, whereas the number M with fixed J runs through the values: $-J, -J+1, \ldots, J-1, J$. From the quantum theoretical laws for the composition of two angular momenta it follows that threefold degeneracy† occurs in the eigenvalue equations (71). Thus, three linearly independent functions $\mathbf{Y}_{JM}(\hat{\mathbf{r}})$ are associated with each pair of numbers J and M. To describe the electromagnetic field we use the vector functions $\mathbf{Y}^{(e)}_{JM}$, $\mathbf{Y}^{(m)}_{JM}$, and $\mathbf{Y}^{(o)}_{JM}$. These functions are constructed out of ordinary spherical functions $Y_{JM}(\hat{\mathbf{r}})$, when $J \neq 0$, as follows:

$$\mathbf{Y}^{(e)}_{JM}(\hat{\mathbf{r}}) \equiv \frac{i}{\sqrt{J(J+1)}}\, r\nabla Y_{JM}(\hat{\mathbf{r}}), \tag{72a}$$

$$\mathbf{Y}^{(m)}_{JM}(\hat{\mathbf{r}}) \equiv \frac{-i}{\sqrt{J(J+1)}}\, (\mathbf{r}\times\nabla) Y_{JM}(\hat{\mathbf{r}}), \tag{72b}$$

$$\mathbf{Y}^{(o)}_{JM}(\hat{\mathbf{r}}) \equiv -i\hat{\mathbf{r}} Y_{JM}(\hat{\mathbf{r}}). \tag{72c}$$

When $J = 0$, only the function $\mathbf{Y}^{(o)}$ is defined. The following relationships emerge from the definitions (72) and the properties of the spherical functions $Y_{JM}(\hat{\mathbf{r}})$:

$$\hat{\mathbf{r}}\cdot\mathbf{Y}^{(e)}_{JM}(\hat{\mathbf{r}}) = 0 = \hat{\mathbf{r}}\cdot\mathbf{Y}^{(m)}_{JM}(\hat{\mathbf{r}}), \tag{73a}$$

$$\hat{\mathbf{r}}\times\mathbf{Y}^{(o)}_{JM}(\hat{\mathbf{r}}) = 0, \tag{73b}$$

† With the exception of the case $J = 0$ when degeneracy does not occur.

$$\hat{\mathbf{r}}\times\mathbf{Y}^{(e)}_{JM}(\hat{\mathbf{r}}) = -\mathbf{Y}^{(m)}_{JM}(\hat{\mathbf{r}}), \tag{73c}$$

$$\hat{\mathbf{r}}\times\mathbf{Y}^{(m)}_{JM}(\hat{\mathbf{r}}) = \mathbf{Y}^{(e)}_{JM}(\hat{\mathbf{r}}), \tag{73d}$$

$$\mathbf{Y}^{(m)}_{JM}(\hat{\mathbf{r}}) = -r\nabla\times\mathbf{Y}^{(e)}_{JM}(\hat{\mathbf{r}}), \tag{73e}$$

$$\mathbf{Y}^{(0)}_{JM}(\hat{\mathbf{r}}) = \frac{1}{\sqrt{J(J+1)}}\,r^2\nabla\times(\nabla\times\mathbf{Y}^{(e)}_{JM}(\hat{\mathbf{r}})). \tag{73f}$$

The vector spherical functions $\mathbf{Y}^{(e)}$, $\mathbf{Y}^{(m)}$, and $\mathbf{Y}^{(0)}$ constitute a complete orthonormal system of functions on the surface of a unit sphere,

$$\int d\Omega\mathbf{Y}^{(\lambda)*}_{JM}(\hat{\mathbf{r}})\cdot\mathbf{Y}^{(\lambda')}_{J'M'}(\hat{\mathbf{r}}) = \delta_{JJ'}\delta_{MM'}\delta_{\lambda\lambda'}, \tag{74a}$$

$$\sum_{JM\lambda}\mathbf{Y}^{(\lambda)}_{JM}(\hat{\mathbf{r}})\mathbf{Y}^{(\lambda)*}_{JM}(\hat{\mathbf{r}}') = \delta_\Omega(\hat{\mathbf{r}},\hat{\mathbf{r}}'), \tag{74b}$$

where λ takes on the values e, m, and 0, and the Dirac function δ_Ω at the surface of the sphere is normalized in the following manner:

$$\int d\Omega\delta_\Omega(\hat{\mathbf{r}},\hat{\mathbf{r}}') = 1.$$

The left-hand side of formula (74b) is a 3×3 matrix whose elements are labelled by the components of both vectors $\mathbf{Y}$. The right-hand side is multiplied by a suitable unit matrix.

Now let us consider the radiation field vector $\mathbf{F}^{\text{rad}}(\mathbf{x}, t)$. By formulae (61) and (7a) it can be expressed in terms of the Fourier transform of the current four-vector:

$$\mathbf{F}^{\text{rad}}(\mathbf{x}, t) = \frac{1}{2}\int d\Gamma\omega\{[\mathbf{n}\times(\mathbf{n}\times\tilde{\mathscr{I}}(\omega\mathbf{n},\omega))-i\mathbf{n}\times\tilde{\mathscr{I}}(\omega\mathbf{n},\omega)]e^{-ik\cdot x}$$
$$+[\mathbf{n}\times(\mathbf{n}\times\tilde{\mathscr{I}}(-\omega\mathbf{n},\ -\omega))-i\mathbf{n}\times\tilde{\mathscr{I}}(-\omega\mathbf{n},\ -\omega)]e^{ik\cdot x}. \tag{75}$$

To obtain the multipole expansion of the field, we shall first expand the Fourier transform of the current into spherical functions,

$$\tilde{\mathscr{I}}(\omega\mathbf{n},\omega) = \frac{4\pi}{\omega^2}\sum_{JM\lambda}(-i)^J\iota_{(\lambda)}(JM\omega)\mathbf{Y}^{(\lambda)}_{JM}(\mathbf{n}). \tag{76}$$

The coefficients $\iota_{(\lambda)}(JM\omega)$ in this expansion are given by the formulae

$$\iota_{(\lambda)}(JM\omega) = \frac{i^J\omega^2}{4\pi}\int d\Omega\mathbf{Y}^{(\lambda)*}_{JM}(\mathbf{n})\cdot\tilde{\mathscr{I}}(\omega\mathbf{n},\omega). \tag{77}$$

When the current expansion (76) is substituted in formula (75) for the radiation field vector, we obtain the multipole expansion of the field.

It turns out that this expansion does not contain the coefficients $\iota_{(0)}(JM\omega)$. Integration with respect to the angular variables in the formula for $\mathbf{F}^{\mathrm{rad}}(\mathbf{x}, t)$ is effected with the aid of the formulae

$$\int d\Omega\, e^{i\mathbf{k}\cdot\mathbf{r}}\mathbf{Y}_{JM}^{(m)}(\mathbf{n}) = 4\pi i^J j_J(kr)\mathbf{Y}_{JM}^{(m)}(\hat{\mathbf{r}}), \tag{78a}$$

$$\int d\Omega\, e^{i\mathbf{k}\cdot\mathbf{r}}\mathbf{Y}_{JM}^{(e)}(\mathbf{n}) = -4\pi i^{J+1}k^{-1}\nabla\times\left(j_J(kr)\mathbf{Y}_{JM}^{(m)}(\hat{\mathbf{r}})\right), \tag{78b}$$

where $j_l(z)$ denotes the spherical Bessel functions

$$j_l(z) = \sqrt{\frac{\pi}{2z}}\, J_{l+\frac{1}{2}}(z).$$

These formulae can be derived by using the definition of the functions $\mathbf{Y}_{JM}^{(m)}$, the relation (73a), and the decomposition of the function exp $(i\mathbf{k}\cdot\mathbf{r})$ into spherical functions,

$$e^{i\mathbf{k}\cdot\mathbf{r}} = 4\pi\sum_{lm}\int_0^\infty i^l j_l(kr)\, Y_{lm}(\hat{\mathbf{r}})\, Y_{lm}^*(\mathbf{n}).$$

In this way we arrive at

$$\mathbf{F}^{\mathrm{rad}}(\mathbf{r}, t) = \sum_{JM}\int_0^\infty \frac{d\omega}{2\pi}$$

$$\times\,[e^{-i\omega t}\left(-\iota_{(m)}(JM\omega)+i\iota_{(e)}(JM\omega)\right)(1+\omega^{-1}\nabla\times)\, j_J(\omega r)\mathbf{Y}_{JM}^{(m)}(\hat{\mathbf{r}})$$

$$+e^{i\omega t}\left(-\iota_{(m)}^*(JM\omega)+i\iota_{(e)}^*(JM\omega)\right)(1-\omega^{-1}\nabla\times)\, j_J(\omega r)\mathbf{Y}_{JM}^{(m)*}(\hat{\mathbf{r}})]. \tag{79}$$

This formula determines the *decomposition of the radiation field into spherical functions*, just as formula (75) does into plane waves. From this we obtain the following multipole expansion of the vectors **E** and **B**:

$$\mathbf{E}^{\mathrm{rad}}(\mathbf{r}, t) = \sum_{JM}\int_0^\infty \frac{d\omega}{2\pi}\, e^{-i\omega t}[\iota_{(e)}(JM\omega)\mathbf{E}_{(e)}(JM\omega;\mathbf{r})$$

$$+\iota_{(m)}(JM\omega)\mathbf{E}_{(m)}(JM\omega;\mathbf{r})]+\text{c.c.}, \tag{80a}$$

$$\mathbf{B}^{\mathrm{rad}}(\mathbf{r}, t) = \sum_{JM}\int_0^\infty \frac{d\omega}{2\pi}\, e^{-i\omega t}[\iota_{(e)}(JM\omega)\mathbf{B}_{(e)}(JM\omega;\mathbf{r})$$

$$+\iota_{(m)}(JM\omega)\mathbf{B}_{(m)}(JM\omega;\mathbf{r})]+\text{c.c.}, \tag{80b}$$

where

$$\mathbf{E}_{(e)}(JM\omega;\mathbf{r}) \equiv 2\omega^{-1}i\nabla\times\big(j_J(\omega r)\mathbf{Y}^{(m)}_{JM}(\hat{\mathbf{r}})\big), \tag{81a}$$

$$\mathbf{B}_{(e)}(JM\omega;\mathbf{r}) \equiv 2j_J(\omega r)\mathbf{Y}^{(m)}_{JM}(\hat{\mathbf{r}}), \tag{81b}$$

$$\mathbf{E}_{(m)}(JM\omega;\mathbf{r}) \equiv -2j_J(\omega r)\mathbf{Y}^{(m)}_{JM}(\hat{\mathbf{r}}), \tag{81c}$$

$$\mathbf{B}_{(m)}(JM\omega;\mathbf{r}) \equiv 2\omega^{-1}i\nabla\times\big(j_J(\omega r)\mathbf{Y}^{(m)}_{JM}(\hat{\mathbf{r}})\big). \tag{81d}$$

The fields $\mathbf{E}^{\text{rad}}$ and $\mathbf{B}^{\text{rad}}$ satisfy the system of free Maxwell equations for every choice of coefficients $\iota_{(e)}(JM\omega)$ and $\iota_{(m)}(JM\omega)$. Two pairs of complex vectors $\big(e^{-i\omega t}\mathbf{E}_{(e,m)}(JM\omega;\mathbf{r}),\ e^{-i\omega t}\mathbf{B}_{(e,m)}(JM\omega;\mathbf{r})\big)$ for an arbitrary system of numbers J, M, ω thus are also solutions of the Maxwell equations without sources.† Each of these complex solutions determines two real solutions.

The coefficients $\iota_{(e)}$ and $\iota_{(m)}$ may, by virtue of formulae (78), be written as integrals over all space:

$$\iota_{(e)}(JM\omega) = i\omega\int d^3x\big(\nabla\times j_J(\omega|\mathbf{x}|)\mathbf{Y}^{(m)*}_{JM}(\hat{\mathbf{x}})\big)\cdot\mathscr{J}_\omega(\mathbf{x}),$$

$$\iota_{(m)}(JM\omega) = \omega^2\int d^3x j_J(\omega|\mathbf{x}|)\mathbf{Y}^{(m)*}_{JM}(\hat{\mathbf{x}})\cdot\mathscr{J}_\omega(\mathbf{x}),$$

where

$$\mathscr{J}_\omega(\mathbf{x}) = \int dt e^{i\omega t}\mathscr{J}(\mathbf{x},t).$$

If the linear dimensions l of the radiating system are much smaller than the wavelength, the argument ωr of the Bessel functions in the formulae above is much smaller than unity in the region in which the function $\mathscr{J}_\omega(\mathbf{x})$ is non-zero. Accordingly, use may be made of the approximate formula

$$j_k(z) \approx \frac{z^k}{(2k+1)!!}.$$

The coefficients $\iota_{(e,m)}$ may then be expressed in terms of the electric and magnetic multipole moments, Q_{JM} and M_{JM}, of the radiating system:

$$Q_{JM}(\omega) \equiv \int d^3x|\mathbf{x}|^J Y^*_{JM}(\hat{\mathbf{x}})\varrho_\omega(\mathbf{x}), \tag{82a}$$

$$M_{JM}(\omega) \equiv \frac{-1}{J+1}\int d^3x|\mathbf{x}|^J Y^*_{JM}(\hat{\mathbf{x}})\big(\nabla\cdot(\mathbf{x}\times\mathscr{J}_\omega(\mathbf{x}))\big), \tag{82b}$$

where $\varrho_\omega(x)$ denotes the density of charge distribution,

$$\varrho_\omega(x) = \frac{1}{i\omega}\nabla\cdot\mathscr{J}_\omega(\mathbf{x}).$$

† This can also be verified by direct calculation.

Inasmuch as the function $|\mathbf{x}|^J Y_{JM}(\hat{\mathbf{x}})$ is a homogeneous polynomial of degree J in the coordinates of the vector $\mathbf{x}$, therefore Q_{JM} represents the J-th moment† of the charge distribution (dipole, when $J = 1$; quadrupole when $J = 2$; etc.), and M_{JM} represents the $(J-1)$-st moment of the magnetic moment $\mathbf{x}\times\mathscr{I}_\omega(\mathbf{x})$.

Using the relationship

$$\nabla\times|\mathbf{x}|^J \mathbf{Y}_{JM}^{(m)}(\hat{\mathbf{x}}) = i\sqrt{\frac{J+1}{J}}\,\nabla(|\mathbf{x}|^J Y_{JM}(\hat{\mathbf{x}})),$$

we obtain

$$\iota_{(e)}(JM\omega) = i\,\frac{\omega^{J+2}}{(2J+1)!!}\sqrt{\frac{J+1}{J}}\,Q_{JM}(\omega),$$

$$\iota_{(m)}(JM\omega) = -i\,\frac{\omega^{J+2}}{(2J+1)!!}\sqrt{\frac{J+1}{J}}\,M_{JM}(\omega).$$

Multipole moments of order J are approximately proportional to the J-th power of the length l. The coefficients $\iota_{(e,m)}$ of the multipole expansion of the current,

$$\iota_{(e,m)}(JM\omega) \sim \frac{(\omega l)^J}{(2J+1)!!},$$

decrease fast as J increases. For small radiating systems, the multipole expansion becomes a power series in the small parameter kl. One may confine oneself to the first few terms of this expansion.

To obtain the multipole expansion of a field radiated by a system of charges and currents (in the region of space-time where these sources do not occur) we must break up the radiation field into two parts: the field emitted by the sources and the field absorbed by the sources. To this end we shall write the spherical Bessel functions in formulae (81) as sums of spherical Hankel functions,

$$j_l(z) = \frac{1}{2}\left(h_l^{(1)}(z)+h_l^{(2)}(z)\right),$$

and we shall split the radiation field described by formulae (80) into two parts containing, respectively, the Hankel functions $h^{(1)}$ and $h^{(2)}$:

$$\mathbf{E}^{\mathrm{rad}}(\mathbf{r}, t) = \mathbf{E}^{(1)}(\mathbf{r}, t)+\mathbf{E}^{(2)}(\mathbf{r}, t),$$
$$\mathbf{B}^{\mathrm{rad}}(\mathbf{r}, t) = \mathbf{B}^{(1)}(\mathbf{r}, t)+\mathbf{B}^{(2)}(\mathbf{r}, t).$$

---

† The integrals

$$\int d^3x\, x^K y^M z^N f(\mathbf{x}), \quad \text{when} \quad K+M+N = J,$$

determine the J-th moment of the function $f(\mathbf{x})$.

To find the physical interpretation of this splitting, we shall assume that the origin of the coordinate system lies inside the region occupied by the sources of radiation and we shall study the behaviour of the functions $\mathbf{E}^{(1,2)}$ and $\mathbf{B}^{(1,2)}$ in the wave zone ($kr \gg 1$). We can then use the asymptotic expansions of the Hankel functions,

$$h_J^{(1,2)}(kr) \underset{r\to\infty}{\sim} (kr)^{-1} e^{\pm i\left(kr - \frac{1}{2}(J+1)\pi\right)},$$

and obtain formulae for the field intensity as a superposition of spherical waves propagating from the radiating system and converging at the radiating system:

$$\mathbf{E}^{(1,2)}(\mathbf{r}, t) \underset{r\to\infty}{\sim} \sum_{JM} (-i)^{J+1} \int_0^\infty \frac{d\omega}{2\pi}$$

$$\times \left[\iota_{(e)}(JM\omega)\, i\omega^{-1}\nabla\times - \iota_{(m)}(JM\omega)\right] \frac{e^{-i\omega(t\mp r)}}{kr}\, \mathbf{Y}_{JM}^{(m)}(\pm\hat{\mathbf{r}}) + \text{c.c.},$$

$$\mathbf{B}^{(1,2)}(\mathbf{r}, t) \underset{r\to\infty}{\sim} \sum_{JM} (-i)^{J+1} \int_0^\infty \frac{d\omega}{2\pi}$$

$$\times \left[\iota_{(e)}(JM\omega) + \iota_{(m)}(JM\omega)\, i\omega^{-1}\nabla\times\right] \frac{e^{-i\omega(t\mp r)}}{kr}\, \mathbf{Y}_{JM}^{(m)}(\pm\hat{\mathbf{r}}) + \text{c.c.}$$

The fields ($\mathbf{E}^{(1)}$, $\mathbf{B}^{(1)}$) and ($\mathbf{E}^{(2)}$, $\mathbf{B}^{(2)}$) are interpreted, respectively, as the *field radiated* and the *field absorbed by the sources*. The total radiation field contains radiated and absorbed waves in equal proportions.

Both the field intensities ($\mathbf{E}^{(1)}$, $\mathbf{B}^{(1)}$) and ($\mathbf{E}^{(2)}$, $\mathbf{B}^{(2)}$) correctly describe the radiated field and the absorbed field only beyond the region in which the sources are contained. The real radiated and absorbed fields must satisfy the inhomogeneous Maxwell equations with current $\mathscr{I}_\mu(x)$ as the source of field, whereas the fields ($\mathbf{E}^{(1,2)}$, $\mathbf{B}^{(1,2)}$) satisfy the free equations everywhere† apart from the origin of the coordinate system. In this way real sources of field have been replaced by equivalent fictitious sources at the point $\mathbf{r} = 0$. These sources can be described by the function $\delta(\mathbf{r})$ and its derivatives. A similar procedure is frequently

† This stems from the fact that Hankel functions satisfy the same kind of differential equations as do Bessel functions.

employed in electrostatics, by replacing the continuous distribution of charge by an equivalent system of multipoles.

The intensities $(\mathbf{E}^{(1)}_{(e)}(JM\omega),\ \mathbf{B}^{(1)}_{(e)}(JM\omega))$ describe the field of an electric 2^J-pole,† for this is a field radiated by an oscillating electric moment of order J. In like manner, the intensities $(\mathbf{E}^{(1)}_{(m)}(JM\omega),\ \mathbf{B}^{(1)}_{(m)}(JM\omega))$ describe the field of a magnetic 2^J-pole.

The multipole expansion plays a very important part in the quantum theory of radiation. In transitions between the quantum states of a radiating system with particular values of angular momentum only certain multipoles are radiated (owing to selection rules).

The method we have described for obtaining the multipole expansion may be applied not only to the radiation field but also to every free electromagnetic field. For we can represent each free field in the form (75), defining the function $\mathscr{I}(\mathbf{k}, \omega)$ with the formula

$$\mathscr{I}(\mathbf{k}, \omega) = -i\mathbf{e}(\mathbf{k})f(\mathbf{k}, +1) + i\mathbf{e}^*(\mathbf{k})f^*(\mathbf{k}, -1) + \mathbf{n}\alpha(\mathbf{k}),$$

where $\alpha(\mathbf{k})$ is an arbitrary function of the vector $\mathbf{k}$.

Just as in determining the energy and momentum radiated by a system it is convenient to use the current Fourier transforms, so in determining the energy and projection of the angular momentum it is convenient to use the multipole expansion of the current. When the expansion (76) of the current Fourier transform is substituted in formulae (58) and (59), we obtain

$$E = \sum_{JM} \int_0^\infty \frac{d\omega}{\pi\omega^2} \left(|\iota_{(e)}(JM\omega)|^2 + |\iota_{(m)}(JM\omega)|^2\right),$$

$$M_{12} = \sum_{JM} M \int_0^\infty \frac{d\omega}{\pi\omega^3} \left(|\iota_{(e)}(JM\omega)|^2 + |\iota_{(m)}(JM\omega)|^2\right).$$

Radiation of a Point Particle

Now we shall consider a simple model of a radiating system, consisting of a charged point particle moving with acceleration. The current Fourier transform for this system is of the form (cf. formula (8.3a))

† Of a dipole ($J = 1$), quadrupole ($J = 2$), octupole ($J = 3$), etc.

$$\tilde{\mathscr{I}}^{\mu}(k) = \int d^4x e^{ik\cdot x} e \int_{-\infty}^{+\infty} d\tau \frac{d\xi^{\mu}(\tau)}{d\tau} \delta^{(4)}(x-\xi(\tau))$$

$$= e \int_{-\infty}^{+\infty} d\tau \frac{d\xi^{\mu}(\tau)}{d\tau} e^{ik\cdot\xi(\tau)}. \tag{83}$$

The current is non-zero along the world line of the particle, hence in an unbounded part of space-time. The considerations of this section have been applied to cases when the current vanished in the remote past and in the far future. They may, however, also be applied to the model in question if it is assumed that in the remote past and in the far future the particle moves uniformly, thus radiating only along a finite segment of the world line. The current modified by the introduction of an exponential damping factor of the form $\exp(-\varepsilon|\tau|)$, where ε is a small, positive number,

$$\tilde{\mathscr{I}}^{\mu}_{\varepsilon}(k) = e \int_{-\infty}^{+\infty} d\tau e^{-\varepsilon|\tau|} \frac{d\xi^{\mu}(\tau)}{d\tau} e^{ik\cdot\xi(\tau)}, \tag{84}$$

is, for sufficiently small ε's, a source of the same kind of radiation as is the current $\tilde{\mathscr{I}}^{\mu}(k)$. The factor $\exp(-\varepsilon|\tau|)$ gives rise to an arbitrarily small† change in the current on that segment of the world line where the particle radiates, but switches off the current in the remote past and in the far future. By q^{μ}/m and p^{μ}/m let us denote the initial and final four-velocity of the particle and suppose that acceleration occurs in the interval $(-\tau_1, \tau_1)$ of the proper time. When partial integration and passage to the limit with $\varepsilon(\varepsilon \to 0)$ are performed formula (84) leads to the following expression for $\tilde{\mathscr{I}}^{\mu}(k)$:

$$\tilde{\mathscr{I}}^{\mu}(k) = ie\left[p^{\mu}\frac{e^{-i\tau_1 k\cdot p/m}}{k\cdot p} - q^{\mu}\frac{e^{i\tau_1 k\cdot q/m}}{k\cdot q}\right]$$

$$+ e\int_{-\tau_1}^{\tau_1} d\tau \frac{d\xi^{\mu}(\tau)}{d\tau} e^{ik\cdot\xi(\tau)}. \tag{85}$$

---

† This is so only in classical theory. In quantum theory, radiation emitted by a particle travelling with arbitrarily small acceleration does indeed carry a negligible quantity of energy but is composed of a vast number of photons. This fact significantly complicates the quantum theory of radiation.

It is seen that $\mathcal{J}^\mu(k)$ is non-zero for four-vectors k of the form $(|\mathbf{k}|, \mathbf{k})$. This means that each acceleration of particles is accompanied by the emission of radiation.

For small† values of k^μ we obtain an approximate formula for $\tilde{\mathcal{J}}^\mu(k)$,

$$\tilde{\mathcal{J}}^\mu(k) = ie\left(\frac{p^\mu}{k\cdot p} - \frac{q^\mu}{k\cdot q}\right) + \begin{array}{l}\text{terms which are}\\ \text{finite when } k\to 0.\end{array} \tag{86}$$

Radiation associated with the values of the current transform $\tilde{\mathcal{J}}^\mu(\mathbf{k}, \omega)$ for small k^μ's is called the *long-wavelength part of the radiation.* It is determined, as shown by formula (86), by the asymptotic values of the momenta of the particle in the past and in the future. In every experiment in which the phenomenon of radiation is studied, there occurs a certain maximum wavelength $\lambda_{\max}$ $\left(\text{to which corresponds a certain minimum frequency } \omega_{\min} = \frac{2\pi}{\lambda_{\max}}\right)$ such that waves with a wavelength greater than $\lambda_{\max}$ escape our observations. Now we shall calculate the energy carried off by this radiation in the case when that radiation is due to the scattering of a point particle. If $\omega_{\min}$ is assumed to be so small that the approximate formula (86) may be used when $|\mathbf{k}| < \omega_{\min}$, we obtain the following expression for the energy sought, which we shall denote by the symbol‡

$$E_{\mathrm{ir}} = -e^2 \int\limits_{\omega<\omega_{\min}} \mathrm{d}\Gamma \left(\frac{p^\mu}{k\cdot p} - \frac{q^\mu}{k\cdot q}\right)^2 \omega(\mathbf{k}). \tag{87}$$

This integral depends on the reference frame, since the condition $\omega > \omega_{\min}$ is not invariant. We shall calculate it in a system of coordinates in which $\mathbf{p}+\mathbf{q} = 0$, obtaining

$$E_{\mathrm{ir}} = \frac{e^2}{4\pi\lambda_{\max}} f\left(\sqrt{1-\frac{4m^2}{t}}\right), \tag{88}$$

† The term small values of k^μ means that $\max|k\cdot\xi(\tau)| \ll 1$, when $|\tau| < \tau_1$.

‡ The index ir stands for infrared.

where the variable t is the momentum transfer, $t = (p-q)^2$, and the function f is of the form

$$f(x) = 2\left[(x+x^{-1})\ln\frac{x+1}{x-1}-2\right].$$

For values of the momentum transfer t which are large in comparison with the squared mass m^2, the function f becomes

$$f\left(\sqrt{1-\frac{4m^2}{t}}\right) = 4\ln\left(-\frac{t}{m^2}\right)+\begin{matrix}\text{terms which are}\\ \text{finite when } t\to\infty.\end{matrix}$$

Hence, this is a function which increases slowly with t. The function f is a factor of proportionality between the energy carried off by radiation with a wavelength greater than λ_{max}, and the energy of the electrostatic interaction of particles of charge e located at a distance λ_{max}. To illustrate what a small fraction of the energy is carried off by long-wavelength radiation in concrete cases, we note that for $\lambda_{\text{max}} = 1$ cm even with momentum transfers, $t \simeq 10^6$ $(\text{GeV}/c)^2$ much greater than the values now achieved in electron accelerators, the energy E_{ir} is of the order 5×10^{-3} eV.

As an illustrative example for our discussion, let us consider a simple case. Suppose that the world line of a particle is described by the formula

$$\xi^\mu(\lambda) = \frac{p^\mu+q^\mu}{2m}\lambda+\frac{p^\mu-q^\mu}{2m}\sqrt{\varrho^2+\lambda^2}+b^\mu,$$

where λ is a parameter of the trajectory† while b^μ specifies the position of the trajectory in space. The parameter ϱ (with the dimension of length) is related to the maximum acceleration of the particle by

$$\left(\frac{d^2\xi}{d\tau^2}\right)^2_{\text{max}} = -\frac{(p-q)^2}{\varrho^2(p+q)^2}.$$

The current transform in this case is of the form

$$\tilde{\mathscr{J}}^\mu(k) = ie\left(\frac{p^\mu}{k\cdot p}-\frac{q^\mu}{k\cdot q}\right)e^{ik\cdot b}\,\frac{\varrho}{m}\sqrt{(p\cdot k)(q\cdot k)}\,K_1\left(\frac{\varrho}{m}\sqrt{(p\cdot k)(q\cdot k)}\right)$$

† The relationship between λ and τ may be determined by integrating the differential equation

$$\frac{d\lambda}{d\tau} = 2m\sqrt{\frac{\varrho^2+\lambda^2}{\varrho^2(p+q)^2+4m^2\lambda^2}}.$$

where K_1 is a modified Bessel function. For small values of the argument x the function $K_1(x)$ behaves as x^{-1}. Accordingly, for small k^{μ}'s, in accordance with the general considerations, we obtain the current Fourier transform in the form of formula (86).

The energy and momentum radiated by a particle moving along the trajectory in question can be calculated from formula (58):

$$P_{\mu}^{\text{rad}} = \frac{p_{\mu}+q_{\mu}}{\sqrt{4m^2-t}} \frac{2e^2}{3\pi^2\varrho}\left(1-\frac{4m^2}{t}\right)^{-1} \int_0^{\infty} dx x^2 K_1^2(x)$$

$$= \frac{p_{\mu}+q_{\mu}}{\sqrt{4m^2-t}} \frac{e^2}{16\varrho}\left(1-\frac{4m^2}{t}\right)^{-1}.$$

The ratio of the energy carried off by the long-wavelength radiation to the total energy radiated is

$$\frac{E_{\text{ir}}}{E_{\text{rad}}} = \frac{\varrho}{\lambda_{\text{max}}} \cdot 4\pi^{-1} f\left(\sqrt{1-\frac{4m^2}{t}}\right)\left(1-\frac{4m^2}{t}\right).$$

For collisions of elementary particles, the parameter ϱ is of the order of the Compton wawelength and the dimensionless parameter $\varrho/\lambda_{\text{max}}$ is very small.

Chapter 4

THE QUANTUM THEORY OF THE ELECTROMAGNETIC FIELD

10. CANONICAL QUANTIZATION OF THE ELECTROMAGNETIC FIELD

THE quantum theory of the electromagnetic field# in a vacuum will be built up by using the method of canonical quantization described in Section 3. In the place of the canonical variables of classical theory, $D_i(\mathbf{x}, t)$ and $B_i(\mathbf{x}, t)$, we shall introduce the *field operators*† $\hat{D}_i(\mathbf{x}, t)$ and $\hat{B}_i(\mathbf{x}, t)$ acting in the Hilbert space of the state vectors of the electromagnetic field. Field operators taken at the same instant of time t satisfy the following *commutation relations*,‡

$$[\hat{B}_i(\mathbf{x}, t)\, \hat{D}_j(\mathbf{y}, t)] = i\hbar\varepsilon_{ijk}\, \partial_k\, \delta(\mathbf{x}-\mathbf{y}), \tag{1}$$

\# The quantum theory of the free field of radiation was given by P. A. M. Dirac (1927). Further references to the literature may be found in the book by Heitler.

† These quantities are customarily referred to as operators in the physical literature. In fact, they are operator-valued distributions just as are the field operators defined in Chapter 2.

‡ The commutation relations (1) can be derived (to within a multiplicative constant) from the following general assumptions:

1. The commutator of the fields $\hat{D}_i(\mathbf{x}, t)$ and $\hat{B}_j(\mathbf{y}, t)$ is a number.

2. All the field operators $\hat{f}_{\mu\nu}(x)$ and $\hat{h}^{\lambda\varrho}(y)$ commute if the points x and y are separated by a space-like interval.

3. There exists a vacuum state which is invariant under Poincaré transformations.

4. The tensors $\hat{f}_{\mu\nu}$ and $\hat{h}^{\lambda\varrho}$ satisfy the classical field equations. Assumptions 2, 3, and 4 imply that

$$(\Omega|[\hat{f}_{\mu\nu}(x), \hat{h}^{\lambda\varrho}(y)]\Omega) = i\int_0^\infty \mathrm{d}M^2\varrho(M^2)\left\{\partial_{[\mu}\delta_{\nu]}{}^{[\lambda}\partial^{\varrho]} - \frac{M^2}{2}\,\delta^{[\lambda}_{[\mu}\delta^{\varrho]}_{\nu]}\right\}\Delta(x-y, M^2),$$

where $\varrho(M^2)$ is a real function. On substituting $x^0 = y^0$, putting $\int_0^\infty \mathrm{d}(M^2)\varrho(M^2) = 1$ and taking account of assumption 1, we obtain the commutation relation (1).

which are obtained by replacing the Poisson brackets by commutators.†

To construct a complete quantum theory of the electromagnetic field, it would be necessary to define the action of the operators $\hat{D}_i$ and $\hat{B}_i$ on the state-vector space, and construct the other components of the field operators $\hat{f}_{\mu\nu}$ and $\hat{h}^{\mu\nu}$ as well as the generators of the Poincaré transformations: the energy and momentum operators $\hat{P}^{\mu}$ and the four-dimensional angular momentum operators $\hat{M}_{\mu\nu}$. Alas, no one has succeeded in accomplishing this in the case of non-linear electrodynamics. How this can be done is known only for Maxwellian electrodynamics. However, before we proceed to discuss this theory, let us describe in the general case the role of Poincaré transformations as symmetry transformations in the quantum theory of the electromagnetic field and the conditions imposed on the theory by the requirement of this symmetry.

The Poincaré Group as a Symmetry Group

Taking electromagnetic field theory as an example, we shall consider in detail the properties of quantum field theory which are associated with the symmetry of the theory with respect to the Poincaré group, that is with the relativistic invariance of the theory. The foundations of the relativistic invariance of quantum theory were given in Section 2, whereas now we shall look at the conditions which operators must satisfy in relativistically invariant theory.

From the requirement that the Poincaré group be a symmetry group of quantum theory it follows that a unitary representation $\mathsf{U}(a, \Lambda)$ of the Poincaré group‡ exists in the Hilbert space of state vectors of the system. The physical quantities describing the electromagnetic field in relativistic theory are the components of tensors or tensor fields (momentum and energy, field intensities, energy-momentum density,

† In this section Plack's constant will be written in all formulae.

‡ In the general case, this is the representation of the universal covering group of the Poincaré group (cf. Appendix I). In the case of electromagnetic field theory, one can confine oneself to representations of the Poincaré group itself since no representations of half-integral spin appear here.

etc.). We require that in quantum theory the average values of physical quantities transform under Poincaré transformations of states of the system just as the corresponding tensor fields do in classical theory. Let us write out these transformation relations for pure states:

$$\begin{aligned}(\mathsf{U}(a, \Lambda)\Psi|\hat{T}^{\mu_1\ldots\mu_k}_{\nu_1\ldots\nu_l}(x)\mathsf{U}(a, \Lambda)\Psi) \\ = \Lambda^{\mu_1}{}_{\lambda_1}\ldots \Lambda^{\mu_k}{}_{\lambda_k}\Lambda_{\nu_1}{}^{\varrho_1}\ldots \Lambda_{\nu_l}{}^{\varrho_l}\left(\Psi|\hat{T}^{\lambda_1\ldots\lambda_k}_{\varrho_1\ldots\varrho_l}(\Lambda^{-1}(x-a))\Psi\right),\end{aligned} \tag{2}$$

where $\hat{T}^{\mu_1\ldots\mu_k}_{\nu_1\ldots\nu_l}$ is an operator (operator-valued distribution) representing a given physical quantity. Since eq. (2) must hold for all states of the system, we arrive at the equation for operators†

$$\begin{aligned}\mathsf{U}^{-1}(a, \Lambda)\hat{T}^{\mu_1\ldots\mu_k}_{\nu_1\ldots\nu_l}\mathsf{U}(a, \Lambda) \\ = \Lambda^{\mu_1}{}_{\lambda_1}\ldots \Lambda^{\mu_k}{}_{\lambda_k}\Lambda_{\nu_1}{}^{\varrho_1}\ldots \Lambda_{\nu_l}{}^{\varrho_l}\hat{T}^{\lambda_1\ldots\lambda_k}_{\varrho_1\ldots\varrho_l}(\Lambda^{-1}(x-a)).\end{aligned} \tag{3}$$

The foregoing equations hold for all tensor quantities representing physical quantities in relativistically invariant theories. For geometrical objects with different transformation properties (for instance, for spinor fields) suitable changes must be made in the expressions on the right-hand side of these equations. Several conclusions may be drawn from eqs. (3), the most important of which are the commutation relations between the Poincaré group generators and other operators. These relations are obtained by considering infinitesimal Poincaré transformations.

---

† The polarization lemma which follows justifies this conclusion. *Polarization lemma*: If $(\Psi|A\Psi) = 0$ for the vectors Ψ constituting a dense linear set D in Hilbert space, then $A = 0$.

By way of proof, let us substitute Ψ in the form of a linear combination $\Psi = \alpha_1\Psi_1+\alpha_2\Psi_2$ where $\Psi_{1,2} \in D$. We then obtain

$$\begin{aligned}|\alpha_1|^2(\Psi_1|A\Psi_1)+|\alpha_2|^2(\Psi_2|A\Psi_2)+\alpha_1^*\alpha_2(\Psi_1|A\Psi_2) \\ +\alpha_1\alpha_2^*(\Psi_2|A\Psi_1) = 0.\end{aligned}$$

The first two terms vanish by the hypothesis of the lemma, whereas the other two must each be separately equal to zero because the choice of the phases of the complex coefficients α_1 and α_2 is arbitrary. Thus, $(\Psi_1|A\Psi_2)$ is zero for any two vectors of the dense set, which means that $A = 0$.

The infinitesimal Poincaré transformations of points of space-time can be parametrized by means of ten parameters $\delta\alpha^{\mu\nu}$ and $\delta\alpha^{\nu}$,

$$x^{\mu} \to x^{\mu} + \delta x^{\mu}, \tag{4a}$$

$$\delta x^{\mu} = -\delta\alpha^{\mu}{}_{\nu} x^{\nu} + \delta\alpha^{\mu}, \tag{4b}$$

$$\delta\alpha_{\mu\nu} = -\delta\alpha_{\nu\mu}. \tag{4c}$$

The *generators of Poincaré transformations* in quantum theory are the operators $\hat{P}_{\mu}$ and $\hat{M}_{\mu\nu}$ which are associated with the expansion of the operators U with respect to the parameters $\delta\alpha^{\mu}$ and $\delta\alpha^{\mu\nu}$ about the unity of the group by the formula:

$$\mathsf{U}(a, \Lambda) \simeq 1 + \frac{i}{\hbar}\,\delta\alpha^{\mu}\hat{P}_{\mu} + \frac{i}{2\hbar}\,\delta\alpha^{\mu\nu}\hat{M}_{\mu\nu}. \tag{5}$$

The laws of composition of Poincaré transformations lead to the following commutation relations for the group generators:

$$\frac{1}{i\hbar}\,[\hat{P}_{\mu}, \hat{P}_{\nu}] = 0, \tag{6a}$$

$$\frac{1}{i\hbar}\,[\hat{M}_{\mu\nu}, \hat{P}_{\lambda}] = -g_{\mu\lambda}\hat{P}_{\nu} + g_{\nu\lambda}\hat{P}_{\mu}, \tag{6b}$$

$$\frac{1}{i\hbar}\,[\hat{M}_{\mu\nu}, \hat{M}_{\lambda\varrho}] = -g_{\mu\lambda}\hat{M}_{\nu\varrho} + g_{\mu\varrho}\hat{M}_{\nu\lambda} - g_{\nu\varrho}\hat{M}_{\mu\lambda} + g_{\nu\lambda}\hat{M}_{\mu\varrho}. \tag{6c}$$

They have the same structure as the Poisson brackets of the corresponding classical quantities (cf. formulae (7.17)).

On substituting expansions (4) and (5) into formula (3) and retaining only linear expressions, because of the arbitrariness of the parameters $\delta\alpha^{\mu}$ and $\delta\alpha^{\mu\nu}$ we then obtain the commutation relations

$$\frac{1}{i\hbar}\,[\hat{P}_{\mu}, \hat{T}^{\mu_1\ldots\mu_k}_{\nu_1\ldots\nu_l}] = -\partial_{\mu}\hat{T}_{\nu_1\ldots\nu_l}^{\mu_1\ldots\mu_k}(x), \tag{7a}$$

$$\frac{1}{i\hbar}\,[\hat{M}_{\mu\nu}, \hat{T}^{\mu_1\ldots\mu_k}_{\nu_1\ldots\nu_l}] = -x_{[\mu}\partial_{\nu]}\hat{T}^{\mu_1\ldots\mu_k}_{\nu_1\ldots\nu_l}(x)$$

$$-\sum_{i=1}^{k}\delta^{\mu_i}_{[\mu}g_{\nu]\lambda}\hat{T}^{\mu_1\ldots\lambda\ldots\mu_k}_{\nu_1\ldots\nu_l}(x) - \sum_{i=1}^{l} g_{\nu_i[\mu}\delta^{\varrho}_{\nu]}\hat{T}^{\mu_1\ldots\mu_k}{}_{\nu_1\ldots\varrho\ldots\nu_l}(x). \tag{7b}$$

If in particular $\hat{T}^{\mu_1\ldots\mu_k}_{\nu_1\ldots\nu_l}(x)$ is replaced by the field tensors $\hat{f}_{\mu\nu}$ and $\hat{h}^{\mu\nu}$, we arrive at

$$\frac{1}{i\hbar}[\hat{P}_\mu, \hat{f}_{\lambda\varrho}(x)] = -\partial_\mu \hat{f}_{\lambda\varrho}(x), \tag{8a}$$

$$\frac{1}{i\hbar}[\hat{P}_\mu, \hat{h}^{\lambda\varrho}(x)] = -\partial_\mu \hat{h}^{\lambda\varrho}(x), \tag{8b}$$

$$\frac{1}{i\hbar}[\hat{M}_{\mu\nu}, \hat{f}_{\lambda\varrho}(x)] = -x_{[\mu}\partial_{\nu]}\hat{f}_{\lambda\partial}(x) - \delta^{\sigma}_{[\lambda}g_{\varrho][\mu}\delta^{\tau}_{\nu]}\hat{f}_{\sigma\tau}(x), \tag{8c}$$

$$\frac{1}{i\hbar}[\hat{M}_{\mu\nu}, \hat{h}^{\lambda\varrho}(x)] = -x_{[\mu}\partial_{\nu]}\hat{h}^{\lambda\varrho}(x) - \delta^{[\lambda}_{\sigma}\delta^{\varrho]}{}_{[\mu}g_{\nu]\tau}\hat{h}^{\sigma\tau}(x). \tag{8d}$$

These commutation relations constitute the conditions for the average values of the field intensities to transform as tensors. Similarly, the commutation relations (6) for Poincaré group generators express the fact that the average energy and momentum form a four-vector, whereas the average four-dimensional angular momentum constitutes an antisymmetrical tensor.

We shall now go on to discuss finite Poincaré transformations. The unitary operators $\mathsf{U}(a, \Lambda)$ forming the representation of a Poincaré group can be expressed in terms of the Poincaré group generators in the following manner (cf. formulae (7.25) and (7.26)),

$$\mathsf{U}(a, \Lambda) = \exp\left(\frac{i}{\hbar}\alpha^\mu \hat{P}_\mu + \frac{i}{2\hbar}\alpha^{\mu\nu}\hat{M}_{\mu\nu}\right). \tag{9}$$

To prove this representation, let us consider a one-parameter subgroup of Poincaré transformations generated by a linear combination of generators. The unitary operators U_λ which constitute the representation of this subgroup are of the form

$$\mathsf{U}_\lambda \equiv \exp\left[\frac{i\lambda}{\hbar}\left(\alpha^\mu \hat{P}_\mu + \frac{1}{2}\alpha^{\mu\nu}\hat{M}_{\mu\nu}\right)\right].$$

On differentiating the expression $\mathsf{U}_\lambda^{-1} T^{\mu_1\ldots\mu_k}_{\nu_1\ldots\nu_l}\mathsf{U}_\lambda$ with respect to the parameter λ, with the aid of the commutation relations (7), we obtain† a differential equation for this expression. The solution of this differential equation for $\lambda = 1$ is of the form

† We used a similar method in the proofs of the relativistic invariance of classical theory (cf. Section 7).

of the right-hand side of eq. (3). Thus, it follows from eqs. (7) that the exponential functions (9) of the Poincaré transformation generators constitute a representation of that group.

Formula (9) is used most frequently when $\Lambda = 0$, that is when the Poincaré transformations are pure translations. In that event we obtain ($\hbar = 1$)

$$e^{-ia\cdot\hat{P}}\hat{T}^{\mu_1\ldots\mu_k}_{\nu_1\ldots\nu_l}(x)e^{ia\cdot\hat{P}} = \hat{T}^{\mu_1\ldots\mu_k}_{\nu_1\ldots\nu_l}(x-a).$$

The aforementioned properties of generators and fields are part of the general programme of quantization which no one has hitherto managed to realize in the case of non-linear electrodynamics. Realization of this programme in linear theory will be taken up in the next section.

11. QUANTIZATION OF THE FREE MAXWELLIAN FIELD

In this section we shall present the quantum theory of the Maxwellian field. To simplify the writing, we shall omit the hat (^) on operators. Planck's constant $\hbar$ will also be omitted in the formulae. In the preceding section, in the general theory of the electromagnetic field we introduced the commutation relations for field operators corresponding to the canonical variables B_i and D_j taken at the same instant of time. In the classical Maxwellian theory, the existence of linear relationships between the fields **E** and **D** and **B** and **H** made it possible to find the values of the Poisson brackets for the components of the field tensor $f_{\mu\nu}$ taken even at arbitrary instants. When the Poisson brackets are replaced by commutators, commutation relations are obtained for the field operator $f_{\mu\nu}$ at two arbitrary points x and y in space-time:

$$\frac{1}{i}[f_{\mu\nu}(x), f_{\lambda\varrho}(y)] = \partial_{[\mu}g_{\nu][\lambda}\partial_{\varrho]}D(x-y). \tag{1}$$

Because of their tensor character (the function $D(x)$ is a scalar) these equations are invariant with respect to Poincaré transformations. Under the transformations of the field operators,

$$f_{\mu\nu}(x) \to {}^{\mathrm{IL}}f_{\mu\nu}(x) = \Lambda_\mu{}^\lambda\Lambda_\nu{}^\varrho f_{\lambda\varrho}(\Lambda^{-1}(x-a)), \tag{2}$$

the commutation relations do not undergo any change, that is

$$[{}^{\mathrm{IL}}f_{\mu\nu}(x), {}^{\mathrm{IL}}f_{\lambda\varrho}(y)] = [f_{\mu\nu}(x), f_{\lambda\varrho}(y)]. \tag{3}$$

The invariance condition (3) is a necessary condition for the commutation relations (1) to be compatible with the requirement that the proper Poincaré group be a symmetry group of the theory. For it follows from this requirement that the operators $f_{\mu\nu}$ and ${}^{IL}f_{\mu\nu}$ are related by the unitary transformation

$$\mathsf{U}^{-1}(a, \Lambda)f_{\mu\nu}(x)\mathsf{U}(a, \Lambda) = {}^{IL}f_{\mu\nu}(x), \tag{4}$$

hence, that they satisfy the same commutation relations.

Now, let us examine the content of equality (1), particularly its relation to the causality principle. From the representation (E.25b) of the function $D(x)$ it follows that this function does not vanish only for arguments which lie on the light cone. The field operators $f_{\mu\nu}(x)$ and $f_{\lambda\varrho}(y)$, therefore, commute if the points x and y cannot be joined by a light signal. This means that measurements of the field intensity in spatially separated regions are independent of each other. This result is compatible with the principle that the velocity of light is the maximum velocity with which perturbations can propagate. Perturbations caused by measurement of the field# at the point x cannot go outside the light cone with apex at x. In the case under consideration, the perturbation propagates along the cone, hence exactly at the speed of light.†

We shall investigate further properties of field operators by using their Fourier representation. As an auxiliary measure, we introduce the field operators $F_{\mu\nu}$ and their conjugates $F^{\dagger}_{\mu\nu}$ which are defined by analogy to the functions named in the same way in classical theory:

$$F_{\mu\nu} \equiv \frac{1}{2}\,(f_{\mu\nu}+i\check{f}_{\mu\nu}), \tag{5a}$$

$$F^{\dagger}_{\mu\nu} \equiv \frac{1}{2}\,(f_{\mu\nu}-i\check{f}_{\mu\nu}). \tag{5b}$$

The theory of measurement of the intensities of the quantum electromagnetic field was given by N. Bohr and L. Rosenfeld (1933).

† It is interesting that this property is possessed by the theory of the electromagnetic field in four-dimensional space-time. For the three-dimensional world (two-dimensional space and time), the function D is non-zero inside the cone as well.

These operators satisfy the commutation relations

$$\frac{1}{i}[F_{\mu\nu}(x), F^{\dagger}_{\lambda\varrho}(y)]$$

$$= \left[\frac{1}{2}\partial_{[\mu}g_{\nu][\lambda}\partial_{\varrho]} + \frac{1}{4}i\varepsilon_{\varkappa\mu\nu[\lambda}\partial_{\varrho]}\partial^{\varkappa} - \frac{1}{4}i\varepsilon_{\varkappa\lambda\varrho[\mu}\partial_{\nu]}\partial^{\varkappa}\right]D(x-y), \tag{6}$$

which follow from the commutation relations (1) for the operators $f_{\mu\nu}(x)$.

The Fourier representations of the field operators $F_{\mu\nu}$ and $F^{\dagger}_{\mu\nu}$ are obtained when the amplitudes $f(\mathbf{k}, \lambda)$ and $f^*(\mathbf{k}, \lambda)$ in formula (9.27) are replaced by the operator-like quantities $c(\mathbf{k}, \lambda)$ and $c^{\dagger}(\mathbf{k}, \lambda)$:

$$F_{\mu\nu}(x) = \int d\Gamma e_{\mu\nu}(\mathbf{k})[c(\mathbf{k}, +1)e^{-ik\cdot x} + c^{\dagger}(\mathbf{k}, -1)e^{ik\cdot x}], \tag{7a}$$

$$F^{\dagger}_{\mu\nu}(x) = \int d\Gamma e^{*}_{\mu\nu}(\mathbf{k})[c(\mathbf{k}, -1)e^{-ik\cdot x} + c^{\dagger}(\mathbf{k}, +1)e^{ik\cdot x}]. \tag{7b}$$

The quantities $c(\mathbf{k}, \lambda)$ and $c^{\dagger}(\mathbf{k}, \lambda)$ may be expected to be operator-valued distributions and not operators since, by virtue of (6), they satisfy the commutation relations with the function δ on the right-hand side

$$[c(\mathbf{k}, \lambda), c^{\dagger}(\mathbf{k}', \lambda')] = \delta_{\lambda\lambda'}\delta_{\Gamma}(\mathbf{k}, \mathbf{k}'), \tag{8a}$$

$$[c(\mathbf{k}, \lambda), c(\mathbf{k}', \lambda')] = 0 = [c^{\dagger}(\mathbf{k}, \lambda), c^{\dagger}(\mathbf{k}', \lambda')]. \tag{8b}$$

These relations are also immediately obtainable from formulae (9.34) if the Poisson brackets for the functions $f(\mathbf{k}, \lambda)$ and $f^*(\mathbf{k}, \lambda)$ are replaced by the commutators of the quantities $c(\mathbf{k}, \lambda)$ and $c^{\dagger}(\mathbf{k}, \lambda)$, which in quantum theory represent the functions f and f^*.

To have operators subsequently in the quantization procedure, distribution-like objects will now be replaced by smoothed-out quantities. To this end we introduce a *complete orthonormal system of wave packet profiles* $f_i(\mathbf{k}, \lambda)$ in momentum space,

$$\sum_{\lambda}\int d\Gamma f_i^*(\mathbf{k}, \lambda) f_j(\mathbf{k}, \lambda) = \delta_{ij}, \tag{9a}$$

$$\sum_{i=1}^{\infty} f_i(\mathbf{k}, \lambda) f_i^*(\mathbf{k}', \lambda') = \delta_{\lambda\lambda'}\delta_{\Gamma},(\mathbf{k}, \mathbf{k}') \tag{9b}$$

and in terms of them we define the quantities

$$c_i \equiv c[f_i] \equiv \sum_\lambda \mathrm{d}\Gamma f_i^*(\mathbf{k}, \lambda) c(\mathbf{k}, \lambda), \tag{10a}$$

$$c_i^\dagger \equiv c^\dagger[f_i] \equiv \sum_\lambda \mathrm{d}\Gamma f_i(\mathbf{k}, \lambda) c^\dagger(\mathbf{k}, \lambda), \tag{10b}$$

which satisfy the commutation relations

$$[c_i, c_j^\dagger] = \delta_{ij}, \tag{11a}$$

$$[c_i, c_j] = 0 = [c_i^\dagger, c_j^\dagger]. \tag{11b}$$

The next step in the procedure for quantizing a free Maxwellian field is to find the representations of the commutation relations above, i.e. to specify the quantities c_i and $c_i^\dagger$ as mutually conjugate linear operators acting in Hilbert space. Let us choose the Fock representation# which has a particularly lucid physical interpretation.† This representation is characterized by the existence in Hilbert space of a distinguished vector‡ Ω on which all operators c_i become zero,

$$c_i\Omega = 0, \quad \text{for all } i\text{'s}. \tag{12}$$

Choosing the Fock representation means identifying the operators c_i and $c_i^\dagger$ with the annihilation and creation operators defined in Section 5. The quantum theory of the free Maxwellian field takes on the form of the quantum theory of a many-boson system. The Fock space, i.e. the Hilbert space in which operators c_i and $c_i^\dagger$ operate, is the direct sum of the vacuum vector, one-particle space, two-particle space, etc.

V. Fock (1932) was the first to give a general theory of creation and annihilation operators. The existence of other canonical representations of the commutation relations was not discovered until the fifties.

† In the quantum theory of systems with an infinite number of degrees of freedom there exists an infinite number of non-equivalent, irreducible representations of canonical commutation relations. The choice of a particular representation is in the nature of an auxiliary postulate. It may well be, however, that only the Fock representation is suitable for describing physical systems.

In the quantum mechanics of systems with a finite number of degrees of freedom all the irreducible representations of the commutation relations are unitarily equivalent. This theorem was proved by J. von Neumann. The precise formulation and modern proof of the von Neumann theorem can be found in the book by Jauch.

‡ For other (non-Fock) representations of canonical commutation relations no vector in Hilbert space has these properties.

In this way, indistinguishable boson particles *photons*,. appear in quantum electromagnetic field theory.

The Fock basis in the Hilbert space of the state vectors of the electromagnetic field† consists of a set of vectors $\Phi_{\{n_k\}} = \Phi_{n_1, \ldots, n_k, \ldots}$ $(\sum_{i=1}^{\infty} n_i < \infty)$, characterized by sequences of occupation numbers $n_1, \ldots,$ $n_k, \ldots$. These vectors are the result of the action of the creation operators $c_i^\dagger$ on the vacuum vector

$$\Phi_{\{n_k\}} = \Big(\prod_{i=1}^{\infty} \frac{1}{\sqrt{n_i!}} (c_i^\dagger)^{n_i}\Big) \Omega .$$

The field operators $f_{\mu\nu}(x)$ are constructed by taking the sum of the operators $F_{\mu\nu}(x)$ and $F_{\mu\nu}^\dagger(x)$ as given by formulae (7):

$$f_{\mu\nu}(x) = f_{\mu\nu}^{(+)}(x) + f_{\mu\nu}^{(-)}(x), \tag{13a}$$

$$f_{\mu\nu}^{(+)}(x) \equiv \int d\Gamma [e_{\mu\nu}(\mathbf{k}) c(\mathbf{k}, +1) + e_{\mu\nu}^*(\mathbf{k}) c(\mathbf{k}, -1)] e^{-ik\cdot x}, \tag{13b}$$

$$f_{\mu\nu}^{(-)}(x) \equiv \int d\Gamma [e_{\mu\nu}^*(\mathbf{k}) c^\dagger(\mathbf{k}, +1) + e_{\mu\nu}(\mathbf{k}) c^\dagger(\mathbf{k}, -1)] e^{ik\cdot x}, \tag{13c}$$

$$[f_{\mu\nu}^{(+)}(x)]^\dagger = f_{\mu\nu}^{(-)}(x). \tag{13d}$$

Operator-valued distributions $c(\mathbf{k}, \lambda)$ and $c^\dagger(\mathbf{k}, \lambda)$ can be correctly defined as in Section 5 by parametrizing the Fock space of photons with sequences of functions $(f_0, f(\mathbf{k}, \lambda), f(\mathbf{k}_1 \lambda_1, \mathbf{k}_2 \lambda_1), \ldots)$. Note, however, that the construction of field operators given by formulae (13a, b and c) differs from that of field operators in non-relativistic many-particle theory. This follows from the requirement that the field operators possess suitable transformation properties with respect to Poincaré transformations.

The field operators $f_{\mu\nu}^{(\pm)}$ are also expressible as linear combinations of the annihilation and creation operators, respectively. To this end let us reverse the relations (10), utilizing the completeness of the set of profiles f_i in doing this,

$$c(\mathbf{k}, \lambda) = \sum_{i=1}^{\infty} f_i(\mathbf{k}, \lambda) c_i, \tag{14a}$$

$$c^\dagger(\mathbf{k}, \lambda) = \sum_{i=1}^{\infty} f_i^*(\mathbf{k}, \lambda) c_i^\dagger . \tag{14b}$$

† Cf. Section 5.

Substitution of these expressions into formulae (13) yields†

$$f_{\mu\nu}^{(+)}(x) = \sum_{i=1}^{\infty} \varphi_{\mu\nu}[x|f_i]c_i, \tag{15a}$$

$$f_{\mu\nu}^{(-)}(x) = \sum_{i=1}^{\infty} \varphi_{\mu\nu}^{*}[x|f_i]c_i^{\dagger}, \tag{15b}$$

where

$$\varphi_{\mu\nu}[x|f_i] \equiv \int d\Gamma[e_{\mu\nu}(\mathbf{k})f_i(\mathbf{k}, +1)+e_{\mu\nu}^{*}(\mathbf{k})f_i(\mathbf{k}, -1)]e^{-ik\cdot x}. \tag{16}$$

The functions $\varphi_{\mu\nu}[x|f_i]$, which we shall call the *tensor wave functions of the photon*, are complex solutions of the classical field equations. The operators $c_i^{\dagger}$ create, and operators c_i annihilate, photons in the state described by the wave function $\varphi_{\mu\nu}[x|f_i]$.

Photon wave functions form a Hilbert space. The scalar product in this space may be easily expressed in terms of wave-packet profiles f,

$$(f_1|f_2) = \sum_{\lambda} \int d\Gamma f_1^{*}(\mathbf{k}, \lambda)f_2(\mathbf{k}, \lambda),$$

but it can also be expressed in terms of the components of wave functions $\varphi_{\mu\nu}$. To this end let us define the expression:

$$(\varphi_1|\varphi_2) \equiv \frac{i}{4\pi}\int d^3x\int d^3y\left\{\mathscr{E}_1^{*}(\mathbf{x}, t)\frac{1}{|\mathbf{x}-\mathbf{y}|}\cdot\nabla\times\mathscr{B}_2(\mathbf{y}, t)-\right.$$
$$\left.-(\nabla\times\mathscr{B}_1^{*}(\mathbf{x}, t))\frac{1}{|\mathbf{x}-\mathbf{y}|}\cdot\mathscr{E}_2(\mathbf{y}, t)\right\}, \tag{17}$$

where $\mathscr{E}$ i $\mathscr{B}$ are complex field vectors constructed out of components of the wave function $\varphi_{\mu\nu}$ in the same way as vectors **E** and **B** are constructed out of components of the field tensor $f_{\mu\nu}$. Inserting the Fourier expansion (16) of the tensor wave functions into the formula (17), we obtain:

$$(\varphi[f_1]|\varphi[f_2]) = (f_1|f_2).$$

The non-local character of the scalar product (17) originates from the lack of symmetry between the coordinate representation and the mo-

† The mathematical sense of the series figuring in formulae (16) and (17) has been discussed in Section 5.

mentum representation for photons. Photon momentum operators exist†, but there are no photon position operators.

Solutions with positive and negative frequencies are orthogonal with respect to the scalar product (17). With the help of this scalar product we can express the creation and anihilation operators by the field operators in the simple form,

$$c[f] = (\varphi[f]|\hat{f}) = -(\hat{f}|\varphi^*[f]), \tag{18a}$$

$$c^\dagger[f] = (\hat{f}|\varphi[f]) = -(\varphi^*[f]|\hat{f}). \tag{18b}$$

To distinguish operators from profiles we have again introduced hats for operators.

Relativistic Invariance

If relativistic invariance is to be ensured, the generators of the Poincaré transformations must be given in the form of self-adjoint operators acting in the Hilbert space of state vectors of the electromagnetic field. In the method of canonical quantization these operators are obtained from the classical expressions (9.33) when the functions $f(\mathbf{k}, \lambda)$ and $f^*(\mathbf{k}, \lambda)$ are replaced by the operators $c(\mathbf{k}, \lambda)$ and $c^\dagger(\mathbf{k}, \lambda)$. Even though the operators c and $c^\dagger$ do not commute, this procedure leads to a unique result, if only we make the assumption—a natural one from the physical point of view—that the vacuum vector is invariant under all Poincaré transformations,

$$\mathsf{U}(a, \Lambda)\Omega = \Omega. \tag{19}$$

Hence we get additional conditions for the operators P_μ and $M_{\mu\nu}$,

$$P_\mu \Omega = 0,$$

$$M_{\mu\nu}\Omega = 0.$$

In order that these conditions be satisfied, it is necessary to choose the normal ordering of creation and annihilation operators in the

† We shall show later in this section that the operators of components of photon momentum act on profiles $f(\mathbf{k}, \lambda)$ as multiplication by k_x, k_y, k_z.

expressions for the energy, momentum, and four-dimensional angular momentum operators,†

$$H = \sum_{\lambda} \int d\Gamma \omega(\mathbf{k}) c^{\dagger}(\mathbf{k}, \lambda) c(\mathbf{k}, \lambda), \tag{20a}$$

$$\boldsymbol{P} = \sum_{\lambda} \int d\Gamma \mathbf{k} c^{\dagger}(\mathbf{k}, \lambda) c(\mathbf{k}, \lambda), \tag{20b}$$

$$\boldsymbol{M} = \sum_{\lambda} \int d\Gamma \left[c^{\dagger}(\mathbf{k}, \lambda) \left(\mathbf{k} \times \frac{1}{i} \mathbf{D} \right) c(\mathbf{k}, \lambda) + \lambda \mathbf{n}(\mathbf{k}) c^{\dagger}(\mathbf{k}, \lambda) c(\mathbf{k}, \lambda) \right], \tag{20c}$$

$$N = \frac{i}{2} \sum_{\lambda} \int d\Gamma \omega(\mathbf{k}) [c^{\dagger}(\mathbf{k}, \lambda) \mathbf{D} c(\mathbf{k}, \lambda) - (\mathbf{D} c(\mathbf{k}, \lambda))^{\dagger} c(\mathbf{k}, \lambda)]. \tag{20d}$$

Generators of conformal transformations can also be constructed in similar fashion.

For expressions which are linear and quadratic in creation and annihilation operators the commutation relations are identical in structure with the Poisson brackets in classical theory. Accordingly, no additional calculations are even necessary to ascertain that all the commutation relations between generators, and between generators and field operators, are of the proper form (10.6) and (10.8) in accordance with the requirements of relativistics invariance of the theory. In particular, we obtain the following set of commutation relations between generators and annihilation operators,

$$[H, c(\mathbf{k}, \lambda)] = -\omega(\mathbf{k}) c(\mathbf{k}, \lambda), \tag{21a}$$

$$[\mathbf{P}, c(\mathbf{k}, \lambda)] = -\mathbf{k} c(\mathbf{k}, \lambda), \tag{21b}$$

$$[\mathbf{M}, c(\mathbf{k}, \lambda)] = -\left[\left(\mathbf{k} \times \frac{1}{i} \mathbf{D} \right) + \lambda \mathbf{n} \right] c(\mathbf{k}, \lambda), \tag{21c}$$

$$[\mathbf{N}, c(\mathbf{k}, \lambda)] = -\omega(\mathbf{k})\, i \mathbf{D} c(\mathbf{k}, \lambda), \tag{21d}$$

† Formulae (20) are of a formal nature since multiplication of the operators $c(\mathbf{k}, \lambda)$ and $c^{\dagger}(\mathbf{k}, \lambda)$, which are operator-valued distributions, is not a well-defined operation. However, if these expressions are used formally, it is possible to obtain well-defined expressions which may be treated as a correct definition of the operators P_{μ} and $M_{\mu\nu}$. In this way, for the four-momentum operator P_{μ} we obtain the formula

$$P_{\mu} c^{\dagger}_{i_1} \ldots c^{\dagger}_{i_n} \Omega = \sum_{j=1}^{n} c^{\dagger}_{i_1} \ldots c^{\dagger}_{i_{j-1}} \sum_{\lambda} \int d\Gamma c^{\dagger}(\mathbf{k}, \lambda) k_{\mu} f_{i_j}(\mathbf{k}, \lambda) c^{\dagger}_{i_{j+1}} \ldots c^{\dagger}_{i_n} \Omega.$$

which may serve as a starting point for calculating further commutators.

Photons

Photons may be figuratively described as elementary excitations of the quantum electromagnetic field. In the particular case of the free field under consideration here, these excitations satisfy the principle of superposition: the sum of the creation operators of photons is also (within a normalization factor) a photon creation operator.

By taking a linear combination of state vectors with various numbers of photons, we obtain state vectors with an undetermined number of photons and we may easily "obliterate" the photon structure of the theory. For it must be realized that the occurrence of photons in the description of a free field is related only to the choice of a convenient representation of the state vectors. This concept does not take on more profound physical meaning until we come to the description of a weak† electromagnetic field interacting with matter.

We shall now show that the profile of a wave packet $f(\mathbf{k}, \lambda)$ is a wave function of a photon (in the momentum representation) created by the operator $c^\dagger[f]$. Namely, we shall demonstrate that the action of the momentum operator on the state vector of photon, $c^\dagger[f]\Omega$, corresponds to multiplication of the function $f(\mathbf{k}, \lambda)$ by the vector $\mathbf{k}$. To this end we shall examine the transformation rule for the state vectors $c^\dagger[f]\Omega$.

The transformation laws (4) for field operators lead to the following ones for the operators $c^\dagger(\mathbf{k}, \lambda)$,

$$\mathsf{U}^{-1}(a, \Lambda)c^\dagger(\mathbf{k}, \lambda)\mathsf{U}(a, \Lambda) = \mathrm{e}^{-i\lambda\theta(\mathbf{k}, \Lambda)}\mathrm{e}^{-ik\cdot a}c^\dagger(\Lambda^{-1}\mathbf{k}, \lambda). \tag{22}$$

From this we obtain the following transformation formula for vectors describing one-photon states,

$$\mathsf{U}(a, \Lambda)c^\dagger[f]\Omega = c^\dagger[{}^{\mathrm{IL}}f]\Omega, \tag{23}$$

where ${}^{\mathrm{IL}}f$ is the transformed function f,

$${}^{\mathrm{IL}}f(\mathbf{k}, \lambda) = \mathrm{e}^{i\lambda\theta(\mathbf{k}, \Lambda)}\mathrm{e}^{ik\cdot a}f(\Lambda^{-1}\mathbf{k}, \lambda). \tag{24}$$

Relations similar to relations (23) may also be given for vectors describing many-photon states.

---

† In describing strong electromagnetic fields in quantum theory an important role is played by the representation of state vectors in terms of coherent state vectors. They will be discussed further on in this section.

The transformation laws of profiles $f(\mathbf{k}, \lambda)$ of wave packets describing photon states are thus the same as those for the functions $f(\mathbf{k}, \lambda)$ which were introduced in Section 9 in order to describe the classical free electromagnetic field.

Now let us examine the implications of formula (23). From this formula it follows in particular that the action of the momentum operator $\mathbf{P}$ on the one-photon state vector corresponds to multiplication of a function $f(\mathbf{k}, \lambda)$ by $\mathbf{k}$. The function $f(\mathbf{k}, \lambda)$ may thus be regarded as a photon wave function in the momentum representation. It also follows from formula (23) that the action of the time-displacement generator, i.e. of the energy operator P_0, on the one-photon state vector corresponds to multiplication of the function $f(\mathbf{k}, \lambda)$ by $\mathbf{k}_0 = \omega(\mathbf{k})$. Photons hence have zero rest mass ($k^2 = 0$) and move at the speed of light. Accordingly, we can imagine that the field perturbations propagating, at the speed of light as we know, are transmitted by photons.

The energy and momentum of a photon in infinite space have continuous spectra. The photon energy and momentum thus are not sharply defined, but the profile of the wave packet may be chosen so that the spread of these quantities about their average values be arbitrarily small. The index λ is the analogue of the spin index σ for particles with non-zero rest mass. For photons, this index specifies the *helicity*† of the photon, i.e. the projection of the angular momentum on the direction of the momentum.

To justify this interpretation of the index λ, let us consider the photon creation operator $c^{\dagger}_{\mathbf{k}_0, \pm}$ with a wave packet profile characterized by an insignificant spread of momentum $\mathbf{k}$ about the value $\mathbf{k}_0$, that is,

$$\int d\Gamma(\mathbf{k}-\mathbf{k}_0)^2 |f_{\mathbf{k}_0}(\mathbf{k}, \lambda)|^2 \ll |\mathbf{k}_0|^2,$$

and with one component corresponding to the value $\lambda = +1$ (or $\lambda = -1$). From the commutation relations (21) it then follows that

$$\frac{\mathbf{k}_0 \cdot \mathbf{M}}{|\mathbf{k}_0|} c^{\dagger}_{\mathbf{k}_0, \pm}\Omega \simeq \pm c^{\dagger}_{\mathbf{k}_0, \pm}\Omega,$$

† Also frequently referred to as the polarization of the photon in view of the connection with classical electromagnetic field theory. A helicity of $+1$ corresponds to left-handed circular polarization and a helicity of -1 to right-handed circular polarization.

where

$$c^{\dagger}_{\mathbf{k},0\pm} = \int \mathrm{d}\Gamma f_{\mathbf{k}_0}(\mathbf{k}, \pm 1) c^{\dagger}(\mathbf{k}, \pm 1).$$

This means that the one-photon state under consideration is an (approximate) eigenstate of the operator of the angular momentum projection on the direction of the average momentum $\mathbf{k}_0$, belonging to the eigenvalue† $+1$ (or -1).

The fact that only two photon polarization states occur, and not three as for particles of spin 1, follows from the zero rest mass of the photon.

In the general case the wave packet profile $f(\mathbf{k}, \lambda)$ describes the state of a photon with undetermined helicity (undetermined polarization). The probabilities of obtaining a helicity of $+1$ or -1 in measurement are equal, respectively, to

$$\int \mathrm{d}\Gamma |f(\mathbf{k}, +1)|^2 \quad \text{and} \quad \int \mathrm{d}\Gamma |f(\mathbf{k}, -1)|^2.$$

The wave-packet profile which, by formulae (10), defines the photon creation and annihilation operators may also be chosen so that the photon created and annihilated have a given total angular momentum J and its projection M. Thus the electromagnetic field associated with these photons will be the field of a particular multipole of the electric or magnetic type. Such a profile, which we shall denote by $f_{(e)JM}(\mathbf{k}, \lambda)$ or $f_{(m)JM}(\mathbf{k}, \lambda)$, must satisfy the condition

$$[\mathbf{e}(\mathbf{k}) f_{(e,m)JM}(\mathbf{k}, +1) - \mathbf{e}^*(\mathbf{k}) f_{(e,m)JM}(\mathbf{k}, -1)] = f(\omega) \mathbf{Y}^{(e,m)}_{JM}(\hat{\mathbf{n}})$$

where $\mathbf{Y}^{(e,m)}_{JM}(\hat{\mathbf{n}})$ is a spherical vector function in momentum space, of the electric or magnetic type, whose properties we have discussed in Section 9, whereas $f(\omega)$ is a properly normalized function of the frequency ω. The corresponding creation operators will be denoted by the symbols $c^{\dagger}_{(e)JM}(\omega)$ and $c^{\dagger}_{(m)JM}(\omega)$.

To the multiple expansion of the classical radiation field given in Section 9 (cf. formulae (9.80)) there corresponds the following decomposition of the operators of the electric field $\mathbf{E}$ and the magnetic induction $\mathbf{B}$ into the annihilation and creation operators, $c_{(e,m)JM}(\omega)$ and

† By the convention adopted, these values are actually equal to $\hbar$ and $-\hbar$.

$c^{\dagger}_{(e,m)JM}(\omega)$:

$$\mathbf{E}(\mathbf{r},t) = \sum_{JM}\left\{\int_0^{\infty}\frac{\mathrm{d}\omega}{2\pi}\,c_{(e)JM}(\omega)\mathrm{e}^{-i\omega t}\mathbf{E}_{(e)}(JM\omega,\mathbf{r})\right.$$

$$\left.+\int_0^{\infty}\frac{\mathrm{d}\omega}{2\pi}\,c_{(m)JM}(\omega)\mathrm{e}^{-i\omega t}\mathbf{E}_{(m)}(JM\omega,\mathbf{r})\right\} + \text{h.c.}$$

$$\mathbf{B}(\mathbf{r},t) = \sum_{JM}\left\{\int_0^{\infty}\frac{\mathrm{d}\omega}{2\pi}\,c_{(e)JM}(\omega)\mathrm{e}^{-i\omega t}\mathbf{B}_{(e)}(JM\omega,\mathbf{r})\right.$$

$$\left.+\int_0^{\infty}\frac{\mathrm{d}\omega}{2\pi}\,c_{(m)JM}(\omega)\mathrm{e}^{-i\omega t}\mathbf{B}_{(m)}(JM\omega,\mathbf{r})\right\} + \text{h.c.}$$

The functions $\mathbf{E}_{(e)}(JM\omega,\mathbf{r})$, $\mathbf{B}_{(e)}(JM\omega,\mathbf{r})$ and $\mathbf{E}_{(m)}(JM\omega,\mathbf{r})$, $\mathbf{B}_{(m)}(JM\omega,\mathbf{r})$, defined by formulae (9.81), describe the radiation field of an electric and a magnetic multipole. Thus, the operators $c_{(e,m)JM}$ annihilate, and the operators $c^{\dagger}_{(e,m)JM}$ create, photons with a particular total angular momentum J and projection of the angular momentum M.

Photon states with particular values of energy and momentum may be obtained in the case of an electromagnetic field enclosed in a cube of volume V, which we have described classically in Section 9. The coefficients in the decomposition of the classical field then are the amplitudes $f_{\mathbf{k},\pm}$ and $f^{*}_{\mathbf{k},\pm}$. Quantizing this theory, we obtain the decomposition of the field operator into the operators $c_{\mathbf{k},\pm}$ and $c^{\dagger}_{\mathbf{k},\pm}$ which satisfy the commutation relations

$$[c_{\mathbf{k},+},c^{\dagger}_{\mathbf{k}',+}] = \delta_{\ \mathbf{k}'} = [c_{\mathbf{k},-},c^{\dagger}_{\mathbf{k}',-}], \tag{25a}$$

$$[c_{\mathbf{k},\pm},c_{\mathbf{k}',-}] = 0 = [c^{\dagger}_{\mathbf{k},\pm},c^{\dagger}_{\mathbf{k}',-}], \tag{25b}$$

obtained when the corresponding Poisson brackets are replaced by commutators. The energy and momentum operators in this case are of the form

$$H = \sum_{\mathbf{k}}\omega(\mathbf{k})\,(c^{\dagger}_{\mathbf{k},+}c_{\mathbf{k},+}+c^{\dagger}_{\mathbf{k},-}c_{\mathbf{k},-}), \tag{26a}$$

$$\mathbf{P} = \sum_{\mathbf{k}}\mathbf{k}\,(c^{\dagger}_{\mathbf{k},+}c_{\mathbf{k},+}+c^{\dagger}_{\mathbf{k},-}c_{\mathbf{k},-}). \tag{26b}$$

Formulae (25) and (26) yield

$$Hc^{\dagger}_{\mathbf{k},\pm}\Omega = \omega(\mathbf{k})c^{\dagger}_{\mathbf{k},\pm}\Omega, \tag{27a}$$

$$\mathbf{P}c^{\dagger}_{\mathbf{k},\pm}\Omega = \mathbf{k}c^{\dagger}_{\mathbf{k},\pm}\Omega. \tag{27b}$$

The operator $c^{\dagger}_{\mathbf{k},\pm}$ acts on the vacuum vector to give a one-photon state vector which is an eigenvector of the energy operator with eigenvalue $\omega(\mathbf{k})$ and the momentum operator with eigenvalue $\mathbf{k}$. Thus, the operator $c^{\dagger}_{\mathbf{k},\pm}$ creates, and the operator $c_{\mathbf{k},\pm}$ annihilates, a photon of momentum $\mathbf{k}$, energy $\omega(\mathbf{k})$, and helicity ± 1.

Coherent States

The vectors $\Phi_{\{n_k\}}$ which constitute the Fock basis are at the same time the eigenvectors of all operators of the numbers of photons N_k corresponding to the occupation numbers n_k. These vectors are, therefore, especially useful for describing processes in which a particular small number of photons takes part. The expectation values of all the components of the electromagnetic field $f_{\mu\nu}(x)$ in the states $\Phi_{\{n_k\}}$ are equal to zero. The operator $f_{\mu\nu}(x)$ consists of two parts: one (with positive frequencies) decreases while the other (with negative frequencies) increases the number of photons by one (cf. formulae (17)). Vectors with different population numbers are orthogonal.

Now let us consider the pure states of the field, known as *coherent states*, which are particularly convenient in describing strong electromagnetic fields. A coherent state of the electromagnetic field is the name we give to a state represented by a vector which is at the same time an eigenvector of all photon annihilation operators† c_i. Coherent state vectors will be denoted by the symbols $\Phi(\alpha_1, \ldots, \alpha_k, \ldots)$ or $\Phi\{\alpha_k\}$ where the symbol $\{\alpha_k\}$ denotes a sequence of complex eigenvalues $\alpha_1, \alpha_2, \ldots$ of the operators $c_1, c_2, \ldots$,

$$c_i\Phi\{\alpha_\kappa\} = \alpha_i\Phi\{\alpha_k\}, \quad \text{for all } i\text{'s}. \tag{28}$$

Consider the solutions of this eigenvalue problem corresponding to the sequences $\{\alpha_k\}$ with only a finite number of non-zero terms.

† The eigenvalue problem for photon creation operators $c^{\dagger}_{i}$ does not have any solution.

A normalized vector constituting such a solution may be written as

$$\Phi\{\alpha_k\} = \sum_{\{n_k\}} \left(\prod_{i=1}^{\infty} \mathrm{e}^{-\frac{1}{2}|\alpha_i|^2} \frac{(\alpha_i)^{n_i}}{\sqrt{n_i!}} \right) \Phi_{\{n_k\}}. \tag{29}$$

To verify that it satisfies the eigenvalue equation (28), we write this vector in the equivalent form

$$\Phi\{\alpha_k\} = \exp\left(-\frac{1}{2}\sum_k |\alpha_k|^2\right)\left(\prod_{i=1}^{\infty}\sum_{n_i=0}^{\infty} \frac{(\alpha_i)^{n_i}}{n_i!} (c_i^\dagger)^{n_i}\right)\Omega.$$

Utilizing the commutativity of the operators c_j and $c_i^\dagger$ for $j \neq i$ and the relations $[c_i,(c_j^\dagger)^n] = n(c_j^\dagger)^{n-1}$, we arrive at the following result for the action of the operator c_j on vector (29),

$$c_j\Phi\{\alpha_k\} = \exp\left(-\frac{1}{2}\sum_k |\alpha_k|^2\right)\left[\prod_{i\neq j}^{\infty}\sum_{n_i=0}^{\infty} \frac{(\alpha_i)^{n_i}}{n_i!} (c_i^\dagger)^{n_i}\right]$$
$$\times\left[\sum_{n_j=1}^{\infty} \frac{(\alpha_j)^{n_j}}{n_j!} n_j(c_j^\dagger)^{n_j-1}\right]\Omega$$
$$= \exp\left(-\frac{1}{2}\sum_k |\alpha_k|^2\right)\alpha_j\left[\prod_i\sum_{n_i=0}^{\infty} \frac{(\alpha_i)^{n_i}}{n_i!} (c_i^\dagger)^{n_i}\right]\Omega = \alpha_j\Phi\{\alpha_k\}.$$

The vector (29) thus is a coherent state vector.

Now let us examine the properties of the particular vectors $\Phi\{\alpha_k\}$ corresponding to such sequences $\{\alpha_k\}$ in which only one eigenvalue (labelled with the index i) is non-zero. These vectors describe a coherent state of the electromagnetic field in which only one mode of vibration has been excited. For simplicity of notation we shall in this case everywhere omit the index i. The vector under consideration will be denoted by the symbol Φ_α,

$$\Phi_\alpha = \mathrm{e}^{-\frac{1}{2}|\alpha|^2}\sum_{n=0}^{\infty}\frac{\alpha^n}{\sqrt{n!}}\Phi_n = \mathrm{e}^{-\frac{1}{2}|\alpha|^2}\left[\sum_{n=0}^{\infty}\frac{\alpha^n}{n!}(c^\dagger)^n\right]\Omega. \tag{30}$$

The symbol Φ_n stands for a Fock basis vector $\Phi_{0,\dots,n,0,\dots}$ with only one occupation number (with index i) different from zero.

The vectors Φ_α and Φ_β belonging to different eigenvalues of the annihilation operator c are not orthogonal. The scalar product of these vectors is given by the formula

$$(\Phi_\alpha|\Phi_\beta) = \exp\left(-\frac{1}{2}|\alpha-\beta^2|+i\,\mathrm{Im}\,\alpha^*\beta\right). \tag{31}$$

This product is easily calculated by representing the vectors Φ_α and Φ_β in the form of a superposition (30) of vectors Φ_n and by utilizing the orthonormality of the latter,

$$(\Phi_\alpha|\Phi_\beta) = \exp\left(-\frac{1}{2}|\alpha|^2-\frac{1}{2}|\beta|^2\right)\sum_{n=0}^{\infty}\frac{(\alpha^*\beta)^n}{n!}$$

$$= \exp\left(-\frac{1}{2}|\alpha|^2-\frac{1}{2}|\beta|^2+\alpha^*\beta\right) = \exp\left(-\frac{1}{2}|\alpha-\beta|^2+i\operatorname{Im}\alpha^*\beta\right).$$

The vectors Φ_α can be obtained by operating on the vacuum vector with the unitary operator $\mathsf{D}(\alpha)$, called the *displacement operator*, which is defined by

$$\mathsf{D}(\alpha) \equiv \exp(\alpha c^\dagger - \alpha^* c). \tag{32}$$

By the Baker–Hausdorff identity (H.5) the displacement operator $\mathsf{D}(\alpha)$ can be written as

$$\mathsf{D}(\alpha) = \exp\left(-\frac{1}{2}|\alpha|^2\right)\exp(\alpha c^\dagger)\exp(-\alpha^* c).$$

And by acting on the vacuum vector with this operator we obtain

$$\mathsf{D}(\alpha)\Omega = \exp\left(-\frac{1}{2}|\alpha|^2\right)\exp(\alpha c^\dagger)\Omega = \Phi_\alpha.$$

The name displacement operator stems from the following property of the operator $\mathsf{D}(\alpha)$, one which follows from the commutation rules for creation and annihilation operators,

$$\mathsf{D}^{-1}(\alpha)c\mathsf{D}(\alpha) = c+\alpha, \tag{33a}$$

$$\mathsf{D}^{-1}(\alpha)c^\dagger\mathsf{D}(\alpha) = c^\dagger+\alpha^*. \tag{33b}$$

Now let us discuss the properties of arbitrary coherent state vectors $\Phi\{\alpha_k\}$. These vectors are not mutually orthogonal. The scalar product of two coherent state vectors is

$$(\Phi\{\alpha_k\}|\Phi\{\alpha'_k\}) = \exp\left\{\sum_i\left(-\frac{1}{2}|\alpha_i-\alpha'_i|^2+i\operatorname{Im}\alpha_i^*\alpha'_i\right)\right\}. \tag{34}$$

The vector $\Phi\{\alpha_k\}$ is the result of the action of the product of the displacement operators on the vacuum vector,

$$\Phi\{\alpha_k\} = \mathsf{D}(\{\alpha_k\})\Omega, \tag{35}$$

where

$$\mathsf{D}(\{\alpha_k\}) \equiv \prod_i \mathsf{D}(\alpha_i) = \exp\left\{\sum_i (\alpha_i c_i^\dagger - \alpha_i^* c_i)\right\}. \tag{36}$$

Because of the property (33) of operators $\mathsf{D}(\alpha_i)$ and the commutativity of operators c_i and $c_j^\dagger$ for $i \neq j$, the following relations hold for all creation and annihilation operators:

$$\mathsf{D}^{-1}(\{\alpha_k\}) c_i \mathsf{D}(\{\alpha_k\}) = c_i + \alpha_i, \tag{37a}$$

$$\mathsf{D}^{-1}(\{\alpha_k\}) c_i^\dagger \mathsf{D}(\{\alpha_k\}) = c_i^\dagger + \alpha_i^*. \tag{37b}$$

The coherent state vectors will now be shown to form a complete set of vectors in the Hilbert space of the field. Namely, we shall demonstrate that any two Fock basis vectors $\Phi_{\{n_k\}}$, and $\Phi_{\{n_k'\}}$, satisfy the equality

$$(\Phi_{\{n_k\}}|\Phi_{\{n_k'\}}) = \int \prod_i d\mu(\alpha_i)\, (\Phi_{\{n_k\}}|\Phi\{\alpha_k\})\, (\Phi\{\alpha_k\}|\Phi_{\{n_k'\}}), \tag{38}$$

where the measure $d\mu(\alpha_i)$ is specified by the formula

$$d\mu(\alpha_i) \equiv \frac{1}{\pi} d^2\alpha = \frac{1}{\pi} d(\operatorname{Re}\alpha) d(\operatorname{Im}\alpha), \tag{39}$$

and the product $\prod_i d\mu(\alpha_i)$ contains the factors $d\mu(\alpha_i)$ which correspond to only those modes i for which the occupation numbers n_i or n_i' are non-zero.†

When the representation (29) of vectors and the orthonormality of the Fock basis vectors are invoked, we get

$$\int \prod_i d\mu(\alpha_i)\, (\Phi_{\{n_k\}}|\Phi\{\alpha_k\})(\Phi\{\alpha_k\}|\Phi_{\{n_k'\}}) = \int \prod_i d\mu(\alpha_i) \exp(-|\alpha_i|^2)\, \frac{\alpha_i^{n_i}(\alpha_i^*)^{n_i'}}{\sqrt{n_i!\, n_i'!}}.$$

Putting $\alpha_i = re^{i\varphi}$, we obtain

$$\int d\mu(\alpha_i) \exp(-|\alpha_i|^2) \frac{\alpha_i^{n_i}(\alpha_i^*)^{n_i'}}{\sqrt{n_i!\, n_i'!}} = \frac{1}{\sqrt{n_i!\, n_i'!}} \int_0^\infty dr\, r^{1+n_i+n_i'} e^{-r^2} \int_0^{2\pi} d\varphi\, e^{i\varphi(n_i-n_i')}$$

$$= \begin{cases} 0, & \text{when} \quad n_i \neq n_i'. \\ 1, & \text{when} \quad n_i = n_i'. \end{cases}$$

---

† Because the occupation numbers $\{n_k\}$ and $\{n_k'\}$ satisfy the conditions $\sum_k n_k < \infty$ and $\sum_k n_k' < \infty$, both sequences $\{n_k\}$ and $\{n_k'\}$ contain only a finite number of non-zero terms.

Thus, we have

$$\int \prod_i \mathrm{d}\mu(\alpha_i)\,(\Phi_{\{n_k\}}|\Phi\{\alpha_k\})\,(\Phi\{\alpha_k\}|\Phi_{\{n_k'\}})$$
$$= \delta_{n_1 n_1'} \ldots \delta_{n_i n_i'} \ldots = (\Phi_{\{n_k\}}|\Phi_{\{n_k'\}}).$$

The set of vectors $\Phi\{\alpha_k\}$ is not linearly independent; for we have the relation

$$\Phi\{\alpha_k\} = \int \prod_i d\mu(\beta_i) \exp\left[\sum_i \left(-\frac{1}{2}|\alpha_i - \beta_i|^2 + i\,\mathrm{Im}\,\beta_i^* \alpha_i\right)\right] \Phi\{\beta_k\}. \quad (40)$$

Coherent state vectors $\Phi\{\alpha_k\}$ thus constitute an overcomplete system in the state-vector Hilbert space of the electromagnetic field.

Simultaneous eigenvectors of all annihilation operators c_i for which infinite sets of eigenvalues $\{\alpha_k\}$ are different from zero also belong to Hilbert space, provided the following condition is satisfied:

$$\sum_{i=1}^{\infty} |\alpha_i|^2 < \infty. \quad (41)$$

Such a vector is the limit of a convergent sequence of the vectors $\Phi\{\alpha_k\}$ for which finite sequences of eigenvalues differ from zero. The n-th term of this sequence, $\Phi^{(n)}$, is a vector $\Phi(\alpha_1, \ldots, \alpha_n, 0, \ldots, 0, \ldots)$. The corresponding set of eigenvalues consists of the first n terms of the infinite set $\{\alpha_k\}$ satisfying the condition (41) and with zeros in the other positions. The set of vectors constructed in this way is convergent† by virtue of condition (41).

Simultaneous eigenvectors of operators c_i, wchich are characterized by infinite sets of eigenvalues $\{\alpha_k\}$ satisfying condition (41), will also be called coherent state vectors and will be denoted by the symbols $\Phi\{\alpha_k\}$.

Coherent state vectors $\Phi\{\alpha_k\}$ are eigenvectors of the part of the field with positive frequencies $f_{\mu\nu}^{(+)}(x)$. The corresponding eigenvalues are linear combinations of the tensor wave functions $\varphi_{\mu\nu}[x|f_i]$ of a photon. For by formulae (17a) and (28) we have

† Associated with every real number ε is a natural number N such that for every $n > N$ and $m > N$

$$||\Phi^{(n)} - \Phi^{(m)}|| < \varepsilon,$$

since the norm $||\Phi^{(n)} - \Phi^{(m)}||$ is (assuming that $n > m$)

$$||\Phi^{(n)} - \Phi^{(m)}|| = \left[2\left(1 - \mathrm{Re}\exp\left(-\sum_{i=m}^{n} \frac{1}{2}|\alpha_i|^2\right)\right)\right]^{\frac{1}{2}}.$$

$$f_{\mu\nu}^{(+)}(x)\Phi\{\alpha_k\} = \Big(\sum_{i=1}^{\infty} \alpha_i \varphi_{\mu\nu}[x|f_i]\Big)\, \Phi\{\alpha_k\}. \tag{42}$$

The expectation value of the field operator $f_{\mu\nu}(x)$ in the coherent state $\Phi\{\alpha_k\}$ is equal to twice the real part of the corresponding eigenvalue of the field $f_{\mu\nu}^{(+)}(x)$. The expectation value of a field in the coherent state $\Phi\{\alpha_k\}$ thus is a real field which constitutes a solution of classical free Maxwell equations:

$$(\Phi\{\alpha_k\}|f_{\mu\nu}(x)\,\Phi\{\alpha_k\}) = \sum_{i=1}^{\infty} (\alpha_i\varphi_{\mu\nu}[x|f_i] + \alpha_i^*\varphi_{\mu\nu}^*[x|f_i]). \tag{43}$$

The average number of type-i photons in the coherent state $\Phi\{\alpha_k\}$ is equal to the squared modulus of the number α_i,

$$(\Phi\{\alpha_k\}|N_i\Phi\{\alpha_k\}) = |\alpha_i|^2. \tag{44}$$

The condition (41) thus means that the vector $\Phi\{\alpha_k\}$ describes the state of the field in which the operator of the total number of photons has a finite expectation value,

$$(\Phi\{\alpha_k\}|N\Phi\{\alpha_k\}) = \sum_{i=1}^{\infty} |\alpha_i|^2 < \infty. \tag{45}$$

Coherence of Electromagnetic Radiation

Let us consider the expectation values in a given state ϱ of the normal products of creation operators $c_i^\dagger$ and annihilation operators c_j. We shall assume, as we have done hitherto, that the operators $c_i^\dagger$ and c_j create and annihilate independent modes of field excitations:

$$\mathscr{G}^{(n,m)}(i_1, \ldots, i_n, i_{n+1}, \ldots, i_{n+m})$$
$$\equiv \mathrm{Tr}\,\{\varrho c_{i_1}^\dagger \ldots c_{i_n}^\dagger c_{i_{n+1}} \ldots c_{i_{n+m}}\} \equiv \langle c_{i_1}^\dagger \ldots c_{i_n}^\dagger c_{i_{n+1}} \ldots c_{n+m}\rangle. \tag{46}$$

The expectation values $\mathscr{G}^{(n,m)}$ will be called *correlation functions*. Since any operator can be decomposed into normal products of creation and annihilation operators, as discussed in Section 5, the expectation value of each linear operator acting in Hilbert space may be expressed in terms of the correlation functions $\mathscr{G}^{(n,m)}$.

The state of the electromagnetic field represented by a density operator

ϱ is said to be# a *fully coherent state* if there exists a sequence of complex numbers $\{z_i\} = z_1, z_2, \ldots$ such that for every value n and for every set of indices $i_1, \ldots, i_n, i_{n+1}, \ldots, i_{2n}$ we have

$$\mathscr{G}^{(n,n)}(i_1, \ldots, i_n, i_{n+1}, \ldots, i_{2n}) = \prod_{k=1}^{n} z_{i_k}^* \prod_{l=n+1}^{2n} z_{i_l}. \tag{47}$$

However, if the correlation functions $\mathscr{G}^{(n,n)}$ possess this property only for $n \leqslant M$, we say that the state of the field has the *M-th order coherence.*

The property (47) of the correlation functions $\mathscr{G}^{(n,n)}$ implies similar properties of the *correlation functions for field intensities.* For example, the correlation functions of electric and magnetic field intensities, $G_E^{(n,n)}$ and $G_B^{(n,n)}$ respectively, satisfy the relations

$$\begin{aligned} &G_E^{(n,n)}(x_1 l_1, \ldots, x_n l_n; x_{n+1} l_{n+1}, \ldots, x_{2n} l_{2n}) \\ &\equiv \langle E_{l_1}^{(-)}(x_1) \ldots E_{l_n}^{(-)}(x_n) E_{l_{n+1}}^{(+)}(x_{n+1}) \ldots E_{l_{2n}}^{(+)}(x_{2n}) \rangle \\ &= \prod_{i=1}^{n} \mathscr{E}_{l_i}^*(x_i) \prod_{j=n+1}^{2n} \mathscr{E}_{l_j}(x_j), \end{aligned} \tag{48a}$$

$$\begin{aligned} &G_B^{(n,n)}(x_1 l_1, \ldots, x_n l_n; x_{n+1} l_{n+1}, \ldots, x_{2n} l_{2n}) \\ &\equiv \langle B_{l_1}^{(-)}(x_1) \ldots B_{l_n}^{(-)}(x_n) B_{l_{n+1}}^{(+)}(x_{n+1}) \ldots B_{l_{2n}}^{(+)}(x_{2n}) \rangle \\ &= \prod_{i=1}^{n} \mathscr{B}_{l_i}^*(x_i) \prod_{j=n+1}^{2n} \mathscr{B}_{l_j}(x_j), \end{aligned} \tag{48b}$$

if the state is fully coherent.

The complex vector fields $\mathscr{E}(x)$ and $\mathscr{B}(x)$ constitute the solution of free Maxwell equations and are constructed in the following manner:

$$\mathscr{E}(x) = \sum_{i=1}^{\infty} z_i \mathscr{E}[x|f_i], \tag{49a}$$

$$\mathscr{B}(x) = \sum_{i=1}^{\infty} z_i \mathscr{B}[x|f_i], \tag{49b}$$

A definition of the full coherence of the quantum electromagnetic field, equivalent to the definition given in the present book, was formulated by R. J. Glauber (1963a, b). A detailed discussion of various aspects of the coherence of electromagnetic fields can be found in the book by Klauder and Sudarshan.

where

$$\mathscr{E}[x|f_i] = i\int d\Gamma\omega(\mathbf{k})[\mathbf{e}(\mathbf{k})f_i(\mathbf{k}, +1) - \mathbf{e}^*(\mathbf{k})f_i(\mathbf{k}, -1)]e^{-ik\cdot x}, \quad (50a)$$

$$\mathscr{B}[x|f_i] = i\int d\Gamma\mathbf{k}\times[\mathbf{e}(\mathbf{k})f_i(\mathbf{k}, +1) - \mathbf{e}^*(\mathbf{k})f_i(\mathbf{k}, -1)]e^{-ik\cdot x}. \quad (50b)$$

The indices l_i in formulae (48) label the vector components and run over the values 1, 2, 3.

First-order coherence, in accordance with the meaning of this concept in classical optics, signifies a capability for interference on the part of radiation from two different regions of space-time. In interference experiments, the energy density is measured as an average over the time interval required for the measurement. Inasmuch as the light used in these experiments is quasi-monochromatic, with a period much shorter than the measuring time, the quantity measured is an average of the intensity $\overline{I}(x)$ (average energy density) given by the formula:

$$\overline{I}(x) = \frac{1}{2}\overline{\langle:(\mathbf{E}^2(x)+\mathbf{B}^2(x)):\rangle}$$

$$= \frac{1}{2}\overline{\langle(\mathbf{E}^{(-)}(x)\cdot\mathbf{E}^{(+)}(x)+\mathbf{B}^{(-)}(x)\cdot\mathbf{B}^{(+)}(x))\rangle}. \quad (51)$$

If, because of the conditions of the experiment, the operators of the field intensities in the region of the measurement can be written† as the sum of operators,

$$\mathbf{E}(\mathbf{x}, t) = \lambda(\mathbf{x}, \mathbf{x}_1)\mathbf{E}(\mathbf{x}_1, t_1) + \lambda(\mathbf{x}, \mathbf{x}_2)\mathbf{E}(\mathbf{x}_2, t_2), \quad (52a)$$

$$\mathbf{B}(\mathbf{x}, t) = \lambda(\mathbf{x}, \mathbf{x}_1)\mathbf{B}(\mathbf{x}_1, t_1) + \lambda(\mathbf{x}, \mathbf{x}_2)\mathbf{B}(\mathbf{x}_2, t_2), \quad (52b)$$

where

$$t_1 = t - |\mathbf{x}-\mathbf{x}_1|, \quad t_2 = t - |\mathbf{x}-\mathbf{x}_2|,$$

and $\lambda(\mathbf{x}, \mathbf{x}_1)$ and $\lambda(\mathbf{x}, \mathbf{x}_2)$ are functions of pairs of points in space-time, the intensity $\overline{I(x)}$ measured at the point $\mathbf{x}$ is the sum of three expressions: the intensity I_1 coming from the point $\mathbf{x}_1$ (that is, the intensity that would be observed if only the first terms appeared on the right-hand side in formulae (52)), the intensity I_2 coming from the point $\mathbf{x}_2$, and

† This is a solution of the boundary-value problem for field operators, based on the corresponding solution for the field functions in the classical diffraction theory. Cf. the book by Jackson.

an interference term constructed out of the first-order correlation functions for the field intensities,

$$\overline{I(x)} = I_1 + I_2 + \lambda^*(\mathbf{x}, \mathbf{x}_1)\,\lambda(\mathbf{x}, \mathbf{x}_2)\,\mathrm{Re}\sum_{k=1}^{3}\overline{[G_E^{(1,1)}(x_1k, x_2k)} \\ \overline{+ G_B^{(1,1)}(x_1k, x_2k)]}. \tag{53}$$

With the conditions of the experiment fixed, i.e. with the functions $\lambda(\mathbf{x}, \mathbf{x}_1)$ and $\lambda(\mathbf{x}, \mathbf{x}_2)$ fixed, the contrast of the interference pattern, that is to say variations in the intensity $\overline{I(x)}$, for small changes of the observation point $\mathbf{x}$ depend on the sum of the moduli of the correlation functions $G_E^{(1,1)}$ and $G_B^{(1,1)}$ figuring in formula (53). For quasi-monochromatic radiation the function $G_B^{(1,1)}$ differs slightly from the function $G_E^{(1,1)}$; the definition of the interference pattern thus depends on the modulus of the correlation function $G_E^{(1,1)}(x_1k, x_2k)$. By virtue of the Schwartz inequality, on the other hand, we have

$$|G_E^{(1,1)}(x_1k, x_2k)|^2 \leqslant G_E^{(1,1)}(x_1k, x_1k)\,G_E^{(1,1)}(x_2k, x_2k). \tag{54}$$

It thus follows that the interference pattern has maximum contrast when

$$|G_E^{(1,1)}(x_1k, x_2k)|^2 = G_E^{(1,1)}(x_1k, x_1k)\,G_E^{(1,1)}(x_2k, x_2k), \tag{55}$$

which means first-order coherence.

The above definition of complete coherence is an extension of the classical concept of coherence. Correlation functions of a higher order characterize the coherence of the radiation, which is revealed in measurements of intensity correlation and not of the actual intensity. An example of experiments which permit some correlation functions of a higher order to be determined are *photon counting* experiments, which give us the probability distribution $p(\{n_i\})$ of photons in a given beam of electromagnetic radiation. The probability that in the ϱ state there are n_1 photons of the first kind, n_2 of the second kind, and in general n_i photons of the i-th kind, is the expectation value in this state of the operator $P_{\{n_i\}}$ which projects on the state vector $\Phi_{\{n_i\}}$ with these very occupation numbers:

$$P_{\{n_i\}}\Phi_{\{n_i\}} = \Phi_{\{n_i\}},$$

$$P_{\{n_i\}}\Phi_{\{n_i'\}} = 0,$$

where the sequence $\{n_i\}$ and $\{n'_i\}$ differ from each other. The projection operator $P_{\{n_i\}}$ decomposes into the normal products of creation and annihilation operators as follows,

$$P_{\{n_i\}} = \prod_{i=1}^{\infty} \frac{1}{n_i!} \sum_{k=0}^{\infty} \frac{(-1)^k}{k!} (c_i^\dagger)^{n_i+k} c_i^{n_i+k}. \tag{56}$$

To prove formula (56) we shall rewrite the projection operator $P_{\{n_i\}}$ as

$$P_{\{n_i\}} = \prod_{i=1}^{\infty} \frac{1}{2\pi} \int_0^{2\pi} d\varphi \, e^{i\varphi(c_i^\dagger c_i - n_i)}.$$

Since the factors in this product commute, let us consider one of them, say the i-th one. The exponential operator $\exp(i\varphi c_i^\dagger c_i)$ will be expressed in the form of a sum of normal products. To this end we shall examine the decomposition of the operator $\exp(\beta c^\dagger c)$,

$$e^{\beta c^\dagger c} = \sum_{n=0}^{\infty} \gamma_n(\beta) (c^\dagger)^n c^n.$$

The coefficients $\gamma_n(\beta)$ for $n \geqslant 1$ satisfy the following differential equations obtained when both sides of the above equality are differentiated and the coefficients at the same terms in the series are compared,

$$\frac{d\gamma_n(\beta)}{d\beta} = \gamma_{n-1}(\beta) + n\gamma_n(\beta).$$

In doing this we have made use of the relation $[c, (c^\dagger)^n] = n(c^\dagger)^{n-1}$. The solution of this differential equation is of the form:

$$\gamma_n(\beta) = \frac{1}{n!} (e^\beta - 1)^n$$

whence it follows that

$$e^{i\varphi c_i^\dagger c_i} = \sum_{n=0}^{\infty} \frac{1}{n!} (e^{i\varphi} - 1)^n (c_i^\dagger)^n c_i^n.$$

For all i's, therefore, we have

$$\frac{1}{2\pi} \int_0^{2\pi} d\varphi \, e^{i\varphi(c_i^\dagger c_i - n_i)} = \sum_{n=0}^{\infty} \frac{1}{n!} (c_i^\dagger)^n c_i^n \frac{1}{2\pi} \int_0^{2\pi} d\varphi \, e^{-i\varphi n_i} (e^{i\varphi} - 1)^n$$

$$= \sum_{n=n_i}^{\infty} \frac{1}{n!} (c_i^\dagger)^n c_i^n \binom{n}{n_i} (-1)^{n-n_i} = \frac{1}{n!} \sum_{k=0}^{\infty} \frac{(-1)^k}{k!} (c_i^\dagger)^{n_i+k} c_i^{n_i+k},$$

which leads to formula (56).

The expectation value $p(\{n_i\})$ of the operator $P_{\{n_i\}}$ in a fully coherent state is:

$$p(\{n_i\}) = \prod_{i=1}^{\infty} p(n_i), \tag{57a}$$

where

$$p(n_i) = e^{-|z_i|^2} \frac{|z_i|^{2n_i}}{n_i!} = e^{-\langle N_i \rangle} \frac{\langle N_i \rangle^{n_i}}{n_i!}. \tag{57b}$$

The symbol $\langle N_i \rangle$ has been used here to denote the average number of photons of the i-th kind, and the numbers z_i form a sequence $\{z_i\}$ which, by the definition of full coherence (47), determines the values of the correlation functions. The probability $p(\{n_i\})$ of finding the state, which has occupation numbers $\{n_i\}$, in the fully coherent state thus is the product of the probabilities for various kinds of photons. The probability distribution $p(n_i)$ for one type of photons is, in the coherent state, a Poisson distribution. The dispersion for such a distribution is given by the formula:

$$\sigma^2(n_i) = \langle N_i \rangle.$$

The difference between the probability distribution of finding $\{n_i\}$ numbers of photons, as observed in the given state, and the Poisson distribution may thus be regarded as a measure of the deviation from full coherence.

An example of a fully coherent state is the coherent state $\Phi\{\alpha_k\}$. The sequence of complex numbers $\{z_i\}$ which corresponds to it by the definition (47) is a sequence of numbers $\{\alpha_i\}$ which are the eigenvalues of the annihilation operators, to which the vector $\Phi\{\alpha_k\}$ belongs. In this state, not only the functions $\mathscr{G}^{(n,n)}$, but also all other correlation functions $\mathscr{G}^{(n,m)}(n \neq m)$ are equal to the products of the corresponding numbers α_i and α_j^*. In view of the properties of the coherent states, for calculating the correlation functions in an arbitrary state it is convenient to decompose the density operator of that state into operators projecting onto coherent states. Such a representation, which is called a *diagonal* or *P representation*, is given by the formula:

$$\varrho = \int \prod_i d\mu(\alpha_i) \mathscr{P}(\{\alpha_k\}) |\Phi\{\alpha_k\}\rangle\langle\Phi\{\alpha_k\}|, \tag{58}$$

where $\mathscr{P}(\{\alpha_k\})$ is a function of the sequence of numbers $\{\alpha_k\}$ and the symbol $|\Phi\{\alpha_k\}\rangle\langle\Phi\{\alpha_k\}|$ denotes the operator projecting onto the coherent state vector $\Phi\{\alpha_k\}$.

The function $\mathscr{P}(\{\alpha_k\})$ is associated in the following manner with the expectation value $T(\{\alpha_k\})$ of the density operator ϱ in the coherent state $\Phi\{\alpha_k\}$,

$$T(\{\alpha_i\}) \equiv (\Phi\{\alpha_k\}|\varrho\Phi\{\alpha_k\})$$
$$= \int \prod_i \mathrm{d}\mu(\beta_i)\mathscr{P}(\{\beta_k\})|(\Phi\{\alpha_k\}|\Phi\{\beta_k\})|^2$$
$$= \int \prod_i \mathrm{d}\mu(\beta_i)\mathscr{P}(\{\beta_k\})\exp\Big(-\sum_{k=1}^{\infty}|\alpha_k-\beta_k|^2\Big).$$

The Fourier transform $\tilde{\mathscr{P}}(\{z_k\})$ of the function $\mathscr{P}(\{\alpha_k\})$ with respect to the real and imaginary parts of the numbers α_k thus is given, because of the properties of convolution, by the formula†

$$\tilde{\mathscr{P}}(\{z_k\}) = \tilde{T}(\{z_k\})\exp\Big(\sum_{k=1}^{\infty}|z_k|^2/4\Big),$$

where $\tilde{T}(\{z_k\})$ is the Fourier transform of the function $T(\{\alpha_k\})$ and $\frac{1}{2}\exp\big(-\sum_{k=1}^{\infty}|z_k|^2/4\big)$ is the Fourier transform of the function $\exp\big(-\sum_{k=1}^{\infty}|\alpha_k|^2\big)$. Since the factor $\exp\big(\sum_k |z_k|^2/4\big)$ appears in the formula for $\tilde{\mathscr{P}}(\{z_k\})$, the function $\tilde{\mathscr{P}}(\{z_k\})$ may for some states not have the inverse Fourier transform. Such states are said not to possess a diagonal representation.

The correlation functions $\mathscr{G}^{(n,n)}(i_1, \ldots, \ldots, i_{2n})$ in the state ϱ which does have a diagonal representation (58) are given by the formula:

$$\mathscr{G}^{(n,n)}(i_1, \ldots, \ldots, i_{2n}) = \int \prod_i \mathrm{d}\mu(\alpha_i)\mathscr{P}(\{\alpha_k\})\prod_{j=1}^{n}\alpha_{i_j}^{*}\prod_{l=n+1}^{2n}\alpha_{i_l}. \quad (59)$$

Now we shall discuss the coherence property of various selected states of the electromagnetic field. An example of a fully coherent mixed state is the so-called *ideal single-mode laser state.* The density operator

---

† Fourier transforms are defined in this case with complete symmetry (the convention given in the introduction concerning the Fourier transform of a space-time function is not applicable here), that is:

$$\tilde{f}(z) = \frac{1}{2\pi}\int \mathrm{d}(\mathrm{Re}\,\alpha)\int \mathrm{d}(\mathrm{Im}\,\alpha)f(\alpha)\mathrm{e}^{i(\mathrm{Re}\,\alpha\mathrm{Re}\,z+\mathrm{Im}\,\alpha\mathrm{Im}\,z)},$$

$$f(\alpha) = \frac{1}{2\pi}\int \mathrm{d}(\mathrm{Re}\,z)\int \mathrm{d}(\mathrm{Im}\,z)\tilde{f}(z)\mathrm{e}^{-i(\mathrm{Re}\,\alpha\mathrm{Re}\,z+\mathrm{Im}\,\alpha\mathrm{Im}\,z)}.$$

ϱ_L of this state possesses a diagonal representation in which the function $\mathscr{P}(\{\alpha_k\})$ is of the form:

$$\mathscr{P}_L(\{\alpha_k\}) = \frac{1}{2|\alpha_j|}\,\delta(|\alpha_j|-r)\prod_{\substack{i=1\\ i\neq j}}^{\infty}\delta^{(2)}(\alpha_i).$$

Thus, the decomposition of the density operator ϱ_L contains operators which project onto the vectors of coherent states with only one type of field excitation, which we have indicated by the index j. The moduli of the numbers α_j are the same and equal to r in this mixture whereas the phases are distributed with a constant probability density throughout the entire range,

$$\varrho_L = \frac{1}{2\pi}\int_0^{2\pi} \mathrm{d}\varphi |\Phi(0, \ldots, 0, re^{i\varphi}, 0, \ldots)\rangle\langle\Phi(0, \ldots, 0, re^{i\varphi}, 0, \ldots)|. \quad (60)$$

The set of numbers $\{z_i\}$ specifying the values of the correlation functions $\mathscr{G}^{(n,n)}$ in this state consists of zeros and the numbers r in the j-th position, $\{z_i\} = (0, \ldots, 0, r, 0, \ldots)$. The functions $\mathscr{G}^{(n,n)}$ calculated in this state are equal to the corresponding correlation functions calculated in any state contained in the mixed state (60). Those correlation functions $\mathscr{G}^{(n,m)}$ for which $n \neq m$, on the other hand, are equal to zero in the state of an ideal single-mode laser. It follows in particular from this that the expectation value of the field operator $f_{\mu\nu}(x)$ vanishes in this state.

The Fock basis vectors $\Phi_{\{n_k\}}$, i.e. the eigenvectors of the photon number operators, describe incoherent states of the electromagnetic field.

Another example of an incoherent state of the field is that of *black-body radiation*. Such radiation is, in a confined volume, in thermal equilibrium with the surrounding walls. It is described by means of a density operator of the form:

$$\varrho_T = \exp(-\beta H)/Tr\{\exp(-\beta H)\}, \quad (61)$$

where $\beta = 1/kT$, T denoting the temperature and k the Boltzmann constant, while the Hamiltonian H is given by the formula

$$H = \sum_{\mathbf{k},\lambda} \omega(\mathbf{k}) c^{\dagger}_{\mathbf{k},\lambda} c_{\mathbf{k},\lambda}.$$

The density operator ϱ_T possesses a diagonal representation,

$$\varrho_T = \int \prod_{\mathbf{k},\lambda} \mathrm{d}\mu(\alpha_{\mathbf{k},\lambda}) \mathscr{P}_T(\{\alpha_{\mathbf{k},\lambda}\}) |\Phi\{\alpha_{\mathbf{k},\lambda}\}\rangle \langle\Phi\{\alpha_{\mathbf{k},\lambda}\}|,$$

where the symbol $|\Phi\{\alpha_{\mathbf{k},\lambda}\}\rangle \langle\Phi\{\alpha_{\mathbf{k},\lambda}\}|$ denotes an operator projecting onto the coherent state vector, i.e. the simultaneous eigenvector of all annihilation operators $c_{\mathbf{k},\lambda}$. The function $\mathscr{P}_T(\{\alpha_{\mathbf{k},\lambda}\})$ is of the form

$$\mathscr{P}_T(\{\alpha_{\mathbf{k},\lambda}\}) = \prod_{\mathbf{k},\lambda} \frac{1}{\langle N_{\mathbf{k},\lambda}\rangle} \exp(-|\alpha_{\mathbf{k},\lambda}|^2/\langle N_{\mathbf{k},\lambda}\rangle),$$

where $\langle N_{\mathbf{k},\lambda}\rangle$ is the average number of photons with momentum $\mathbf{k}$ and helicity λ,

$$\langle N_{\mathbf{k},\lambda}\rangle = (e^{\beta\omega(\mathbf{k})} - 1)^{-1}.$$

The probability distribution $p(\{n_{\mathbf{k},\lambda}\})$ of finding in the blackbody radiation $n_{\mathbf{k},\lambda}$ photons with momenta $\mathbf{k}$ and polarizations λ is given by the formula:

$$p(\{n_{\mathbf{k},\lambda}\}) = \prod_{\mathbf{k},\lambda} p_{\mathbf{k},\lambda}(n_{\mathbf{k},\lambda}),$$

$$p_{\mathbf{k},\lambda}(n) = \langle N_{\mathbf{k},\lambda}\rangle^n/(1+\langle N_{\mathbf{k},\lambda}\rangle)^{n+1}.$$

The dispersion for this distribution is greater than for the Poisson distribution, amounting as it does to

$$\sigma^2(n_{\mathbf{k},\lambda}) = \langle N_{\mathbf{k},\lambda}\rangle (\langle N_{\mathbf{k},\lambda}\rangle + 1).$$

12. THE INTERACTION OF THE QUANTUM ELECTROMAGNETIC FIELD WITH EXTERNAL SOURCES

Let us now consider the interaction of the quantized electromagnetic field with charges and currents. The charges and currents will be assumed to be described in the classical (non-quantum) manner by numerical functions: the components of the current four-vector $\mathscr{I}^\mu(x)$. Furthermore, we shall assume that the influence of the electromagnetic field on the distribution of the charges and currents is so small as to be negligible. As in classical theory, such sources are called *external*; their distribution and motion are given beforehand. We shall consider sources which vanish in the remote past ($x^0 < T_1$) and in the far future ($x^0 > T_2$).

In Section 9 we presented the classical theory of the interaction of the electromagnetic field with external sources. On the basis of this we shall construct a theory for the interaction of the quantum field. We assume that the field operators $f_{\mu\nu}(x)$ satisfy the following equations:

$$\partial_\mu f^{\mu\nu}(x) = \mathscr{I}^\nu(x), \tag{1a}$$

$$\partial_\mu \check{f}^{\mu\nu}(x) = 0, \tag{1b}$$

and the asymptotic conditions:

$$f_{\mu\nu}(x) = \begin{cases} f^{\text{in}}_{\mu\nu}(x) & \text{for } x^0 < T_1, & (2a) \\ f^{\text{out}}_{\mu\nu}(x) & \text{for } x^0 > T_2, & (2b) \end{cases}$$

where $f^{\text{in}}_{\mu\nu}$ and $f^{\text{out}}_{\mu\nu}$ satisfy the source-free Maxwell equations. The above formula is frequently interpreted by the statement that the operator $f_{\mu\nu}(x)$ *interpolates* between the operators $f^{\text{in}}_{\mu\nu}(x)$ and $f^{\text{out}}_{\mu\nu}(x)$. Since they are linear, eqs. (1) for operators may be solved just as equations for functions, the result being a relation between the operators $f^{\text{in}}_{\mu\nu}(x)$ and $f^{\text{out}}_{\mu\nu}(x)$, as in Section 9 (cf. formulae (9.54) and (9.56)):

$$f^{\text{out}}_{\mu\nu}(x) = f^{\text{in}}_{\mu\nu}(x) + \partial_{[\mu} \int \mathrm{d}^4y\, D(x-y)\mathscr{I}_{\nu]}(y). \tag{3}$$

From the rules of canonical quantization it follows that the operators $f^{\text{in}}_{\mu\nu}(x)$ and $f^{\text{out}}_{\mu\nu}(x)$ satisfy the commutation relations (11.1) for a free field. Using the Fourier integral decomposition of the fields $f^{\text{in}}_{\mu\nu}(x)$ and $f^{\text{out}}_{\mu\nu}(x)$ we determine, as described in the preceding section, two sets of operators $c^{\text{in}}(\mathbf{k}, \lambda)$ and $c^{\text{out}}(\mathbf{k}, \lambda)$. To construct the two sets of photon annihilation operators c^{in}_i and c^{out}_i, we make use of the same complete orthonormal system of wave packet profiles $f_i(\mathbf{k}, \lambda)$. Since the operators $f^{\text{in}}_{\mu\nu}$ and $f^{\text{out}}_{\mu\nu}$ satisfy the commutation relations for a free field, the operators c^{in}_i and $c^{\dagger\text{in}}_j$ as well as the operators c^{out}_i and $c^{\dagger\text{out}}_j$ satisfy the commutation relations for boson annihilation and creation operators. Equation (3) implies the following relation between the annihilation operators c^{in}_k and c^{out}_k:

$$c^{\text{out}}_k = c^{\text{in}}_k + i\iota_k, \tag{4}$$

where the coefficients ι_k are obtained when the functions $\iota(\mathbf{k}, \lambda)$ defined in Section 9 are integrated with the wave packet profiles $f_k(\mathbf{k}, \lambda)$,

$$\iota_k \equiv \sum_\lambda \int \mathrm{d}\Gamma \bar{f}_k(\mathbf{k}, \lambda)\,\iota(\mathbf{k}, \lambda). \tag{5}$$

With the help of the two sets of IN and OUT photon creation operators we form two bases in the state vector space of the field constructed out of the vectors $\Phi^{\text{in}}_{\{n_k\}}$ and the vectors $\Phi^{\text{out}}_{\{n_k\}}$, respectively. The states described by these vectors are characterized by giving the number of photons whose states are labelled with i in the remote past, and the far future, respectively. Each of these bases contains one vacuum vector. These vectors are defined by the conditions:

$$c^{\text{in}}_k \Omega^{\text{in}} = 0, \quad \text{for all } k\text{'s} \tag{6a}$$

$$c^{\text{out}}_k \Omega^{\text{out}} = 0, \quad \text{for all } k\text{'s}. \tag{6b}$$

In the past, the field operator $f_{\mu\nu}(x)$ becomes the operator $f^{\text{in}}_{\mu\nu}(x)$ and for that reason the basis $\Phi^{\text{in}}_{\{n_k\}}$ is most convenient for a state characterization which refers to the past. Similarly, the basis $\Phi^{\text{out}}_{\{n_k\}}$ is particularly convenient for describing the properties which characterize the state of the system in the future.

In using the vectors of these two bases, we define the *transition amplitude* $A(\{n_i\}, \{n'_i\})$,

$$A(\{n_i\}, \{n'_i\}) \equiv A(n_1, \ldots, n_k, \ldots; n'_1, \ldots, n'_k, \ldots)$$

$$\equiv (\Phi^{\text{out}}_{\{n_i\}} | \Phi^{\text{in}}_{\{n'_i\}}). \tag{7}$$

The transition amplitude $A(\{n_i\}, \{n'_i\})$ is the amplitude of the probability that in the state characterized in the past by the occupation numbers $n'_1, \ldots, n'_k, \ldots$ we detect in the future a state characterized by the occupation numbers $n_1, \ldots, n_k, \ldots$. In brief, we frequently call A the amplitude of transition from the state $\Phi^{\text{in}}_{\{n'\}}$ to the state $\Phi^{\text{out}}_{\{n_i\}}$. Sometimes, an even more figurative term is used, by calling A the probability amplitude for a process consisting in photons undergoing transition from the state $\{n'_i\}$ to the state $\{n_i\}$. Expressions of this kind should be used with caution, however, bearing in mind that we are concerned with a quantum theory in which the concept of a physical process signifies only a correlation between measurements on a system.

The assumption that the vectors $\Phi^{\text{in}}_{\{n_i\}}$ and $\Phi^{\text{out}}_{\{n_i\}}$ form two complete orthonormal systems of vectors in the state-vector Hilbert space leads to the unitarity conditions for the transition amplitudes. These conditions make it possible to determine a unitary operator S, whose definition and properties will be discussed in the next section.

All the transition amplitudes can be expressed, owing to the existence of the relations (4) between the operators c_k^{out} and c_k^{in}, in terms of the *vacuum-to-vacuum transition amplitude* $(\Omega^{\text{out}}|\Omega^{\text{in}})$ and the coefficients ι_k.

Now let us calculate the vacuum-to-vacuum transition amplitude which we can determine up to the phase by using the probability-conservation law which is a result of the unitarity condition for the amplitudes. It follows from this law that, among other things, the sum of transition probabilities from the vacuum state to all possible final states is equal to unity. By formulae (4) and (6) the amplitude of the transition from the vacuum state to an arbitrary final state is of the form

$$A(n_1 \ldots n_k \ldots; 0 \ldots 0 \ldots) = \left(\prod_{k=1}^{\infty} \frac{1}{\sqrt{n_k!}} (c_k^{\dagger\text{out}})^{n_k} \Omega^{\text{out}} \middle| \Omega^{\text{in}}\right)$$

$$= (n_1! \ldots n_k! \ldots)^{-\frac{1}{2}} (\Omega^{\text{out}}|\Omega^{\text{in}}) \prod_{k=1}^{\infty} (i\iota_k)^{n_k}. \tag{8}$$

Probability conservation gives us

$$1 = |(\Omega^{\text{out}}|\Omega^{\text{in}})|^2 \sum_{n_1 \ldots n_k \ldots} (n_1! \ldots n_k! \ldots)^{-1} \prod_{k=1}^{\infty} |\iota_k|^{2n_k}$$

$$= |(\Omega^{\text{out}}|\Omega^{\text{in}})|^2 \exp\left(\sum_{k=1}^{\infty} |\iota_k|^2\right). \tag{9}$$

From this we obtain the value of the vacuum-to-vacuum transition amplitude determined to within the phase factor $\exp(i\alpha)$, where α is a real number which we shall find in Section 13,

$$(\Omega^{\text{out}}|\Omega^{\text{in}}) = e^{i\alpha} \exp\left(-\frac{1}{2} \sum_{k=1}^{\infty} |\iota_k|^2\right). \tag{10}$$

The expression in the exponent in eq. (10) has a straightforward physical interpretation. It can be expressed in terms of the average number $\bar{N}$ of photons radiated by the source when the initial state is a vacuum.

We shall demonstrate that

$$\bar{N} = \sum_{k=1}^{\infty} |\iota_k|^2 = -\int \mathrm{d}\Gamma \tilde{\mathscr{I}}_\mu^*(k)\tilde{\mathscr{I}}^\mu(k)$$

$$= -\frac{1}{2}\int \mathrm{d}^4x\,\mathrm{d}^4y\,\mathscr{I}_\mu(x)\,D^{(1)}(x-y)\,\mathscr{I}^\mu(y), \tag{11}$$

where

$$D^{(1)}(x) = \int \mathrm{d}\Gamma(\mathrm{e}^{-ik\cdot x}+\mathrm{e}^{ik\cdot x}). \tag{12}$$

As an auxiliary measure, let us calculate the average number $\bar{N}_k$ of photons in the k-th state, emitted by the source in a process in which no radiation is incident. The number $\bar{N}_k$ is equal to the sum of the probabilities of the transitions to the state in which there are, among others, n_k photons in the k-th state, multiplied by the number n_k,

$$\bar{N}_k = \sum_{n_1 \dots n_k \dots} n_k |(\Phi^{\text{out}}_{n_1 \dots n_k \dots}|\Phi^{\text{in}})|^2. \tag{13}$$

Utilizing the fact that the vectors $\Phi^{\text{out}}_{\{n_k\}}$ form a complete set, we write formula (13) as

$$\bar{N}_k = (\Omega^{\text{in}}|N_k^{\text{out}}\Omega^{\text{in}}), \tag{14}$$

where N_k^{out} is the operator of the number of photons in the k-th state and is constructed from the operators c_k^{out} and $c_k^{\dagger\text{out}}$. When we use the relation (4) between the operators c_k^{out} and c_k^{in}, we arrive at:

$$\bar{N}_k = |\iota_k|^2. \tag{15}$$

The average number $\bar{N}$ of all photons emitted is the sum of the numbers $\bar{N}_k$. In this way we have demonstrated the validity of formula (11).

The same method can be used to calculate the average energy, average momentum, and average angular momentum radiated by the source, on the assumption that the initial state is a vacuum. These quantities are equal to the analogous quantities calculated in the classical theory (cf. formulae (9.58) and (9.59)). As an example, we give the formula for the four-vector of the average momentum

$$\bar{P}_\mu = (\Omega^{\text{in}}|P_\mu^{\text{out}}\Omega^{\text{in}}) = -\int \mathrm{d}\Gamma k_\mu \tilde{\mathscr{I}}_\nu^*(k)\tilde{\mathscr{I}}^\nu(k). \tag{16}$$

Now let us go on to an examination of the statistical law governing the emission of photons of various types in the case when the initial state is a vacuum. The probability $p_{n_1 \dots n_k \dots}$ of the emission of n_1

photons of the first kind, n_2 photons of the second kind, etc., is given by virtue of the equalities (8), (10), and (15) by the following formula:

$$p_{n_1 \ldots n_k \ldots} \equiv p_{\{n_k\}} = \prod_{i=1}^{\infty} \frac{1}{n_i!} (\bar{N}_i)^{n_i} e^{-\bar{N}_i}. \tag{17}$$

It follows from this formula that:

1. Processes of the emission of photons of various kinds are independent phenomena. The total probability $p_{n_1 \ldots n_k \ldots}$ is the product of the emission probabilities of the particular types of photons.

2. The emission probability of a particular (i-th) type of photons is governed by the Poisson distribution,

$$p_n = \frac{1}{n!} (\bar{N}_i)^n e^{-\bar{N}_i}. \tag{18}$$

The emission probability distribution of photons with occupation numbers $\{n_k\}$ thus is of the same form as the probability distribution of finding the occupation numbers of photons in a fully coherent state. We shall return to this subject in the next section. The Poisson distribution characterizes processes consisting in certain completely random events being able to occur an arbitrary number of times, with only the average value $\bar{n}$ being known. The Poisson distribution governs, for instance, the number of connections made per unit time by a telephone exchange if we know the average number of connections and each connection is a random event, independent of the others. From the fact that the Poisson distribution governs the emission of photons of a particular type and that the processes of emission of photons of various types are independent we deduce that the individual photon emissions by an external source are statistically independent events. This result is a consequence of the assumption that the electromagnetic field does not affect the source.

We now go on to study the transition amplitude in the case when the initial state is one with a well-defined number of photons. The general formula for the transition amplitude has a complicated structure. These complications are due to the fact that not all photons which we detect when making measurements in the far future were emitted by the sources. Some of the photons turned up there because they had been present in the past and they entered the recording instrument without ever interacting with the radiation sources. This problem is illustrated with the example of the amplitude of the transition from an initial state, containing n photons in some state chosen to be the first, to a final state in which there are $n+m$ photons of the same kind. The probability amplitude of this process is given by the formula

$$A(n+m; n) \equiv A(n+m, 0, \ldots; n, 0, \ldots)$$

$$= [(n+m)!\, n!]^{-\frac{1}{2}} \big((c_1^{\dagger\text{out}})^{n+m} \Omega^{\text{out}} | (c_1^{\dagger\text{in}})^n \Omega^{\text{in}}\big)$$

$$= [(n+m)!\, n!]^{-\frac{1}{2}} (\Omega^{\text{out}}|\Omega^{\text{in}}) \Big[(i\iota_1)^{n+m} (i\iota_1^*)^n$$

$$+ (i\iota_1)^{n+m-1} (i\iota_1^*)^{n-1} (n+m) n$$

$$\ldots + (i\iota_1)^{n+m-k} (i\iota_1^*)^{n-k} \binom{n+m}{k} \binom{n}{k} k! + \ldots + \big((n+m)!/m!\big) (i\iota_1)^m \Big]$$

$$= \big(n!/(n+m)!\big)^{\frac{1}{2}} (i\iota_1)^m L_n^m(|\iota_1|^2) (\Omega^{\text{out}}|\Omega^{\text{in}}), \tag{19}$$

where $L_n^m(|\iota_1|^2)$ is a Laguerre polynomial. This formula enables us to interpret the transition amplitude $A(n+m; n)$ as the sum of the probability amplitudes of the processes in which the source absorbs and emits a particular number of photons. The first term in the sum describes the process in which the source absorbs all the photons of the incident beam and emits $n+m$. The second term describes the process in which the source absorbs $n-1$ photons and emits $n+m-1$; one of the photons from the incident beam is transmitted without interacting with the source. The last term describes the process in which the source does not absorb any photons, while emitting m photons. The appearance of numerical factors in the particular terms in the sum is due to the indistinguishability of photons. The factor in front of the k-th term is of the form $\binom{n+m}{k}\binom{n}{k} k!$. For out of n photons in the initial state we can select k photons, which do not interact with the sources, in $\binom{n}{k}$ different ways. Similarly, k photons out of $n+m$ photons in the final state can be chosen in $\binom{n+m}{k}$ ways. The factor $k!$ comes from the fact that we do not distinguish processes differing in the way k photons not interacting with the sources are arranged. It is seen, therefore, that the concept of a photon is useful for describing the phenomena of emission and absorption of radiation. Almost all processes of interaction of a quantized electromagnetic field with sources can be pictured as consisting of elementary acts of photon emission and absorption.

The formula (19) derived for the transition amplitude $A(n+m; n)$ will now be utilized to discuss *stimulated emission.* In the simplified case, when the sources are weak enough so that the multiple emission of a photon may be neglected, the sum figuring in formula (19) reduces to the last term. In order for the last term but one (and also all the preceding terms) to be negligible in comparison with the last term, the following condition must be satisfied:

$$|\iota_1|^2 \frac{n}{m+1} \ll 1. \tag{20}$$

The characteristic parameter of the expansion thus is the product of the intensity of the radiation sources† $|\iota|^2$ and the incident beam intensity n (measured by the number of photons). If condition (20) is satisfied, it is sufficient to take only the photon emission by the source into account, while absorption and re-emission may be neglected.

In the simplest case, when $m = 1$, we then have

$$A(n+1; n) \simeq i\sqrt{n+1}\,(\Omega^{\text{out}}|\Omega^{\text{in}})\,\iota_1. \tag{21}$$

Let us compare this amplitude with the transition amplitude for an initial state of n photons of the first kind and a final state in which there are n photons of the first kind and one photon of the k-th kind ($k \neq 1$).

---

† As an example, note that in the case when the source of radiation is a point particle of charge e, moving slowly ($v/c \ll 1$), by virtue of the considerations in Section 9, we obtain

$$|\iota|^2 \sim 10^{-3} \frac{e^2}{\hbar c} \left(\frac{v}{c}\right)^2 c^3 \left(\frac{\Delta k_x}{\omega}\right)\left(\frac{\Delta k_y}{\omega}\right)\left(\frac{\Delta k_z}{\omega}\right),$$

where Δk characterizes the spread of the photon wave vectors.

We denote this amplitude by the symbol $A(n, 1; n)$ and calculate it in the weak-source approximation, i.e. under the assumption that condition (20) is satisfied. In this approximation the amplitude is equal to the probability amplitude of the source emitting a photon in a hitherto unoccupied one-particle state. We thus have

$$A(n, 1; n) \simeq i(\Omega^{\text{out}}|\Omega^{\text{in}})\iota_k. \tag{22}$$

The probabilities of the processes described above are equal to the squared moduli of the corresponding amplitudes,

$$|A(n+1;n)|^2 \simeq (n+1)p_1, \tag{23a}$$

$$|A(n, 1; n)|^2 \simeq p_k, \quad (k \neq 1), \tag{23b}$$

where

$$p_i \equiv |(\Omega^{\text{out}}|\Omega^{\text{in}})|^2|\iota_i|^2. \tag{24}$$

Thus, we see that in the weak-source approximation the probability of a photon being emitted to a state n-fold occupied is $n+1$ greater than the probability of emission to an unoccupied state. The probability $|A(n+1; n)|^2$ is the sum of the probability p_1 of *spontaneous emission*, which is independent of the state of the radiation incident on the source, and the probability np_1 of *stimulated emission*, which is proportional to the intensity (number of photons) of the incident radiation. The resultant effect of emitted photons being "attracted" by photons present in the incident beam is related, as follows from our considerations, to the fact that photons are subject to Bose–Einstein statistics. The corpuscular picture of radiation phenomena proves useful in explaining this effect. The stimulated emission effect is the principle underlying the operation of quantum amplifiers of electromagnetic radiation: masers and lasers.

An arbitrary transition amplitude from a state $\Phi^{\text{in}}_{\{n'_i\}}$ to a state $\Phi^{\text{out}}_{\{n_i\}}$ may also be presented, just as the amplitude $A(n+m; n)$, in the form of a sum of terms, each of which describes a process with a particular number of photons emitted and absorbed. We shall give only the first term of this sum. It describes the absorption of all photons from the initial state and the emission of all photons which we observe in the final state. The other terms may be presented as the sum of amplitudes of transition between states with a smaller number of photons,

$$A(n_1 \ldots n_k \ldots; n_1' \ldots n_k' \ldots) = (n_1! \ldots n_k! \ldots n_1'! \ldots n_k'! \ldots)^{-\frac{1}{2}}$$

$$\times (\Omega^{\text{out}}|\Omega^{\text{in}}) \Big[\prod_{k=1}^{\infty} (i\iota_k)^{n_k} (i\iota_k^*)^{n_k'} + \ldots\Big]. \tag{25}$$

13. THE FORMULATION OF THE QUANTUM THEORY OF THE MAXWELLIAN FIELD WITH THE AID OF POTENTIALS

In the theory of the free quantum electromagnetic field and in the quantum theory of the electromagnetic field with external sources there is no need to introduce the concept of a potential operator. In the preceding section we have shown how the probability amplitudes of physical processes taking place in the presence of external sources can be calculated by a method employing only field strength operators. It is difficult, however, to avoid introducing potentials of the electromagnetic field when describing the influence of the field on sources. The classical theory of the mutual interaction between an electromagnetic field and charged particles moving in that field is usually constructed on the basis of a Lagrangian which is the sum of the Lagrangian of the field, the Lagrangian of the particles, and an expression describing the interaction; the interaction Lagrangian is of the form

$$-e \sum_{A=1}^{N} \int_A d\tau \, \frac{d\xi_A^\mu}{d\tau} A_\mu(\xi_A(\tau)).$$

The potential A_μ is an auxiliary quantity in this theory. It does not figure in the equations of motion of the particles, nor in the formulae for the dynamical quantities (cf. Section 8). The introduction of this concept does, however, simplify the theory. Subsequent chapters will deal with the quantum mechanics of systems of particles in an external electromagnetic field and then with the complete quantum theory of mutually interacting electrons and photons. It is possible to formulate these theories without employing the concept of the potential, but the theory thus obtained has an extremely complicated form.# To facilitate the construction of complete quantum electrodynamics and to build a bridge between it and the theory under consideration here, we shall formulate in this section the theory of the electromagnetic field with external sources, using potential operators to do so. Employing the concept of the potential operator, we shall construct the S operator in the form of a chronological exponential operator. This will enable us to compare the theory in question with the theory of a vector field with mass. The S operator expressed as a chronological exponential operator will furthermore serve to carry out a discussion of the problem of the relativistic invariance of the theory and the causality condition.

In introducing the potential operators A_μ in the quantum theory of the Maxwellian field, we must give up the tensor notation of the equations of this theory. The operator $A_\mu(x)$ is not a vector field## in quantum theory. To specify a potential uniquely, one condition would have to be imposed on the four components of the potential. No such condition, even the *Lorentz condition*

$$\partial^\mu A_\mu^L(x) = 0,$$

which has an apparently tensor form, can be made compatible with the commutation relations for field operators and the assumption that the components $A_\mu^L(x)$ form a four-vector.

This is seen most easily upon considering the simplest quantum electromagnetic field theory: the Maxwellian theory of a source-free field. With the assumption that the vacuum state vector Ω is invariant under Poincaré transformations and that the field strength operators and potential operator A_μ^L obey the tensor transformation laws under Poincaré transformations, we obtain the following expression for the vacuum expectation value for the commutator of the potential components,

$$(\Omega|[A_\mu^L(x), A_\nu^L(y)]\Omega) = i\lambda\partial_\mu\partial_\nu D(x-y).$$

Several formulations of quantum electrodynamics without potentials have been given. Recently, such a formulation has been given by Mandelstam (1962).

The potential operator does retain its vector character in formulations of quantum electrodynamics given by E. Fermi (1932) and S. N. Gupta (1950) and K. Bleuler (1950). In both cases, however, this is accomplished at the price of introducing spaces of state vectors which are not Hilbert spaces. In the Fermi formulation the state vectors cannot be normalized, while in the Gupta–Bleuler formulation the length of the vectors may assume negative values or become zero.

To get this expression we have made use of the fact that the components of the potential operator A_μ^L satisfy the d'Alembert equation and the Lorentz condition. The formula so obtained is incompatible with the commutation relations (11.1) for the components of the field operator $f_{\mu\nu}(x)$.

The mere introduction of the potential operator in quantum field theory means discarding the tensor notation of the theory, so we choose the condition for the potential components in a form that is dependent on the reference frame but is convenient for subsequent consideration. Namely, we shall make use of the potential in the *radiation gauge*, sometimes also called the *Coulomb gauge*. The potential operator in the radiation gauge satisfies the condition

$$\nabla \cdot \mathbf{A}(x) = 0. \tag{1}$$

Before formulating the quantum field theory in the radiation gauge we shall make several remarks on the classical electromagnetic theory in the radiation gauge.

THE CLASSICAL THEORY

It follows from Maxwell's equations that the components of a radiation-gauge potential satisfy the system of equations:

$$\Box \mathbf{A}(x) = \mathscr{I}^{\text{tr}}(x), \tag{2a}$$

$$-\Delta A^0(x) = \mathscr{I}^0(x), \tag{2b}$$

where the symbol $\mathscr{I}^{\text{tr}}$ denotes the transverse (divergence-free) part of the current vector $\mathscr{I}$,

$$\mathscr{I}^{\text{tr}}(\mathbf{x}, t) \equiv \mathscr{I}(\mathbf{x}, t) - \nabla\Delta^{-1}(\nabla \cdot \mathscr{I}(\mathbf{x}, t)). \tag{3}$$

The operator Δ^{-1} is the inverse integral operator of the Laplacian,

$$\Delta^{-1} f(\mathbf{x}) \equiv -\int \frac{d^3k}{(2\pi)^3} \frac{\tilde{f}(\mathbf{k})}{|\mathbf{k}|^2} e^{i\mathbf{k}\cdot\mathbf{x}} = -\frac{1}{4\pi}\int d^3x' \frac{f(\mathbf{x}')}{|\mathbf{x}-\mathbf{x}'|}. \tag{4}$$

From eq. (2b) we deduce that the time component of the radiation-gauge potential does not depend on the quantities describing independent degrees of freedom of the radiation field. It is determined uniquely by the charge density $\mathscr{I}^0(x)$. In the radiation gauge the radiation field is described only by the spatial component of the potential vector.

The equations connecting the field intensity with the potential can, in the radiation gauge, be solved for the potential components, yielding

$$A^0(\mathbf{x}, t) = \frac{1}{4\pi}\int d^3y \frac{1}{|\mathbf{x}-\mathbf{y}|} \mathscr{I}^0(\mathbf{y}, t), \tag{5a}$$

$$\mathbf{A}(\mathbf{x}, t) = -\int_{-\infty}^{t} dt \mathbf{E}(\mathbf{x}, t), \tag{5b}$$

or

$$\mathbf{A}(\mathbf{x}, t) = -\frac{1}{4\pi}\int d^3y \frac{1}{|\mathbf{x}-\mathbf{y}|} \nabla\times\mathbf{B}(\mathbf{y}, t). \tag{5c}$$

It follows from these formulae that in the absence of sources the potential vector is constructed in the following manner from the functions $f(\mathbf{k}, \lambda)$ which describe the independent degrees of freedom of the field,

$$\mathbf{A}(x) = \int d\Gamma[\mathbf{e}(\mathbf{k})f(\mathbf{k}, +1)-\mathbf{e}^*(\mathbf{k})f(\mathbf{k}, -1)]e^{-ik\cdot x}+\text{c.c.} \tag{6}$$

The Quantum Theory of the Free Field

The radiation-gauge potential operator in the quantum theory of the free electromagnetic field is the name given to the operator $\mathbf{A}(x)$, as defined by formula (5b) or (5c), in which the symbols $\mathbf{E}$ and $\mathbf{B}$ now denote operators constructed out of the components of the operator $f_{\mu\nu}(x)$. The operator $\mathbf{A}(x)$ is related to the operators $c(\mathbf{k}, \lambda)$ by the formula:

$$\mathbf{A}(x) = \int d\Gamma[\mathbf{e}(\mathbf{k})c(\mathbf{k}, +1)-\mathbf{e}^*(\mathbf{k})c(\mathbf{k}, -1)]e^{-ik\cdot x}+\text{h.c.}, \tag{7}$$

where h.c. denotes the Hermitian conjugate of the first operator. The operator $\mathbf{A}(x)$ may also be expressed in terms of the photon annihilation and creation operators c_i and $c_i^\dagger$,

$$\mathbf{A}(x) = \mathbf{A}^{(+)}(x)+\mathbf{A}^{(-)}(x), \tag{8a}$$

$$\mathbf{A}^{(+)}(x) = \sum_{i=1}^{\infty} c_i\mathbf{e}[x|f_i], \tag{8b}$$

$$\mathbf{A}^{(-)}(x) = \sum_{i=1}^{\infty} c_i^\dagger\mathbf{e}^*[x|f_i], \tag{8c}$$

where

$$\mathbf{e}[x|f_i] \equiv \int \mathrm{d}\Gamma[\mathbf{e}(\mathbf{k})f_i(\mathbf{k}, +1) - \mathbf{e}^*(\mathbf{k})f_i(\mathbf{k}, -1)]\mathrm{e}^{-ik\cdot x}. \tag{9}$$

The function $\mathbf{e}[x|f_i]$ is a *vector wave function of the photon* in the radiation gauge of the vectors ε_μ. In general, by the photon vector wave function we shall understand the expression $\varphi_\mu[x|f]$,

$$\varphi_\mu[x|f] \equiv \int \mathrm{d}\Gamma f_\mu(\mathbf{k})\mathrm{e}^{-ik\cdot x}, \tag{10}$$

where

$$f_\mu(\mathbf{k}) \equiv \varepsilon_\mu(\mathbf{k})f(\mathbf{k}, +1) - \varepsilon_\mu^*(\mathbf{k})f(\mathbf{k}, -1). \tag{11}$$

The vector wave function $\varphi_\mu[x|f]$ thus depends not only on the choice of wave-packet profile $f(\mathbf{k}, \lambda)$ but also on the choice of gauge for the vectors $\varepsilon_\mu(\mathbf{k})$. In every gauge, however, the following relations are satisfied:

$$k^\mu f_\mu(\mathbf{k}) = 0, \tag{12a}$$

$$\partial^\mu \varphi_\mu[x|f] = 0, \tag{12b}$$

$$\Box \varphi_\mu[x|f] = 0. \tag{12c}$$

The scalar product of two wave-packet profiles is expressible in terms of the vector wave functions f_μ by the formula

$$(f_1|f_2) = -\int \mathrm{d}\Gamma f_\mu^{1*}(\mathbf{k}) f^{2\mu}(\mathbf{k}).$$

This same scalar product may also be written in terms of the vector wave functions φ_μ. With this in mind, let us define the expression

$$(\varphi[f_1]\|\varphi[f_2]) \equiv i\int \mathrm{d}\sigma_\mu\{(\partial^{[\mu}\varphi^{\nu]}[x|f_1])^*\varphi_\nu[x|f_2] - \varphi_\nu^*[x|f_1]\partial^{[\mu}\varphi^{\nu]}[x|f_2]\}. \tag{12d}$$

The integral in this formula does not depend on the choice of the hypersurface σ by virtue of the field equations satisfied by the functions φ_μ. If we take this hypersurface to be the hyperplane t = const, we get

$$(\varphi[f_1]\|\varphi[f_2]) = (f_1|f_2).$$

The notation used here for the scalar product of vector wave functions is the same as that used for the scalar product of tensor wave functions (cf. formula (11.17)). To avoid confusion wherever both of these scalar products appear next to each other, we shall distinguish between them by providing the wave functions with the superscript T or V.

The scalar product expressed in terms of vector wave functions of the photon, in contrast to the corresponding expression for tensor functions, is in the form of a space integral, which is typical of scalar products in quantum mechanics. The price we pay for giving such a local form to the scalar product in the coordinate representation is that gauge-dependent quantities make their appearance. The entire integrand in formula (12d) depends on the choice of the gauge. The gauge-dependence disappears, however, when the integration is performed.

The functions $f^i_\mu(\mathbf{k})$ constructed from the complete orthonormal system of profiles $f_i(\mathbf{k}, \lambda)$ satisfy the following completeness and orthonormality conditions:

$$\int d\Gamma f^{i*}_\mu(\mathbf{k}) f^{j\mu}(\mathbf{k}) = -\delta_{ij}, \tag{13a}$$

$$\sum_{i=1}^{\infty} k_{[\mu} f^i_{\nu]}(\mathbf{k}) k'_{[\lambda} f^{i*}_{\varrho]}(\mathbf{k}') = -k_{[\mu} g_{\nu][\lambda} k_{\varrho]} \delta_\Gamma(\mathbf{k}, \mathbf{k}'). \tag{13b}$$

In the radiation gauge, the completeness condition for the set of functions $\mathbf{f}^i(\mathbf{k})$ is of the form:

$$\sum_{i=1}^{\infty} f^i_k(\mathbf{k}) f^{i*}_l(\mathbf{k}') = \left(\delta_{kl} - \frac{k_k k_l}{|\mathbf{k}|^2}\right) \delta_\Gamma(\mathbf{k}, \mathbf{k}'). \tag{14}$$

The following completeness condition emerges from it for the set of functions $\mathbf{e}[x|f_i]$:

$$\sum_{i=1}^{\infty} e_k[x|f_i]\,(e_l[y|f_i])^* = \frac{1}{i}\,(\delta_{kl} - \Delta^{-1}\partial_k \partial_l)\, D^{(+)}(x-y), \tag{15}$$

where

$$D^{(+)}(x) \equiv i \int d\Gamma e^{-ik\cdot x}. \tag{16}$$

The function $D^{(+)}(x)$ is that part of the function $D(x)$ which contains only positive frequencies. The part containing only negative frequencies will be denoted by the symbol $D^{(-)}(x)$,

$$D^{(-)}(x) \equiv -i \int d\Gamma e^{ik\cdot x}, \tag{17}$$

$$D(x) = D^{(+)}(x) + D^{(-)}(x). \tag{18}$$

The commutation relations for the creation and annihilation operators and the completeness condition for the set of functions $\mathbf{e}[x|f_i]$ lead to the following commutation relations for the potential operators:

$$[A_k(x), A_l(y)] = \frac{1}{i}\,(\delta_{kl} - \Delta^{-1}\partial_k \partial_l)\, D(x-y). \tag{19}$$

In the particular case when $x^0 = y^0 = t$, we have

$$[A_k(\mathbf{x}, t), A_l(\mathbf{y}, t)] = 0 = [\partial_t A_k(\mathbf{x}, t), \partial_t A_l(\mathbf{y}, t)], \tag{20a}$$

$$[A_k(\mathbf{x}, t), \partial_t A_l(\mathbf{y}, t)] = i\left(\delta_{kl}\,\delta(\mathbf{x}-\mathbf{y}) + \partial_k\partial_l \frac{1}{4\pi}\frac{1}{|\mathbf{x}-\mathbf{y}|}\right). \tag{20b}$$

The potential operators $A_k(x)$ and $\partial_t A_l(x)$ do not commute also when the points x and y are spatially separated. This fact is not incompatible with the causality principle, since the potential operators do not represent measurable quantities (only the field intensities are measurable).

The Quantum Theory of the Electromagnetic Field with External Sources

In the case when external sources do appear, the potential operator defined by formulae (5b) and (5c) satisfies the d'Alembert equation with sources (2a). This operator goes over in the remote past into the operator $\mathbf{A}^{\text{in}}$, and in the far future into the operator $\mathbf{A}^{\text{out}}$,

$$\mathbf{A}(\mathbf{x}, t) = \begin{cases} \mathbf{A}^{\text{in}}(\mathbf{x}, t), & \text{when} \quad t < T_1, \\ \mathbf{A}^{\text{out}}(\mathbf{x}, t), & \text{when} \quad t > T_2. \end{cases} \tag{21}$$

The asymptotic operators $\mathbf{A}^{\text{in}}$ and $\mathbf{A}^{\text{out}}$ are also defined by formulae (5b) and (5c), but in these formulae $\mathbf{E}$ or $\mathbf{B}$ must be replaced, respectively, by $\mathbf{E}^{\text{in}}$ or $\mathbf{B}^{\text{in}}$ and $\mathbf{E}^{\text{out}}$ or $\mathbf{B}^{\text{out}}$.

As in many-particle theory, the S operator is defined as follows:

$$(\Phi^{\text{in}}_{\{n_i\}}|\mathsf{S}\Phi^{\text{in}}_{\{n'_i\}}) \equiv (\Phi^{\text{out}}_{\{n_i\}}|\Phi^{\text{in}}_{\{n'_i\}}) = A(\{n_i\}; \{n'_i\}). \tag{22}$$

It will be shown that this operator may be written in the form of a chronological exponential operator:

$$\mathsf{S} = T\exp\left(-i\int \mathrm{d}t\left[-\int \mathrm{d}^3x\,\mathscr{I}(\mathbf{x}, t)\cdot\mathbf{A}^{\text{in}}(\mathbf{x}, t) + \frac{1}{8\pi}\int \mathrm{d}^3x\,\mathrm{d}^3y\,\mathscr{I}^0(\mathbf{x}, t)\frac{1}{|\mathbf{x}-\mathbf{y}|}\mathscr{I}^0(\mathbf{y}, t)\right]\right). \tag{23}$$

In order to verify the validity of this formula, we shall demonstrate that the operator defined by the formula takes IN operators into OUT operators. To begin with, let us consider the essential part of this operator, postponing the discussion of the numerical $\mathscr{I}^0$-dependent phase. Applying the formulae given in Appendix H and the commutation relations (19) for the operators $\mathbf{A}^{\text{in}}(x)$, we obtain:

$$T\exp\left(i\int d^4x\,\mathbf{A}^{\text{in}}(x)\cdot\mathscr{I}(x)\right)$$

$$=\exp\left[\frac{1}{2}i\int d^4x d^4y\,\mathscr{I}_k(x)\,(\delta_{kl}-\Delta^{-1}\partial_k\partial_l)\,D_F(x-y)\,\mathscr{I}_l(y)\right]$$

$$\times\exp\left(i\int d^4x\,\mathbf{A}^{(-)\text{in}}(x)\cdot\mathscr{I}(x)\right)\exp\left(i\int d^4x\,\mathbf{A}^{(+)\text{in}}(x)\cdot\mathscr{I}(x)\right), \tag{24}$$

where D_F is the *Feynman Green's function* for the d'Alembert equation,

$$D_F(x)\equiv\theta(x^0)\,D^{(+)}(x)+\theta(-x^0)\,D^{(+)}(-x)$$
$$=\theta(x^0)\,D^{(+)}(x)-\theta(-x^0)\,D^{(-)}(x). \tag{25}$$

The function D_F has real and imaginary parts which are given by the formula (cf. Appendix E):

$$D_F(x)=\frac{1}{2}\,\varepsilon(x)\,D(x)+\frac{i}{2}\,D^{(1)}(x).$$

Thus, when the phase factor due to the real part of the function D_F is omitted, the operator (24) can be written as (using formula (H.5))

$$\exp\left(i\int d^4x\,\mathbf{A}^{\text{in}}(x)\cdot\mathscr{I}(x)\right).$$

This is a displacement operator† since the operators appearing as exponents in the formula in question may be expressed in the form of the following sums of creation and annihilation operators:

$$\int d^4x\,\mathbf{A}^{(-)\text{in}}(x)\cdot\mathscr{I}(x)=\sum_{k=1}^{\infty}\iota_k c_k^{\dagger\text{in}}=\sum_{\lambda}\int d\Gamma\,\iota(\mathbf{k},\lambda)\,c^{\dagger\text{in}}(\mathbf{k},\lambda),$$

$$\int d^4x\,\mathbf{A}^{(+)\text{in}}(x)\cdot\mathscr{I}(x)=\sum_{k=1}^{\infty}\iota_k^* c_k^{\text{in}}=\sum_{\lambda}\int d\Gamma\,\iota^*(\mathbf{k},\lambda)\,c^{\text{in}}(\mathbf{k},\lambda).$$

† Cf. formula (11.32)

The operator given by formula (23) can, therefore, be rewritten as

$$\mathsf{S} = e^{i\alpha}\exp\Big(-i\sum_{k=1}^{\infty}(\iota_k c_k^{\dagger\mathrm{in}} + \iota_k^* c_k^{\mathrm{in}})\Big), \tag{26}$$

whence it follows that

$$\mathsf{S}^\dagger c_k^{\mathrm{in}}\mathsf{S} = c_k^{\mathrm{in}} + i\iota_k, \tag{27}$$

whereby the operator (23) is the S operator for a quantum electromagnetic field interacting with the given sources.

As already mentioned, the S operator given by formula (26) is, up to the phase factor, the displacement operator which we defined in Section 11. It thus follows that a state which in the past was a vacuum state will, under the influence of external radiation sources, be a coherent state in the future. The eigenvalues of the operators c_k^{out} are the coefficients $i\iota_k$. No wonder, therefore, that the probability distribution obtained in Section 12 for the number of photons emitted by an external source when the initial state is a vacuum is the same as for a fully coherent state.

In the representation of the S operator as a chronological exponential operator the expression in the exponent is the operator of the interaction energy of the electromagnetic field and the external sources. The first term describes the interaction energy of the field (photons) with an external current, whereas the second describes the Coulomb energy of the charges.

We shall now show that the S operator defined above is a relativistic invariant and that it satisfies the Bogolyubov causality condition.

Collecting the parts dependent on the products of the currents in formulae (23) and (24), we obtain

$$\frac{1}{2}i\int d^4x\, d^4y\, \mathcal{J}^\mu(x)\mathcal{D}^{\mathrm{F}}_{\mu\nu}(x-y, n)\mathcal{J}^\nu(y), \tag{28}$$

where

$$\mathcal{D}^{\mathrm{F}}_{\mu\nu}(x, n) \equiv \left[-g_{\mu\nu} + \frac{\partial_\mu\partial_\nu - n_\mu(n\cdot\partial)\partial_\nu - n_\nu(n\cdot\partial)\partial_\mu}{\Box - (n\cdot\partial)^2}\right] D_{\mathrm{F}}(x), \tag{29}$$

and the vector n_μ has the components $(1, 0, 0, 0)$. Since the current satisfies the continuity equation, the expression (28) does not in effect depend on the vector n_μ. The function $\mathcal{D}^{\mathrm{F}}_{\mu\nu}(x, n)$ can thus be replaced in this expression by the function $D^{\mathrm{F}}_{\mu\nu}(x)$:

$$D^{\mathrm{F}}_{\mu\nu}(x) \equiv -g_{\mu\nu}D_{\mathrm{F}}(x). \tag{30}$$

The final formula for the S operator thus has an explicitly relativistically invariant form:

$$\mathsf{S} = \exp\left[-\frac{1}{2}i\int d^4x d^4y \mathscr{I}_\mu(x) D_F(x-y)\mathscr{I}^\mu(y)\right] \tag{31}$$

$$\times \exp\left(i\sum_\lambda \int d\Gamma \iota(\mathbf{k}, \lambda) c^{\dagger\text{in}}(\mathbf{k}, \lambda)\right) \exp\left(i\sum_\lambda \int d\Gamma \iota^*(\mathbf{k}, \lambda) c^{\text{in}}(\mathbf{k}, \lambda)\right).$$

The creation and annihilation operators appear in this formula in the normal order. The first exponential function is the amplitude of the vacuum-to-vacuum transition

$$(\Omega^{\text{out}}|\Omega^{\text{in}}) = \exp\left[-\frac{1}{2}i\int d^4x d^4y \mathscr{I}_\mu(x) D_F(x-y)\mathscr{I}^\mu(y)\right]. \tag{32}$$

The squared modulus of this expression may be written as

$$|(\Omega^{\text{out}}|\Omega^{\text{in}})|^2 = \exp\left[\frac{1}{2}\int d^4x d^4y \mathscr{I}_\mu(x) D^{(1)}(x-y)\mathscr{I}^\mu(y)\right]$$

in accordance with the formula given in the preceding section.

The S operator which we have constructed as a chronological exponential operator satisfies the causality condition, just as does the S operator in the many-particle theory (cf. p. 66). This condition cannot, however, be expressed by using the functional derivatives of the S operator with respect to the function $\mathscr{I}^\mu(x)$, since these functions are not independent. The current $\mathscr{I}^\mu(x)$ satisfies the continuity equation. Instead of the derivatives with respect to the function $\mathscr{I}^\mu(x)$, we shall use the derivatives with respect to the independent components of the tensor $f^{\text{ext}}_{\mu\nu}$, defined by the formula

$$\partial^\mu f^{\text{ext}}_{\mu\nu}(x) = \mathscr{I}_\nu(x). \tag{33}$$

In doing this, we shall make use of the equality

$$\frac{\delta\mathscr{I}^\mu(x)}{\delta f^{\text{ext}}_{\lambda\varrho}(y)} = \partial^{[\lambda}g^{\varrho]\mu}\delta^{(4)}(x-y),$$

the identities given in Appendix H for exponential operators, and the relation

$$D_F(x) + D^{(-)}(x) = D_R(x). \tag{34}$$

As a result, we obtain the equality

$$\mathsf{S}^{\dagger}\frac{1}{i}\frac{\delta \mathsf{S}}{\delta f^{\text{ext}}_{\lambda\varrho}(x)} = f^{\text{in}\lambda\varrho}(x) + \partial^{[\lambda}\int d^4y\, D_{\text{R}}(x-y)\mathscr{J}^{\varrho]}(y),$$

wherefrom the *Bogolyubov causality condition* follows in the form

$$\frac{\delta}{\delta f^{\text{ext}}_{\mu\nu}(x)}\left(\mathsf{S}^{\dagger}\frac{\delta \mathsf{S}}{\delta f^{\text{ext}}_{\lambda\varrho}(y)}\right) = 0, \quad \text{when} \quad x \gtrsim y. \tag{35}$$

The relation $x \gtrsim y$ means that the point x lies outside the lower half of the light cone with apex at the point y. This condition is met by a suitable choice of the S-operator phase. This choice was made when we took the S operator in the form (23).

The Poincaré Group as the Symmetry Group of Amplitudes

Now we shall study how the Poincaré group acts in the set of objects determining the quantum theory of the electromagnetic field interacting with external sources. The Poincaré group is not a symmetry group of this theory in the sense explained in Sections 2 and 10. For it is not possible to define a representation of this group $\mathsf{U}(a, \Lambda)$ which would satisfy the condition (10.2) for all field operators. The existence of the external currents $\mathscr{J}^{\mu}(x)$ destroys the homogeneity and isotropy of space-time. We shall show, however, that if both the sources and state vectors are simultaneously transformed, the transition amplitudes will not suffer change.

In the theory being considered here, we can define two groups of unitary operators $\mathsf{U}^{\text{in}}(a, \Lambda)$ and $\mathsf{U}^{\text{out}}(a, \Lambda)$ which constitute a unitary representation of the proper Poincaré group in the set of state vectors Φ^{in} and Φ^{out}, respectively. In acting on an arbitrary basis vector Φ^{in}, the operator $\mathsf{U}^{\text{in}}(a, \Lambda)$ produces from the former a vector ${}^{\text{I}\text{L}}\Phi^{\text{in}}$ characterized as follows: all wave-packet profiles $f_i(\mathbf{k}, \lambda)$ specifying one-particle photon states in the remote past in the state Φ^{in} will be replaced by the profiles ${}^{\text{I}\text{L}}f_i(\mathbf{k}, \lambda)$ transformed according to formula (9.38). Similarly, we define the action of the operator $\mathsf{U}^{\text{out}}(a, \Lambda)$ on the vector Φ^{out}. The operators $\mathsf{U}^{\text{in}}(a, \Lambda)$ and $\mathsf{U}^{\text{out}}(a, \Lambda)$ satisfy the relations

$$\mathsf{U}^{\text{in}-1}(a, \Lambda) f^{\text{in}}_{\mu\nu}(x)\, \mathsf{U}^{\text{in}}(a, \Lambda) = {}^{\text{I}\text{L}}f^{\text{in}}_{\mu\nu}(x), \tag{36a}$$

$$\mathsf{U}^{\text{out}-1}(a, \Lambda) f^{\text{out}}_{\mu\nu}(x)\, \mathsf{U}^{\text{out}}(a, \Lambda) = {}^{\text{I}\text{L}}f^{\text{out}}_{\mu\nu}(x). \tag{36b}$$

The S operator, treated as the functional $\mathsf{S}[\mathscr{J}]$ of the current $\mathscr{J}^\mu(x)$, has the property

$$\mathsf{U}^{\mathrm{in}-1}(a, \Lambda)\,\mathsf{S}[{}^{\mathrm{IL}}\mathscr{J}]\,\mathsf{U}^{\mathrm{in}}(a, \Lambda) = \mathsf{S}[\mathscr{J}], \tag{37}$$

where

$${}^{\mathrm{IL}}\mathscr{J}^\mu(x) = \Lambda^\mu{}_\nu \mathscr{J}^\nu\big(\Lambda^{-1}(x-a)\big). \tag{38}$$

To prove formula (37), use should be made of formulae expressing the products $\mathbf{A}^{(\pm)}, \mathscr{J}$ in terms of the creation and annihilation operators, the transformation formulae (11.22) for the annihilation and creation operators, formula (9.65), and the following property of the expression $\iota(\mathbf{k}, \lambda)$:

$${}^{\mathrm{IL}}\iota(\mathbf{k}, +1) = -\int d^4x\varepsilon_\mu^*(\mathbf{k})\,{}^{\mathrm{IL}}\mathscr{J}^\mu(x)e^{ik\cdot x} = e^{ik\cdot a}e^{i\theta(\mathbf{k}, \Lambda)}\iota(\Lambda^{-1}\mathbf{k}, +1),$$

$${}^{\mathrm{IL}}\iota(\mathbf{k}, -1) = e^{ik\cdot a}e^{-i\theta(\mathbf{k}, \Lambda)}\iota(\Lambda^{-1}\mathbf{k}, -1).$$

Calculating the matrix elements of both sides of the equality (37) between any two vectors Φ^{in}, we obtain a relation with a straightforward physical interpretation

$$({}^{\mathrm{IL}}\Phi_f^{\mathrm{in}}|\mathsf{S}[{}^{\mathrm{IL}}\mathscr{J}]\,{}^{\mathrm{IL}}\Phi_i^{\mathrm{in}}) = (\Phi_f^{\mathrm{in}}|\mathsf{S}[\mathscr{J}]\Phi_i^{\mathrm{in}}). \tag{39}$$

This equality means that the transition amplitude does not change if we simultaneously perform the Poincaré transformation of the external current $\mathscr{J}^\mu(x)$ and the Poincaré transformation of all photon wave functions specifying the asymptotic states of the system.

Equality (39) will serve as a starting point for the generalization of the symmetry group to the case of the quantum theory in which external fields or external sources appear. The group G shall be said to act in the set of transition amplitudes $(\Phi_f^{\mathrm{out}}|\Phi_i^{\mathrm{in}})$ in the presence of external fields or external sources if in the sets of the vectors Φ^{in} and Φ^{out} there act unitary or anti-unitary representations of this group, $T^{\mathrm{in}}(g)$ and $T^{\mathrm{out}}(g)$, related to each other by the compatibility condition

$$T^{\mathrm{in}\dagger}(g)\,\mathsf{S}\,T^{\mathrm{out}}(g) = \mathsf{S}, \tag{40}$$

and in the set of external fields or sources there acts the representation $\mathscr{T}_g$ of this group.

The group G acting in the set of transition amplitudes is a *symmetry group* of these amplitudes if the equalities

$$\mathscr{T}_g\big(T^{\mathrm{out}}(g)\Phi_f^{\mathrm{out}}|T^{\mathrm{in}}(g)\Phi_i^{\mathrm{in}}\big) = (\Phi_f^{\mathrm{out}}|\Phi_i^{\mathrm{in}}) \tag{41}$$

are satisfied for the unitary operators $T^{\text{in}}(g)$ and $T^{\text{out}}(g)$, or

$$\mathscr{T}_g\big(T^{\text{out}}(g)\Phi_f^{\text{out}}|T^{\text{in}}(g)\Phi_i^{\text{in}}\big) = (\Phi_i^{\text{out}}|\Phi_f^{\text{in}}) \tag{42}$$

for the anti-unitary operators $T^{\text{in}}(g)$ and $T^{\text{out}}(g)$. The transformation $\mathscr{T}_g$ acts on the external fields and sources on which the transition amplitude depends. The interchange of the indices i and f in formula (42) means that the state Φ_i^{out} is characterized in the future in the same way as the state Φ_i^{in} was in the past. If use is made of the compatibility condition (40), the conditions (41) and (42) may be rewritten as

$$\mathscr{T}_g\big(T^{\text{in}}(g)\Phi_f^{\text{in}}|\mathsf{S}\,T^{\text{in}}(g)\Phi_i^{\text{in}}\big) = (\Phi_f^{\text{in}}|\mathsf{S}\Phi_i^{\text{in}}), \tag{43a}$$

or

$$\mathscr{T}_g\big(T^{\text{in}}(g)\Phi_f^{\text{in}}|\mathsf{S}\,T^{\text{in}}(g)\Phi_i^{\text{in}}\big) = (\Phi_i^{\text{in}}|\mathsf{S}\Phi_f^{\text{in}}). \tag{43b}$$

The proper Poincaré group acts in the set of transition amplitudes in the quantum theory of an electromagnetic field interacting with an external current. It is a symmetry group of amplitudes by the equality (39). The full Poincaré group may be shown also to be a symmetry group of the set of amplitudes.

A Simple Representation of Transition Amplitudes

We shall now give an important relation between the probability amplitude of an arbitrary radiative process and the vacuum-to-vacuum transition amplitude. To derive this relation, we shall use the external potential method which will be employed frequently further on. This method is formally analogous to the method of an arbitrary external potential $U(\mathbf{x}, t)$ used in the discussion of the scattering of particles in non-relativistic quantum mechanics, for, we shall assume that in addition to the quantum electromagnetic field there is also an external electromagnetic field described by the potential $\mathscr{A}_\mu(x)$. The potential $\mathscr{A}_\mu$ will be treated as an additional, auxiliary variable of the theory. In the final formulae, of course, $\mathscr{A}_\mu$ must always be fixed to describe the actual field occurring in the experiment. In the simplified version of the theory, when the sources of the field are not governed by dynamical laws, we most often put $\mathscr{A}_\mu = 0$. However, when the motion of the sources depends on the external field, the potential $\mathscr{A}_\mu$ is fixed so that it describes the field of a magnet, atomic nucleus, electromagnetic wave, etc., which is external to the quantum system under study.

In Maxwellian linear electrodynamics, to which we have confined our discussion, no coupling appears between the given field $\partial_{[\mu}\mathscr{A}_{\nu]}$ and the quantum field when the influence of the quantum field on the source is neglected. Only the phase of the S operator will experience some modification since the energy of the interaction of the quantum field with the given current and the Coulomb energy of the charges should now be augmented by the energy H_{ex} of the interaction between the given current $\mathscr{I}^\mu$ and the external field of potential $\mathscr{A}_\mu$. This energy is given by the formula (cf. also formula (23)):

$$H_{\text{ex}}(t) = -\int \mathrm{d}^3x\, \mathscr{A}(\mathbf{x}, t)\cdot \mathscr{I}(\mathbf{x}, t)$$

$$-\frac{1}{4\pi}\int \mathrm{d}^3x\mathrm{d}^3y\, \Delta\mathscr{A}_0(\mathbf{x}, t)\frac{1}{|\mathbf{x}-\mathbf{y}|}\,\mathscr{I}_0(\mathbf{y}, t)$$

$$= \int \mathrm{d}^3x\, \mathscr{A}_\mu(\mathbf{x}, t)\mathscr{I}^\mu(\mathbf{x}, t). \tag{44}$$

As a result, the operator $\mathsf{S}[\mathscr{A}]$ in the presence of an external field differs only by a phase factor from the S operator defined previously,

$$\mathsf{S}[\mathscr{A}] = \mathsf{S}\exp\left(-i\int \mathrm{d}^4x\, \mathscr{A}_\mu(x)\mathscr{I}^\mu(x)\right). \tag{45}$$

Such a phase factor does not affect the transition probabilities, but it does enable us to introduce functional derivatives with respect to the potential $\mathscr{A}_\mu$ and to derive new formulae which not only are neat and lucid in form but also have an interesting physical interpretation.

Note first of all that the operation of functional differentiation $\delta/\delta\mathscr{A}_\mu(x)$ may be substituted for multiplication by $-i\mathscr{I}^\mu(x)$ in the operation on the phase factor of the operator $\mathsf{S}[\mathscr{A}]$. The operator $\mathsf{S}[\mathscr{A}]$ can thus be presented, in view of formulae (45) and (31), in the form

$$\mathsf{S}[\mathscr{A}] = \exp\left(\sum_k c_k^{+\,\text{in}} D^*[f_k]\right)\exp\left(\sum_k c_k^{\text{in}} D[f_k]\right)\mathscr{C}[\mathscr{A}], \tag{46}$$

where $\mathscr{C}[\mathscr{A}]$ is the amplitude of the vacuum-to-vacuum transition in the presence of the external potential,

$$\mathscr{C}[\mathscr{A}] = \exp\left[-\frac{1}{2}i\int d^4x d^4y \mathscr{I}_\mu(x) D_F(x-y)\mathscr{I}^\mu(y)\right]\exp\left(-i\int d^4x \mathscr{A}_\lambda \mathscr{I}^\lambda\right)$$

$$= \exp\left[-\frac{1}{2}i\int d^4x d^4y \frac{\delta}{\delta\mathscr{A}_\mu(x)} D^F_{\mu\nu}(x-y)\frac{\delta}{\delta\mathscr{A}_\nu(y)}\right]$$

$$\times \exp\left(-i\int d^4x \mathscr{A}_\lambda(x)\mathscr{I}^\lambda(x)\right), \tag{47}$$

whereas $D[f]$ is the functional derivative "in the direction of" the photon wave function,

$$D[f] \equiv \int d^4x \mathbf{e}[x|f]\cdot\frac{\delta}{\delta\mathscr{A}(x)} = \int d^4x \varphi_\mu[x|f]\frac{\delta}{\delta\mathscr{A}_\mu(x)}. \tag{48}$$

In the foregoing equality we have utilized the invariance of the theory with respect to gauge transformations. Any photon wave function φ_μ differs from its corresponding radiation-gauge function by the term $\partial_\mu\lambda(x)$ (cf. formulae (11) and (9.63)). On the other hand, by virtue of the continuity equation for the current, the condition

$$\partial_\mu\frac{\delta}{\delta\mathscr{A}_\mu(x)}\mathscr{C}[\mathscr{A}] = 0 \tag{49}$$

is satisfied, which makes it possible to eliminate the term dependent on the gauge function $\lambda(x)$.

The condition (49) should hold not only for the amplitude of the vacuum-to-vacuum transition but also for every other functional $F[\mathscr{A}]$ which, just as $\mathscr{C}[\mathscr{A}]$, is invariant under gauge transformation. For under infinitesimal gauge transformations we have

$$0 = F[\mathscr{A}+\partial\delta\lambda]-F[\mathscr{A}] = -\int d^4x\delta\lambda(x)\partial_\mu\frac{\delta}{\delta\mathscr{A}_\mu(x)}F[\mathscr{A}].$$

The general formula for the transition amplitude is obtained by enclosing the S operator between the basis vectors Φ_f^{in} and Φ_i^{in},

$$(\Phi_f^{\text{out}}|\Phi_i^{\text{in}}) = (\Phi_f^{\text{in}}|\mathrm{S}\Phi_i^{\text{in}})$$

$$= \left(\Phi_f^{\text{in}}\middle|\exp\Big(\sum_k c_k^{\dagger\text{in}}D^*[f_k]\Big)\exp\Big(\sum_k c_k^{\text{in}}D[f_k]\Big)\Phi_i^{\text{in}}\right)\mathscr{C}[\mathscr{A}]. \tag{50}$$

In the simplest cases of emission oı absorption of a single photon, the amplitudes are given by the formulae

$$D^*[f_k]\mathscr{C}[\mathscr{A}] \quad \text{or} \quad D[f_k]\mathscr{C}[\mathscr{A}].$$

In the general case, the transition amplitude for states with particular photon occupation numbers is in the form of a polynomial in the operations D and D^* acting on the vacuum-to-vacuum transition amplitude.

The transition amplitude for coherent states has an interesting form. Suppose that the vectors Φ_i^{in} and Φ_f^{in} describe coherent states, i.e. are eigenvectors of the operators c_k^{in} with some non-vanishing eigenvalues,

$$c_k^{\text{in}}\Phi_i^{\text{in}} = \beta_k\Phi_i^{\text{in}}, \qquad k = 1, \dots, m,$$
$$c_k^{\text{in}}\Phi_f^{\text{in}} = \alpha_k\Phi_f^{\text{in}}, \qquad k = 1, \dots, n.$$

In this case the transition amplitude is of the form

$$(\Phi_f^{\text{out}}|\Phi_i^{\text{in}}) =$$
$$\left(\Phi_f^{\text{in}}\middle|\exp\left(\sum_{k=n+1}^{\infty} c_k^{\dagger\text{in}}D^*[f_k]\right)\exp\left(\sum_{k=m+1}^{\infty} c_k^{\text{in}}D[f_k]\right)\Phi_i^{\text{in}}\right)\mathscr{C}[\mathscr{A}+\varphi], \tag{51}$$

where

$$\varphi_\mu(x) \equiv \sum_{k=1}^{m}\beta_k\varphi_\mu[x|f_k]+\sum_{k=1}^{n}\alpha_k^*\varphi_\mu^*[x|f_k]. \tag{52}$$

To derive this formula we made use of the equality

$$\exp\int\left(\mathrm{d}^4x\varphi_\mu(x)\frac{\delta}{\delta\mathscr{A}_\mu(x)}\right)\mathscr{C}[\mathscr{A}] = \mathscr{C}[\mathscr{A}+\varphi].$$

Formula (51) for the transition amplitude offers still one more confirmation of the almost classical character of coherent states. With the aid of this formula, one can also get a better understanding of the concept of external field in quantum theory. In the general case, the components of the potential $\varphi_\mu(x)$ are complex functions, but if $\alpha_k = \beta_k$, then $\varphi_\mu(x)$ is a real potential of the average electromagnetic field in the coherent state considered. Such a real field is thus obtained when we investigate the probability amplitude of a coherent state remaining unchanged. If the average field intensity in such a coherent state is large relative to the field intensity of radiation emitted spontaneously by the given system,† then only the diagonal elements of the S operator

† In the case under consideration this means that

$$|\alpha_k| \gg |\iota_k|.$$

with respect to that coherent state of the field are of practical significance. In an experiment we can never distinguish an initial coherent state with an enormous average number of photons from a state in which a few more photons occur.

Our fundamental formula (50) for the transition amplitude in an external field may thus be regarded as the diagonal matrix element of the S operator in the coherent state of the radiation. The average field in this state is described by the potential $\mathscr{A}_\mu$.

A characteristic feature of the formulae arrived at is that the transition amplitudes are separated into two parts: the part $\mathscr{C}[\mathscr{A}]$ which is dependent only on the radiation sources and independent of the initial and final states of the quantum electromagnetic field, and a part which is dependent only on the states of the field. All the information about the sources of radiation is contained in the vacuum-to-vacuum transition amplitude, provided that we know this amplitude for an arbitrary external potential. The fact that the amplitudes of all radiative transitions, i.e. transitions with the emission and/or absorption of photons, may be obtained from the amplitude of the non-radiative transition, i.e. the amplitude of the vacuum-to-vacuum transition, treated as a functional of the external potential, is an expression of the important *principle of universality of electromagnetic interactions.* The coupling of field sources to an external field is of the same form as the coupling of the quantum field to sources. The principle of the universality of electromagnetic interactions will be used in subsequent considerations.

Note, finally, the striking similarity between formula (47) for $\mathscr{C}[\mathscr{A}]$ with functional derivatives with respect to $\mathscr{A}_\mu$, and formula (5.77) derived in the theory of interacting particles. The exponential operation

$$\exp\left(\frac{1}{2i}\int\frac{\delta}{\delta\mathscr{A}}\,D^{\mathrm{F}}\,\frac{\delta}{\delta\mathscr{A}}\right)$$

introduces an interaction due to the occurrence of the quantum electromagnetic field into the theory of a classical external field interacting with a classical current. The role of the potential of the mutual interaction of currents is played by the function $D^{\mathrm{F}}_{\mu\nu}(x-y)$. In accordance with the theory of relativity, this potential does not transmit an instantaneous interaction but describes the retardation effects due to the finite propagation velocity of interaction. This interaction is of a relativistically invariant character since the function $D_F(x-y)$ is a relativistic scalar,

$$D^F_{\mu\nu}(x) = -g_{\mu\nu}\left(\frac{1}{4\pi}\,\delta(x^2)+\frac{1}{4\pi i}\,\frac{1}{x^2}\right).$$

We shall return to these problems again when discussing the complete quantum electrodynamics.

Induced Processes in Intense Photon Beams

Amplitudes for induced emission and absorption have a simple representation when the intensity of incoming radiation is large.# To derive this representation we will use formula (12.19) expressing the transition amplitudes $A(n+m;n)$ in terms of the associated Laguerre polynomials $L_n^m(|\iota|^2)$. For large values of the index n these polynomials can be approximated by Bessel functions according to the formula

$$\frac{n!}{(n+m)!}\,e^{z/2}L_n^m(z) = 2^n\,\frac{J_m(2\sqrt{Nz})}{(2\sqrt{Nz})^m} + \begin{matrix}\text{terms of higher}\\ \text{order in } z/\sqrt{N}\end{matrix}, \tag{53}$$

where

$$N = n+\frac{m+1}{2}.$$

Next we use the integral representation for the Bessel functions

$$(iw)^m|w|^{-m}J_m(2|w|) = \frac{1}{2\pi}\int_0^{2\pi} d\varphi\, e^{-im\varphi}\exp(iwe^{i\varphi}+iw^*e^{-i\varphi}). \tag{54}$$

From eqs. (53) and (54), making again use of the condition $n \gg m$, we obtain the following approximate formula for the induced emission amplitude

$$A(n+m;n) \simeq \frac{1}{(2\pi)}\int_0^{2\pi} d\varphi\, e^{-im\varphi}\exp(i\sqrt{N}\iota e^{i\varphi}+i\sqrt{N}\iota^* e^{-i\varphi}). \tag{55a}$$

In a similar way we can obtain an approximate formula for the induced absorption amplitude,

$$A(n;n+m) \simeq \frac{1}{2\pi}\int_0^{2\pi} d\varphi\, e^{im\varphi}\exp(i\sqrt{N}\iota e^{i\varphi}+i\sqrt{N}\iota^* e^{-i\varphi}). \tag{55b}$$

This problem has been studied in our recent paper (I. Białynicki-Birula, Z. Białynicka-Birula (1973). We refer the reader to this paper for all the details.

With the help of eqs. (12.5) and (9.67) we can transform expressions (55a) and (55b) into the form

$$\left.\begin{matrix} A(n+m;n) \\ A(n;n+m) \end{matrix}\right\} \simeq \frac{1}{2\pi}\int_0^{2\pi} \mathrm{d}\varphi\, \mathrm{e}^{\mp im\varphi} \exp\left[-i\int \mathrm{d}^3x\,\mathrm{d}t\, \mathscr{A}_\mu^{(\varphi)}(\mathbf{x},t)\mathscr{I}^\mu(\mathbf{x},t)\right], \quad (56)$$

where $\mathscr{A}_\mu^{(\varphi)}(\mathbf{x},t)$ is the electromagnetic field potential constructed out of the photon wave function, defined by formulae (10) and (11),

$$\mathscr{A}_\mu^{(\varphi)}(\mathbf{x},t) \equiv \sqrt{N}\varphi_\mu[x|f]\mathrm{e}^{-i\varphi} + \sqrt{N}\varphi_\mu^*[x|f]\mathrm{e}^{i\varphi}. \quad (57)$$

Formula (57) has a simple physical interpretation, since the exponential function $\exp(-i\int \mathrm{d}^4x\mathscr{A}\cdot\mathscr{I})$ represents the transition amplitude (cf. formula (45)) for a (trivial!) physical system consisting of a *given* external current $\mathscr{I}^\mu$ and a *given* classical electromagnetic field $\mathscr{A}_\mu$.

Thus, induced emission and absorption amplitudes are approximately equal to the Fourier expansion coefficients for the transition amplitude in an external electromagnetic field treated as a function of the phase of the field.

Similar results can also be obtained in full quantum electrodynamics when quantized sources interact with the quantized electromagnetic field. Only in this case does it make sense (owing to the energy conservation) to consider transitions in which the number of photons in an intense beam increases or decreases by a small number m.

Many-Photon Propagators

It is convenient to use propagators to describe the quantized electromagnetic field treated as a system of many photons. In as simple a case as the interaction of the quantized field with the external current, this method may appear to be too sophisticated. The method of propagators, however, will play such an important role in the formulation of full quantum electrodynamics that we have decided to introduce it here in order to explain how this method works in the simplest possible case.

Many-photon propagators $T[z_1 \ldots z_k|\mathscr{A},\mathscr{I}]$ are defined as follows,

$$T_{\mu_1\ldots\mu_k}[z_1 \ldots z_k|\mathscr{A},\mathscr{I}]$$
$$\equiv \exp\left(\frac{1}{2i}\int \frac{\delta}{\delta\mathscr{A}} D^{\mathrm{F}} \frac{\delta}{\delta\mathscr{A}}\right) \mathscr{A}_{\mu_1}(z_1)\ldots\mathscr{A}_{\mu_k}(z_k)\exp\left(-i\int \mathscr{I}\cdot\mathscr{A}\right). \quad (58)$$

They can be also represented as multiple functional derivatives with respect to the external current of the vacuum-to-vacuum transition amplitude †

$$T_{\mu_1 \ldots \mu_k}[z_1 \ldots z_k|\mathscr{A}, \mathscr{J}] = i^k \frac{\delta^k}{\delta \mathscr{J}^{\mu_1}(z_1) \ldots \delta \mathscr{J}^{\mu_k}(z_k)} \mathscr{C}[\mathscr{A}].$$

Two examples of the simplest propagators are given below,

$$T_\mu[z|\mathscr{A}, \mathscr{J}] = \mathscr{C}[\mathscr{A}]\mathfrak{A}_\mu(z), \tag{59a}$$

$$T_{\mu\nu}[z_1 z_2|\mathscr{A}, \mathscr{J}] = \mathscr{C}[\mathscr{A}](\mathfrak{A}_\mu(z_1)\mathfrak{A}_\nu(z_2) - iD^{\mathrm{F}}_{\mu\nu}(z_1 - z_2)), \tag{59b}$$

where

$$\mathfrak{A}_\mu(z) \equiv \mathscr{A}_\mu(z) - \int \mathrm{d}z' \, D^{\mathrm{F}}_{\mu\nu}(z - z')\mathscr{J}^\nu(z'). \tag{60}$$

Many-photon propagators play the role of wave functions for many-photon systems. They describe photon propagation between points $z_1, \ldots, z_k$ in the presence of the external field $\mathscr{A}$ and the external current $\mathscr{J}$. Each propagator contains a part describing free propagation governed by the propagator $D^{\mathrm{F}}_{\mu\nu}$ as well as a part describing photon emission and absorption by external sources.

The relationship between propagators and the wave functions ψ_R and ψ_A which describe scattering in non-relativistic quantum mechanics is particularly close. For, we can express the photon transition amplitudes as scalar products (calculated in the future or in the past) of the propagators and free photon wave functions. To illustrate this we shall consider the one-photon propagator $T_\mu[z] \equiv T_\mu[z|\mathscr{A}, \mathscr{J}]$.

In the remote past and in the far future in regions free of external fields and sources all photon propagators obey free Maxwell equations in each variable. It follows from properties of the propagator D^{F} that

$$T_\mu[z] = i \int \mathrm{d}\Gamma \mathrm{e}^{-ikx} \tilde{\mathscr{J}}_\mu(k)\mathscr{C}[\mathscr{A}], \text{ when } z^0 > T_2, \tag{61a}$$

$$T_\mu[z] = i \int \mathrm{d}\Gamma \mathrm{e}^{ik \cdot x} \tilde{\mathscr{J}}^*_\mu(k)\mathscr{C}[\mathscr{A}], \text{ when } z^0 < T_1. \tag{61b}$$

These expressions are of the form of the photon wave function and its complex conjugate (cf. formula (10)). On calculating the scalar products of these two functions with given photon wave function $\varphi_\mu[z|f]$ we

† In calculating functional derivatives we treat all components of the current vector as independent functions. The continuity equation $\partial_\mu \mathscr{J}^\mu = 0$ is imposed after all the functional differentiations have been performed.

obtain the familiar formulae for photon emission and photon absorption,

$$(\varphi[f]|T)_{t>T_2} = -i\int \mathrm{d}\Gamma f_\mu^*(k)\tilde{\mathscr{I}}^\mu(k)\,(\Omega^{\mathrm{out}}|\Omega^{\mathrm{in}}), \tag{62a}$$

$$(T^*|\varphi[f])_{t<T_1} = -i\int \mathrm{d}\Gamma f_\mu(k\;\tilde{\mathscr{I}}^{\mu *}(k)\,(\Omega^{\mathrm{out}}|\Omega^{\mathrm{in}}). \tag{62b}$$

The two-photon propagator $T_{\mu\nu}[z_1 z_2|\mathscr{A},\mathscr{I}]$ describes all three processes in which two photons take part: the emission or the absorption of two photons and the emission of one photon with the absorption of the other photon. Transition amplitudes for these processes calculated as appropriate scalar products of this propagator with photon wave functions are equal to the expressions obtained before.

This procedure can be extended to many-photon propagators. We return to this problem in Chapter 6, where we define many-photon propagators in full quantum electrodynamics.

14. THE PROCA THEORY

In this section we shall give the classical and the quantum theory of a vector field with mass μ. It will be shown that the transition amplitudes have a limit when we make the transition $\mu \to 0$ and that the results obtained in the limiting case coincide with the results yielded previously by the Maxwellian theory.

The Classical Theory of a Vector Field with Mass

The classical vector field with mass is described in terms of a real vector field $A_\mu(x)$. The Lagrangian density for this field is assumed to be given by the formula

$$\mathscr{L}(x) = -\frac{1}{4} f_{\mu\nu}(x) f^{\mu\nu}(x) + \frac{1}{2}\mu^2 A_\mu(x) A^\mu(x), \tag{1}$$

where

$$f_{\mu\nu}(x) \equiv \partial_{[\mu} A_{\nu]}(x). \tag{2}$$

This theory will be referred to as the *Proca theory*.# Field equations, called the *Proca equations*,

$$\partial_\mu f^{\mu\nu}(x) + \mu^2 A^\nu(x) = 0, \tag{3}$$

\# Proca is usually thought to have been the first to give the classical theory of a vector field with mass (Proca, 1936), but actually these equations were discovered earlier by Lanczos (1929), although in a different context.

follow from the Lagrangian. These equations may be presented in a somewhat different, equivalent form:

$$K_x A_\mu(x) \equiv (\Box + \mu^2) A_\mu(x) = 0, \tag{4a}$$

$$\partial^\mu A_\mu(x) = 0. \tag{4b}$$

In quantum theory, eq. (4a)—called the *Klein–Gordon equation*—states that particles associated with the field $A_\mu(x)$ have a mass μ, whereas eq. (4b) (the Lorentz condition) states that these particles have only spin 1. The solutions of the classical field equations (4) may be written in terms of the Fourier integral:

$$A_\mu(x) = \int \mathrm{d}\Gamma(\mu)[f_\mu(\mathbf{k})\mathrm{e}^{-ik\cdot x} + f_\mu^*(\mathbf{k})\mathrm{e}^{ik\cdot x}], \tag{5}$$

where

$$f_\mu(\mathbf{k}) = \sum_{r=0,+1,-1} \varepsilon_\mu(\mathbf{k}, r) f(\mathbf{k}, r), \tag{6}$$

and the three vectors $\varepsilon_\mu(\mathbf{k}, \pm 1)$ and $\varepsilon_\mu(\mathbf{k}, 0)$ in the fixed reference frame may be constructed in the following manner:

$$\varepsilon^\mu(\mathbf{k}, 0) = \frac{\sqrt{\mu^2 + \mathbf{k}^2}}{\mu|\mathbf{k}|}\left(\frac{|\mathbf{k}|^2}{\sqrt{\mu^2+\mathbf{k}^2}}, \mathbf{k}\right), \tag{7a}$$

$$\varepsilon^\mu(\mathbf{k}, +1) = (0, \mathbf{e}(\mathbf{k})), \tag{7b}$$

$$\varepsilon^\mu(\mathbf{k}, -1) = (0, -\mathbf{e}^*(\mathbf{k})). \tag{7c}$$

The symbol $\mathrm{d}\Gamma(\mu)$ denotes an invariant measure on the hyperboloidal surface—called the mass hyperboloid—defined by the equation $k^2 = \mu^2$, normalized in the following manner:

$$\mathrm{d}\Gamma(\mu) \equiv \frac{\mathrm{d}^3 k}{2(2\pi)^3 \sqrt{\mu^2 + \mathbf{k}^2}}. \tag{8}$$

The symbol $\int \mathrm{d}\Gamma(\mu)$ always will be used to denote integration over the upper sheet of the hyperboloid. The vectors $\varepsilon_\mu(\mathbf{k}, r)$ satisfy the orthonormality relations

$$g^{\mu\nu}\varepsilon_\mu^*(\mathbf{k}, r)\varepsilon_\nu(\mathbf{k}, s) = -\delta_{rs}, \quad r, s = 0, -1, +1, \tag{9}$$

and the completeness condition (in a subspace orthogonal to k_μ)

$$\sum_{r=0,-1,+1} \varepsilon_\mu(\mathbf{k}, r)\varepsilon_\nu^*(\mathbf{k}, r) = -g_{\mu\nu} + \mu^{-2}k_\mu k_\nu. \tag{10}$$

The coefficients $f(\mathbf{k}, r)$ are arbitrary functions describing independent degrees of freedom of the field.

By varying the Lagrangian with respect to the metric tensor (cf. Appendix F), we obtain the following formula for the energy-momentum tensor $T^{\mu\nu}(x)$:

$$T^{\mu\nu}(x) = f^{\mu}{}_{\lambda}(x) f^{\lambda\nu}(x) + \mu^2 A^{\mu}(x) A^{\nu}(x)$$
$$+ g^{\mu\nu}\left(\frac{1}{4} f_{\lambda\varrho}(x) f^{\lambda\varrho}(x) - \frac{1}{2}\mu^2 A_{\lambda}(x) A^{\lambda}(x)\right). \tag{11}$$

The Proca theory in canonical formulation is simpler than the canonical Maxwellian field theory, since it is closer to the canonical formulation of mechanics. The field $\boldsymbol{\pi}(x)$ canonically conjugate to the field $\mathbf{A}(x)$ is defined as:

$$\boldsymbol{\pi}(x) \equiv \frac{\partial \mathscr{L}(x)}{\partial \dot{\mathbf{A}}(x)} = \partial_0 \mathbf{A}(x) + \nabla A_0(x). \tag{12}$$

The time component of the field $A_0(x)$ has no field canonically conjugate to it since the derivative of the Lagrangian density with respect to $\dot{A}_0(x)$ is equal to zero. This does not pose any difficulties since the component of the field $A_0(x)$ is not independent of the other canonical variables because by virtue of the field equations it can be expressed in terms of the function $\boldsymbol{\pi}(x)$ in the following manner:

$$A_0(x) = \mu^{-2} \nabla \cdot \boldsymbol{\pi}(x). \tag{13}$$

The three spatial components of the field $\mathbf{A}$ in canonical formulation will be denoted by $\boldsymbol{\phi}$. The independent degrees of freedom of the field are thus described by six canonical variables $\boldsymbol{\phi}(x)$ and $\boldsymbol{\pi}(x)$. Out of the components of the tensor $T^{\mu\nu}(x)$ we construct expressions representing the energy H, the momentum $\mathbf{P}$, and the angular momentum $\mathbf{M}$ of the field. They are expressed in terms of the fields $\boldsymbol{\phi}(x)$ and $\boldsymbol{\pi}(x)$:

$$H = \frac{1}{2}\int d^3x \left[(\boldsymbol{\pi}(x))^2 + \mu^{-2}(\nabla \cdot \boldsymbol{\pi}(x))^2 + (\nabla \times \boldsymbol{\phi}(x))^2 + \mu^2 (\boldsymbol{\phi}(x))^2\right], \tag{14a}$$

$$\mathbf{P} = \int d^3x \sum_{i=1}^{3} \phi_i(x) \nabla \pi_i(x), \tag{14b}$$

$$\mathbf{M} = \int d^3x \left[\mathbf{x} \times \sum_{i=1}^{3} \phi_i(x) \nabla \pi_i(x) + \boldsymbol{\phi}(x) \times \boldsymbol{\pi}(x)\right]. \tag{14c}$$

On writing the fields $\boldsymbol{\phi}(x)$ and $\boldsymbol{\pi}(x)$ in terms of the coefficients $f(\mathbf{k}, r)$ and substituting them into the formulae (14), we obtain expressions for the energy, momentum, and angular momentum (analogous to the expressions obtained in Maxwellian theory):

$$H = \sum_{r=0,-1,+1} \int d\Gamma(\mu) \sqrt{\mu^2+\mathbf{k}^2}\, f^*(\mathbf{k}, r) f(\mathbf{k}, r), \tag{15a}$$

$$\mathbf{P} = \sum_{r=0,-1,+1} \int d\Gamma(\mu) \mathbf{k} f^*(\mathbf{k}, r) f(\mathbf{k}, r), \tag{15b}$$

$$\mathbf{M} = \sum_{r=0,-1,+1} \int d\Gamma(\mu) \left[f^*(\mathbf{k}, r) \left(\mathbf{k} \times \frac{1}{i} \mathbf{D} \right) f(\mathbf{k}, r) + r\mathbf{n}(\mathbf{k}) f^*(\mathbf{k}, r) f(\mathbf{k}, r) \right]. \tag{15c}$$

If the Poisson brackets of two arbitrary functionals $\mathscr{F}[\boldsymbol{\phi}, \boldsymbol{\pi}]$ and $\mathscr{G}[\boldsymbol{\phi}, \boldsymbol{\pi}]$ which are dependent on the canonical fields $\boldsymbol{\phi}(x)$ and $\boldsymbol{\pi}(x)$, taken at the instant of time t, are defined as

$$\{\mathscr{F}[\boldsymbol{\phi}, \boldsymbol{\pi}], \mathscr{G}[\boldsymbol{\phi}, \boldsymbol{\pi}]\} \equiv \sum_{i=1}^{3} \int d^3x \left[\frac{\delta \mathscr{F}}{\delta \phi_i(\mathbf{x}, t)} \cdot \frac{\delta \mathscr{G}}{\delta \pi_i(\mathbf{x}, t)} - \frac{\delta \mathscr{F}}{\delta \pi_i(\mathbf{x}, t)} \cdot \frac{\delta \mathscr{G}}{\delta \phi_i(\mathbf{x}, t)} \right], \tag{16}$$

the field equations may be written in the canonical form:

$$\dot{\boldsymbol{\phi}}(x) = \{\boldsymbol{\phi}(x), H\} = \boldsymbol{\pi}(x) - \mu^{-2} \nabla(\nabla \cdot \boldsymbol{\pi}(x)), \tag{17a}$$

$$\dot{\boldsymbol{\pi}}(x) = \{\boldsymbol{\pi}(x), H\} = \Delta \boldsymbol{\phi}(x) - \nabla(\nabla \cdot \boldsymbol{\phi}(x)) - \mu^2 \boldsymbol{\phi}(\mathbf{x}). \tag{17b}$$

The Poisson brackets of the canonical variables $\boldsymbol{\phi}(x)$ and $\boldsymbol{\pi}(x)$ do not contain derivatives of the Dirac delta function:

$$\{\phi_k(\mathbf{x}, t), \pi_l(\mathbf{y}, t)\} = \delta_{kl} \delta(\mathbf{x}-\mathbf{y}), \tag{18a}$$

$$\{\phi_k(\mathbf{x}, t), \phi_l(\mathbf{y}, t)\} = 0 = \{\pi_k(\mathbf{x}, t), \pi_l(\mathbf{y}, t)\}. \tag{18b}$$

Just as we have done in Chapter 3, so now it may be shown that the canonical formulation of the Proca theory is covariant with respect to Poincaré transformations.

In the presence of sources of the field $A_\mu(x)$, represented by the current four-vector $\mathscr{J}^\mu(x)$, the field is described by a Lagrangian which differs from the Lagrangian (1) by the expression $-A_\mu(x)\mathscr{I}^\mu(x)$. The field equations

$$-K^{\mu\nu}A_\nu(x) \equiv [(\Box+\mu^2)g^{\mu\nu}-\partial^\mu\partial^\nu]A_\nu(x) = \mathscr{I}^\mu(x) \tag{19}$$

then lead to the following relation between the time component of the field $A_0(x)$ and the field $\boldsymbol{\pi}(x)$ canonically conjugate to $\boldsymbol{\phi}(x)$:

$$A_0(x) = \mu^{-2}(\nabla\cdot\boldsymbol{\pi}(x)+\mathscr{I}_0(x)). \tag{20}$$

The energy density $T^{00}(x)$ differs from the expression in the integrand in formula (14a) by the term $A_\mu(x)\mathscr{I}^\mu(x)$. As a result of these changes, the expression describing the energy of the field assumes the form

$$H = \frac{1}{2}\int d^3x[(\boldsymbol{\pi}(x))^2+\mu^{-2}(\nabla\cdot\boldsymbol{\pi}(x)+\mathscr{I}_0(x))^2+(\nabla\times\boldsymbol{\phi}(x))^2$$
$$+\mu^2(\boldsymbol{\phi}(x))^2]-\int d^3x\boldsymbol{\mathscr{I}}(x)\cdot\boldsymbol{\phi}(x). \tag{21}$$

This expression differs from the one describing the energy in the absence of sources by the term

$$\int d^3x\left[\frac{1}{2}\mu^{-2}(\mathscr{I}_0(x))^2+\mu^{-2}(\nabla\cdot\boldsymbol{\pi}(x))\mathscr{I}_0(x)-\boldsymbol{\mathscr{I}}(x)\cdot\boldsymbol{\phi}(x)\right], \tag{22}$$

which we interpret as the energy of interaction of the field A_μ with external sources and the interaction energy of the sources.

The Quantum Theory of a Vector Field with Mass

Quantization of a vector field consists in replacing the fields $\boldsymbol{\phi}(x)$ and $\boldsymbol{\pi}(x)$ by the operators $\boldsymbol{\phi}(x)$ and $\boldsymbol{\pi}(x)$, and the Poisson brackets (18) by the commutators

$$[\phi_k(\mathbf{x},t),\pi_l(\mathbf{y},t)] = i\delta_{kl}\delta(\mathbf{x}-\mathbf{y}), \tag{23a}$$

$$[\phi_k(\mathbf{x},t),\phi_l(\mathbf{y},t)] = 0 = [\pi_k(\mathbf{x},t),\pi_l(\mathbf{y},t)]. \tag{23b}$$

The annihilation and creation operators c and $c\dagger$ are defined by means of the decomposition of the operators $\boldsymbol{\phi}$ and $\boldsymbol{\pi}$ into plane waves. The operators $\boldsymbol{\phi}(x)$ and $\boldsymbol{\pi}(x)$ are associated with the operators $c(\mathbf{k},r)$ and $c^\dagger(\mathbf{k},r)$ in the same way as the functions $\boldsymbol{\phi}(x)$ and $\boldsymbol{\pi}(x)$ were related to the functions $f(\mathbf{k},r)$ and $f^*(\mathbf{k},r)$. The operators $c(\mathbf{k},r)$ and $c^\dagger(\mathbf{k},r)$

satisfy the commutation relations

$$[c(\mathbf{k}, r), c^\dagger(\mathbf{k}', s)] = \delta_{rs}\delta_\Gamma(\mathbf{k}, \mathbf{k}'). \tag{24}$$

After multiplication by the wave-packet profiles $f(\mathbf{k}, r)$ and integration, these operators are interpreted as the annihilation and creation operators of vector particles with mass μ; we denote them by the symbols $c[f]$ and $c^\dagger[f]$. The indices r and s label different states of polarization. In addition to the polarizations $+1$ and -1, particles with mass may also have polarization 0. Out of the operators $c(\mathbf{k}, r)$ we can construct a vector annihilation operator $c_\mu(\mathbf{k})$ according to the formula

$$c_\mu(\mathbf{k}) \equiv \sum_{r=0,-1,+1} \varepsilon_\mu(\mathbf{k}, r)\, c(\mathbf{k}, r). \tag{25}$$

From the commutation relations (23) and from the field equations we can obtain commutation relations for fields at two arbitrary points of space-time. It is most convenient to use the Fourier representation and the relations (24) for this purpose. As a result we obtain:

$$[A_\mu(x), A_\nu(y)] = \frac{1}{i}\, \Delta_{\mu\nu}(x-y), \tag{26}$$

where

$$\Delta_{\mu\nu}(x) \equiv -(g_{\mu\nu}+\mu^{-2}\partial_\mu\partial_\nu)\Delta(x), \tag{27}$$

and the function $\Delta(x)$ is an invariant solution of the Klein–Gordon equation:

$$\Delta(x, \mu) \equiv \Delta^{(+)}(x, \mu)+\Delta^{(-)}(x, \mu), \tag{28a}$$

$$\Delta^{(\pm)}(x, \mu) \equiv \pm i \int \mathrm{d}\Gamma(\mu)\, \mathrm{e}^{\mp ik\cdot x}. \tag{28b}$$

In this section and subsequently wherever this does not lead to ambiguity, we shall not indicate the mass-dependence of the function Δ. With the signs $(+)$ and $(-)$ we have denoted functions containing, respectively, only positive and only negative frequencies. The concept of positive and negative frequencies is defined as in the case of the d'Alembert equation (cf. remarks on p. 127).

Now let us go on to discuss the problem of the interaction of a quantum field $A_\mu(x)$ with an external current $\mathscr{J}^\mu(x)$. The operator $A_\mu(x)$ then satisfies the field equations

$$\Box A^\nu(x)+\mu^2 A^\nu(x)-\partial^\nu\partial_\mu A^\mu(x) = \mathscr{J}^\nu(x). \tag{29}$$

We use the continuity equation for current to rewrite this as a system of equations:

$$(\Box + \mu^2) A_\mu(x) = \mathscr{I}_\mu(x), \tag{30a}$$

$$\partial^\mu A_\mu(x) = 0. \tag{30b}$$

We introduce two bases Ψ^{in} and Ψ^{out} into the state-vector space of the system. In the asymptotic source-free regions of space-time, the operators $A_\mu(x)$ go over into the operators $A_\mu^{\text{in}}(x)$ and $A_\mu^{\text{out}}(x)$:

$$A_\mu(x) = \begin{cases} A_\mu^{\text{in}}(x), & \text{when} \quad x^0 < T_1, \\ A_\mu^{\text{out}}(x), & \text{when} \quad x^0 > T_2. \end{cases} \tag{31}$$

$$A_\mu^{\substack{\text{in}\\ \text{out}}}(x) = \sum_{i=1}^{\infty} \left(\varphi_\mu[x|f_i]\, c_i^{\substack{\text{in}\\ \text{out}}} + (\varphi_\mu[x|f_i])^* c_i^{\substack{\text{in}\\ \text{out}}\dagger}\right)$$

$$= A_\mu^{\substack{\text{in}\\ \text{out}}(+)} + A_\mu^{\substack{\text{in}\\ \text{out}}(-)}, \tag{32}$$

where

$$\varphi_\mu[x|f_i] = \sum_{r=0,-1,+1} \int \mathrm{d}\Gamma(\mu)\, \varepsilon_\mu(\mathbf{k}, r) f_i(\mathbf{k}, r) \mathrm{e}^{-ik\cdot x} \tag{33}$$

and the functions $f_i(\mathbf{x}, r)$ form a complete orthonormal set of wave-packet profiles:

$$\sum_{i=1}^{\infty} f_i(\mathbf{k}, r) f_i^*(\mathbf{k}', s) = \delta_{rs} \delta_\Gamma(\mathbf{k}, \mathbf{k}'), \tag{34a}$$

$$\sum_{r=0,-1,+1} \int \mathrm{d}\Gamma(\mu) f_i^*(\mathbf{k}, r) f_j(\mathbf{k}, r) = \delta_{ij}. \tag{34b}$$

The scalar product for the functions $\varphi_\mu[x|f_i]$ is of the form

$$i \int \mathrm{d}\sigma_\mu((\varphi^{\mu\nu}[x|f_i])^* \varphi_\nu[x|f_j] - (\varphi_\nu[x|f_i])^* \varphi^{\mu\nu}[x|f_j])$$

$$= \sum_r \int \mathrm{d}\Gamma(\mu) f_i^*(\mathbf{k}, r) f_j(\mathbf{k}, r), \tag{35}$$

where

$$\varphi_{\mu\nu}[x|f_i] \equiv \partial_{[\mu} \varphi_{\nu]}[x|f_i]. \tag{36}$$

By the Proca equations, the integral over the hypersurface σ does not depend on the choice of σ. We can calculate this integral by taking σ in the form of the hypersurface t = const. With the aid of formula (35), the annihilation operators can be expressed in the following manner in terms of the field operators:

$$c_i^{\substack{\text{in}\\ \text{out}}} = i\int d\sigma_\mu\big((\varphi^{\mu\nu}[x|f_i])^* A_\nu^{\substack{\text{in}\\ \text{out}}}(x) - (\varphi_\nu[x|f_i])^* f^{\mu\nu}_{\substack{\text{in}\\ \text{out}}}(x)\big). \tag{37}$$

The time component of the potential operator, which is not an independent quantity, may be eliminated from expression (37) by means of the field equations.

The S Operator in the Presence of External Sources

As in Maxwell's theory, we shall construct the S operator in Proca theory in the form of a chronological exponential operator. Such construction will ensure the causality and unitarity of the S operator. We expect this operator to be of the form

$$S = T\exp\Big(-i\int d^4x$$

$$\times\Big[-\boldsymbol{\phi}^{\text{in}}(x)\cdot\boldsymbol{\mathscr{I}}(x) + \mu^{-2}\nabla\cdot\boldsymbol{\pi}^{\text{in}}(x)\mathscr{I}_0(x) + \frac{1}{2}\mu^{-2}(\mathscr{I}_0(x))^2\Big]\Big). \tag{38}$$

The brackets contain the energy density operator of the interaction of the particles with the sources (cf. formula (22)). The validity of formula (38) may be proved by the method used in Maxwellian theory and in many-particle theory. Here, we shall give a somewhat different proof, after previously rearranging formula (38) into a relativistically invariant form. Using the commutation relations for the operators $\boldsymbol{\phi}^{\text{in}}$ and $\boldsymbol{\pi}^{\text{in}}$ and formula (H.13), we obtain

$$S = \exp\Big[\frac{1}{2}i\int d^4x\, d^4y \mathscr{I}^\mu(x)\Delta^F_{\mu\nu}(x-y)\mathscr{I}^\nu(y)\Big]$$

$$\times\exp\Big[-i\int d^4x A_\mu^{\text{in}(-)}(x)\mathscr{I}^\mu(x)\Big]\exp\Big[-i\int d^4x A_\mu^{\text{in}(+)}(x)\mathscr{I}^\mu(x)\Big], \tag{39}$$

where $\Delta^F_{\mu\nu}(x)$ is the *Feynman Green's function for the Proca equation*

$$\Delta^F_{\mu\nu}(x) \equiv -(g_{\mu\nu}+\mu^{-2}\partial_\mu\partial_\nu)\Delta_F(x), \tag{40}$$

and $\Delta_F(x)$ is the *Feynman Green's function for the Klein–Gordon equation*

$$\Delta_F(x) \equiv \theta(x^0)\Delta^{(+)}(x) - \theta(-x^0)\Delta^{(-)}(x). \tag{41}$$

Let us now outline the proof of formula (39). The operator defined by formula (39) satisfies the relation

$$[A_\lambda^{\text{in}}(x), S] = S\int d^4y \Delta_{\lambda\varrho}(x-y)\mathscr{I}^\varrho(y), \tag{42}$$

by virtue of (H.1). By the field equations, the operators $A_\lambda^{\text{out}}(x)$ and $A_\lambda^{\text{in}}(x)$ are related by

$$A_\lambda^{\text{out}}(x) = A_\lambda^{\text{in}}(x)+\int d^4y\Delta_{\lambda\varrho}(x-y)\mathscr{J}^\varrho(y). \tag{43}$$

Equations (39) and (43) lead to the relation

$$A_\lambda^{\text{in}}(x)\mathsf{S} = \mathsf{S}A_\lambda^{\text{out}}(x), \tag{44}$$

which means that the operator defined by formula (39) or the equivalent formula (38) is the S operator.

We shall now go to the limit $\mu \to 0$ and show that, with respect to the description of all photon emission and absorption phenomena, the theory so obtained is equivalent to the quantum Maxwellian theory with sources.† We make the assumption that the external current satisfies the continuity equation. Then, in formula (39) the function $\Delta_{\mu\nu}^{\text{F}}(x-y)$ can be replaced by the function $-g_{\mu\nu}\Delta_{\text{F}}(x-y)$, which tends to $-g_{\mu\nu}D_{\text{F}}(x-y)$ in the limit, when $\mu \to 0$. The expression in the exponent of the second exponential function in formula (39) goes over in this limit,‡ to the expression

$$-i\sum_{r=\pm 1}\int d\Gamma\varepsilon_\lambda(\mathbf{k}, r)\tilde{\mathscr{J}}^\lambda(\mathbf{k}, \omega)c^\dagger(\mathbf{k}, r)|_{\mu=0}. \tag{45}$$

Similar expressions containing the annihilation operators $c(\mathbf{k}, r)$ are obtained as an exponent of the third exponential function. Since the vectors $\varepsilon_\lambda(\mathbf{k}, \pm 1)$ go over in the limit into the vectors $\varepsilon_\lambda(\mathbf{k})$ and $-\varepsilon_\lambda^*(\mathbf{k})$ in the radiation gauge, the operator S for vector particles with mass μ goes over in the limit, when $\mu \to 0$, into the S operator for Maxwellian theory. Particles whose spin projection on the direction of the momentum is equal to zero do not interact with the sources in the limit when $\mu \to 0$. For, in this limit the S operator does not depend on the operators $c(\mathbf{k}, 0)$ and $c^\dagger(\mathbf{k}, 0)$. This result is a consequence of the assumption that the current satisfies the continuity equation. Thus we have gained a new way of describing the quantum theory of the electromagnetic field, one that is entirely equivalent to the description given earlier.

† The field operators $A_\mu(x)$ do not, however, have a limit when $\mu \to 0$.

‡ For, we have

$$\varepsilon_\lambda(\mathbf{k}, 0)\tilde{\mathscr{J}}^\lambda(\mathbf{k}, \omega) = \frac{\mu}{|\mathbf{k}|}\tilde{\mathscr{J}}^0(\mathbf{k}, \omega) \to 0.$$

Objects representing particles with zero polarization do appear in this description, it is true, but as follows from the given theory, these particles cease to be emitted or absorbed by the sources in the limit, when $\mu \to 0$. They then move freely, without affecting the processes in which the other particles participate. Compared with Maxwellian theory, the theory in which the Proca field and the transition to the limit when $\mu \to 0$ appear has the advantage that there exists a tensor description of this theory with the use of potentials. In the theory which describes photons with mass, the field $A_\mu(x)$ is a vector field. Various experimental data available at present# indicate that had we wanted to describe the electromagnetic field in terms of a vector field with mass μ, that mass would have to be many orders of magnitude smaller than the electron mass. The most stringent restriction on μ is obtained from satellite measurements of the terrestrial magnetic field: μ must be less than 10^{-20} times the electron mass.

The best laboratory measurements## permitting the upper limit of the photon mass to be determined consist in verifying Coulomb's law with extreme precision. A hypothetical modification of Coulomb's law is written in the form $1/r^{1+q}$ and the upper limit for the parameter q is then found. The upper limit for μ ($\mu < 8\times 10^{-15}$ eV) is evaluated on the basis of the experimental value of q ($q < (2.7\pm 3.1)\times 10^{-16}$).

15. THE INFRARED CATASTROPHE

In this section we shall investigate radiation from a charged point particle moving with acceleration. A classical description of such a radiating system was given in Section 9. Now we shall demonstrate that in quantum theory the description of such a system in terms of photons entails difficulties which are referred to as the *infrared catastrophe*.###

Cf. the review article by A. Goldhaber and M. M. Nieto (1971).

E. R. Williams, J. E. Faller, and H. A. Hill (1971).

The first solution of the problem of the infrared catastrophe was given by F. Bloch and A. Nordsieck (1936). A description of the method due to Bloch and Nordsieck can be found in the book by Hamilton. The infrared catastrophe in complete relativistic electrodynamics has been dealt with in many works. The classic work in this field is the extensive paper by D. R. Yennie, S. Frautchi, and H. Suura (1961).

These difficulties follow from the fact that photons have zero rest mass and can, therefore, have arbitrarily small energy.

As follows from the discussion in Section 9, the Fourier transform of the current $\mathscr{I}^{\mu}(x)$ generated by a single point particle for small values of k^{μ} is of the form

$$\tilde{\mathscr{I}}^{\mu}(k) = -ie\left(\frac{p^{\mu}}{k \cdot p} - \frac{q^{\mu}}{k \cdot q}\right) + \tilde{\mathscr{I}}^{\mu}_{\text{reg}}(k), \tag{1}$$

where p^{μ} and q^{μ} are asymptotic four-vectors of the momentum of the particle in the past and in the future and $\mathscr{I}^{\mu}_{\text{reg}}$ is a regular function at the point $k = 0$. On substituting this current transform into formula (13.31), we get the following expression for the S operator:

$$\mathsf{S} = \exp\left\{ie^2 \int \frac{\mathrm{d}^4 k}{(2\pi)^4} \frac{1}{k^2 + i\varepsilon}\left[\left(\frac{p^{\lambda}}{k \cdot p} - \frac{q^{\lambda}}{k \cdot q}\right)\left(\frac{p_{\lambda}}{k \cdot p} - \frac{q_{\lambda}}{k \cdot q}\right) + R(k)\right]\right\} \exp\left(i \sum_{n=1}^{\infty} \iota_n c_n^{\dagger \text{in}}\right) \exp\left(i \sum_{n=1}^{\infty} \iota_n^* c_n^{\text{in}}\right), \tag{2}$$

where the residual term $R(k)$ tends more slowly to infinity for $k \to 0$ than does the term written out explicitly. The imaginary part of the integral in the exponent (the real part gives only an unobservable phase factor in the S operator) diverges logarithmically for small values of k^{μ}. It may be written as

$$\frac{1}{2} \int \mathrm{d}\Gamma \left[\left(\frac{p^{\lambda}}{k \cdot p} - \frac{q^{\lambda}}{k \cdot q}\right)\left(\frac{p_{\lambda}}{k \cdot p} - \frac{q_{\lambda}}{k \cdot q}\right) + R(k)\right]. \tag{3}$$

The S operator thus contains the factor $e^{-\infty}$, which means that all the matrix elements of the S operator in the Fock basis are zero.

The divergent integral in the exponent is, by virtue of formula (12.11), half the average number of photons radiated. Since the integrand has a non-integrable singularity $1/\omega$ at the point $\omega = 0$, the number of long-wavelength photons radiated is infinite.

The source of the infrared catastrophe lies in idealization being carried out too far in the description of experiments with charged particles. All measurements are always performed in a finite time, whereby the time over which the sources of electromagnetic field act is never in reality infinitely long. The considered simple model of current produced by a single charged point particle† is a source of an infinite number of radiated photons because the duration of the charge is infinite.

To illustrate this assertion, let us once again make use (cf. Section 9) of a formal procedure wherein the current produced by the charged particle is switched on and off by means of the exponential switching factor $\exp(-\varepsilon|\tau|)$. The value of the Fourier transform of the modified current at the point $k = 0$ is given by

$$\tilde{\mathscr{I}}^{\mu}_{\varepsilon}(0) = \frac{e}{\varepsilon}\left(\frac{p^{\mu}}{m} - \frac{q^{\mu}}{m}\right) + \begin{array}{l}\text{terms which are}\\ \text{finite when } \varepsilon \to 0.\end{array}$$

This expression is finite, whence the infrared catastrophe is completely eliminated.

The introduction of the switching factor and the preservation of a finite value of ε in the final formulae does prove the existence of a direct relation between the time over which the sources act and the infrared catastrophe, but is nevertheless not a satisfactory operation. For finite values of ε, $\tilde{\mathscr{I}}^{\mu}_{\varepsilon}(\nu)$ is finite, it is true, but the current does not satisfy the continuity equation since the switching factor causes charge to come into being and vanish. The equation of continuity for current, on the other hand, is a condition for the internal consistency of the Maxwell equations and violation of this condition can by no means be permitted.

A solution of the infrared catastrophe problem that is free of these shortcomings will be obtained by analysing the measuring capabilities of actual instruments. The finite duration of every measuring process, to which we drew attention earlier, imposes obvious restrictions on measurements of very small frequencies. Similarly, the finite spatial dimensions of measuring instruments impose restrictions on measurements of very long wavelengths. In quantum theory, the energies and momenta of photons replace the vibration frequencies and wavelengths, and the restrictions on measurements of these quantities are usually characterized by giving the resolution of the instrument. If this resolution is ΔE in the case of energy measurements, systems of photons with a total energy less than ΔE escape observation. The

† The infrared catastrophe also occurs for particles with a continuous distribution of charge, provided that the total charge is not zero.

situation is similar with resolution in measurements of momentum. Photons with low energies, and hence with long wavelengths, are called *soft photons.*

It will now be shown that in the considered simple model with given sources, the undetectability of photons with frequencies below a certain cut-off frequency ω_{min} eliminates from all formulae for the probability of real processes that part of the Fourier transform of the current $\tilde{\mathscr{I}}^{\mu}(\mathbf{k}, \omega)$ which corresponds to the values $|\mathbf{k}| = \omega < \omega_{\text{min}}$.

Consider any measurement in which we do not get information concerning soft photons. The projection operator P representing such a measurement (question) is a unit operator in the subspace of long-wavelength (soft) photons,

$$[c(\mathbf{k}), P] = 0, \quad \text{when} \quad |\mathbf{k}| < \omega_{\text{min}}. \tag{4}$$

The probability p of obtaining an affirmative answer to the question represented by the operator P, when the electromagnetic field is in the state described in the past by the density operator ϱ_i, is given by

$$p = \operatorname{Tr}\{P\mathsf{S}\varrho_i\mathsf{S}^{\dagger}\} = \operatorname{Tr}\{\mathsf{S}^{\dagger}P\mathsf{S}\varrho_i\}. \tag{5}$$

The operator $\mathsf{S}^{\dagger}P\,\mathsf{S}$ figuring in this formula no longer displays the infrared catastrophe. For we have

$$\mathsf{S}^{\dagger}[\mathscr{I}]P\mathsf{S}[\mathscr{I}] = \mathsf{S}^{\dagger}[\mathscr{I}_{\text{h}}]P\mathsf{S}[\mathscr{I}_{\text{h}}], \tag{6}$$

where $\mathsf{S}[\mathscr{I}]$ denotes the S operator with external current, whereas $\mathscr{I}_{\text{h}}$ denotes the *hard part of the current,* which we define by the formulae

$$\mathscr{I}^{\mu}_{\text{h}}(x) \equiv \mathscr{I}^{\mu}(x) - \mathscr{I}^{\mu}_{\text{s}}(x), \tag{7a}$$

$$\mathscr{I}^{\mu}_{\text{s}}(x) \equiv \int\limits_{\substack{k_0<\omega_{\text{min}} \\ |\mathbf{k}|<\omega_{\text{min}}}} \mathrm{d}^4k\,\mathrm{e}^{-ik\cdot x}\tilde{\mathscr{I}}^{\mu}(k). \tag{7b}$$

The value of the Fourier transform of the current in the neighbourhood of the point $k = 0$ thus does not at all affect the results of real measurements.

The separation of the current into a hard part $\mathscr{I}_{\text{h}}$ and a soft part $\mathscr{I}_{\text{s}}$ can be given a simple physical interpretation which we shall use subsequently. To this end let us consider the current produced by the motion of a point particle and let us separate the space-time density of the charge into the aforementioned parts:

$$\mathscr{I}^{\mu}(x) = \int \mathrm{d}\xi^{\mu} e\delta^{(4)}(x-\xi) = \int \mathrm{d}\xi^{\mu}\big(\varrho_{\mathrm{h}}(x-\zeta)+\varrho_{\mathrm{s}}(x-\xi)\big), \tag{8}$$

where

$$\varrho_{\mathrm{s}}(x) \equiv e \int\limits_{\substack{k_0<\omega_{\min}\\ |\mathbf{k}|<\omega_{\min}}} \mathrm{d}^4k\, \mathrm{e}^{-ik\cdot x}. \tag{9}$$

The function $\varrho_{\mathrm{s}}(x)$ determines the space-time distribution of the charge associated with the soft part of the current. This is an analytical function in the variables $\mathbf{x}$ and t, with a maximum value $\varrho_{\mathrm{s}}(0)$ of $e(2\omega_{\min})^4$. The charge concentrated about the point $x = 0$ in a region much smaller than $\left(\frac{1}{\omega_{\min}}\right)^4$ is only a tiny fraction of the full charge e. The total charge determined by the soft part is, however, equal to the full charge of the particle,

$$\int \mathrm{d}^4x\varrho_{\mathrm{s}}(x) = e.$$

From the properties of the soft part described it follows that the hard part of the current,

$$\mathscr{I}^{\mu}_{\mathrm{h}}(x) = \int \mathrm{d}\xi^{\mu}\big(e\delta^{(4)}(x-\xi)-\varrho_{\mathrm{s}}(x-\xi)\big),$$

may be regarded as the difference between the original current flowing along the world line of the particle and the screening current flowing in all space. The density of the screening current is insignificant, and yet the screening is complete. In actual experiments, screening charges and screening currents always occur and for this reason the hard part of the current describes the motion of charges more faithfully than does the original current $\mathscr{I}^{\mu}(x)$. Instead of starting by taking the current $\mathscr{I}^{\mu}(x)$ caused by the motion of a point particle and then eliminating the infrared catastrophe by restricting the resolution of the instruments, one can thus from the very beginning consider the screened current $\mathscr{I}^{\mu}_{\mathrm{h}}(x)$ which does not interact at all with photons of too low frequencies.

The separation of the current into a hard and a soft part is not relativistically invariant and hence is inconvenient for analysing the infrared catastrophe in the full quantum electrodynamics. The principal purpose of this separation, viz. to eliminate photons with very low frequencies, can also be achieved in a different way, if from the outset, use is made of the theory of vector particles with rest mass μ, as formulated in the preceding section. To conclude the considerations of the infrared catastrophe in the theory with external currents, we shall use just that formulation of the theory.

For a field $A_\mu(x)$ with mass μ the S operator is not burdened with the infrared catastrophe since the divergent integral in formula (2) may be replaced by the integral

$$-\int \frac{\mathrm{d}^4k}{(2\pi)^4} \frac{1}{\mu^2-k^2-i\varepsilon} \left[\left(\frac{p^\lambda}{k\cdot p}-\frac{q^\lambda}{k\cdot q}\right)\left(\frac{p_\lambda}{k\cdot p}-\frac{q_\lambda}{k\cdot q}\right)+R(k)\right],$$

in which the integrand for $k = 0$ has an integrable singularity. As an illustration, let us consider a state which is the vacuum in the remote past. The state vector Ω^{in} may be expressed in terms of the S operator by the formula

$$\Omega^{\text{in}} = \mathrm{S}\Omega^{\text{out}} = \mathrm{e}^{i\varphi}\exp\left[\frac{1}{2}\int \mathrm{d}\Gamma(\mu)\tilde{\mathscr{J}}^*_\lambda(k)\tilde{\mathscr{J}}^\lambda(k)\right]$$
$$\times \exp\left[-i\int \mathrm{d}\Gamma(\mu)\, c_\nu^{\dagger\text{out}}(\mathbf{k})\tilde{\mathscr{J}}^\nu(x)\right]\Omega^{\text{out}}, \qquad (10)$$

where $\mathrm{e}^{i\varphi}$ is a phase factor dependent on $\tilde{\mathscr{J}}^\mu(k)$. If the components of the current $\tilde{\mathscr{J}}^\mu(k)$ behave like $1/|\mathbf{k}|$ in the vicinity of $k = 0$, then in the limit when $\mu \to 0$ all the components of the vector Ω^{in} in the Fock basis OUT tend to zero (this means the infrared catastrophe). However, if the hard part of the current is used instead of $\mathscr{J}^\mu(x)$, then in accordance with what has been said above (cf. p. 220), in the limit when $\mu \to 0$ we obtain the same results as in Maxwellian quantum electrodynamics.

CHAPTER 5

RELATIVISTIC QUANTUM MECHANICS OF ELECTRONS

16. THE DIRAC EQUATION AND THE SYMMETRIES OF ITS SOLUTIONS

IN RELATIVISTIC quantum mechanics the *wave function of an electron* in an external electromagnetic field $\mathscr{A}_\mu(x)$ is a set of four complex functions $\psi_a(x)$ ($a = 1, \ldots, 4$) satisfying the *Dirac equation.* The Dirac equation may be written in matrix form

$$[-i\gamma^\mu \partial_\mu + m + e\gamma^\mu \mathscr{A}_\mu(x)]\psi(x) = 0, \tag{1}$$

where we have denoted a one-column, four-row matrix constructed from the functions $\psi_a(x)$ by $\psi(x)$. The four functions $\psi_a(x)$ form a *bispinor field.* The wave function $\psi(x)$ is often called a *bispinor.*† By the symbol γ^μ we have denoted an arbitrary set of fourth-order matrices satisfying the following conditions:

$$\gamma^\mu\gamma^\nu + \gamma^\nu\gamma^\mu = 2g^{\mu\nu}. \tag{2}$$

The four matrices γ^μ and their products can be used to construct a complete set of 16 linearly-independent matrices $\Gamma_A (A = 1, \ldots, 16)$,

$$I, \gamma^\mu, \gamma^{\mu\nu}, \gamma_5\gamma^\mu, \gamma_5, \tag{3}$$

where I is a fourth-order unit matrix,

$$\gamma^{\mu\nu} \equiv \begin{cases} \gamma^\mu\gamma^\nu & (\mu \neq \nu), \\ 0 & (\mu = \nu), \end{cases} \tag{4}$$

$$\gamma_5 \equiv i\gamma^0\gamma^1\gamma^2\gamma^3. \tag{5}$$

All the matrices Γ_A (with the exception of I) have zero traces.

---

† The elements of the theory of spinors and bispinors are given in Appendix I.

Most problems involving the Dirac equation can be solved without resorting to a particular representation# of the matrices Γ_A. At times, however, knowledge of the specific form of these matrices proves useful. The Dirac, Majorana, and spinor representations are the ones used most often.

Below we give the form of five γ matrices in these representations.

The Dirac representation

$$\gamma^0 = \begin{pmatrix} 1 & 0 & 0 & 0 \\ 0 & 1 & 0 & 0 \\ 0 & 0 & -1 & 0 \\ 0 & 0 & 0 & -1 \end{pmatrix}, \quad \gamma^1 = \begin{pmatrix} 0 & 0 & 0 & 1 \\ 0 & 0 & 1 & 0 \\ 0 & -1 & 0 & 0 \\ -1 & 0 & 0 & 0 \end{pmatrix},$$

$$\gamma^2 = \begin{pmatrix} 0 & 0 & 0 & -i \\ 0 & 0 & i & 0 \\ 0 & i & 0 & 0 \\ -i & 0 & 0 & 0 \end{pmatrix}, \quad \gamma^3 = \begin{pmatrix} 0 & 0 & 1 & 0 \\ 0 & 0 & 0 & -1 \\ -1 & 0 & 0 & 0 \\ 0 & 1 & 0 & 0 \end{pmatrix} \tag{6}$$

$$\gamma_5 = \begin{pmatrix} 0 & 0 & 1 & 0 \\ 0 & 0 & 0 & 1 \\ 1 & 0 & 0 & 0 \\ 0 & 1 & 0 & 0 \end{pmatrix}.$$

The Majorana representation†

$$\gamma^0 = \begin{pmatrix} 0 & 0 & i & 0 \\ 0 & 0 & 0 & i \\ -i & 0 & 0 & 0 \\ 0 & -i & 0 & 0 \end{pmatrix}, \quad \gamma^1 = \begin{pmatrix} 0 & 0 & 0 & -i \\ 0 & 0 & -i & 0 \\ 0 & -i & 0 & 0 \\ -i & 0 & 0 & 0 \end{pmatrix},$$

$$\gamma^2 = \begin{pmatrix} i & 0 & 0 & 0 \\ 0 & i & 0 & 0 \\ 0 & 0 & -i & 0 \\ 0 & 0 & 0 & -i \end{pmatrix}, \quad \gamma^3 = \begin{pmatrix} 0 & 0 & -i & 0 \\ 0 & 0 & 0 & i \\ -i & 0 & 0 & 0 \\ 0 & i & 0 & 0 \end{pmatrix}, \tag{7}$$

A general theory of the γ matrices (without use of a concrete representation of them) was given by W. Pauli (1936).

† The representation we have given differs from that given by Majorana. We shall nevertheless refer to it as the Majorana representation because it has a characteristic feature of that given by Majorana, viz. the elements of all the matrices γ^μ are imaginary.

$$\gamma_5 = \begin{pmatrix} 0 & 0 & 0 & -i \\ 0 & 0 & i & 0 \\ 0 & -i & 0 & 0 \\ i & 0 & 0 & 0 \end{pmatrix}.$$

The spinor representation

$$\gamma^0 = \begin{pmatrix} 0 & 0 & 1 & 0 \\ 0 & 0 & 0 & 1 \\ 1 & 0 & 0 & 0 \\ 0 & 1 & 0 & 0 \end{pmatrix}, \quad \gamma^1 = \begin{pmatrix} 0 & 0 & 0 & -1 \\ 0 & 0 & -1 & 0 \\ 0 & 1 & 0 & 0 \\ 1 & 0 & 0 & 0 \end{pmatrix},$$

$$\gamma^2 = \begin{pmatrix} 0 & 0 & 0 & i \\ 0 & 0 & -i & 0 \\ 0 & -i & 0 & 0 \\ i & 0 & 0 & 0 \end{pmatrix}, \quad \gamma^3 = \begin{pmatrix} 0 & 0 & -1 & 0 \\ 0 & 0 & 0 & 1 \\ 1 & 0 & 0 & 0 \\ 0 & -1 & 0 & 0 \end{pmatrix}, \tag{8}$$

$$\gamma_5 = \begin{pmatrix} 1 & 0 & 0 & 0 \\ 0 & 1 & 0 & 0 \\ 0 & 0 & -1 & 0 \\ 0 & 0 & 0 & -1 \end{pmatrix}.$$

Each representation of the matrices γ^μ can be obtained from any other representation by means of a similarity transformation;†

$$\gamma^\mu \to S\gamma^\mu S^{-1}.$$

For the representations given here we have the following similarity relations. The spinor representation can be obtained from the Dirac

† This conclusion follows from the theory of representations of finite groups as applied to a group whose elements are 32 matrices of the form $\pm(\gamma^0)^i(\gamma^1)^j(\gamma^2)^k(\gamma^3)^l$, where i, j, k, l may be equal to 0 or 1. The number of classes in this group is 17. For finite groups, on the other hand, the number of irreducible, non-equivalent representations is equal to the number of classes, and the sum of the squared dimensions of all these representations is equal to the number of elements of the group. It is thus seen that there exists only one representation of this group in terms of fourth-order matrices. The other 16 representations are one-dimensional. By Schur's lemma it follows that two representations of the matrices γ^μ determine, to within a factor, the matrix S connecting them.

representation by means of a unitary similarity transformation with the aid of the matrix S_1:

$$S_1 = \frac{1}{\sqrt{2}} \begin{pmatrix} 1 & 0 & 1 & 0 \\ 0 & 1 & 0 & 1 \\ 1 & 0 & -1 & 0 \\ 0 & 1 & 0 & -1 \end{pmatrix}.$$

The Majorana representation, on the other hand, can be obtained from the spinor representation by means of a unitary similarity transformation with the aid of the matrix S_2:

$$S_2 = \frac{1}{\sqrt{2}} \begin{pmatrix} 1 & 0 & 0 & 1 \\ 0 & 1 & -1 & 0 \\ 0 & -i & -i & 0 \\ i & 0 & 0 & -i \end{pmatrix}.$$

In studies of the Dirac equation symmetries, which will constitute the content of a subsequent part of this section, an important role is played by the following transformations of the set of matrices γ^μ:

$$\gamma^\mu \to \gamma^{\mu\dagger}, \tag{9a}$$

$$\gamma^\mu \to -\gamma^{\mu *}. \tag{9b}$$

These are similarity transformations since they lead from a set of matrices satisfying condition (2) to another set which also satisfies this condition. Transformations (9) can be expressed in terms of the matrices A and C which are defined to within multiplicative factors by the relations

$$\mathrm{A}\gamma^\mu \mathrm{A}^{-1} = \gamma^{\mu\dagger}, \tag{10a}$$

$$\mathrm{C}^{-1}\gamma^\mu \mathrm{C} = -\gamma^{\mu *}. \tag{10b}$$

It follows from these relations that the product $\mathrm{A}^\dagger\mathrm{A}^{-1}$ and $\mathrm{C}^*\mathrm{C}$ commute with all the matrices γ^μ and hence, by virtue of Schur's lemma, are proportional to the unit matrix. This property may be expressed as follows:

$$\mathrm{A} = a\mathrm{A}^\dagger, \tag{11a}$$

$$\mathrm{C}^* = c\mathrm{C}^{-1}, \tag{11b}$$

where a and c are factors which satisfy the conditions

$$a^*a = 1,$$
$$c^* = c.$$

By multiplying the matrix A by a suitable phase factor we can always bring equality (11a) to the form

$$\mathrm{A} = \mathrm{A}^\dagger. \tag{12}$$

The matrix C in the Majorana representation may be chosen as the unit matrix. In an arbitrary representation it may be chosen in the form

$$\mathrm{C} = S^*S^{-1}, \tag{13}$$

where S is the matrix of the similarity transformation which takes the Majorana representation into the given representation. The matrix C given by formula (13) satisfies the relation

$$\mathrm{C}^* = \mathrm{C}^{-1}. \tag{14}$$

It is thus seen that the factor c is positive and that all matrices C satisfying the condition (10b) may be normalized so that relation (14) is satisfied. The matrices A and C can be used to construct a matrix B,

$$\mathrm{B} \equiv \mathrm{C}^\dagger \mathrm{A}, \tag{15}$$

with the properties

$$\mathrm{B}\gamma^\mu \mathrm{B}^{-1} = -\gamma^{\mu T}. \tag{16}$$

The matrix B is antisymmetric.

Just as matrix B, matrix B^T satisfies equation (16); accordingly, the relation

$$\mathrm{B}^T = \pm \mathrm{B}$$

must hold. The plus sign in this equation would mean that out of the 16 linearly-independent matrices $\Gamma_A\mathrm{B}$, 10 matrices ($\gamma^\mu\mathrm{B}$ and $\gamma^{\mu\nu}\mathrm{B}$) are antisymmetric. This possibility must be excluded since there are at most 6 antisymmetric matrices among the linearly-independent matrices of rank four.

The fact that B is antisymmetric leads to the following relation for C and A:

$$\mathrm{C}^\dagger \mathrm{A} = -\mathrm{A}^T \mathrm{C}^*. \tag{17}$$

The form of the matrices A and C depends on the representation of the matrices γ^μ. Below we give the matrices A and C in the Dirac, Majorana, and spinor representations.†

The Dirac representation

$$A = \begin{pmatrix} 1 & 0 & 0 & 0 \\ 0 & 1 & 0 & 0 \\ 0 & 1 & -1 & 0 \\ 0 & 0 & 0 & -1 \end{pmatrix}, \quad C = \begin{pmatrix} 0 & 0 & 0 & 1 \\ 0 & 0 & -1 & 0 \\ 0 & -1 & 0 & 0 \\ 1 & 0 & 0 & 0 \end{pmatrix}.$$

The Majorana representation

$$A = \begin{pmatrix} 0 & 0 & i & 0 \\ 0 & 0 & 0 & i \\ -i & 0 & 0 & 0 \\ 0 & -i & 0 & 0 \end{pmatrix}, \quad C = \begin{pmatrix} 1 & 0 & 0 & 0 \\ 0 & 1 & 0 & 0 \\ 0 & 0 & 1 & 0 \\ 0 & 0 & 0 & 1 \end{pmatrix}.$$

The spinor representation

$$A = \begin{pmatrix} 0 & 0 & 1 & 0 \\ 0 & 0 & 0 & 1 \\ 1 & 0 & 0 & 0 \\ 0 & 1 & 0 & 0 \end{pmatrix}, \quad C = \begin{pmatrix} 0 & 0 & 0 & 1 \\ 0 & 0 & -1 & 0 \\ 0 & -1 & 0 & 0 \\ 1 & 0 & 0 & 0 \end{pmatrix}.$$

In these representations the matrix A is equal to the matrix γ^0.

The matrix A may be chosen to be equal to γ^0 in all representations which are unitarily equivalent to the spinor representation and we shall confine ourselves in what follows only to such representations.

The matrix A will now be used to define the scalar product of two solutions of the Dirac equation. We introduce the notation

$$\overline{\psi} \equiv \psi^\dagger A. \tag{18}$$

From the properties (10a) of the matrix A it follows that if the function ψ satisfies the Dirac equation, the function $\overline{\psi}$ satisfies the conjugate Dirac equation:

$$\overline{\psi}(x)[i\gamma^\mu \overleftarrow{\partial}_\mu + m + e\gamma^\mu \mathscr{A}_\mu(x)] = 0. \tag{19}$$

It thus follows that for any two solutions ψ_1 and ψ_2 of the Dirac equation, the expression

$$\mathscr{S}^\mu(x) \equiv \overline{\psi}_1(x)\gamma^\mu \psi_2(x)$$

† The aforementioned conditions do not determine matrices A and C uniquely. The choice of a concrete form for these matrices thus is an additional convention.

satisfies the continuity equation

$$\partial_\mu \mathscr{S}^\mu(x) = 0.$$

The scalar product of two solutions of the Dirac equation has the form

$$(\psi_1|\psi_2) \equiv \int_\sigma d\sigma_\mu \bar{\psi}_1(x)\gamma^\mu\psi_2(x). \tag{20}$$

By virtue of the continuity equation for $\mathscr{S}^\mu(x)$, the scalar product does not depend on the choice of spacelike hypersurface σ. If in particular we choose the hypersurface defined by the equation x^0 = const, we obtain the scalar product in a form resembling the scalar product of non-relativistic quantum mechanics,

$$(\psi_1|\psi_2) = \int d^3x\psi_1^\dagger(\mathbf{x}, x^0)\psi_2(\mathbf{x}, x^0).$$

It follows from this formula that the scalar product of two solutions of the Dirac equation is positive definite.

Now let us examine the *symmetries of the Dirac equation*. We shall show that the representation† of the full Poincaré group acts in the set of geometrical objects $\psi(x)$ and $\mathscr{A}_\mu(x)$, related by the Dirac equation. The *transformation laws of the electromagnetic potential* are known from classical electrodynamics. Fields transformed by means of a translation and proper Lorentz transformation ${}^{\mathrm{IL}}\mathscr{A}_\mu(x)$, spatially reflected ${}^{\mathrm{P}}\mathscr{A}_\mu(x)$, and time-reversed ${}^{\mathrm{T}}\mathscr{A}_\mu(x)$, are defined by the formulas

$${}^{\mathrm{IL}}\mathscr{A}_\mu(x) = \Lambda_\mu{}^\nu\mathscr{A}_\nu(\Lambda^{-1}(x-a)), \tag{21a}$$

$${}^{\mathrm{P}}\mathscr{A}_\mu(x) = \mathscr{P}_\mu^\nu\mathscr{A}_\nu(\mathscr{P}x), \tag{21b}$$

$${}^{\mathrm{T}}\mathscr{A}_\mu(x) = \mathscr{P}_\mu^\nu\mathscr{A}_\nu(-\mathscr{P}x), \tag{21c}$$

where the matrix $\mathscr{P}_\mu^\nu$ was defined in Section 6. Bispinor fields transformed by means of a translation and a proper Lorentz transformation, spatially reflected, and time-reversed, will be denoted by the symbols ${}^{\mathrm{IL}}\psi(x)$, ${}^{\mathrm{P}}\psi(x)$, and ${}^{\mathrm{T}}\psi(x)$, respectively. The invariance of the form of the

† To be more exact, this is the representation of an inhomogeneous group $SL(2, C)$ (cf. Appendix I). The term representation is used here in a somewhat more general sense than usual, for we also apply it to the assignment of antilinear transformations to elements of the group.

Dirac equation under transformations of the full Poincaré group is ensured by the following *transformation laws of the bispinor field*:

$$^{\mathrm{IL}}\psi(x) = \mathrm{S_L}\psi(\Lambda^{-1}(x-a)), \tag{22a}$$

$$^{\mathrm{P}}\psi(x) = \mathrm{P}\psi(\mathscr{P}x), \tag{22b}$$

$$^{\mathrm{T}}\psi(x) = (\mathrm{T}\psi(-\mathscr{P}x))^*, \tag{22c}$$

where the matrices $\mathrm{S_L}$, P, and T satisfy the relations

$$\mathrm{S_L^{-1}}\gamma^\mu \mathrm{S_L} = \Lambda_\nu{}^\mu\gamma^\nu, \tag{23a}$$

$$\mathrm{P^{-1}}\gamma^\mu \mathrm{P} = \mathscr{P}^\mu_\nu\gamma^\nu, \tag{23b}$$

$$\mathrm{T^{-1}}\gamma^{\mu *} \mathrm{T} = \mathscr{P}^\mu_\nu\gamma^\nu. \tag{23c}$$

Condition (23a) is satisfied by the matrix $\mathrm{S_L}$ in the form

$$\mathrm{S_L} = \pm\exp\left(\tfrac{1}{4}\, i\alpha_{\mu\nu}\sigma^{\mu\nu}\right), \tag{24}$$

where

$$\sigma^{\mu\nu} \equiv i\gamma^{\mu\nu}, \tag{25}$$

and the coefficients $\alpha_{\mu\nu}$ are related to the matrix $\Lambda_\mu{}^\nu$ by formulae (7.26). Let us verify that matrix (24) does obey condition (23a); we do so by expanding the exponential function as a series in the coefficients $\alpha_{\mu\nu}$ according to the formula (H.1) and making use of the commutation relations for the matrices γ^μ. Conditions (23b) and (23c) are satisfied by matrices P and T of the form:

$$\mathrm{P} = \gamma^0, \tag{26a}$$

$$\mathrm{T} = \mathrm{C}^{-1}\gamma^1\gamma^2\gamma^3. \tag{26b}$$

It may be verified by direct calculation that the transformed bispinors $^{\mathrm{IL}}\psi(x)$, $^{\mathrm{P}}\psi(x)$, and $^{\mathrm{T}}\psi(x)$, which we shall denote by the abbreviated symbol $'\psi(x)$, satisfy the Dirac equation which contains the functions $\mathscr{A}_\mu(x)$, appropriately transformed, i.e. the functions $^{\mathrm{IL}}\mathscr{A}_\mu(x)$, $^{\mathrm{P}}\mathscr{A}_\mu(x)$ and $^{\mathrm{T}}\mathscr{A}_\mu(x)$ (abbreviated to the symbol $'\mathscr{A}_\mu(x)$). This means that

$$[-i\gamma^\mu\partial_\mu + m + e\gamma^\mu{}'\mathscr{A}_\mu(x)]'\psi(x) = 0. \tag{27}$$

Formulae (24) and (26a), as well as the properties of the matrices A and γ^μ, lead moreover to the relations

$$S_L^\dagger A = A S_L^{-1}, \tag{28a}$$

$$P^\dagger A = A P^{-1}, \tag{28b}$$

which ensure that the scalar product of two solutions of the Dirac equation is invariant with respect to orthochronous Poincaré transformations. This follows from the fact that the bilinear form $\overline{\psi}_1(x)\gamma^\mu\psi_2(x)$ transforms like a vector under translations, proper Lorentz transformations, and spatial inversion, and from the fact that the scalar product is independent of the choice of the hypersurface σ.

The matrix T given by formula (26b) satisfies the relation

$$T^* = -T^{-1} \tag{29}$$

and, by formulae (10) and (14), the condition

$$T^T A = -A^* T. \tag{30}$$

This latter property of the matrix T can also be proved by analogy with the proof of the antisymmetry of the matrix B, without use of definition (26b). With relations (29) and (30) it may be shown that under time reversal, the scalar product transforms in the following manner:

$$({}^T\psi_1|{}^T\psi_2) = (\psi_2|\psi_1). \tag{31}$$

An important role in electron theory is played by geometrical objects constructed as bilinear expressions from the components of the bispinor $\psi(x)$ and its conjugate $\psi^\dagger(x)$. The 16 independent expressions of this type which can be constructed are coordinates of the following geometrical objects (under orthochronous Poincaré transformations of the bispinors):

$$\text{scalar} \quad \overline{\psi}(x)\psi(x), \tag{32a}$$

$$\text{vector} \quad \overline{\psi}(x)\gamma^\mu\psi(x), \tag{32b}$$

$$\text{tensor} \quad i\overline{\psi}(x)\gamma^{\mu\nu}\psi(x), \tag{32c}$$

$$\text{pseudovector} \quad \overline{\psi}(x)\gamma_5\gamma^\mu\psi(x), \tag{32d}$$

$$\text{pseudoscalar} \quad i\overline{\psi}(x)\gamma_5\psi(x). \tag{32e}$$

All of the bilinear expressions listed above are real. Under time reversal these quantities transform, by virtue of (23c), (29), and (30), in the following manner: scalar like scalar, vector like pseudovector, tensor like pseudotensor, pseudovector like pseudovector, and pseudoscalar like pseudoscalar.

Apart from the symmetries associated with the properties of the space-time, the Dirac equation has two further important symmetries: *symmetry with respect to charge conjugation*, and *symmetry with respect to gauge transformation.* The charge conjugation transformation leads from the solution of the Dirac equation $\psi(x)$ to the bispinor ${}^{\mathrm{C}}\psi(x)$,

$$ {}^{\mathrm{C}}\psi(x) \equiv \mathrm{C}\psi^*(x). \tag{33}$$

The *charge-conjugated bispinor* is a solution of the Dirac equation in which the charge e is replaced by its opposite,

$$[-i\gamma^{\mu}\partial_{\mu}+m-e\gamma^{\mu}\mathscr{A}_{\mu}(x)]{}^{\mathrm{C}}\psi(x) = 0. \tag{34}$$

The pictures of all physical processes (e.g. the trajectory of an electron in the electromagnetic field) remain unchanged, if we go over from electrons described by the bispinor $\psi(x)$ to electrons described by the bispinor ${}^{\mathrm{C}}\psi(x)$ and at the same time change the direction of all currents producing the external field. It follows from the properties of the C matrix that the scalar product of two solutions of the Dirac equation is invariant under charge conjugation. The probabilities of two processes described, respectively, by the bispinor $\psi(x)$ and ${}^{\mathrm{C}}\psi(x)$ are therefore equal to each other.

The composition of the P, C, and T transformations of the bispinor leads to a very straightforward transformation law,

$$ {}^{\mathrm{PCT}}\psi(x) = {}^{\mathrm{PC}}(\mathrm{T}\psi(-\mathscr{P}x))^* = {}^{\mathrm{P}}(\mathrm{CT}\psi(-\mathscr{P}x)) = \mathrm{PCT}\,\psi(-x). \tag{35}$$

The product PCT of the matrices has the property

$$(\mathrm{PCT})^{-1}\gamma^{\mu}\mathrm{PCT} = -\gamma^{\mu}. \tag{36}$$

This property is possessed by the matrix γ_5 multiplied by an arbitrary factor. Using T and P matrices in the form (26), we obtain

$$\mathrm{PCT} = -i\gamma_5. \tag{37}$$

The Dirac equation remains unchanged if, simultaneously with the gauge transformation of the external potential

$$\mathscr{A}_{\mu}(x) \to {}^{\mathrm{G}}\mathscr{A}_{\mu}(x) = \mathscr{A}_{\mu}(x)+\partial_{\mu}\Lambda(x), \tag{38}$$

where $\Lambda(x)$ is an arbitrary real function, we perform the gauge transformation of the electron wave function according to the formula

$$\psi(x) \to {}^{G}\psi(x) = e^{-ie\Lambda(x)}\psi(x). \tag{39}$$

The gauge transformation of the bispinor consists in changing its phase; it does not alter the scalar product.

From the Dirac equation, by acting on the left-hand side with the operation $(i\gamma^\mu\partial_\mu - e\gamma^\mu\mathscr{A}_\mu + m)$, we obtain a second-order equation for the bispinor ψ in the form

$$\left[(\partial_\mu + ie\mathscr{A}_\mu)(\partial^\mu + ie\mathscr{A}^\mu) + m^2 - \frac{ie}{2} f_{\mu\nu}\gamma^{\mu\nu}\right]\psi(x) = 0. \tag{40}$$

In the spinor representation, the matrices $\gamma^{\mu\nu}$ have the completely reduced form

$$\gamma^{\mu\nu} = \begin{pmatrix} S^{\mu\nu} & 0 \\ 0 & \tilde{S}^{\mu\nu} \end{pmatrix},$$

and, in consequence, eq. (40) may be represented as two independent equations for two spinors ψ and $\tilde{\varphi}$ composing the bispinor ψ. We shall write only the first:

$$\left[(\partial_\mu + ie\mathscr{A}_\mu)(\partial^\mu + ie\mathscr{A}^\mu) + m^2 - \frac{ie}{2} f_{\mu\nu}S^{\mu\nu}\right]\psi(x) = 0. \tag{41}$$

This equation is equivalent to the Dirac equation.

To prove this equivalence, it is sufficient to define the second half of the bispinor by means of the formula

$$\tilde{\varphi} \equiv \frac{i}{m}\sigma^\mu(\partial_\mu + ie\mathscr{A}_\mu)\psi$$

and to make use of the relations between the matrices γ, σ, and $\tilde{\sigma}$ and the properties of the matrices σ and $\tilde{\sigma}$, which are given in Appendix I.

Though somewhat more awkward to apply,# the second-order equation (41) has a more lucid physical interpretation. In the absence of the electromagnetic field, this is the Klein-Gordon equation for both components of the spinor ψ, and hence the condition for the particle to

Some properties of the spinor field are more easily described by using second-order equations (cf., for instance, Brown (1958) and Białynicki-Birula (1964)).

have a particular mass. Interaction with the electromagnetic field is described by two types of coupling. Coupling with the charge of the particle is introduced by replacing the ordinary derivatives ∂_μ by the derivatives D_μ in the presence of the field

$$D_\mu = \partial_\mu + ie\mathscr{A}_\mu .$$

This is the so-called *minimal electromagnetic coupling.* Coupling with the magnetic moment is described by the last term in eq. (41).

The Properties of Solutions of the Dirac Equation without Potential

We shall now examine the properties of the bispinor $\chi(x)$ which is the wave function of a free electron. It satisfies the Dirac equation without potential

$$D_x \chi(x) \equiv (-i\gamma\cdot\partial + m)\chi(x) = 0, \tag{42}$$

where

$$\gamma\cdot\partial = \gamma^\mu \partial_\mu .$$

The components $\chi_a(x)$ of the bispinor $\chi(x)$ are solutions of the Klein–Gordon equation with mass m. This follows from the equality

$$(i\gamma\cdot\partial + m)(-i\gamma\cdot\partial + m) = (\Box + m^2)I.$$

Each solution of the Klein–Gordon equation may be written as a superposition of plane waves, i.e. the functions $\exp(\pm ip\cdot x)$, where $p\cdot x = E_p x^0 - \mathbf{p}\cdot\mathbf{x}$, and

$$E_p \equiv \sqrt{m^2 + \mathbf{p}^2}.$$

This fact will be used in order to find the general solution of the Dirac equation without potential. By analogy with the solutions of the Klein–Gordon equation, the solution of the Dirac equation, whose time-dependence is expressed by the factor $\exp(-i\omega x^0)$, will be called a *solution with positive frequency* if $\omega > 0$, and a *solution with negative frequency* if $\omega < 0$. In non-relativistic quantum mechanics the sign of the frequency is simply the sign of the energy. Each solution of the Dirac equation without potential may be rewritten in the form

$$\chi(x) = \sum_r \int d\Gamma(m)[u(\mathbf{p}, r)e^{-ip\cdot x}g^+(\mathbf{p}, r) + v(\mathbf{p}, r)e^{ip\cdot x}g^{-*}(\mathbf{p}, r)], \tag{43}$$

where r is an index assuming two values; we shall discuss its meaning further on. The occurrence of a complex conjugate function in the second term will be justified in Section 17. The components of the bispinors $u(\mathbf{p}, r)$ and $v(\mathbf{p}, r)$ satisfy the systems of algebraic equations,

$$(\gamma \cdot p - m)u(\mathbf{p}, r) = 0, \tag{44a}$$

$$(\gamma \cdot p + m)v(\mathbf{p}, r) = 0, \tag{44b}$$

where $\gamma \cdot p = \gamma^\mu p_\mu$. Equations (44) are the eigenvalue equations of the matrix $\gamma \cdot p$. The eigenvalues of this matrix, m and $-m$, are doubly degenerate. Accordingly, the sets of eigenfunctions $u(\mathbf{p}, r)$ and $v(\mathbf{p}, r)$ form two two-dimensional subspaces in the four-dimensional space of the bispinors. The phases of the bispinors u and v are chosen so that the following relations hold between the bispinors u and v and the bispinors ${}^{\mathrm{C}}u$ and ${}^{\mathrm{C}}v$ charge-conjugated with them:

$$v(\mathbf{p}, r) = {}^{\mathrm{C}}u(\mathbf{p}, r) \equiv \mathbf{C}u^*(\mathbf{p}, r), \tag{45a}$$

$$u(\mathbf{p}, r) = {}^{\mathrm{C}}v(\mathbf{p}, r) \equiv \mathbf{C}v^*(\mathbf{p}, r). \tag{45b}$$

As basis vectors in the subspaces u and v we can take, for instance, the eigenvectors of the helicity operator $\hat{\lambda}$:

$$\hat{\lambda}(p) \equiv i\gamma_5 \gamma^\mu s_\mu(p) \tag{46}$$

where

$$s^\mu(p) \equiv \left(\frac{|\mathbf{p}|}{m}, \frac{p_0 \mathbf{p}}{m|\mathbf{p}|}\right). \tag{47}$$

The definition of $s^\mu(p)$ implies the following properties of this expression:

$$s_\mu s^\mu = -1, \tag{48a}$$

$$s_\mu(p) p^\mu = 0. \tag{48b}$$

The operator $\hat{\lambda}$ has two eigenvalues, $+1$ and -1. It is evident from condition (48b) that the matrices $\hat{\lambda}(p)$ and $\gamma \cdot p$ commute; thus, we may look for their simultaneous eigenvectors. The operator $\hat{\lambda}$ depends on the choice of the reference system. Under Lorentz transformations, this operator undergoes changes. Those solutions of eqs. (44) which are simultaneously also eigenvectors of the helicity operator will be denoted by the symbols $u(\mathbf{p}, \pm 1)$ and $v(\mathbf{p}, \pm 1)$:

$$\hat{\lambda}(p)u(\mathbf{p}, \pm 1) = \pm u(\mathbf{p}, \pm 1), \tag{49a}$$

$$\hat{\lambda}(p)v(\mathbf{p}, \pm 1) = \mp v(\mathbf{p}, \mp 1). \tag{49b}$$

The index r in formula (43) thus takes on the values $+1$ and -1. Reversal of the plus and minus signs in eq. (49b) is in accordance with the convention (45). The bispinors u and v can be normalized in the following manner,

$$\bar{u}(\mathbf{p}, r)u(\mathbf{p}, s) = 2m\,\delta_{rs}, \tag{50a}$$

$$\bar{v}(\mathbf{p}, r)v(\mathbf{p}, s) = -2m\,\delta_{rs}, \tag{50b}$$

if the sign of the matrix A is chosen suitably. Such normalization will be used henceforth. The bispinors u and v moreover satisfy the orthogonality condition

$$\bar{u}(\mathbf{p}, r)v(\mathbf{p}, s) = 0. \tag{51}$$

In the Dirac representation, the solutions of eqs. (44) and (49) which satisfy conditions (50) and (45) are of the form

$$u(\mathbf{p}, r) = \sqrt{E_p+m}\begin{pmatrix} w(\mathbf{p}, r) \\ \dfrac{\boldsymbol{\sigma}\cdot\mathbf{p}}{E_p+m}\,w(\mathbf{p}, r)\end{pmatrix}, \tag{52a}$$

$$v(\mathbf{p}, r) = \sqrt{E_p+m}\begin{pmatrix} \dfrac{\boldsymbol{\sigma}\cdot\mathbf{p}}{E_p+m}\,\hat{\varepsilon}w^*(\mathbf{p}, r) \\ \hat{\varepsilon}w^*(\mathbf{p}, r)\end{pmatrix}. \tag{52b}$$

In these formulae, $\boldsymbol{\sigma}$ is the spin vector, with the Pauli matrices as its components:

$$\boldsymbol{\sigma} \equiv \left(\begin{pmatrix}0 & 1\\ 1 & 0\end{pmatrix}, \begin{pmatrix}0 & -i\\ i & 0\end{pmatrix}, \begin{pmatrix}1 & 0\\ 0 & -1\end{pmatrix}\right);$$

the matrix $\hat{\varepsilon}$ is of the form

$$\hat{\varepsilon} \equiv \begin{pmatrix}0 & -1\\ 1 & 0\end{pmatrix}, \tag{53}$$

and $w(\mathbf{p}, r)$ may be chosen as normalized eigenvectors of the projection of the spin operator $\boldsymbol{\sigma}$ onto the momentum direction

$$\frac{\boldsymbol{\sigma}\cdot\mathbf{p}}{|\mathbf{p}|}\,w(\mathbf{p}, \pm 1) = \pm w(\mathbf{p}, \pm 1), \tag{54a}$$

$$w^\dagger(\mathbf{p}, r)\,w(\mathbf{p}, s) = \delta_{rs}. \tag{54b}$$

The phases of these vectors are chosen so that

$$-i\hat{\varepsilon}w^*(\mathbf{p}, \pm 1) = w(\mathbf{p}, \mp 1). \tag{54c}$$

The functions $g^+(\mathbf{p}, r)$ and $g^-(\mathbf{p}, r)$ figuring in formula (43) are coefficients in the expansion of the general solution of the Dirac equation without potential into the functions $\chi^{(+)}(x; \mathbf{p}, r)$ and $\chi^{(-)}(x; \mathbf{p}, r)$ defined by the formulae

$$\chi^{(+)}(x; \mathbf{p}, r) \equiv u(\mathbf{p}, r)\mathrm{e}^{-ip\cdot x}, \tag{55a}$$

$$\chi^{(-)}(x; \mathbf{p}, r) \equiv v(\mathbf{p}, r)\mathrm{e}^{ip\cdot x}. \tag{55b}$$

The functions $\chi^{(+)}(x; \mathbf{p}, r)$ and $\chi^{(-)}(x; \mathbf{p}, r)$ are solutions of the Dirac equation without potential which represent plane waves. The function $\chi^{(+)}$ is a solution with positive frequency, and the function $\chi^{(-)}$ is one with negative frequency. The solution of the Dirac equation which is a superposition of plane waves with only positive frequencies will be denoted by the symbol $\chi^{(+)}(x)$, and the solution containing only negative frequencies by the symbol $\chi^{(-)}(x)$:

$$\chi^{(+)}[x|g] = \sum_r \int \mathrm{d}\Gamma(m)\chi^{(+)}(x; \mathbf{p}, r)g(\mathbf{p}, r), \tag{56a}$$

$$\chi^{(-)}[x|g] = \sum_r \int \mathrm{d}\Gamma(m)\chi^{(-)}(x; \mathbf{p}, r)g^*(\mathbf{p}, r). \tag{56b}$$

The function $g(\mathbf{p}, r)$ will be called the *wave-packet profile*.

The scalar product of two solutions χ_1 and χ_2 of the Dirac equation without potential is expressed with the aid of the coefficients of the expansion of these functions into plane waves, which we shall denote by the symbols f^+ and f^- and g^+ and g^-, respectively:

$$(\chi_1|\chi_2) = \sum_r \int \mathrm{d}\Gamma(m)[f^{+*}(\mathbf{p}, r)g^+(\mathbf{p}, r) + f^-(\mathbf{p}, r)g^{-*}(\mathbf{p}, r)]. \tag{57}$$

It is evident from this formula that the scalar product of two solutions of the Dirac equation without potential does not depend on time. The scalar products of the functions f and g which appear on the right-hand side of formula (57) will be denoted by the symbol $(f|g)$:

$$(f|g) \equiv \sum_r \int \mathrm{d}\Gamma(m)f^*(\mathbf{p}, r)g(\mathbf{p}, r). \tag{58}$$

The bispinors describing wave functions with positive frequencies, $\chi^{(+)}(x; \mathbf{p}, r)$, and with negative frequencies, $\chi^{(-)}(x; \mathbf{p}, r)$, satisfy the following *completeness relations*,

$$\sum_r \int \mathrm{d}\Gamma(m)\, \chi^{(+)}(x; \mathbf{p}, r)\, \overline{\chi^{(+)}}(y; \mathbf{p}, r) = \frac{1}{i}\, S^{(+)}(x-y), \tag{59a}$$

$$\sum_r \int \mathrm{d}\Gamma(m)\, \chi^{(-)}(x; \mathbf{p}, r)\, \overline{\chi^{(-)}}(y; \mathbf{p}, r) = \frac{1}{i}\, S^{(-)}(x-y), \tag{59b}$$

where

$$S^{(\pm)}(x) \equiv (i\gamma \cdot \partial + m)\Delta^{(\pm)}(x, m). \tag{60}$$

The proof of eqs. (59) is based on the relations

$$\sum_r u(\mathbf{p}, r)\bar{u}(\mathbf{p}, r) = \gamma \cdot p + m, \tag{61a}$$

$$\sum_r v(\mathbf{p}, r)\bar{v}(\mathbf{p}, r) = \gamma \cdot p - m, \tag{61b}$$

which result from eqs. (44) satisfied by the bispinors u and v and from the normalization condition (50) of these bispinors. The proof of these relations is given in Appendix J. The completeness relations (59) are conditions for the completeness of the set of the solutions of the Dirac equation. The solutions $\chi_n^{(+)}(x)$ or $\chi_n^{(-)}(x)$ will be said to form *complete sets of solutions* if they satisfy the completeness relations

$$\sum_n \chi_n^{(+)}(x)\, \overline{\chi_n^{(+)}}(y) = \frac{1}{i}\, S^{(+)}(x-y), \tag{62a}$$

$$\sum_n \chi_n^{(-)}(x)\, \overline{\chi_n^{(-)}}(y) = \frac{1}{i}\, S^{(-)}(x-y). \tag{62b}$$

A set of wave packets with positive frequencies, $\chi^{(+)}[x|g_n]$, whose profiles $g_n(\mathbf{p}, r)$ form a complete set of functions, i.e. satisfy the condition

$$\sum_n g_n(\mathbf{p}, r) g_n^*(\mathbf{q}, s) = (2\pi)^3 2E_p\, \delta(\mathbf{p}-\mathbf{q})\, \delta_{rs},$$

constitutes a complete set of positive-frequency solutions of the Dirac equation without potential. Each solution of this equation, containing only positive frequencies, may be written as a linear combination of the bispinors $\chi^{(+)}[x|g_n]$. Similarly, the set of wave packets with negative frequencies $\chi^{(-)}[x|g_n]$, forms a complete set of negative-frequency solutions of the Dirac equation without potential if the profiles of these packets constitute a complete set of functions.

To conclude this section, we shall give the *transformation laws* under Poincaré transformations for the *wave-packet profiles* $g^{\pm}(\mathbf{p}, r)$ which characterize the solutions of the Dirac equation without potential. In doing this, we shall make use of the following identities which are satisfied by the bispinors describing plane waves:

$$\chi^{(\pm)}(x-a; \mathbf{p}, r) = e^{\pm ip\cdot a}\chi^{(\pm)}(x; \mathbf{p}, r), \tag{63a}$$

$$\mathrm{S_L}\chi^{(\pm)}(\Lambda^{-1}x; \mathbf{p}, r) = \sum_s \Lambda^{(\pm)}(\mathbf{p}; s, r)\chi^{(\pm)}(x; \Lambda\mathbf{p}, s), \tag{63b}$$

$$\mathrm{P}\chi^{(\pm)}(\mathscr{P}x; \mathbf{p}, r) = \pm\chi^{(\pm)}(x; -\mathbf{p}, r), \tag{63c}$$

$$(\mathrm{T}\chi^{(\pm)}(-\mathscr{P}x; \mathbf{p}, r))^* = \chi^{(\pm)}(x; -\mathbf{p}, -r), \tag{63d}$$

where the matrices $\Lambda^{(\pm)}(\mathbf{p}; r, s)$ satisfy the relations

$$\mathrm{S_L}u(\mathbf{p}, r) = \sum_s \Lambda^{(+)}(\mathbf{p}; s, r)u(\Lambda\mathbf{p}, s), \tag{64a}$$

$$\mathrm{S_L}v(\mathbf{p}, r) = \sum_s \Lambda^{(-)}(\mathbf{p}; s, r)v(\Lambda\,\mathbf{p}, s). \tag{64b}$$

The existence of linear relations between the bispinors $\mathbf{S_L}u(\mathbf{p}, r)$ and $u(\mathbf{p}, s)$ and between $\mathbf{S_L}v(\mathbf{p}, r)$ and $v(\mathbf{p}, s)$ follows from eqs. (44) and the properties of the matrix $\mathbf{S_L}$. Identities (63) may be proved by using the knowledge of the bispinors u and v in the Dirac representation.

The transformation properties of the profiles $g^{\pm}(\mathbf{p}, r)$ are read from the relations

$$'\chi(x) = \sum_r \int d\Gamma(m)$$
$$\times [\chi^{(+)}(x; \mathbf{p}, r)'g^{+}(\mathbf{p}, r)+\chi^{(-)}(x; \mathbf{p}, r)'g^{-*}(\mathbf{p}, r)],$$

if $'\chi(x)$ is replaced, respectively, by the translated bispinor $^{\mathrm{Tr}}\chi(x)$, subjected to the proper Lorentz transformation bispinor $^{\mathrm{L}}\chi(x)$, spatially reflected bispinor ${}^{P}\chi(x)$, and time-reversed bispinor ${}^{T}\chi(x)$, and then identities (63) are applied. As a result we obtain

$$ {}^{Tr}g^{\pm}(\mathbf{p}, r) = e^{ip\cdot a}g^{\pm}(\mathbf{p}, r), \tag{65a} $$

$$ {}^{L}g^{\pm}(\mathbf{p}, r) = \sum_{s} \Lambda(\mathbf{p}; r, s)g^{\pm}(\Lambda^{-1}\mathbf{p}, s), \tag{65b} $$

$$ {}^{P}g^{\pm}(\mathbf{p}, r) = \pm g^{\pm}(-\mathbf{p}, r), \tag{65c} $$

$$ {}^{T}g^{\pm}(\mathbf{p}, r) = (g^{\pm}(-\mathbf{p}, -r))^{*}, \tag{65d} $$

where

$$ \Lambda(\mathbf{p}; r, s) = \Lambda^{(+)}(\mathbf{p}; r, s) = \Lambda^{(-)*}(\mathbf{p}; r, s). \tag{66} $$

This latter equality follows from the adopted convention (45) concerning the choice of the phase of the bispinors u and v.

The scalar product of the functions $g^{\pm}(\mathbf{p}, r)$ does not change under transformations (65a), (65b) and (65c). Under the influence of transformation (65d) the scalar product of these functions changes into its complex conjugate.

17. ELECTRON SCATTERING IN THE ELECTROMAGNETIC FIELD

In this section we shall discuss one of the more difficult problems of relativistic quantum mechanics, the physical interpretation of the wave function, and particularly the physical interpretation of negative-frequency solutions of the Dirac equation. The application of the Dirac equation to the solution of certain problems in which a static potential occurs (e.g. to describe the hydrogen atom) does not require solution of the problem under consideration. In the case of static potentials there is no need to use solutions which are superpositions at the same time of functions with positive and negative frequencies. This is why it was possible to verify experimentally the applicability of the Dirac equation to the description of the electron even before the proper interpretation of its solutions was discovered. In the case of scattering in a time-dependent field, regardless of its strength, the Dirac equation does not have solutions with a particular sign of frequency simultaneously in both asymptotic regions of space-time, i.e. in the remote past and far future. The interpretation of solutions with negative frequencies thus is related to the scattering problem, to which this section is devoted.#

To simplify subsequent considerations, we shall assume that the electromagnetic potential $\mathscr{A}_\mu(x)$ vanishes beyond the region of space-time bounded by the two hyperplanes $x^0 = T_1$ and $x^0 = T_2$. The other two regions of space-time, $x^0 < T_1$ and $x^0 > T_2$, will be referred to in brief as the *past* and the *future*.

Our analysis of electron scattering in an external electromagnetic field described by means of the potential $\mathscr{A}_\mu(x)$ will begin with a discussion of the Born approximation. The scattering amplitude in this approximation will be patterned after the analogous expression in non-relativistic quantum mechanics (cf. formula (4.25)). Two kinds of wave packets exist in relativistic quantum mechanics, viz. functions with positive frequencies $\chi^{(+)}[x|g]$ and functions with negative frequencies $\chi^{(-)}[x|g]$. Choosing four different combinations of these functions, we obtain four scattering amplitudes in the Born approximation, the meaning of which we explain below:

$$S_{\mathrm{B}}^{++}[f,g] = (\chi^{(+)}[f]|\chi^{(+)}[g]) - ie\int \mathrm{d}^4x\,\overline{\chi^{(+)}}[x|f]\gamma\cdot\mathscr{A}(x)\chi^{(+)}[x|g], \tag{1a}$$

$$S_{\mathrm{B}}^{-+}[f,g] = ie\int \mathrm{d}^4x\,\overline{\chi^{(-)}}[x|f]\gamma\cdot\mathscr{A}(x)\chi^{(+)}[x|g], \tag{1b}$$

$$S_{\mathrm{B}}^{--}[f,g] = (\chi^{(-)}[f]|\chi^{(-)}[g]) + ie\int \mathrm{d}^4x\,\overline{\chi^{(-)}}[x|f]\gamma\cdot\mathscr{A}(x)\chi^{(-)}[x|g], \tag{1c}$$

$$S_{\mathrm{B}}^{+-}[f,g] = -ie\int \mathrm{d}^4x\,\overline{\chi^{(+)}}[x|f]\gamma\cdot\mathscr{A}(x)\chi^{(-)}[x|g], \tag{1d}$$

where $\gamma\cdot\mathscr{A} = \gamma^\mu\mathscr{A}_\mu$. Formulae (1b) and (1d) do not contain the first terms of the expansion, also called the zeroth-order terms since the solutions of the Dirac equation without potential with different signs

---

The treatment of the problems of relativistic mechanics in this book is patterned after the papers of R. P. Feynman (1949a, 1949b). Readers are ardently advised to study these epoch-making papers in the original.

Another interpretation of solutions with negative frequencies (called the sea theory or the hole theory) was given by P. A. M. Dirac (1930) (cf. also the book by Dirac).

of frequency are orthogonal to each other. The choice of signs in front of first-order terms was dictated by the following considerations. In formulae (1a) and (1b), this sign was chosen to obtain the correspondence with the non-relativistic quantum mechanics. For, as will become clear further on in this section, the amplitudes (1a) and (1c) describe the scattering of particles of opposite charge. In formulae (1b) and (1d), the sign of the first-order term in one-electron theory is not physically meaningful in view of the absence of the zeroth-order term. We adopted the plus and minus signs in these expressions with an eye to the symmetry of subsequent formulae.

A corresponding S matrix can be defined for every amplitude $S[f, g]$, just as in non-relativistic quantum mechanics. In this way we obtain four distributions $S^{\pm\pm}(\mathbf{p}, r; \mathbf{q}, s)$ which, for all suitably regular functions f and g, satisfy the relations:

$$S^{++}[f, g] = \int \mathrm{d}\Gamma_p(m)\mathrm{d}\Gamma_q(m) \sum_r \sum_s f^*(\mathbf{p}, r) S^{++}(\mathbf{p}, r; \mathbf{q}, s) g(\mathbf{p}, r), \tag{2a}$$

$$S^{-+}[f, g] = \int \mathrm{d}\Gamma_p(m)\mathrm{d}\Gamma_q(m) \sum_r \sum_s f(\mathbf{p}, r) S^{-+}(\mathbf{p}, r; \mathbf{q}, s) g(\mathbf{p}, r), \tag{2b}$$

$$S^{--}[f, g] = \int \mathrm{d}\Gamma_p(m)\mathrm{d}\Gamma_q(m) \sum_r \sum_s f(\mathbf{p}, r) S^{--}(\mathbf{p}, r; \mathbf{q}, s) g^*(\mathbf{p}, r), \tag{2c}$$

$$S^{+-}[f, g] = \int \mathrm{d}\Gamma_p(m)\mathrm{d}\Gamma_q(m) \sum_r \sum_s f^*(\mathbf{p}, r) S^{+-}(\mathbf{p}, r; \mathbf{q}, s) g^*(\mathbf{p}, r). \tag{2d}$$

In the Born approximation, the following four S matrices correspond to the four amplitudes $S_{\mathrm{B}}[f, g]$ given by formulae (1):

$$S_{\mathrm{B}}^{++}(\mathbf{p}, r; \mathbf{q}, s) = \delta_{\Gamma(m)}(\mathbf{p}, \mathbf{q})\, \delta_{rs} - ie\bar{u}(\mathbf{p}, r)\gamma \cdot \tilde{\mathscr{A}}(p-q) u(\mathbf{q}, s), \tag{3a}$$

$$S_{\mathrm{B}}^{-+}(\mathbf{p}, r; \mathbf{q}, s) = ie\bar{v}(\mathbf{p}, r)\gamma \cdot \tilde{\mathscr{A}}(-p-q) u(\mathbf{q}, s), \tag{3b}$$

$$S_{\mathrm{B}}^{--}(\mathbf{p}, r; \mathbf{q}, s) = \delta_{\Gamma(m)}(\mathbf{p}, \mathbf{q})\, \delta_{rs} + ie\bar{v}(\mathbf{p}, r)\gamma \cdot \tilde{\mathscr{A}}(-p+q) v(\mathbf{q}, s), \tag{3c}$$

$$S_{\mathrm{B}}^{+-}(\mathbf{p}, r; \mathbf{q}, s) = -ie\bar{u}(\mathbf{p}, r)\gamma \cdot \tilde{\mathscr{A}}(p+q) v(\mathbf{q}, s), \tag{3d}$$

where the symbol $\tilde{\mathscr{A}}_\mu(p)$ denotes the Fourier transform of the potential $\mathscr{A}_\mu(x)$, and the four-momenta p and q satisfy the following conditions:

$$p^2 = m^2, \quad q^2 = m^2, \quad p^0 > 0, \quad q^0 > 0,$$
$$\delta_{\Gamma(m)}(\mathbf{p}, \mathbf{q}) \equiv (2\pi)^3 2\sqrt{m^2+\mathbf{p}^2}\,\delta(\mathbf{p}-\mathbf{q}). \tag{4}$$

Now we shall give the four amplitudes S_B a physical interpretation. For this purpose, we assume, as in non-relativistic quantum mechanics, that the argument of the Fourier transform of the potential represents the momentum and energy transferred by the external field to the quantum system. The amplitude $S_B^{++}(\mathbf{p}, r; \mathbf{q}, s)$ thus describes the scattering of an electron with initial momentum $\mathbf{q}$ and initial energy E_q, and final momentum and energy $\mathbf{p}$ and E_p, whereas the amplitude $S_B^{--}(\mathbf{p}, r; \mathbf{q}, s)$ describes the scattering of an electron with initial momentum and energy $\mathbf{p}$ and E_p, and final momentum and energy $\mathbf{q}$ and E_q. In accordance with this principle, we must further assume that the amplitude $S_B^{+-}(\mathbf{p}, r; \mathbf{q}, s)$ describes a process as a result of which the field transfers to the quantum system a momentum $\mathbf{p}+\mathbf{q}$ and energy E_p+E_q. This process is known as *pair creation.* At the expense of the external electromagnetic field, two particles are created: one with momentum $\mathbf{p}$ and energy E_p, and another with momentum $\mathbf{q}$ and energy E_q. If the principle of charge conservation is to be satisfied, we must also assume that the particles created by the external field have opposite charges. In similar fashion we conclude that the amplitude S_B^{-+} describes the process of *pair annihilation*, in which the momentum and energy of two particles of opposite charge are transferred to the external electromagnetic field. The transition amplitudes S_B^{+-} and S_B^{-+} describing the processes of creation and annihilation contain pairs of functions with positive and negative frequencies. We thus adopt the principle that the solutions of the Dirac equation without potential which have positive frequencies describe negative electrons, i.e. *negatons*, while those with negative frequencies describe positive electrons, i.e. *positons.*† Solutions with positive frequencies are constructed from the bispinors u, and

† The negaton-positon pair is an example of a particle-antiparticle pair. All charged particles occurring in nature, as well as some neutral ones, have corresponding antiparticles.

solutions with negative frequencies from the bispinors v. The amplitude S_B^{++} thus describes the scattering of a negaton, and the amplitude S_B^{--}, the scattering of a positon. Moreover, formulae (3) imply the following rule: functions describing the initial state of a negaton or the final state of a positon are written on the right-hand side of the expression for the amplitude, whereas functions describing the final state of a negaton or initial state of a positon are written on the left-hand side.

By the convention adopted in Section 16, viz. that the coefficient in the expansion of the function χ into plane waves with negative frequencies is a conjugate function (cf. formula (16.43)), the transition amplitudes are linear functionals of the wave-packet profiles describing the initial states and antilinear functionals of the wave-packet profiles describing the final states, for negatons and positons alike.

The processes of negaton scattering (a), pair annihilation (b), positon scattering (c), and pair creation (d) may be depicted graphically as in Fig. 17.1. These are the simplest diagrams which we construct as follows. Imagine that the time axis is directed upwards. The state of an electron (or electrons) before interaction is represented by a line segment with an arrow (or two line segments with arrows) in the lower part of the drawing. The arrow points upwards if the electron is a negaton, and downwards if it is a positon. Turning points represent interactions with the potential. A segment (or segments) with arrow in the upper

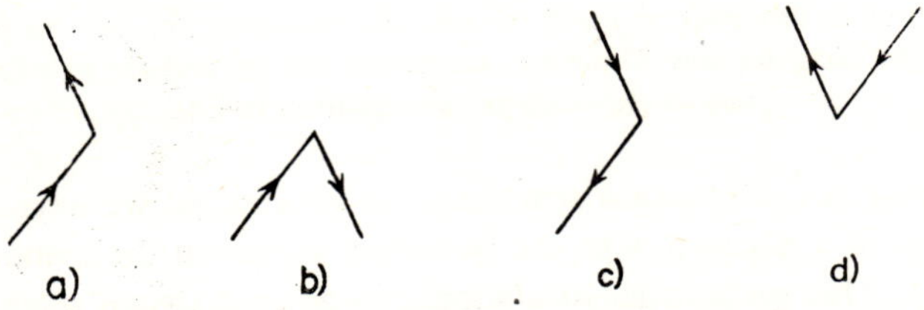

FIG. 17.1. Four elementary processes: (a) negaton scattering, (b) pair annihilation, (c) positon scattering, and (d) pair creation.

part of the drawing represents the state of the system after the interaction. We see that every pair of lines shown in Fig. 17.1 has a definite orientation indicated consistently by both arrows. This fact is the result of charge conservation.

The full scattering amplitude will be constructed in the form of a perturbation series. We shall make the natural assumption that the amplitudes in the higher orders of perturbation theory can be constructed out of the four elementary amplitudes represented by the diagrams in Fig. 17.1, with due account for the principles of conservation of charge, momentum, and energy. The diagrams in Fig. 17.2 represent† all possible processes in second-order perturbation theory, obtained in accordance with the principle we have adopted.‡ Diagrams (a) and (b) in Fig. 17.2

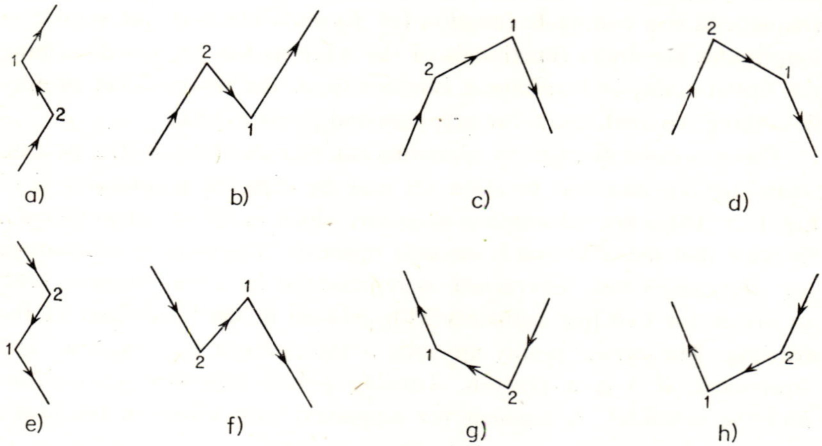

FIG. 17.2. Processes: (a) and (b) negaton scattering, (c) and (d) pair annihilation, (e) and (f) positon scattering, and (g) and (h) pair creation in second-order perturbation theory.

depict processes of negaton scattering. In process (a) we have two-fold interaction of a negaton with the potential, at first at the instant t_2 and then at t_1. This process has its analogue in non-relativistic quantum me-

† As in Section 4, we give here a symbolic "picture" of scattering processes, one which is inseparably tied in with perturbation theory. This picture is helpful in heuristic considerations leading to the formulation of relativistic electron theory.

‡ Subsequently in this chapter, however, we shall demonstrate that because the electrons are indistinguishable account must also be taken of additional diagrams composed (in this order of perturbation theory) of two parts: an open electron line and a closed line.

chanics. In process (b), three particles, a negaton and a pair, occur in the intermediate state in the interval of time between two interactions with the potential. The course of this process is imagined to be as follows. At the instant t_1 a pair of particles is created by the external potential. Both particles of this pair, together with the incident negaton, move freely until the instant t_2 when the positon from the pair annihilates with the incident negaton. From the instant t_1 the negaton from the pair moves freely and is observed as the final particle. Diagrams (c) and (d) depict the annihilation of a pair. In these processes, at first one of the particles of the pair is scattered once by the potential and the pair then suffers annihilation. Diagrams (e) and (f) depict the scattering of a positon, and diagrams (g) and (h) the creation of a pair. In processes (a), (c), (f) and (g), in the time interval (t_2, t_1) between interactions a particle described by a wave function with positive frequencies moves freely from the point x_2 to the point x_1. Since this motion may occur in any state, the integration in the scattering amplitude should be performed over all momentum states and the summation over all possible spin states. Thus, the function $\frac{1}{i} S^{(+)}(x_1 - x_2)$, which has the representation (16.59a), will appear. The function $\frac{1}{i} S^{(-)}(x_1 - x_2)$ given by formula (16.59b) appears in the amplitudes describing processes (b), (d), (e), and (h), in which a particle described by a function with negative frequencies occurs in the time interval (t_1, t_2).

Now we shall write out the formulae for the scattering amplitudes corresponding to processes of negaton scattering as depicted by diagrams (2a) and (2b):

$$S_2^{(a)}[f, g] = (-ie)^2 \sum_r \int \mathrm{d}\Gamma(m) \int_{t_1 > t_2} \mathrm{d}^4x_1 \mathrm{d}^4x_2$$

$$\times \overline{\chi^{(+)}}[x_1|f]\, \gamma \cdot \mathscr{A}(x_1) \chi^{(+)}(x_1; \mathbf{p}, r) \overline{\chi^{(+)}}(x_2; \mathbf{p}, r) \gamma \cdot \mathscr{A}(x_2) \chi^{(+)}[x_2|g], \tag{5a}$$

$$S_2^{(b)}[f, g] = (-ie)(ie) \sum_r \int \mathrm{d}\Gamma(m) \int_{t_2 > t_1} \mathrm{d}^4x_1 \mathrm{d}^4x_2$$

$$\times \overline{\chi^{(+)}}[x_1|f] \gamma \cdot \mathscr{A}(x_1) \chi^{(-)}(x_1; \mathbf{p}, r) \overline{\chi^{(-)}}(x_2; \mathbf{p}, r) \gamma \cdot \mathscr{A}(x_2) \chi^{(+)}[x_2|g]. \tag{5b}$$

The negaton scattering amplitude in second-order perturbation theory, $S_Q^{++}[f, g]^{(2)}$, which is the sum of both these expressions, may be rewritten as

$$S_Q^{++}[f, g]^{(2)} = ie^2 \int d^4x_1 \, d^4x_2$$
$$\times \overline{\chi^{(+)}[x_1|f]}\gamma \cdot \mathscr{A}(x_1) S_F(x_1 - x_2)\gamma \cdot \mathscr{A}(x_2)\chi^{(+)}[x_2|g] \tag{6}$$

by introducing the *Feynman propagator* $S_F(x)$:

$$S_F(x) \equiv \theta(x^0)S^{(+)}(x) - \theta(-x^0)S^{(-)}(x). \tag{7}$$

This amplitude has been provided with the lower suffix Q and we shall refer to it as the *quasi-amplitude*, to distinguish it from the proper amplitude of negaton scattering in an external field. (Cf. the second footnote on p. 252).

Formulae can be obtained in the same way for the transition amplitudes corresponding to the other diagrams in second-order perturbation theory. The quasi-amplitude of each of the other three processes (pair annihilation, positon scattering, and pair creation) is also a sum of two expressions which can be written as one term by using the function S_F:

$$S_Q^{-+}[f, g]^{(2)} = -ie^2 \int d^4x_1 d^4x_2$$
$$\times \overline{\chi^{(-)}[x_1|f]}\gamma \cdot \mathscr{A}(x_1) S_F(x_1 - x_2)\gamma \cdot \mathscr{A}(x_2)\chi^{(+)}[x_2|g], \tag{8a}$$

$$S_Q^{\mp -}[f, g]^{(2)} = \mp ie^2 \int d^4x_1 \, d^4x_2$$
$$\times \overline{\chi^{(\mp)}[x_1|f]}\gamma \cdot \mathscr{A}(x_1) S_F(x_1 - x_2)\gamma \cdot \mathscr{A}(x_2)\chi^{(-)}[x_2|g]. \tag{8b}$$

In the n-th order of perturbation theory the quasi-amplitude $S_Q[f, g]^{(n)}$ describing each of the possible processes is the sum of 2^{n-1} terms. To each of these terms there corresponds a time-ordered diagram, which may be constructed out of the elementary diagrams, Fig. 17.1, depicting the scattering of particles, and creation and annihilation of pairs. On summing 2^{n-1} terms of each quasi-amplitude, we obtain the expression,

$$\pm(-1)^n ie^n \int d^4x_1 \ldots d^4x_n \overline{\chi^{(\cdot)}[x_1|f]}\gamma \cdot \mathscr{A}(x_1) S_F(x_1 - x_2)$$
$$\times \gamma \cdot \mathscr{A}(x_2) \ldots S_F(x_{n-1} - x_n)\gamma \cdot \mathscr{A}(x_n)\chi^{(\cdot)}[x_n|g], \tag{9}$$

where the symbols $\chi^{(\cdot)}$ represent wave packets with positive or negative frequencies depending on the amplitude in question, the plus sign in front of the whole expression referring to the amplitude $S_Q^{+\pm}$ and the minus sign to the amplitude $S_Q^{-\pm}$.

The full quasi-amplitudes $S_Q[f|g]$ are sums of perturbation series constructed from the terms having the form (9). The formulae for these amplitudes may be rewritten in compact form if the functions $\psi_F^{\pm}$ are introduced.

The functions $\psi_F^{+}[g]$ and $\psi_F^{-}[g]$ are chosen so that the quasi-amplitudes $S_Q[f, g]$ are represented as scalar products of the functions $\chi^{(\pm)}$ and the functions $\psi_F^{\pm}$. This may be achieved if the functions $\psi_F^{\pm}$ are defined in the form of the following perturbation series:

$$\psi_F^{\pm}[x|g] \equiv \chi^{(\pm)}[x|g] - e\int d^4z\, S_F(x-z)\gamma\cdot\mathscr{A}(z)\chi^{(\pm)}[z|g]$$
$$+e^2\int d^4z_1\, d^4z_2\, S_F(x-z_1)\gamma\cdot\mathscr{A}(z_1)S_F(z_1-z_2)\gamma\cdot\mathscr{A}(z_2)\chi^{(\pm)}[z_1|g] + \dots \tag{10}$$

The quasi-amplitudes may then be rewritten as

$$S_Q^{++}[f, g] = (\chi^{(+)}[f]|\psi_F^{+}[g])_{\sigma_F}, \tag{11a}$$

$$S_Q^{-+}[f, g] = (\chi^{(-)}[f]|\psi_F^{+}[g])_{\sigma_P}, \tag{11b}$$

$$S_Q^{--}[f, g] = (\chi^{(-)}[f]|\psi_F^{-}[g])_{\sigma_P}, \tag{11c}$$

$$S_Q^{+-}[f, g] = (\chi^{(+)}[f]|\psi_F^{-}[g])_{\sigma_F}, \tag{11d}$$

where σ_P denotes an arbitrary hypersurface lying in the past in a field-free region, whereas σ_F denotes an arbitrary hypersurface lying in the future in a field-free region.

In proving formulae (11), one must make use of the relations

$$\int d^3x\,\overline{\chi^{(\pm)}}[x|g]S^{(\pm)}(x-y) = \frac{1}{i}\,\overline{\chi^{(\pm)}}[y|g],$$

$$\int d^3x\,\overline{\chi^{(\pm)}}[x|g]S^{(\mp)}(x-y) = 0.$$

In view of the simplicity of formulae (11) and their similarity to corresponding formulae in non-relativistic theory, one may hope that the *Feynman functions* $\psi_F^{\pm}$ will be convenient objects for describing electrons in relativistic theory. It should be emphasized that the same wave function ψ_F^{+} describes two processes: the scattering of a negaton and the annihilation of a pair, whereas the same function ψ_F^{-} describes the scattering of a positon and the creation of a pair.

In drawing the diagrams and writing out the quasi-amplitudes and the Feynman wave functions, we have not hitherto taken account of the indistinguishability of the particles and the Pauli exclusion principle. It is true that the initial and final states were, in all the cases considered, states with at most one negaton and at most one positon, but in the space-time region in which the electromagnetic field acts, systems of many identical particles, negatons or positons, may occur as a result of pair creation and annihilation processes.

In the second-order perturbation theory, this kind of situation is shown in Fig. 17.2 by diagrams (b) and (f). Diagram (b) represents a process in which two negatons coexist during the time interval from t_1 to t_2, whereas diagram (f) depicts the analogous situation for positons.

It turns out that taking account of the indistinguishability of the particles and the resulting antisymmetrization of the wave functions of the system of particles does not change the shape of the Feynman function; this function is multiplied by a constant factor $C[\mathscr{A}]$ (independent of x, but dependent on the electromagnetic field):

$$\psi_{\mathrm{F}} \xrightarrow[\text{antisymmetrization}]{} \psi = C[\mathscr{A}]\psi_{\mathrm{F}}.$$

This surprising result is of enormous importance for the theory of systems of indistinguishable particles. For the time being, let us confine ourselves to demonstrating the aforementioned property only in second-order perturbation theory. The general proof requires certain supplementary considerations, and we shall give it at the end of this section.

The wave functions $\psi_{\mathrm{F}}^{\pm}$ can be represented figuratively by means of the same kind of time-ordered diagrams as we used in constructing the quasi-amplitudes. To focus our attention, let us consider the function $\psi_{\mathrm{F}}^{+}[x|g]$. From the structure of diagrams (a), (b), (c), and (d) which represent contributions to the function ψ_{F}^{+} in second-order perturbation theory it is seen that problems associated with the coexistence of identical particles will not appear until we come to the process described by the function $\psi_{\mathrm{F}}^{(b)}$, corresponding to diagram (b),

$$\psi_{\mathrm{F}}^{(b)}[x|g] = -e^2 \int\limits_{t_2>t_1} \mathrm{d}^4 z_1 \mathrm{d}^4 z_2 S^{(+)}(x-z_1)\gamma\cdot\mathscr{A}(z_1)$$
$$\times S^{(-)}(z_1-z_2)\gamma\cdot\mathscr{A}(z_2)\chi^{(+)}[z_2|g].$$

The function $\psi_F^{(b)}[x|g]$ describes a process in which two negatons coexist during the interval from t_1 to t_2. The function $S^{(\pm)}$ appearing in the integrand in the definition of $\psi_F^{(t)}$ will now be expressed in terms of the free-electron wave functions $\chi^{(\pm)}(x; \mathbf{p}, r)$, the result being:

$$\psi_F^{(b)}[x|g] = -e^2 \sum_r \sum_s \int \mathrm{d}\Gamma_1(m)\mathrm{d}\Gamma_2(m) \int_{t_2>t_1} \mathrm{d}^4z_1 \mathrm{d}^4z_2$$
$$\times \chi^{(+)}(x; \mathbf{p}_1, r)\bar{\chi}^{(+)}(z_1; \mathbf{p}_1, r)\gamma \cdot \mathscr{A}(z_1)\chi^{(-)}(z_1; \mathbf{p}_2, s)$$
$$\times \bar{\chi}^{(-)}(z_2; \mathbf{p}_2, s)\gamma \cdot \mathscr{A}(z_2)\chi^{(+)}[z_2|g].$$

It is seen from this representation that the state of the two coexisting negatons is described by the wave function $\chi^{(+)}(x; \mathbf{p}_1, r)\ \chi^{(+)}[z_2|g]$, which is not antisymmetric in the coordinates of the two particles. We can, however, without difficulty meet the requirements of the theory of indistinguishable particles by adding to the function $\psi_F^{(b)}$ the function $\psi_F^{(b')}$ which differs from it as to sign and permutation of the negaton variables,

$$\psi_F^{(b')}[x|g] = e^2 \sum_r \sum_s \int d\Gamma_1(m)\mathrm{d}\Gamma_2(m) \int_{t_2>t_1} \mathrm{d}^4z_1 \mathrm{d}^4z_2$$
$$\times \chi^{(+)}[x|g]\bar{\chi}^{(+)}(z_1; \mathbf{p}_1, r)\gamma \cdot \mathscr{A}(z_1)\chi^{(-)}(z_1; \mathbf{p}_2, s)$$
$$\times \bar{\chi}^{(-)}(z_2; \mathbf{p}_2, s)\gamma \cdot \mathscr{A}(z_2)\chi^{(+)}(z_2; \mathbf{p}_1, r) = -\chi^{(+)}[x|g]e^2 \int_{t_2>t_1} \mathrm{d}^4z_1 \mathrm{d}^4z_2$$
$$\times \mathrm{Tr}\{S^{(+)}(z_2-z_1)\gamma \cdot \mathscr{A}(z_1)S^{(-)}(z_1-z_2)\gamma \cdot \mathscr{A}(z_2)\}$$
$$= \chi^{(+)}[x|g]C^{(2)}[\mathscr{A}].$$

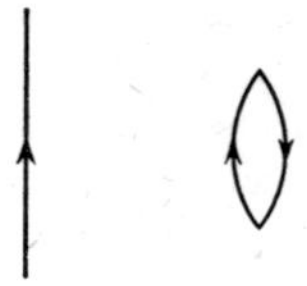

FIG. 17.3. Diagram representing the function $\psi_F^{(b')}$.

The function $\psi_F^{(b')}$ is represented by the diagram in Fig. 17.3. The process depicted by this diagram has an interpretation as lucid as that of the processes portrayed in Fig. 17.2. This time the negaton moves without interacting with the electromagnetic field, just as in zeroth-order perturbation theory. Under the influence of the electromagnetic field, in addition to the moving negaton and completely independent of its motion there is a process in which a pair is created and annihilated.

Since it neither depends on the space-time coordinates nor possesses bispinor indices, the amplitude $C^{(2)}[\mathscr{A}]$ may be treated as a characteristic—independent of the initial and final states of the electrons—of the processes occurring in the external electromagnetic field. In second-order perturbation theory, only one such process will appear: the creation of a pair and its annihilation. As we shall subsequently show, diverse and more complicated processes appear in higher orders of perturbation theory. All such processes, taking place independently of the initial and final states of the particles, will be called *vacuum-to-vacuum processes.* The sum $C[\mathscr{A}]$ of all amplitudes for such processes will be referred to as the *amplitude of the vacuum-to-vacuum transition*, or simply the *vacuum-to-vacuum amplitude.*

In zeroth-order perturbation theory, the vacuum-to-vacuum amplitude is simply equal to 1 since no process can take place without an external field so that the vacuum remains the vacuum.

The vacuum-to-vacuum amplitude is equal to unity also in the case of a static electromagnetic field. The static field cannot transfer energy to the negaton–positon pair and as a result creation processes do not occur. This explains the success Dirac had in applying his equation to the problem of the hydrogen atom before a correct, full interpretation of the theory based on this equation had been discovered.

It is worth noting the close connection between the form of the vacuum-to-vacuum amplitude and the principle of probability conservation. The process of pair creation, which competes with the vacuum-to-vacuum process, already appears in the first-order perturbation theory. It follows from the principle of probability conservation that the sum of the probabilities of all processes which can occur with the vacuum as the initial state must be equal to 1. On the basis of our considerations thus far, we can verify whether this assertion is true in the lowest non-trivial order of perturbation theory.

In the Born approximation, the probability W_P of the creation of any pair is given by the formula:

$$W_P = \sum_{r,s} \int \mathrm{d}\Gamma_p(m)\,\mathrm{d}\Gamma_q(m)\,|S_B^{+-}(\mathbf{p}, r; \mathbf{q}, s)|^2$$

$$= -e^2 \int \mathrm{d}^4z_1\,\mathrm{d}^4z_2\,\mathrm{Tr}\{S^{(+)}(z_1-z_2)\gamma\cdot\mathscr{A}(z_2)S^{(-)}(z_2-z_1)\gamma\cdot\mathscr{A}(z_1)\}.$$

On the other hand, the probability W_V of the vacuum-to-vacuum process in this order of perturbation theory is given by the formula:

$$W_V = 1+2\operatorname{Re} C^{(2)}[\mathscr{A}] = 1+2C^{(2)}[\mathscr{A}].$$

On adding, we get:

$$W_P + W_V = 1.$$

Just as all amplitudes hitherto considered, so the vacuum-to-vacuum amplitude may be expressed in terms of the function S_F. To do so it is sufficient to utilize the definition (17.7) of the function S_F and the invariance of the trace with respect to cyclic permutations,

$$C^{(2)}[\mathscr{A}] = \tfrac{1}{2}e^2 \int d^4z_1\, d^4z_2 \operatorname{Tr}\{S_F(z_1-z_2)\gamma\cdot\mathscr{A}(z_2)\, S_F(z_2-z_1)\gamma\cdot\mathscr{A}(z_1)\}. \tag{12}$$

The corrected wave function, in which the wave functions have been antisymmetrized, is represented—to within second-order terms—as the product of the vacuum-to-vacuum amplitude and the Feynman function,

$$\psi^+[x|g] = (1+C^{(2)}[\mathscr{A}])\psi_F^+[x|g] + \text{terms of a higher order.}$$

It is left to the reader to verify that the corrected function ψ^-, to within second-order terms, is also in the form of a product with the same universal factor $1+C^{(2)}[\mathscr{A}]$. The correct amplitudes $S[f, g]$ can be written as the scalar products of the free solutions $\chi^{(\pm)}[f]$ of the Dirac equation and the corrected wave functions $\psi^{\pm}[g]$.

It follows from the foregoing considerations that in order to calculate them, the amplitudes of negaton and positon scattering and pair creation or annihilation need not be broken up into components which differ as to the time ordering of the vertices $x_1, x_2, \ldots, x_n$ of the relevant diagrams. When the contributions from all processes are added up, the scattering amplitudes may be expressed in terms of the functions S_F which figure in the integrands of integrals taken over *all* space-time. The decomposition of the amplitudes into parts corresponding to various diagrams ordered in time is not only unnecessary but also has no absolute meaning since the sequence of the times of two spatially separated events may differ in various reference systems.

The joint consideration of the processes of scattering, creation, and annihilation makes it possible to link up non-relativistic fragments into a relativistic whole.

Electron Wave Functions

We shall now discuss the properties of the Feynman wave functions ψ_F. Since the wave functions corrected by taking the Pauli exclusion principle into account differ from the functions ψ_F only by the factor $C[\mathscr{A}]$, information about the functions ψ_F can be used directly to describe real processes.

Instead of the two functions ψ_F^+ and ψ_F^- we shall introduce one general Feynman function ψ_F which is the sum of these two,

$$\psi_F(x) = \psi_F[x|f, g] = \psi_F^+[x|g] + \psi_F^-[x|f]. \qquad (13)$$

Setting $f = 0$ or $g = 0$ permits the original functions $\psi_F^+(x)$ and $\psi_F^-(x)$ to be obtained back from $\psi_F(x)$.

The definition of $\psi_F^\pm$ implies that the function ψ_F satisfies the following integral equation,

$$\psi_F(x) = \chi_F(x) - e\int d^4z S_F(x-z)\,\gamma\cdot\mathscr{A}(z)\,\psi_F(z), \qquad (14)$$

where

$$\chi_F(x) = \chi^{(+)}[x|g] + \chi^{(-)}[x|f].$$

The integral equation (14) serves as a more precise definition of the function ψ_F than the perturbation series, since it may admit solutions even when the perturbation series diverges.

The integral equation (14) is equivalent to the Dirac equation for ψ_F,

$$[-i\gamma\cdot\partial + m + e\gamma\cdot\mathscr{A}(x)]\,\psi_F(x) = 0, \qquad (15)$$

together with the following *asymptotic conditions* in the past and in the future. In the remote past, in a region without field, the part of the function $\psi_F[x|f, g]$ with positive frequencies is equal to $\chi^{(+)}[x|g]$. In the far future, in a region without field, the part of the function $\psi_F[x|f, g]$ with negative frequencies is equal to $\chi^{(-)}[x|f]$.

First, we shall show that the Dirac equation (15) and the asymptotic conditions follow from the integral equation (14).

The Dirac equation for ψ_F follows from the fact that the function S_F is the Green's function (fundamental solution) of the Dirac equation.

$$D_x S_F(x-y) = \delta^{(4)}(x-y). \tag{16}$$

To simplify the notation, on the right-hand side of this equation we have omitted the four-dimensional unit matrix in bispinor space. The proof of eq. (16) can be obtained directly from the definition of the function S_F:

$$\begin{aligned} D_x S_F(x-y) &= -i\gamma^0\delta(t-t')S^{(+)}(x-y) - i\gamma^0\delta(t'-t)S^{(-)}(x-y) \\ &= \gamma^0\delta(t-t')\int d\Gamma(m)\,[e^{-ip\cdot(x-y)}(\gamma\cdot p+m)+e^{ip\cdot(x-y)}(\gamma\cdot p-m)] \\ &= I\delta^{(4)}(x-y). \end{aligned}$$

The asymptotic conditions are derived from the integral equation by making use of the fact that for $t < T_1$ $(t > T_2)$ in the field-free region the function S_F in the integrand of eq. (14) may be replaced by the function $-S^{(-)}(x-y)$ $(S^{(+)}(x-y))$ which, as a function in the variable x, contains only negative (positive) frequencies.

Now, we shall give the proof that the integral equation follows from the Dirac equation and the asymptotic conditions.

Suppose that the function ψ_F satisfying this equation has in the remote past a positive-frequency part equal to $\chi^{(+)}[x|g]$ whereas in the far future it has a negative-frequency part equal to $\chi^{(-)}[x|f]$. We multiply both sides of the Dirac equation by $S_F(y-x)$, integrate with respect to x over the region of space-time bounded by the hypersurfaces σ_F and σ_P, and make use of Gauss' theorem and the equation

$$S_F(x-y)(m+i\gamma\cdot\overleftarrow{\partial}) = \delta^{(4)}(x-y),$$

proof of which proceeds by analogy with the proof of eq. (16). The result is written in the form

$$\begin{aligned} \psi_F(y) &= i\int_{\sigma_F} d\sigma_\mu S_F(y-x)\,\gamma^\mu\psi_F(x) - i\int_{\sigma_P} d\sigma_\mu S_F(y-x)\,\gamma^\mu\psi_F(x) \\ &\quad - e\int_{\sigma_P}^{\sigma_F} d^4x\, S_F(y-x)\,\gamma\cdot\mathscr{A}(x)\,\psi_F(x). \end{aligned}$$

It will be assumed that the hypersurfaces σ_F and σ_P lie in the remote past and the far future in field-free regions. The integrals over the hypersurfaces σ describing the boundary (surface) terms in Gauss' theorem will be calculated by making use of the asymptotic conditions. By way of example, let us calculate the first of these integrals. Since the hypersurface σ_F lies in a field-free region, the functions $S_F(y-x)$ for all fixed y's may be replaced by $-S^{(-)}(y-x)$. The first integral may thus be written as:

$$\int_{\sigma_F} d\sigma_\mu \sum_r \int d\Gamma(m)\,\chi^{(-)}(y;\mathbf{p},r)\,\bar{\chi}^{(-)}(x;\mathbf{p},r)\,\gamma^\mu\psi_F(x).$$

It follows from the asymptotic conditions that

$$\int_{\sigma_F} d\sigma_\mu \bar{\chi}^{(-)}(x;\mathbf{p},r)\,\gamma^\mu\psi_F(x) = f^*(\mathbf{p},r),$$

and, hence, the first integral is equal to $\chi^{(-)}[y|f]$. Similarly, the second integral may be shown to be equal to $\chi^{(+)}[y|g]$. The final integral can be extended over all space-time because the field $\mathscr{A}_\mu(x)$ vanishes beyond a finite region. In this manner, we obtain the integral equation (14).

The solution $\psi_F[x|f, g]$ of the Dirac equation thus is determined by giving the profiles of wave packets with negative frequencies in the future and positive frequencies in the past. Earlier we demonstrated that just such a characterization of the solutions of the Dirac equation is useful for describing scattering and pair creation and annihilation. It differs from the characterization typical of classical electrodynamics, in which we give the asymptotic conditions only in the past (retarded solution) or only in the future (advanced solution). In quantum theory, in addition to the Feynman solution characterized by giving a part with positive frequencies in the past and one with negative frequencies in the future, use may also be made of an *anti-Feynman solution* which is the conjugate of the Feynman solution; to do this, in the integral equation we replace the function S_F by the function $S_{\mathscr{A}}$,

$$S_{\mathscr{A}}(x) \equiv \theta(x^0)S^{(-)}(x) - \theta(-x^0)S^{(+)}(x). \tag{17}$$

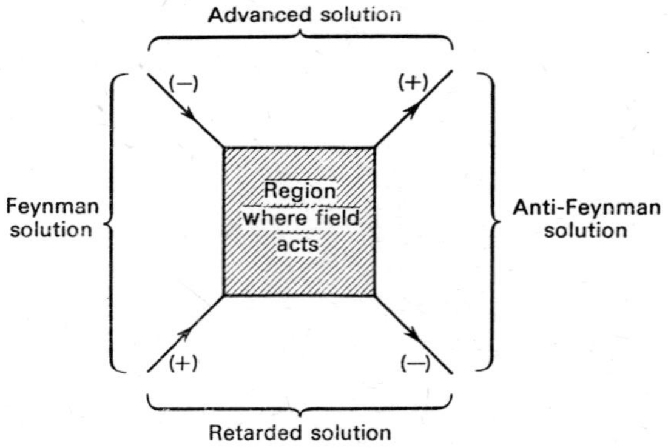

FIG. 17.4. Four types of solution of the Dirac equation.

The four types of solution have been schematically shown in Fig. 17.4. The four wave-packet profiles in the past and in the future cannot, of course, be all given arbitrarily, nor is it even possible arbitrarily to give both wave-packet profiles with positive frequencies in the past and in the future, or both wave-packet profiles with negative frequencies in the past and in the future.

The functions ψ_F^+ and ψ_F^- which are special cases of the Feynman solutions have the simplest interpretation. The solution $\psi_F^+[x|g]$ is determined by giving the profile of a wave packet with positive frequencies in the past. In the far future the part of the function $\psi_F^+[x|g]$ with negative frequencies is equal to zero. Both the part with positive frequencies in the future and the part with negative frequencies in the past can be calculated from knowledge of the potential $\mathscr{A}_\mu(x)$. Subsequently in this section we shall show that they can be expressed in terms of S matrices for negaton scattering and for pair annihilation. Similarly, the solution $\psi_F^-[x|g]$ is determined by giving the profile of a wave with negative frequencies in the future. Its part with positive frequencies in the past is equal to zero, whereas the parts with negative frequencies in the past and positive frequencies in the future may be expressed

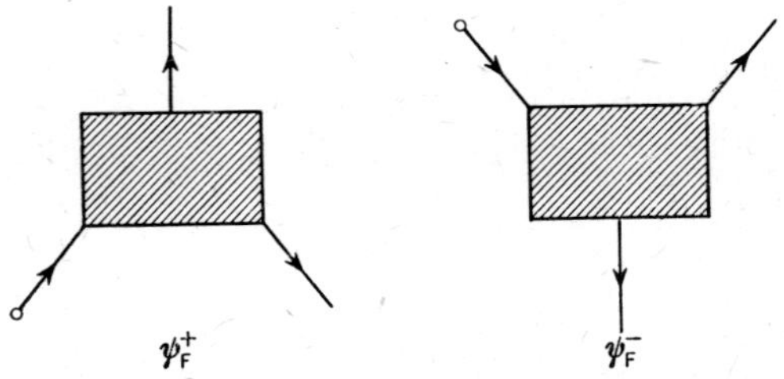

FIG. 17.5. The Feynman functions ψ_F^+ and ψ_F^-.

in terms of the S matrices for pair creation and for positon scattering, respectively.

The solutions $\psi_F^\pm$ may be depicted by means of the diagrams in Fig. 17.5, where we have used circles to denote those parts of the solutions which are given beforehand. In these diagrams an analogy can be seen with the phenomena of propagation of waves in space when they encounter an obstacle. In this case, the electromagnetic field would be an obstacle. The reflected wave is "reversed in time", and hence the occurrence of frequencies opposite to those which appeared in the incident wave. This interesting analogy, first discovered by Wheeler and Feynman, has not hitherto received a full explanation. We do not know whether the description of antiparticles as particles whose travel has been time-reversed is only a formal mathematical operation, or whether some important physical meaning is concealed behind it.

The iterative solution of the integral equation for ψ_F may be expressed in terms of χ_F, the inhomogeneous term in this equation, by means of the *propagator* K_F,

$$K_F[x, y|\mathscr{A}] \equiv S_F(x-y)-e\int d^4z\, S_F(x-z)\gamma\cdot\mathscr{A}(z)\, S_F(z-y)$$
$$+e^2\int d^4z_1 d^4z_2\, S_F(x-z_1)\gamma\cdot\mathscr{A}(z_1)\, S_F(z_1-z_2)\gamma\cdot\mathscr{A}(z_2)\, S_F(z_2-y)+\ldots \tag{18}$$

The perturbation series defining the propagator K_F can be obtained as the iterative solution of the integral equation:

$$K_F[x, y|\mathscr{A}] = S_F(x-y)-e\int d^4z\, S_F(x-z)\gamma\cdot\mathscr{A}(z)\, K_F[z, y|\mathscr{A}]. \tag{19}$$

It follows from this equation that the propagator K_F satisfies the inhomogeneous Dirac equation with electromagnetic field:

$$[-i\gamma\cdot\partial_x+m+e\gamma\cdot\mathscr{A}(x)]\, K_F[x, y|\mathscr{A}] = \delta^{(4)}(x-y).$$

From the perturbation expansion (10) of the function ψ_F it is seen that

$$\psi_F(x) = \chi_F(x)-e\int d^4z K_F[x, z|\mathscr{A}]\gamma\cdot\mathscr{A}(z)\psi_F(z). \tag{20}$$

In view of the important role played by the propagator K_F and related propagators in our formulation of electrodynamics, let us examine the properties of these objects in what follows.

The anti-Feynman function will be denoted by the symbol $\psi_{\mathfrak{A}}$. It may be defined as the solution of the integral equation:

$$\psi_{\mathfrak{A}}(x) = \chi_{\mathfrak{A}}(x)-e\int d^4z S_{\mathfrak{A}}(x-z)\,\gamma\cdot\mathscr{A}(z)\psi_{\mathfrak{A}}(z), \tag{21}$$

where

$$\chi_{\mathfrak{A}}(x) = \chi^{(+)}[x|f]+\chi^{(-)}[x|g],$$

or else in terms of a suitable perturbation series. The asymptotic conditions satisfied by the function $\psi_{\mathfrak{A}}$ are determined by the properties of the free solution of the Dirac equation $\chi_{\mathfrak{A}}(x)$. The positive-frequency part $\chi_{\mathfrak{A}}^{(+)}$ determines the part with the positive frequencies of the function $\psi_{\mathfrak{A}}$ in the future, whereas the negative-frequency part $\chi_{\mathfrak{A}}^{(-)}$

determines the part with negative frequencies of the function in the past. Diagrams depicting these boundary conditions are given in Fig. 17.6. The integral equation for $\psi_{\mathcal{A}}$ may also be written in an equivalent form (cf. formula (14)) containing the Feynman propagator,

$$\bar{\psi}_{\mathcal{A}}(x) = \bar{\chi}_{\mathcal{A}}(x) - e\int \mathrm{d}^4 z \bar{\psi}_{\mathcal{A}}(z)\gamma \cdot \mathscr{A}(z)\, S_{\mathrm{F}}(z-x).$$

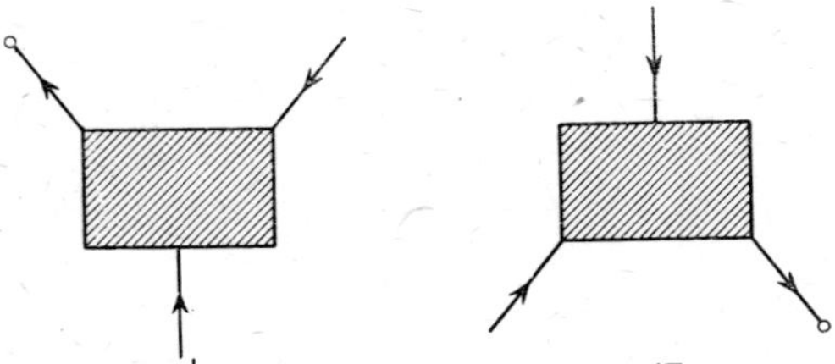

FIG. 17.6. The anti-Feynman functions $\psi_{\mathcal{A}}^{+}$ and $\psi_{\mathcal{A}}^{-}$.

The conjugate anti-Feynman function may also be expressed in terms of the propagator K_{F},

$$\bar{\psi}_{\mathcal{A}}(x) = \bar{\chi}_{\mathcal{A}}(x) - e\int \mathrm{d}^4 z \bar{\chi}_{\mathcal{A}}(z)\gamma \cdot \mathscr{A}(z)\, K_{\mathrm{F}}[z, x|\mathscr{A}].$$

THE GREEN'S FUNCTIONS FOR THE DIRAC EQUATION

The functions S_{F} and K_{F} given above play an important role in the relativistic theory of spin $\frac{1}{2}$ particles, both in quantum mechanics and in quantum field theory. These functions are the Green's functions for the Dirac equation without field and with external electromagnetic field, respectively. In addition to the Feynman Green's functions S_{F} and K_{F} we shall also consider three other types of Green's functions for the Dirac equation corresponding to the other three characteristics of the solutions of the Dirac equation (cf. Fig. 17.4). These will be the *anti-Feynman functions* $S_{\mathcal{A}}$ and $K_{\mathcal{A}}$, the *retarded functions* S_{R} and K_{R}, and the *advanced functions* S_{A} and K_{A}. The properties of the Green's functions for the Dirac equation without potential have been listed in Appendix E. The fundamental solutions of the Dirac equation and other wave equations are, strictly speaking, distributions and not ordinary functions. Nevertheless, in keeping with tradition we shall refer to them as functions.

The Green's functions for the Dirac equation without potential will be expressed in terms of the *Green's functions for the Klein–Gordon equation,* which we denote by the symbols Δ_F, $\Delta_{\bar{F}}$, Δ_R, and Δ_A and define† by means of the following Fourier integral representations:

$$\Delta_F(x) \equiv \int \frac{d^4p}{(2\pi)^4} \frac{1}{m^2 - p^2 - i\varepsilon} e^{-ip\cdot x}, \tag{22a}$$

$$\Delta_{\bar{F}}(x) \equiv \int \frac{d^4p}{(2\pi)^4} \frac{1}{m^2 - p^2 + i\varepsilon} e^{-ip\cdot x}, \tag{22b}$$

$$\Delta_R(x) \equiv \int \frac{d^4p}{(2\pi)^4} \frac{1}{m^2 - p^2 - i\varepsilon \operatorname{sgn} p_0} e^{-ip\cdot x}, \tag{22c}$$

$$\Delta_A(x) \equiv \int \frac{d^4p}{(2\pi)^4} \frac{1}{m^2 - p^2 + i\varepsilon \operatorname{sgn} p_0} e^{-ip\cdot x}. \tag{22d}$$

The four functions above will be denoted by the common symbol Δ_I. They satisfy the inhomogeneous Klein–Gordon equation,

$$(\Box + m^2)\Delta_I(x) = \delta^{(4)}(x). \tag{23}$$

These functions differ in that each satisfies different asymptotic conditions. The latter can be read from the integral representations (22) when integration with respect to the variable p_0 has been carried out by the residue method. Depending on the sign of x^0, the contour of integration in the complex plane p_0 can be closed in the upper or lower half-plane. On integrating, we obtain the following conditions,

$$\Delta_F(x) = \begin{cases} \Delta^{(+)}(x), & x^0 > 0, \\ -\Delta^{(-)}(x), & x^0 < 0, \end{cases} \tag{24a}$$

$$\Delta_{\bar{F}}(x) = \begin{cases} \Delta^{(-)}(x), & x^0 > 0, \\ -\Delta^{(+)}(x), & x^0 < 0, \end{cases} \tag{24b}$$

$$\Delta_R(x) = \begin{cases} \Delta(x), & x^0 > 0, \\ 0, & x^0 < 0, \end{cases} \tag{24c}$$

$$\Delta_A(x) = \begin{cases} 0, & x^0 > 0, \\ -\Delta(x), & x^0 < 0. \end{cases} \tag{24d}$$

† As we shall show subsequently, definition (22a) of the function Δ_F is equivalent to definition (14.41) given earlier.

The functions S_F, $S_{\mathscr{A}}$, S_R, and S_A, which we shall denote by the common symbol S_I, will be defined† by the formula:

$$S_I(x) \equiv (i\gamma \cdot \partial + m)\Delta_I(x). \tag{25}$$

From the properties of the matrices γ^μ and from eq. (23) it follows that the functions S_I satisfy the inhomogeneous Dirac equation without potential,

$$(-i\gamma \cdot \partial + m)S_I(x) = \delta^{(4)}(x). \tag{26}$$

The propagators $S_I(\mathbf{x}, t)$ are distributions with respect to the spatial variables $\mathbf{x}$ and discontinuous functions of the time variable t at the point $t = 0$. This discontinuity is the same for all four propagators:

$$\lim_{\varepsilon \to 0} \big(S_I(\mathbf{x}, \varepsilon) - S_I(\mathbf{x}, -\varepsilon)\big) = i\gamma^0 \delta(\mathbf{x}). \tag{27}$$

This relation is obtained by integrating both sides of eq. (26) with respect to t in the interval $(+\varepsilon, -\varepsilon)$.

The functions $K_{\mathscr{A}}$, K_R, and K_A are defined in a manner similar to that in which K_F was defined, i.e. by means of a perturbation series which we obtain by substituting the functions $S_{\mathscr{A}}$, S_R, and S_A, respectively, for the function S_F everywhere in series (18). It follows from these definitions that all four functions, K_F, $K_{\mathscr{A}}$, K_R, and K_A (which we denote by the common symbol K_I), satisfy the integral equations,

$$K_I[x, y|\mathscr{A}] = S_I(x-y) - \int d^4z S_I(x-z) e\gamma \cdot \mathscr{A}(z) K_I[z, y|\mathscr{A}], \tag{28a}$$

$$K_I[x, y|\mathscr{A}] = S_I(x-y) - \int d^4z K_I[x, z|\mathscr{A}] e\gamma \cdot \mathscr{A}(z) S_I(z-y). \tag{28b}$$

The iterative solution of each of these equations is the defining perturbation series of a function K_I. These equations may have solutions, however, even when the perturbation series is divergent. It follows from the integral equations (28) that the functions K_I are the Green's

† For the function S_F this definition is equivalent to definition (7). The integral representations (14.28b) of the function $\Delta^{(\pm)}(x)$ can be used to demonstrate that the equality

$$\partial_0[\theta(x^0)\Delta^{(+)}(x) - \theta(-x^0)\Delta^{(-)}(x)] = \theta(x^0)\partial_0\Delta^{(+)}(x) - \theta(-x^0)\,\partial_0\Delta^{(-)}(x)$$

is satisfied, whereby the equivalence of the two definitions follows.

functions for the Dirac equation with electromagnetic field. Upon acting on both sides of equations (28) with the Dirac operators D_x and $\overleftarrow{D}_y$, we arrive at the equations

$$[-i\gamma\cdot\partial_x+m+e\gamma\cdot\mathscr{A}(x)]K_{\rm I}[x,y|\mathscr{A}] = \delta^{(4)}(x-y), \tag{29a}$$

$$K_{\rm I}[x,y|\mathscr{A}][i\gamma\cdot\overleftarrow{\partial}_y+m+e\gamma\cdot\mathscr{A}(y)] = \delta^{(4)}(x-y). \tag{29b}$$

The functions $K_{\rm F}$, $K_{\bar{\rm F}}$, $K_{\rm R}$, and $K_{\rm A}$ differ from each other as to the kind of asymptotic conditions they satisfy. For the functions $K_{\rm R}$ and $K_{\rm A}$ these conditions are

$$K_{\rm R}[x,y|\mathscr{A}] = 0, \quad \text{when } x^0 < y^0, \tag{30a}$$

$$K_{\rm A}[x,y|\mathscr{A}] = 0, \quad \text{when } x^0 > y^0. \tag{30b}$$

The asymptotic conditions for the functions $K_{\rm F}[x,y|\mathscr{A}]$ and $K_{\bar{\rm F}}[x,y|\mathscr{A}]$ are more complicated. If $x^0 > y^0$, then for $x^0 > T_2$ the function $K_{\rm F}[x,y|\mathscr{A}]$ as a function of the variable x is the positive-frequency solution of the Dirac equation without potential, and for $y^0 < T_1$, as a function of the variable y it is the negative-frequency solution. If $x^0 < y^0$, then for $x^0 < T_1$ the function $K_{\rm F}[x,y|\mathscr{A}]$ as a function of the variable x is the negative-frequency solution, and for $y^0 > T_2$ as a function of the variable y it is the positive-frequency solution. The asymptotic conditions for the function $K_{\bar{\rm F}}[x,y|\mathscr{A}]$ are formulated in like manner by interchanging the terms *positive-frequency* and *negative-frequency* in the statements above. The asymptotic conditions obeyed by the functions $K_{\rm I}$ emerge from the definition of these functions in terms of perturbation series and from the definition of the functions $S_{\rm I}$.

The discontinuities of the propagators $S_{\rm I}$ are transferred to the propagators $K_{\rm I}$. Just as for $S_{\rm I}$, it may be proved that

$$\lim_{\varepsilon\to 0}(K_{\rm I}[\mathbf{x},t+\varepsilon;\mathbf{y},t|\mathscr{A}]-K_{\rm I}[\mathbf{x},t-\varepsilon;\mathbf{y},t|\mathscr{A}]) = i\gamma^0\delta(\mathbf{x}-\mathbf{y}), \tag{31a}$$

$$\lim_{\varepsilon\to 0}(K_{\rm I}[\mathbf{x},t;\mathbf{y},t+\varepsilon|\mathscr{A}]-K_{\rm I}[\mathbf{x},t;\mathbf{y},t-\varepsilon|\mathscr{A}]) = -i\gamma^0\delta(\mathbf{x}-\mathbf{y}). \tag{31b}$$

These discontinuities must be borne in mind and the values of the propagators for equal times should be given a meaning by specifying the limiting procedure.

Just as propagators in non-relativistic theory, the propagators $K_{\rm I}$ split up into the product of $K_{\rm I}$'s by functional differentiation:

$$\frac{\delta K_I[x, y|\mathscr{A}]}{\delta \mathscr{A}_\mu(z)} = -eK_{\rm I}[x, z|\mathscr{A}]\gamma^\mu K_{\rm I}[z, y|\mathscr{A}]. \tag{32}$$

This relation is proved in the same way as formula (5.73), by functionally differentiating the differential equation for $K_{\rm I}$ and solving the resultant equation for $\delta K_{\rm I}/\delta\mathscr{A}$ by means of the propagators $K_{\rm I}$. The asymptotic conditions play an essential role in this proof.

With the aid of the Green's functions $K_{\rm I}$ the solution of the Dirac equation $\psi(x)$ in the electromagnetic field may be expressed by means of a suitably chosen set of asymptotic solutions of the free equation. To do this we multiply both sides of equation (29a) by $\psi(y)$, integrate with respect to y over the region of space-time bounded by the hypersurfaces $\sigma_<$ and $\sigma_>$, and apply Gauss' theorem. If $\psi(y)$ satisfies the Dirac equation with electromagnetic field, then the final formula will contain only surface terms:

$$\psi(x) = i\int_{\sigma_>} d\sigma_\mu K_{\rm I}[x, y|\mathscr{A}]\gamma^\mu\psi(y) - i\int_{\sigma_<} d\sigma_\mu K_{\rm I}[x, y|\mathscr{A}]\gamma^\mu\psi(y). \tag{33}$$

From this we obtain a generalization of formula (31b). Taking the limit when $\sigma_> \to \sigma(x) \leftarrow \sigma_<$, where $\sigma(x)$ denotes the hypersurface passing through the point x, we get

$$\lim_{\sigma_>\to\sigma(x)\leftarrow\sigma_<} (K_{\rm F}[x, y_>|\mathscr{A}] - K_{\rm F}[x, y_<|\mathscr{A}])\,\gamma^\mu = -i\delta^\mu(x, y),$$

where $\delta^\mu(x, y)$ is the three-dimensional function δ on the hypersurface.

For retarded and advanced propagators, only one of the integrals in formula (33) is different from zero:

$$\psi(x) = -i\int_{\sigma_<} d\sigma_\mu K_{\rm R}[x, y|\mathscr{A}]\gamma^\mu\psi(y), \tag{34a}$$

$$\psi(x) = i\int_{\sigma_>} d\sigma_\mu K_{\rm A}[x, y|\mathscr{A}]\gamma^\mu\psi(y). \tag{34b}$$

The first formula is valid for x's later than $\sigma_<$, whereas the second holds for x's earlier than $\sigma_>$. As in classical field theory, the retarded and advanced Green's functions propagate the initial conditions forward and backward, respectively, in time. On introduction of the function

$$K[x, y|\mathscr{A}] \equiv K_R[x, y|\mathscr{A}] - K_A[x, y|\mathscr{A}],$$

which satisfies the homogeneous Dirac equation with potential, the two formulae (34) may be replaced by one which is valid for any position of the point x with respect to the hypersurface σ,

$$\psi(x) = -i \int_\sigma d\sigma_\mu K[x, y|\mathscr{A}] \gamma^\mu \psi(y). \tag{35}$$

For Feynman and anti-Feynman propagators in the general case, both integrals over the hypersurfaces will appear in formula (33). Writing the solutions of the Dirac equations as the sum of these two integrals over the hypersurfaces corresponds to decomposing the Feynman function into the parts ψ_F^+ and ψ_F^-. Let us consider an arbitrary solution of the Dirac equation, expressed in terms of the Feynman boundary conditions (cf. formula (13)),

$$\psi_F[x|f, g] = \psi_F^+[x|g] + \psi_F^-[x|f].$$

We shall show that

$$\psi_F[x|f, g] = -i \int_{\sigma_<} d\sigma_\mu K_F[x, y|\mathscr{A}] \gamma^\mu \psi_F^+[y|g]$$
$$+ i \int_{\sigma_>} d\sigma_\mu K_F[x, y|\mathscr{A}] \gamma^\mu \psi_F^-[y|g], \tag{36}$$

where the hypersurfaces $\sigma_<$ and $\sigma_>$ lie earlier and later, respectively, than the point x.

For the proof, we shall exploit the fact that the integral $\int_\sigma d\sigma_\mu K_F[x, y|\mathscr{A}]\gamma^\mu \psi(y)$ over the hypersurface σ is independent of the choice of that hypersurface, provided that it does not intersect the point x. This independence stems from the fact that K_F and ψ satisfy, with regard to the variable y, the mutually conjugate Dirac equation when $x \neq y$. On removing the hypersurface beyond the region wherein the field acts and on making use of the boundary conditions satisfied by the propagator K_F and by the solutions $\psi^\pm$, we obtain:

$$\int_{\sigma_<} d\sigma_\mu K_F \gamma^\mu \psi^- = 0 = \int_{\sigma_>} d\sigma_\mu K_F \gamma^\mu \psi^+.$$

When we remove the hypersurfaces $\sigma_<$ and $\sigma_>$, respectively, into the remote past and far future beyond the region where the field acts, we arrive at:

$$\psi_F[x|f, g] = -i \int_{\sigma_P} d\sigma_\mu K_F[x, y|\mathscr{A}]\, \gamma^\mu \chi^{(+)}[y|g] + i \int_{\sigma_F} d\sigma_\mu K_F[x, y|\mathscr{A}]\, \gamma^\mu \chi^{(-)}[y|f], \tag{37}$$

where σ_P denotes the hypersurface in the remote past, and σ_F the hypersurface in the far future. Without affecting the values of the integrals, in this formula $\chi^{(+)}$ and $\chi^{(-)}$ can be replaced by their sum, the function χ_F, since the added terms vanish. On applying Gauss' theorem and using the free Dirac equation, we thus get:

$$\psi_F[x|f, g] = \int_{\sigma_P}^{\sigma_F} d^4y\, K_F[x, y|\mathscr{A}]\, \overleftarrow{D}_y \chi_F(y), \tag{38}$$

where

$$\overleftarrow{D}_y \equiv i\gamma \cdot \overleftarrow{\partial}_y + m.$$

The hypersurfaces σ_P and σ_F can be removed to infinity since the integrand vanishes beyond the region where the field occurs. In the field-free region, the propagator K_F satisfies the free Dirac equation. Formula (38) is a different form of the familiar formula for the general solution of the Dirac equation, expressed in terms of the Feynman boundary conditions (cf. formula (10)).

All the Green's functions K_I are invariant under orthochronous Poincaré transformations and under gauge transformation, provided that at the same time the arguments x and y, the spinor indices, and the field $\mathscr{A}$ are subject to the transformations. The results obtained in Section 16 may be used to demonstrate that the following relations are satisfied:

$$K_I[x, y|\mathscr{A}] = K_I[x+a, y+a|^{Tr}\mathscr{A}], \tag{39a}$$

$$K_I[x, y|\mathscr{A}] = S_L^{-1} K_I[\Lambda x, \Lambda y|^{L}\mathscr{A}] S_L, \tag{39b}$$

$$K_I[x, y|\mathscr{A}] = P^{-1} K_I[\mathscr{P}x, \mathscr{P}y|^{P}\mathscr{A}] P, \tag{39c}$$

$$K_I[x, y|\mathscr{A}] = e^{ie\Lambda(x)} K_I[x, y|^{G}\mathscr{A}] e^{-ie\Lambda(y)}. \tag{39d}$$

Since the asymptotic conditions for various functions K_{I} distinguish the direction in which time runs, the functions K_{I} are not invariant with respect to time reversal. They satisfy the relation

$$K_{T(\mathrm{I})}[x, y|\mathscr{A}] = \mathrm{T}^{-1} K_{\mathrm{I}}^{*}[-\mathscr{P}x, -\mathscr{P}y|^{\mathrm{T}}\mathscr{A}]\,\mathrm{T}, \tag{40}$$

where

$$T(\mathrm{R}) = \mathrm{A}, \quad T(\mathrm{A}) = \mathrm{R}, \quad T(\mathrm{F}) = \text{Ⅎ}, \quad T(\text{Ⅎ}) = \mathrm{F}.$$

The functions K_{R} and K_{A} are invariant under the charge-conjugation transformation, but the functions K_{F} and $K_{\text{Ⅎ}}$ go over into each other under this transformation since in formulating the asymptotic conditions for them we distinguished positive frequencies. Under charge conjugation the electromagnetic field changes sign:

$$K_{C(\mathrm{I})}[x, y|\mathscr{A}] = \mathrm{C}K_{\mathrm{I}}^{*}[x, y|-\mathscr{A}]\mathrm{C}^{-1}, \tag{41}$$

where

$$C(\mathrm{R}) = \mathrm{R}, \quad C(\mathrm{A}) = \mathrm{A}, \quad C(\mathrm{F}) = \text{Ⅎ}, \quad C(\text{Ⅎ}) = \mathrm{F}.$$

On combining the P, C, and T transformations, we obtain for the propagators a simple relation which no longer contains the conjugation operation:

$$K_{CT(\mathrm{I})}[x, y|\mathscr{A}] = \gamma_5 K_{\mathrm{I}}[-x, -y|^{\mathrm{PCT}}\mathscr{A}]\,\gamma_5^{-1}, \tag{42}$$

where

$${}^{\mathrm{PCT}}\mathscr{A}(z) = -\mathscr{A}(-z),$$

whereas the function $CT(\mathrm{I})$ is a combination of the functions C and T defined above,

$$CT(\mathrm{R}) = \mathrm{A}, \quad CT(\mathrm{A}) = \mathrm{R}, \quad CT(\mathrm{F}) = \mathrm{F}, \quad CT(\text{Ⅎ}) = \text{Ⅎ}.$$

In addition to satisfying the relations emerging from the invariance of the Dirac equation for the wave function, the Green's functions K_{I} also satisfy one more additional relation resulting from the existence of the matrix A which effects the Hermitian conjugation of the matrices γ^{μ}:

$$K_{\mathrm{I}}[x, y|\mathscr{A}] = \mathrm{A}^{-1}K_{\mathrm{I}}^{\dagger}[y, x|\mathscr{A}]\,\mathrm{A}. \tag{43}$$

When we combine this last transformation with time reversal, we obtain conditions for each Green's function separately,

$$K_{\mathrm{I}}[x, y|\mathscr{A}] = (\mathrm{A}^{\mathrm{T}}\mathrm{T})^{-1}K_{\mathrm{I}}^{\mathrm{T}}[-\mathscr{P}y, -\mathscr{P}x|^{\mathrm{T}}\mathscr{A}]\mathrm{A}^{\mathrm{T}}\mathrm{T}. \tag{44}$$

The foregoing relation is more useful for discussing the properties of all sorts of transition amplitudes with respect to time reversal than is relation (40) which contains a complex conjugation of propagators.

In conclusion, we shall give one more relation of this type which lends itself to application in calculations. To get it, we combine the transformation C and the Hermitian conjugation,

$$K_{CT(\mathrm{I})}[x, y|\mathscr{A}] = (\mathrm{C}^\dagger \mathrm{A})^{-1} K_{\mathrm{I}}^{\mathrm{T}}[y, x|-\mathscr{A}]\mathrm{C}^\dagger \mathrm{A}. \tag{45}$$

The above relations between the propagators K_{I} take on a particularly simple form when there is no electromagnetic field. We shall write these relations out for the Feynman propagators used most frequently:

$$\begin{aligned} S_{\mathrm{F}}(x) &= \mathrm{S}_{\mathrm{L}} S_{\mathrm{F}}(\Lambda^{-1}x)\mathrm{S}_{\mathrm{L}}^{-1} \\ &= \mathrm{P} S_{\mathrm{F}}(\mathscr{P}x)\mathrm{P}^{-1} \\ &= \mathrm{T}^{-1} S_{\mathrm{A}}^{*}(-\mathscr{P}x)\mathrm{T}, \end{aligned} \tag{46}$$

$$S_{\mathrm{F}}(x) = \mathrm{C} S_{\mathrm{A}}^{*}(x)\mathrm{C}^{-1}, \tag{47}$$

$$S_{\mathrm{F}}(x) = \gamma_5 S_{\mathrm{F}}(-x)\gamma_5^{-1}, \tag{48}$$

$$\begin{aligned} S_{\mathrm{F}}(x) &= \mathrm{A}^{-1} S_{\mathrm{A}}^{\dagger}(-x)\mathrm{A} \\ &= (\mathrm{A}^{\mathrm{T}}\mathrm{T}\mathrm{P})^{-1} S_{\mathrm{F}}^{\mathrm{T}}(x)\mathrm{A}^{\mathrm{T}}\mathrm{T}\mathrm{P} \\ &= \mathrm{B}^{-1} S_{\mathrm{F}}^{\mathrm{T}}(-x)\mathrm{B}. \end{aligned} \tag{49}$$

The Electron in a Static Electromagnetic Field

Now we shall give the construction of the Feynman propagator for the case of a static electromagnetic field and examine this object in regard to some properties which will be needed in later sections.

In these considerations, it is convenient to use the Dirac equation written in a form similar to that of the Schrödinger equation,

$$i\frac{\partial}{\partial t}\psi(\mathbf{x}, t) = H_{\mathrm{D}}\psi(\mathbf{x}, t), \tag{50}$$

where the operator H_{D}, called the *Dirac Hamiltonian*,† is defined by the formula:

$$H_{\mathrm{D}} \equiv \frac{1}{i}\boldsymbol{\alpha}\cdot\nabla - e\boldsymbol{\alpha}\cdot\mathscr{A}(\mathbf{x}) + e\alpha_0\mathscr{A}_0(\mathbf{x}) + \beta m. \tag{51}$$

† H_{D} is not the energy operator since it simultaneously describes evolution forward in time for particles and backward in time for antiparticles.

The Hermitian matrices α^μ and β are defined as

$$\alpha^\mu \equiv \mathrm{A}\gamma^\mu,$$
$$\beta \equiv \mathrm{A}.$$

In the Dirac, spinor, and Majorana representations, $\alpha^0 = 1$, and $\beta = \gamma^0$.

With certain assumptions as to the regularity of the potential $\mathscr{A}_\mu$, the operator H_D is self-adjoint in the Hilbert space of bispinors with scalar product (16.20). In the general case this operator has both a continuous and a discrete spectrum. The continuous spectrum lies on the real axis from $-\infty$ to $-m$ and from m to ∞. The discrete spectrum lies in the energy gap from $-m$ to m. Each square-integrable bispinor may be decomposed into the eigenfunctions φ_n and the generalized eigenfunctions $\varphi^{(\pm)}$ of the operator H_D. The normalized eigenfunctions, once the exponential time dependence $\exp(-iEt)$ has been separated, will be denoted by $\varphi_n(\mathbf{x})$,

$$H_\mathrm{D}\varphi_n(\mathbf{x}) = E_n\varphi_n(\mathbf{x}).$$

These include eigenfunctions belonging to both positive and negative eigenvalues.

The generalized eigenfunctions belonging to the continuous spectrum will be labelled with the components of the vector $\mathbf{p}$ and the polarization index r which refer to the asymptotic behaviour of the solutions

$$H_\mathrm{D}\varphi^{(\pm)}(\mathbf{x}, \mathbf{p}, r) = \pm E_p\varphi^{(\pm)}(\mathbf{x}, \mathbf{p}, r).$$

Just as in non-relativistic theory, so here the functions $\varphi^{(\pm)}$ satisfying the given boundary conditions will be defined by means of integral equations. We shall consider only solutions corresponding to diverging waves,† i.e. solutions with positive frequency will be asymptotically of the form e^{ipr}/r, whereas solutions with negative frequency will be asymptotically of the form e^{-ipr}/r. The integral equations for the functions $\varphi^{(\pm)}$ are of the form

$$\varphi^{(+)}(\mathbf{x}, \mathbf{p}, r) = \mathrm{e}^{i\mathbf{p}\cdot\mathbf{x}}u(\mathbf{p}, r)$$
$$-\left(E_p+\beta m+\frac{1}{i}\boldsymbol{\alpha}\cdot\boldsymbol{\nabla}\right)\frac{1}{4\pi}\int d^3y\,\frac{\mathrm{e}^{ip|\mathbf{x}-\mathbf{y}|}}{|\mathbf{x}-\mathbf{y}|}\,\alpha\cdot\mathscr{A}(\mathbf{y})\varphi^{(+)}(\mathbf{y}, \mathbf{p}, r), \tag{52a}$$

$$\varphi^{(-)}(\mathbf{x}, \mathbf{p}, r) = \mathrm{e}^{-i\mathbf{p}\cdot\mathbf{x}}v(\mathbf{p}, r)$$
$$-\left(-E_p+\beta m+\frac{1}{i}\boldsymbol{\alpha}\cdot\boldsymbol{\nabla}\right)\frac{1}{4\pi}\int d^3y\,\frac{\mathrm{e}^{-ip|\mathbf{x}-\mathbf{y}|}}{|\mathbf{x}-\mathbf{y}|}\,\alpha\cdot\mathscr{A}(\mathbf{y})\varphi^{(-)}(\mathbf{y}, \mathbf{p}, r). \tag{52b}$$

---

† Solutions corresponding to converging waves (i.e. the relativistic analogues of the functions φ^- defined by eq. (4.19b)) can be derived in like manner. Note: Do not confuse the solutions $\varphi^{(-)}$ which correspond to a negative eigenvalue of the operator H_D with the solutions φ^- introduced in Section 4 in the description of converging waves.

The functions $\varphi^{(+)}$ and $\varphi^{(-)}$ are normalized in the same way as the corresponding inhomogeneous terms in the foregoing integral equations. For example:

$$\int d^3x\varphi^{(+)\dagger}(\mathbf{x}, \mathbf{p}, r)\varphi^{(+)}(\mathbf{x}, \mathbf{q}, s) = \delta_{rs}2E_p(2\pi)^3\delta(\mathbf{p}-\mathbf{q}).$$

The proof of this property is similar to that for the corresponding formulae in non-relativistic theory.†

With such a normalization adopted for the functions $\varphi^{(+)}$ and $\varphi^{(-)}$, the resolution of unity associated with the operator H_D may be written as:

$$\sum_n \varphi_n(\mathbf{x})\varphi_n^\dagger(\mathbf{y})+\sum_r\int\frac{d^3p}{2E_p(2\pi)^3}\varphi^{(+)}(\mathbf{x}, \mathbf{p}, r)\varphi^{(+)\dagger}(\mathbf{y}, \mathbf{p}, r)$$

$$+\sum_r\int\frac{d^3p}{2E_p(2\pi)^3}\varphi^{(-)}(\mathbf{x}, \mathbf{p}, r)\,\varphi^{(-)\dagger}(\mathbf{y}, \mathbf{p}, r) = I\delta(\mathbf{x}-\mathbf{y}), \tag{53}$$

where I is a unit 4×4 matrix. This matrix will be dropped from formulae henceforth.

The solutions φ^n and $\varphi^{(\pm)}$ can be used to construct the Feynman propagator by first building the propagators $K^{(\pm)}$ which are the counterparts of the functions $S^{(\pm)}$,

$$K^{(\pm)}[\mathbf{x}t, \mathbf{y}t'|\mathscr{A}] \equiv \sum_n \theta(\pm E_n)\varphi_n(\mathbf{x})\,e^{-iE(t-t')}\overline{\varphi}_n(\mathbf{y})$$

$$+\sum_r\int\frac{d^3p}{2E_p(2\pi)^3}\,\varphi^{(\pm)}(\mathbf{x}, \mathbf{p}, r)\,e^{\mp iE_p(t-t')}\,\overline{\varphi^{(\pm)}}(\mathbf{y}, \mathbf{p}, r), \tag{54}$$

and then forming the propagator K_F from them,

$$K_F[\mathbf{x}t, \mathbf{y}t'|\mathscr{A}] \equiv \theta(t-t')K^{(+)}[\mathbf{x}t, \mathbf{y}t'|\mathscr{A}]-\theta(t'-t)K^{(-)}[\mathbf{x}t, \mathbf{y}t'|\mathscr{A}]. \tag{55}$$

The propagator K_F so defined satisfies the differential eqs. (29) and the Feynman asymptotic conditions (positive frequencies for $t > t'$ and negative frequencies for $t < t'$). In proving eqs. (29), we should use the completeness condition (53) for the functions φ_n, $\varphi^{(+)}$, and $\varphi^{(-)}$.

Since they depend only on the difference between the time t and t', the functions $K^{(\pm)}$ and K_F may be written as single Fourier integrals,

$$K_I[\mathbf{x}t, \mathbf{y}t'|\mathscr{A}] = \int_{-\infty}^{+\infty}\frac{dE}{2\pi}\,e^{-iE(t-t')}\,K_I[\mathbf{x}, \mathbf{y}; E|\mathscr{A}].$$

Comparison of this formula with the definitions of the functions $K^{(\pm)}$ and K_F reveals that

$$K^{(\pm)}[\mathbf{x}, \mathbf{y}; E|\mathscr{A}] = \sum_n \theta(\pm E_n)\delta(E-E_n)\,\varrho_n(\mathbf{x}, \mathbf{y})+\varrho_E^{(\pm)}(\mathbf{x}, \mathbf{y}), \tag{56a}$$

† Cf. the footnote on p. 31.

$$K_F[\mathbf{x}, \mathbf{y}; E|\mathscr{A}] = \sum_n \frac{1}{E_n - E}\varrho_n(\mathbf{x}, \mathbf{y})$$

$$+\int_m^\infty \frac{dE'}{E'-E-i\varepsilon}\varrho_{E'}^{(+)}(\mathbf{x}, \mathbf{y}) + \int_{-\infty}^{-m} \frac{dE'}{E'-E+i\varepsilon}\varrho_{E'}^{(-)}(\mathbf{x}, \mathbf{y}), \tag{56b}$$

where

$$\varrho_n(\mathbf{x}, \mathbf{y}) \equiv \varphi_n(\mathbf{x})\overline{\varphi_n}(\mathbf{y}),$$

$$\varrho_E^{(\pm)}(\mathbf{x}, \mathbf{y}) \equiv \theta(\pm E - m)\frac{\sqrt{E^2-m^2}}{2(2\pi)^3}\sum_r \int d\Omega_n$$

$$\times \varphi^{(\pm)}(\mathbf{x}, \sqrt{E^2-m^2}\,\mathbf{n}, r)\,\varphi^{(\pm)}(\mathbf{y}, \sqrt{E^2-m^2}\,\mathbf{n}, r).$$

The Fourier transform of K_F as a function of the variable E can be extended analytically to the entire complex plane. To this end, in formula (56b) we replace the real value of E by the complex number z ($\operatorname{Im} z \neq 0$),

$$K_F[\mathbf{x}, \mathbf{y}; z|\mathscr{A}] \equiv \sum_n \frac{1}{E_n - z}\varrho_n(\mathbf{x}, \mathbf{y})$$

$$+\int_m^\infty \frac{dE}{E-z}\varrho_E^{(+)}(\mathbf{x}, \mathbf{y}) + \int_{-\infty}^{-m} \frac{dE}{E-z}\varrho_E^{(-)}(\mathbf{x}, \mathbf{y}). \tag{57}$$

The propagator K_F as a function of the complex variable has branch points on the real axis at $z = \pm m$. Hence, this is a multiple-valued function (having more than one Riemann sheet). Formula (57) defines the function K_F on the so-called *physical sheet*. On this sheet the function K_F has poles at the points $z = E_n$ and a discontinuity along the real axis from $-\infty$ to $-m$ and from m to ∞. Off the real axis (to be more accurate, beyond the spectrum of the operator H_D) the function K_F is analytical on the physical sheet. The Fourier transform of the propagator K_F is obtained from the function $K_F[\mathbf{x}, \mathbf{y}; z|\mathscr{A}]$ by going onto the real axis from above for positive values of E and from below for negative values of E.

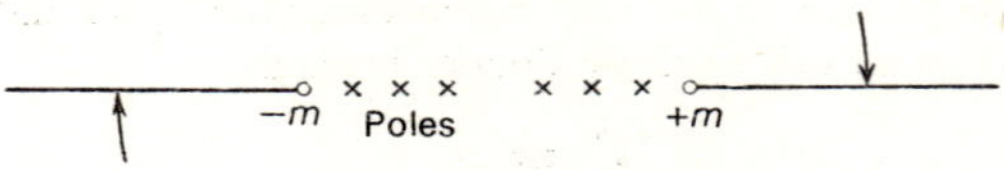

FIG. 17.7. Singularities of the function $K_F[\mathbf{x}, \mathbf{y}; z|\mathscr{A}]$.

The distribution of the singularities and the prescription for obtaining the transform of the Feynman propagator is illustrated in Fig. 17.7.

The function $K_F[\mathbf{x}, \mathbf{y}; z|\mathscr{A}]$ is closely related to the operator of the resolvent of the Hamiltonian H_D. It is simply the coordinate representation of that operator. In the Dirac notation, generalized to the case of bispinors, we can write:

$$(K_F[\mathbf{x}, \mathbf{y}; z|\mathscr{A}])_{ab} = \langle \mathbf{x}, a|(H_D - z)^{-1}|\mathbf{y}, c\rangle \mathrm{A}_{cb},$$

where a and b are bispinor indices labelling the components of the function K_F understood as a matrix in bispinor space.

To prove this formula, it is sufficient to use the spectral resolution of unity associated with the operator H_D and to use the definition (57) of the function $K_F[\mathbf{x}, \mathbf{y}; z|\mathscr{A}]$.

The time-dependent propagator K_F can also be associated with a resolvent. To do this, we shall employ the contour of integration C_F in the plane of the complex vari-

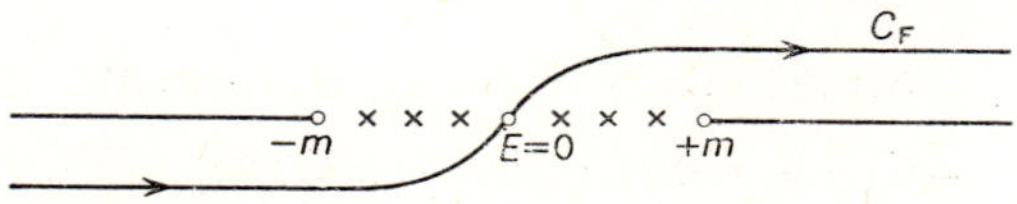

FIG. 17.8. The contour of integration C_F.

able z shown in Fig. 17.8. The propagator $K_F[x, y|\mathscr{A}]$ is the coordinate representation of the operator $K_F[t-t'|\mathscr{A}]$, defined by the formula:

$$K_F[t-t'|\mathscr{A}] \equiv \frac{1}{2\pi}\int_{C_F} dz \frac{e^{-iz(t-t')}}{H_D - z},$$

$$(K_F[x, y|\mathscr{A}])_{ab} = \langle \mathbf{x}, a|K_F[t-t'|\mathscr{A}]|\, \mathbf{y}, c\rangle A_{cb}.$$

The results obtained have their counterpart in non-relativistic mechanics. The retarded propagator $K_R[x, y|U]$ defined there is the coordinate representation of the operator

$$K_R[t-t'|U] = i\theta(t-t')\, e^{-iH(t-t')} = \frac{1}{2\pi}\int_{C_R} dz \frac{e^{-iz(t-t')}}{H-z},$$

where the contour C_R runs from $-\infty$ to ∞ above the real axis.

THE PROPERTIES OF THE S MATRIX FOR AN ELECTRON IN AN EXTERNAL FIELD

Next we shall examine those properties of the amplitudes of scattering and pair creation and annihilation in an external electromagnetic field which will be needed further on in the book. In particular, we shall derive new representations of the amplitudes, discuss the transformation properties of the S matrix elements, and introduce the concept of crossing symmetry.

To simplify the notation and to get more symmetrical formulae, we introduce the following notations for the four components of the S matrix, for the wave functions $\chi^{(\pm)}$, and for the bispinors u and v. We shall denote them by the symbols $S^{\lambda\lambda'}$, $S(\mathbf{p}, r, \lambda; \mathbf{q}, s, \lambda')$, $\chi(x; \mathbf{p}, r, \lambda)$,

$u(\mathbf{p}, r, \lambda)$, etc., where the parameters λ and λ' assume two values, $+1$ and -1. These parameters will be substituted for the indices $+$ and $-$ used previously. The symbol $u(\mathbf{p}, r, +1)$ stands for $u(\mathbf{p}, r)$ and $u(\mathbf{p}, r, -1)$ stands for $v(\mathbf{p}, r)$. To illustrate this new convention, we shall give a formula for the S matrix, in the Born approximation, which is equivalent to the four formulae (3),

$$S(\mathbf{p}, r, \lambda, \mathbf{q}, s, \lambda') = \delta_{\lambda\lambda'}\delta_{rs}\delta_{\Gamma}(\mathbf{p}, \mathbf{q})$$
$$-i\lambda e\bar{u}(\mathbf{p}, r, \lambda)\gamma\cdot\mathscr{A}(\lambda p - \lambda' q)\, u(\mathbf{q}, s, \lambda').$$

Now we shall rewrite the transformation amplitudes in a new form containing the Feynman propagator K_{F}. Formulae (11) for the amplitudes will serve as the starting point. Let us take the indistinguishability of identical particles into account, by replacing the functions $\psi_{\mathrm{F}}^{\pm}$ by the functions $C[\mathscr{A}]\psi_{\mathrm{F}}^{\pm}$:

$$S^{++}[f, g] = C[\mathscr{A}](\chi^{(+)}[f]\|\psi_{\mathrm{F}}^{+}[g])_{\sigma_{\mathrm{F}}}, \tag{58a}$$

$$S^{-+}[f, g] = C[\mathscr{A}](\chi^{(-)}[f]\|\psi_{\mathrm{F}}^{+}[g])_{\sigma_{\mathrm{P}}}, \tag{58b}$$

$$S^{--}[f, g] = C[\mathscr{A}](\chi^{(-)}[f]\|\psi_{\mathrm{F}}^{-}[g])_{\sigma_{\mathrm{P}}}, \tag{58c}$$

$$S^{+-}[f, g] = C[\mathscr{A}](\chi^{(+)}[f]\|\psi_{\mathrm{F}}^{-}[g])_{\sigma_{\mathrm{F}}}. \tag{58d}$$

These formulae can be written in a uniform manner with the aid of the anti-Feynman functions,

$$S^{\lambda\lambda'}[f, g] = C[\mathscr{A}]\,(\psi_{\bar{\mathrm{a}}}^{\lambda}[f]\|\,\psi_{\mathrm{F}}^{\lambda'}[g]). \tag{59}$$

The hypersurface labels could be dropped from this formula because the scalar product of two solutions of the Dirac equation in a field does not depend on σ. On substituting the Feynman functions in the form of integrals over hypersurfaces (36), we obtain

$$S^{\pm+}[f, g] = -iC[\mathscr{A}]\int_{\sigma_x} d\sigma_\mu \int_{\sigma_y} d\sigma_\nu \overline{\psi_{\bar{\mathrm{a}}}^{\pm}}[x|f]\gamma^\mu K_{\mathrm{F}}[x, y|\mathscr{A}]\,\gamma^\nu\psi_{\mathrm{F}}^{+}[y|g], \tag{60a}$$

$$S^{\pm-}[f, g] = iC[\mathscr{A}]\int_{\sigma_x} d\sigma_\mu \int_{\sigma_y} d\sigma_\nu \overline{\psi_{\bar{\mathrm{a}}}^{\pm}}[x|f]\,\gamma^\mu K_{\mathrm{F}}[x, y|\mathscr{A}]\gamma^\nu\psi_{\mathrm{F}}^{-}[y|g]. \tag{60b}$$

In the first formula σ_x lies in the future in relation to σ_y, whereas in the second, σ_y lies in the future in relation to σ_x. If we choose these hypersurfaces suitably and utilize the asymptotic conditions for the wave functions and propagators, we can write the amplitudes in terms of only the propagators and the solutions of the free Dirac equations:

$$S^{++}[f, g] = -iC[\mathscr{A}] \int_{\sigma_F} d\sigma_\mu \int_{\sigma_P} d\sigma_\nu$$
$$\times \overline{\chi^{(+)}}[x|f]\gamma^\mu K_F[x, y|\mathscr{A}]\gamma^\nu \chi^{(+)}[y|g], \tag{61a}$$

$$S^{-+}[f, g] = -iC[\mathscr{A}] \int_{\sigma_P} d\sigma_\mu \int_{\sigma_P} d\sigma_\nu$$
$$\times \overline{\chi^{(-)}}[x|f]\gamma^\mu K_F[x, y|\mathscr{A}]\gamma^\nu \chi^{(+)}[y|g], \tag{61b}$$

$$S^{--}[f, g] = iC[\mathscr{A}] \int_{\sigma_P} d\sigma_\mu \int_{\sigma_F} d\sigma_\nu$$
$$\times \overline{\chi^{(-)}}[x|f]\gamma^\mu K_F[x, y|\mathscr{A}]\gamma^\nu \chi^{(-)}[y|g], \tag{61c}$$

$$S^{+-}[f, g] = iC[\mathscr{A}] \int_{\sigma_F} d\sigma_\mu \int_{\sigma_F} d\sigma_\nu$$
$$\times \overline{\chi^{(+)}}[x|f]\gamma^\mu K_F[x, y|\mathscr{A}]\gamma^\nu \chi^{(-)}[y|g]. \tag{61d}$$

These formulae very suggestively explain the relation between the propagator K_F and the scattering amplitudes. On the basis of these formulae, it might be said that for t's and t''s lying beyond the region of interaction, the expression $\pm C[\mathscr{A}]K_F[\mathbf{x}t, \mathbf{y}t'|\mathscr{A}]$ is an S matrix in the coordinate representation (Cf. also Appendix M).

A shortcoming of the formulae above is that they contain the hypersurfaces σ_F and σ_P, which constitute an external element and violate the explicit relativistic invariance. The dependence on these hypersurfaces is, of course, only apparent; they can be completely eliminated from the formulae by application of Gauss' theorem. For this purpose, let us go back to formulae (58) and write all the scalar products in them in terms of integrals over space-time. Gauss' theorem implies the relations

$$(\chi^{(\lambda)}[f]\|\psi^{(\lambda')}[g])_{\sigma_F} - (\chi^{(\lambda)}[f]\|\psi^{(\lambda')}[g])_{\sigma_P} = i\int_{\sigma_P}^{\sigma_F} d^4x\, \overline{\chi^{(\lambda)}}[x|f]D_x\psi^{(\lambda')}[x|g].$$

By virtue of formula (38), the integral on the right-hand side may be replaced by a double integral over space-time,

$$i\int_{\sigma_P}^{\sigma_F} d^4x \int_{\sigma_P}^{\sigma_F} d^4y\,\overline{\chi^{(\lambda)}}[x|f]\,D_x K_F[x, y|\mathscr{A}]\overleftarrow{D}_y\,\chi^{(\lambda')}[y|g].$$

Removing the hypersurfaces σ_F and σ_P to infinity (cf. the remarks after formula (38)) and utilizing the asymptotic conditions satisfied by the functions $\psi^{\pm}[g]$, we finally arrive at:

$$S^{\lambda\lambda'}[f, g] = C[\mathscr{A}]S_Q^{\lambda\lambda'}[f, g] = C[\mathscr{A}]\,\delta_{\lambda\lambda'}(f|g)$$

$$+i\lambda C[\mathscr{A}]\int d^4x d^4y\,\overline{\chi^{(\lambda)}}[x|f]\,D_x K_F[x, y|\mathscr{A}]\overleftarrow{D}_y\,\chi^{(\lambda)}[y|g]. \qquad (62)$$

When both sides of this equality are stripped of the wave-packet profiles f and g, what is left is the following representation of the four components of the S matrix,

$$S(\mathbf{p}, r, \lambda; \mathbf{q}, s, \lambda') = C[\mathscr{A}]\,\delta_{\Gamma(m)}(\mathbf{p}, \mathbf{q})\,\delta_{\lambda\lambda'}\,\delta_{rs}$$

$$+i\lambda C[\mathscr{A}]\int d^4x d^4y\,\overline{\chi}(x; \mathbf{p}, r, \lambda)\,D_x K_F[x, y|\mathscr{A}]\overleftarrow{D}_y\,\chi(y; \mathbf{q}, s, \lambda'). \quad (63)$$

The asymptotic behaviour of the functions $\psi_F^{\pm}$ can be expressed in terms of the components of the S matrix (as already mentioned earlier). The following relations are satisfied:

$$\psi_F^{+}[x|g]_{t>T_2} = \sum_{r,s}\int d\Gamma_p(m)\,d\Gamma_q(m)\,\chi^{(+)}(x; \mathbf{p}, r)$$

$$\times S_Q^{++}(\mathbf{p}, r; \mathbf{q}, s)\,g(\mathbf{q}, s),$$

$$\psi_F^{+}[x|g]_{t<T_1} = \sum_{r,s} d\Gamma_p(m)\,d\Gamma_q(m)[\chi^{(+)}(x; \mathbf{p}, r)\,\delta_{\Gamma(m)}(\mathbf{p}, \mathbf{q})\,\delta_{rs}$$

$$+\chi^{(-)}(x; \mathbf{p}, r)\,S_Q^{-+}(\mathbf{p}, r; \mathbf{q}, s)]\,g(\mathbf{q}, s),$$

$$\psi_F^{-}[x|f]_{t<T_1} = \sum_{r,s}\int d\Gamma_p(m)\,d\Gamma_q(m)\,\chi^{(-)}(x; \mathbf{p}, r)$$

$$\times S_Q^{--}(\mathbf{p}, r; \mathbf{q}, s)\,f^*(\mathbf{q}, s),$$

$$\psi_F^{-}[x|f]_{t>T_2} = \sum_{r,s}\int d\Gamma_p(m)\,d\Gamma_q(m)[\chi^{(+)}(x; \mathbf{p}, r)\,S_Q^{+-}(\mathbf{p}, r; \mathbf{q}, s)$$

$$+\chi^{(-)}(x; \mathbf{p}, r)\,\delta_{\Gamma(m)}(\mathbf{p}, \mathbf{q})\,\delta_{rs}]\,f^*(\mathbf{q}, s).$$

The functions $\psi^+[x|g]$ and $\psi^-[x|f]$, as functions defined over all space-time, contain more information than do the four amplitudes $S^{\lambda\lambda'}[f, g]$. It seems, however, that this additional information can not be used to compare the results of theory with experiment. The point is that in experiments we have situations in which states prepared by a source and those recorded by a counter are states of free particles, i.e. are described by the asymptotic form of wave functions. A wave function defined in all space-time interpolates between asymptotic values in the past and in the future.

In our considerations thus far, and particularly in the derivation of formula (62), we made the assumption that the electromagnetic field vanishes in the remote past and the far future. One can weaken this assumption and consider the electromagnetic field $\mathscr{A}_\mu(x)$ which in the past and in the future goes over into a static field described by the potential $\mathscr{A}_\mu^{\text{st}}(\mathbf{x})$. An arbitrary solution of the Dirac equation $\psi(x)$ in the remote past and far future can then be written as a superposition of solutions of the Dirac equation in the static field. Among such solutions in general there will be some describing bound states in addition to those describing scattering states. Under the influence of a time-varying field, in addition to scattering and pair creation and annihilation processes, transitions will also occur between bound states and between bound and scattering states. The probability amplitudes of these processes may still be expressed in terms of the Feynman propagator K_{F} which satisfies the inhomogeneous Dirac equation in the electromagnetic field and appropriately modified asymptotic conditions. This propagator, treated as a function in each of two variables x and y, goes over asymptotically into a superposition of solutions of the Dirac equation in the static field with, respectively, positive and negative frequencies. The formulae for the transition amplitudes resemble formula (62), except that Dirac solutions in the static field will appear instead of the function $\chi^{(\lambda)}$.

The transformation properties of the Green's function $K_{\text{I}}[x, y|\mathscr{A}]$ discussed in this section imply that the full Poincaré group acts in the set of amplitudes $S^{\lambda\lambda'}[f, g]$ and that it is a symmetry group. We do not yet know, of course, the properties of the amplitude $C[\mathscr{A}]$ which multiplies all the amplitudes $S_{\text{Q}}^{\lambda\lambda'}[f, g]$. At this point, however, we shall make use of information which will be proved at the end of this section, viz. that the amplitude $C[\mathscr{A}]$ is invariant under Poincaré transformations of the potential, and also unde rgauge transformations, and charge conjugation,

$$C[\mathscr{A}] = C[{}^{\mathrm{Tr}}\mathscr{A}] = C[{}^{\mathrm{L}}\mathscr{A}] = C[{}^{\mathrm{P}}\mathscr{A}] = C[{}^{\mathrm{T}}\mathscr{A}] = C[{}^{\mathrm{G}}\mathscr{A}] = C[-\mathscr{A}].$$

The functional dependence of the scattering amplitudes on the external potential will be denoted by use of the symbol $S^{\lambda\lambda'}[f, g; \mathscr{A}]$. Properties (39) and (40) of the Green's function and the transformation properties of the solutions $\chi^{(\lambda)}[x|f]$ of the Dirac equation lead to the following transformation properties of the four amplitudes $S^{\lambda\lambda'}[f, g; \mathscr{A}]$ under translation, homogeneous Lorentz transformation, spatial inversion, and time reversal.

$$S^{\lambda\lambda'}[f, g; \mathscr{A}] = S^{\lambda\lambda'}[{}^{\mathrm{Tr}}f, {}^{\mathrm{Tr}}g; {}^{\mathrm{Tr}}\mathscr{A}], \tag{64a}$$

$$S^{\lambda\lambda'}[f, g; \mathscr{A}] = S^{\lambda\lambda'}[{}^{\mathrm{L}}f, {}^{\mathrm{L}}g; {}^{\mathrm{L}}\mathscr{A}], \tag{64b}$$

$$S^{\lambda\lambda'}[f, g; \mathscr{A}] = S^{\lambda\lambda'}[{}^{\mathrm{P}}f, {}^{\mathrm{P}}g; {}^{\mathrm{P}}\mathscr{A}], \tag{64c}$$

$$S^{\lambda\lambda'}[f, g; \mathscr{A}] = S^{\lambda\lambda'}[{}^{\mathrm{T}}g, {}^{\mathrm{T}}f; {}^{\mathrm{T}}\mathscr{A}]. \tag{64d}$$

These equations mean that the full Poincaré group is a symmetry group in the set of amplitudes. The transformation properties of the components of the matrices $S^{\lambda\lambda'}[\mathbf{p}, r; \mathbf{q}, s|\mathscr{A}]$ follow from those of the amplitudes $S^{\lambda\lambda'}[f, g; \mathscr{A}]$. As an example we shall now give the transformation properties of the component $S^{++}[\mathbf{p}, r; \mathbf{q}, s|\mathscr{A}]$:

$$S^{++}[\mathbf{p}, r; \mathbf{q}, s|\mathscr{A}] = e^{-i(p-q)\cdot a} S^{++}[\mathbf{p}, r; \mathbf{q}, s|{}^{\mathrm{Tr}}\mathscr{A}], \tag{65a}$$

$$S^{++}[\mathbf{p}, r; \mathbf{q}, s|\mathscr{A}] = \sum_{r's'} S^{++}[\Lambda\mathbf{p}, r'; \Lambda\mathbf{q}, s'|{}^{\mathrm{L}}\mathscr{A}] \times \left(\Lambda^{(+)}(\Lambda\mathbf{p}, r', r)\right)^* \Lambda^{(+)}(\Lambda\mathbf{q}, s', s), \tag{65b}$$

$$S^{++}[\mathbf{p}, r; \mathbf{q}, s|\mathscr{A}] = S^{++}[-\mathbf{p}, r; -\mathbf{q}, s|{}^{\mathrm{P}}\mathscr{A}], \tag{65c}$$

$$S^{++}[\mathbf{p}, r; \mathbf{q}, s|\mathscr{A}] = S^{++}[-\mathbf{q}, -s; -\mathbf{p}, -r|{}^{\mathrm{T}}\mathscr{A}]. \tag{65d}$$

The amplitudes $S^{\lambda\lambda'}[f, g; \mathscr{A}]$ are also invariant under gauge transformations of the external field. Under gauge transformations the propagator K_{F} changes according to formula (39d). We assume that the gauge function $\Lambda(x)$, just as the potential itself, vanishes in the remote past and in the far future. It then follows from the representation (61) of the transition amplitudes that they do not suffer change under gauge transformations of the potential,

$$S^{\lambda\lambda'}[f, g; \mathscr{A}] = S^{\lambda\lambda'}[f, g; {}^{G}\mathscr{A}]. \tag{66}$$

The following relation between the scattering amplitudes of particles and antiparticles emerges from the properties (45) of the propagator and the properties (16.45) of the bispinors:

$$S^{++}[\mathbf{p}, r; \mathbf{q}, s|\mathscr{A}] = S^{--}[\mathbf{q}, s; \mathbf{p}, r|-\mathscr{A}]. \tag{67}$$

This relation gives precise meaning to the notion of symmetry between particles and antiparticles. It means that the scattering amplitude of an antiparticle in the electromagnetic field is equal to the scattering amplitude of a particle with the same asymptotic momenta and polarizations in the electromagnetic field of opposite sign.

In turn, the properties (39c) and (44) of the propagator K_F lead to a relation between the amplitudes of pair creation and annihilation,

$$S^{+-}[\mathbf{p}, r; \mathbf{q}, s|\mathscr{A}] = S^{-+}[\mathbf{q}, -s; \mathbf{p}, -r|^{TP}\mathscr{A}]. \tag{68}$$

The meaning of this relation is that the creation amplitude of a pair is equal to the annihilation amplitude of a pair with opposite polarizations by the electromagnetic field reflected in space-time.

In conclusion, we shall discuss the relations between the S matrix elements which follow from the so-called *crossing symmetry*. The starting point for these considerations will be formula (63) for the S matrix, but rewritten with the use of Fourier transforms:

$$S(\mathbf{p}, r, \lambda; \mathbf{q}, s, \lambda') = C[\mathscr{A}]\,\delta_{\Gamma(m)}(\mathbf{p}, \mathbf{q})\,\delta_{\lambda\lambda'}\,\delta_{rs} + i\lambda\bar{u}(\mathbf{p}, r, \lambda)\,\mathscr{M}(\lambda p, \lambda' q)u(\mathbf{q}, s, \lambda'), \tag{69}$$

where

$$\mathscr{M}(p, q) \equiv (m-\gamma\cdot p)C[\mathscr{A}]\tilde{K}_F[p, q|\mathscr{A}](m-\gamma\cdot q), \tag{70}$$

whereas $\tilde{K}_F[p, q|\mathscr{A}]$ is the double Fourier transform of the propagator,

$$\tilde{K}_F[p, q|\mathscr{A}] \equiv \int d^4x\,d^4y\,e^{ip\cdot x}K_F[x, y|\mathscr{A}]\,e^{-iq\cdot y}.$$

At first glance, it might seem that multiplication of the matrix $\mathscr{M}(p, q)$ on both sides by the bispinors $\bar{u}$ and u yields zero since these bispinors satisfy the equations

$$(m-\lambda\gamma\cdot p)u(\mathbf{p}, r, \lambda) = 0 = \bar{u}(\mathbf{p}, r, \lambda)(m-\lambda\gamma\cdot p).$$

This is not the case, however, because the Fourier transform of the propagator has poles at the points $p^2 = m^2$ and $q^2 = m^2$. Multiplication by $m-\gamma\cdot p$ and by $m-\gamma\cdot q$ eliminates these poles and in consequence the function $\mathscr{M}(p, q)$ is regular at these points. Since in formula (70) we are dealing with an expression of the type 0/0, the transition to the so-called *mass shell* $p^2 = m^2 = q^2$ should be performed in the product defining the function $\mathscr{M}$, and not in each factor separately.

We shall demonstrate now that the Fourier transform of the propagator has poles on the mass shell. The integral equation (19) determining the propagator K_F leads to the following integral equation for the Fourier transform of this propagator:

$$\tilde{K}_F[p, q|\mathscr{A}] = \frac{1}{m-\gamma\cdot p}(2\pi)^4\delta^{(4)}(p-q)$$
$$-\frac{1}{m-\gamma\cdot p}\int\frac{d^4p'}{(2\pi)^4}e\gamma\cdot\tilde{\mathscr{A}}(p-p')\tilde{K}_F[p', q|\mathscr{A}].$$

The matrix $(m-\gamma\cdot p)^{-1}$ can be written for $(m+\gamma\cdot p)(m^2-p^2)^{-1}$, whence K_F has poles on the mass shell.

Using also the second integral eq. (28b) for the propagator K_F, we obtain the following symmetrical integral representation for $\mathscr{M}$:

$$\mathscr{M}(p, q) = C[\mathscr{A}]\Big\{(m-\gamma\cdot p)(2\pi)^4\delta^{(4)}(p-q) - e\gamma\cdot\tilde{\mathscr{A}}(p-q)$$
$$+\int\frac{d^4p'}{(2\pi)^4}\frac{d^4q'}{(2\pi)^4}e\gamma\cdot\tilde{\mathscr{A}}(p-p')\tilde{K}_F[p', q'|\mathscr{A}]e\gamma\cdot\tilde{\mathscr{A}}(q'-q)\Big\}. \quad (71)$$

The first term on the right-hand side makes no contribution to the scattering amplitudes, the second gives the Born approximation, whereas the third contains all the higher-order corrections.

The matrix $\mathscr{M}(p, q)$, which is dependent on two four-vectors p and q, is a universal quantity describing all four processes which correspond to various values of the indices λ and λ'. Depending on the process chosen, the values of the arguments p and q must be chosen appropriately. For example, in the description of negaton scattering these arguments are equal to the initial and final momenta of the negaton, whereas in the description of pair annihilation the vector p is equal to the positon momentum taken with a minus sign, and the vector q

is equal to the negaton momentum. Thus, in the general case the aforementioned universality of description is illusory since the domains of the arguments p and q for all four processes are disjoint. The domains of variation of the time components of the vectors p and q are given in Fig. 17.9. It may happen, however, that the four branches of the func-

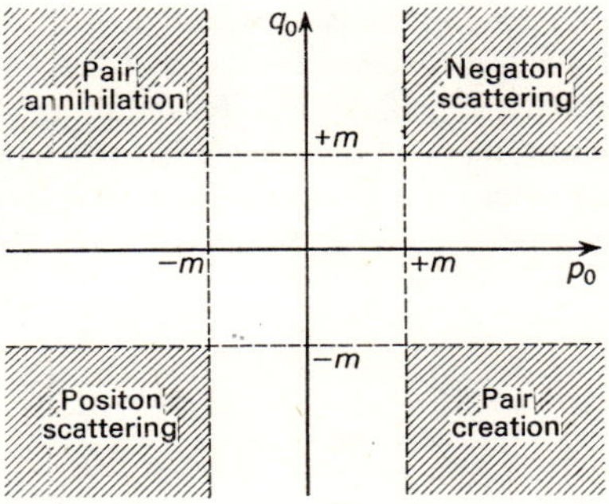

FIG. 17.9. The domains of the components p_0 and q_0 for the four processes described by the amplitude $\mathscr{M}(p, q)$.

tion $\mathscr{M}(p, q)$ in various domains of the arguments constitute a single analytical function. The universality of the function $\mathscr{M}(p, q)$ then takes on a non-trivial meaning and one may speak of crossing symmetry. Such a situation occurs first and foremost in the complete quantum electrodynamics in the absence of an external electromagnetic field and to a more modest extent in the case of scattering of a particle in a static field. The latter example will now be discussed more precisely.

In the case of a static field, neither creation nor annihilation of a pair is possible since the static field cannot supply or receive energy. The amplitude $C[\mathscr{A}]$ thus is equal to unity. Formula (71) for the matrix $\mathscr{M}$ can now be rewritten as:

$$\mathscr{M}(p, q) = (m-\gamma\cdot p)(2\pi)^4\, \delta^{(4)}(p-q) - 2\pi\delta(p_0-q_0)\mathscr{T}(\mathbf{p}, \mathbf{q}; p_0),$$

where

$$\mathscr{T}(\mathbf{p}, \mathbf{q}; E) = e\gamma\cdot\tilde{\mathscr{A}}(\mathbf{p}-\mathbf{q})$$
$$-\int \frac{d^3p'}{(2\pi)^3}\frac{d^3q'}{(2\pi)^3}\, e\gamma\cdot\tilde{\mathscr{A}}(\mathbf{p}-\mathbf{p}')\tilde{K}_F[\mathbf{p}', \mathbf{q}'; E|\mathscr{A}]e\gamma\cdot\tilde{\mathscr{A}}(\mathbf{q}'-\mathbf{q}).$$

The function $\tilde{K}_F[\mathbf{p}, \mathbf{q}; E|\mathscr{A}]$ is the Fourier transform of the function $K_F[\mathbf{x}, \mathbf{y}; E|\mathscr{A}]$ and as a function of the variable E accordingly has the same analytical properties as $K_F[\mathbf{x}, \mathbf{y}; E|\mathscr{A}]$. Depending on whether $E > m$ or $E < -m$, the values of the functions $\tilde{K}_F[\mathbf{p}, \mathbf{q}; E|\mathscr{A}]$ can be obtained from the momentum representation of the resolvent operator $\tilde{K}_F[\mathbf{p}, \mathbf{q}; z|\mathscr{A}]$,

$$(\tilde{K}_F[\mathbf{p}, \mathbf{q}; z|\mathscr{A}])_{ab} = \langle \mathbf{p}, a|(H_D - z)^{-1}|\mathbf{q}, c\rangle \, A_{cb},$$

as the boundary value of this analytical function on the right or left cut. The passage to real values of E is governed by the prescription shown in Fig. 17.7.

The two processes occurring in the case of a static electromagnetic field thus are described by a single analytical function $\mathscr{T}(\mathbf{p}, \mathbf{q}; z)$ whose boundary values on the cut give the negaton scattering amplitude ($E > m$) and the positon scattering amplitude ($E < -m$). The statement that we use the same function to describe both processes, merely inserting different values of the arguments, thus has a non-trivial meaning in this case. The two processes are said to be linked by *crossing symmetry*. The name of this symmetry is associated with the diagram method. On the diagrams, antiparticle scattering may be obtained from particle scattering by means of the operation of line crossing as shown in Fig. 17.10.

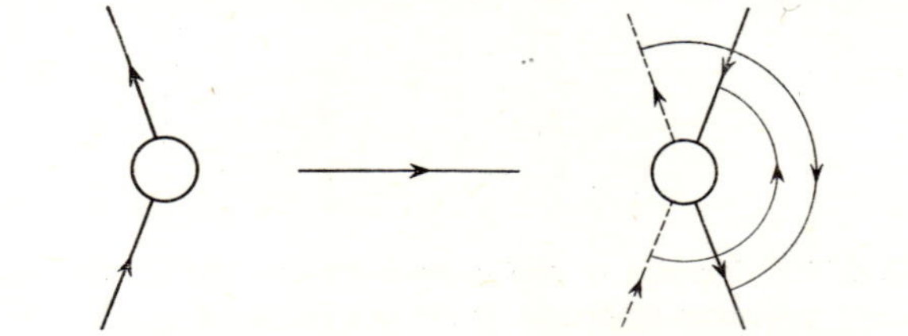

FIG. 17.10. The operation of line crossing takes a particle-scattering diagram into an antiparticle-scattering diagram.

The Indistinguishability of Particles and the Vacuum-to-Vacuum Transition Amplitude

Now we shall give the previously mentioned proof that taking account of the indistinguishability of particles in all orders of perturbation theory results, in consequence, in the Feynman wave functions being multiplied by the x-independent factor $C[\mathscr{A}]$. To carry out the proof we shall employ the Fredholm method of solving integral equations. This method will be applied to the integral equation (19) for the propagator K_F, in terms of which we then express the Feynman wave function ψ_F.

In our discussion thus far, we solved integral equations by using the method of successive approximations, which yields the perturbation series. By the Fredholm method we obtain a solution in the form of the quotient of two perturbation series. Thus, we get

$$K_F[x, y|\mathscr{A}] = \frac{C[x, y|\mathscr{A}]}{C[\mathscr{A}]} \tag{72}$$

and we shall show that the propagator $C[x, y|\mathscr{A}]$ figuring in the numerator contains contributions only from suitably antisymmetrized products of wave functions. Taking antisymmetrization into account therefore is tantamount to multiplying K_F by $C[\mathscr{A}]$. In order to determine $C[x, y|\mathscr{A}]$ and $C[\mathscr{A}]$, we substitute K_F in the form of quotient (72) into the integral equation (19). The result is the following integral equation for $C[x, y|\mathscr{A}]$:

$$\begin{aligned} C[x, y|\mathscr{A}] &= C[\mathscr{A}]S_F(x-y) \\ &-e\int d^4z S_F(x-z)\gamma\cdot\mathscr{A}(z)C[z, y|\mathscr{A}]. \end{aligned} \tag{73}$$

The solution of this equation will be sought in the form of perturbation series (power series in e):

$$C[\mathscr{A}] = 1+C^{(1)}+C^{(2)}+\ldots$$

$$C[x, y|\mathscr{A}] = S_F(x-y)+C^{(1)}[x, y|\mathscr{A}]+C^{(2)}[x, y|\mathscr{A}]+\ldots$$

The Fredholm method gives the following integral expressions for the successive terms of the expansion,†

$$\begin{aligned} C^{(n)}[x, y|\mathscr{A}] = \frac{e^n}{n!}\int &dz_1 \ldots dz_n S_F(xz_1 \ldots z_n, yz_1 \ldots z_n) \\ &\times \gamma(z_1)\cdot\mathscr{A}(z_1) \ldots \gamma(z_n)\cdot\mathscr{A}(z_n), \end{aligned} \tag{74a}$$

$$\begin{aligned} C^{(n)}[\mathscr{A}] = \frac{e^n}{n!}\int &dz_1 \ldots dz_n S_F(z_1 \ldots z_n, z_1 \ldots z_n) \\ &\times \gamma(z_1)\cdot\mathscr{A}(z_1) \ldots \gamma(z_n)\cdot\mathscr{A}(z_n), \end{aligned} \tag{74b}$$

† Knowledge of the Fredholm method is not necessary for understanding the deductions which follow. It will be subsequently shown by direct calculation that the expressions given below satisfy the integral equation (73). For a fuller understanding of the results arrived at, however, the reader is advised to become acquainted with the elements of Fredholm theory from any good textbook on integral equations.

In using the Fredholm method, we shall at first treat the propagators $S_F(x)$ as if they were ordinary functions. Further on in this section we shall deal with the difficulties caused by the fact that the S_F's are distributions.

where

$$S_F(x_1 \dots x_n, y_1 \dots y_n) = S_F^{(n)} \equiv \mathrm{Det}\begin{vmatrix} S_F(x_1-y_1)\, S_F(x_1-y_2) \dots S_F(x_1-y_n) \\ S_F(x_2-y_1)\, S_F(x_2-y_2) \dots S_F(x_2-y_n) \\ \vdots \qquad\qquad \vdots \qquad\qquad \vdots \\ S_F(x_n-y_1)\, S_F(x_n-y_2) \dots S_F(x_n-y_n) \end{vmatrix}$$

$$= \sum_{\mathrm{perm}\, i} \varepsilon_p S_F(x_1-y_{i_1}) \dots S_F(x_n-y_{i_n})$$

$$= \sum_{\mathrm{perm}\, i} \varepsilon_p S_F(x_{i_1}-y_1) \dots S_F(x_{i_n}-y_n). \tag{75}$$

In formulae (74) we have applied several conventions which simplify the notation. First of all, we have dropped the superscript 4 in the volume elements d^4z. We also use a new convention concerning summation over the bispinor indices. The function $S_F^{(n)}$ defined by determinant (75) is a $4n \times 4n$ matrix since multiplication of propagators in formula (75) denotes external multiplication of matrices in bispinor space. The matrices γ have been endowed with the arguments z_i so as to show the order in which multiplication should be performed in bispinor space. Multiplication of the matrix $S_F^{(n)}$ by $\gamma(z_i)$ is taken to mean that the propagator S_F in which z_i appears as an argument is multiplied by γ. The argument z_i may occur twice in the matrix $S_F^{(n)}$, once in the first group of arguments and again in the second. In that event, both multiplications should be carried out.

For example:

$$S_F(x, y)\gamma^\mu(y) = S_F(x-y)\gamma^\mu,$$

$$S_F(x, y)\gamma^\mu(x) = \gamma^\mu S_F(x-y),$$

$$S_F(xz, yz)\gamma^\mu(z) = S_F(x-y)\,\mathrm{Tr}\{\gamma^\mu S_F(z-z)\} - S_F(x-z)\gamma^\mu S_F(z-y)$$

where the expressions on the right-hand side stand for ordinary matrix products.

Formulae (74) contain the propagators $S_F^{(n)}$, whose arguments z_i appear twice. The determinant defining $S_F^{(n)}$ will then have ill-determined expressions: the values of the propagators S_F at zero. It is seen from the structure of expressions (74a) and (74b) that $S_F(0)$ always occurs in the combination $\mathrm{Tr}\{\gamma^\mu S_F(0)\}$. Such meaningless expressions have appeared because, in applying the Fredholm method, we have inadvertently manipulated distributions, treating them as ordinary functione. In this way, however, we have obtained convenient formulae for propagators and the awkward expressions can be eliminated from them by applying, for instance, the procedure called regularization, which is discussed in Appendix K. Suppose that in the integral equation (73) the propagator S_F has been replaced by a sequence S_F^{Reg} of functions regular at the point $x = 0$, this sequence tending to the distribution S_F; further suppose that the sequence is constructed from functions with the same invariance properties as the propagator S_F (cf. formula (46)). Each function in this sequence accordingly satisfies the condition:

$$\operatorname{Tr}\{\gamma^\mu S_F^{Reg}(x)\} = i\partial^\mu \Delta_F^{Reg}(x) = x^\mu f(x^2).$$

From the assumption of regularity at the point $x = 0$ it follows that this expression is equal to zero at this point. On this basis we shall regard the expressions $\operatorname{Tr}\{\gamma^\mu S_F(0)\}$ which figure in formulae (74) as also being equal to zero. We arrive at the same result if we define the undetermined expression $\operatorname{Tr}\{\gamma^\mu S_F(0)\}$ as the following symmetrical limit,

$$\operatorname{Tr}\{\gamma^\mu S_F(0)\} = \frac{1}{2} \lim_{\substack{\eta\to 0\\ \eta^2>0,\ \eta^0>0}} \operatorname{Tr}\{\gamma^\mu(S_F(\eta)+S_F(-\eta))\}.$$

In what follows, whenever a propagator with two identical arguments occurs we shall take such an expression to mean a symmetrical limit, e.g.

$$\operatorname{Tr}\{\gamma^\mu K_F[x, x|\mathscr{A}]\}$$
$$= \frac{1}{2} \lim_{\substack{\eta\to 0\\ \eta^2>0,\ \eta^0>0}} \operatorname{Tr}\{\gamma^\mu(K_F[x+\eta, x|\mathscr{A}]+K_F[x-\eta, x|\mathscr{A}])\}. \tag{76}$$

For readers not familiar with the Fredholm method, we shall now show that the solutions (74) obtained with it satisfy the integral equations (73). Expanding the determinant (75) in terms of the first row, we get:

$$S_F(xz_1 \ldots z_n, yz_1 \ldots z_n) = S_F(x-y)S_F(z_1 \ldots z_n, z_1 \ldots z_n)$$
$$-\sum_{k=1}^{n} S_F(x-z_k)\, S_F(z_k\, z_1 \ldots \check{z}_k \ldots z_n, y\, z_1 \ldots \check{z}_k \ldots z_n).$$

When we multiply this identity by $\frac{e^n}{n!}\gamma(z_1)\cdot\mathscr{A}(z_1)\ldots\gamma(z_n)\cdot\mathscr{A}(z_n)$, then integrate with respect to $z_1,\ldots,z_n$ over all space-time, and utilize formulae (74), we arrive at the recursion equation:

$$C^{(n)}[x,y|\mathscr{A}] = C^{(n)}[\mathscr{A}]S_F(x-y) - e\int dz S_F(x-z)\gamma\cdot\mathscr{A}(z)C^{(n-1)}[z,y|\mathscr{A}]. \tag{77}$$

This equation guarantees that the power series constructed from the terms $C^{(n)}$ is a solution of the original integral equation.

Now let us study the wave function $\psi[x|f,g]$ obtained by multiplying the Feynman function by $C[\mathscr{A}]$,

$$\psi[x|f,g] = C[\mathscr{A}]\psi_F[x|f,g] = C[\mathscr{A}]\chi_F[x|f,g] - \int d^4y C[x,y|\mathscr{A}]e\gamma\cdot\mathscr{A}(y)\chi_F[y|f,g]. \tag{78}$$

The Fredholm solution may be exploited in order to write the functions $\psi^{(n)}$ in the n-th order perturbation theory in the following form:

$$\psi^{(n)}[x|f,g] = \frac{e^n}{n!}\int dz_1\ldots dz_n\gamma(z_1)\cdot\mathscr{A}(z_1)\ldots\gamma(z_n)\cdot\mathscr{A}(z_n)$$
$$\times\{\chi_F[x|f,g]S_F(z_1\ldots z_n, z_1\ldots z_n) - \chi_F[z_1|f,g]S_F(xz_2\ldots z_n, z_1z_2\ldots z_n)$$
$$+\ldots+(-1)^k\chi_F[z_k|f,g]S_F(xz_1\ldots\check{z}_k\ldots z_n, z_1z_2\ldots z_k\ldots z_n)$$
$$+\ldots+(-1)^n\chi_F[z_n|f,g]S_F(xz_1\ldots z_{n-1}, z_1z_2\ldots z_n)\}.$$

Each function $S_F^{(n)}$ can be decomposed into a sum of products of the functions $S^{(+)}$ and $S^{(-)}$. Let us consider the terms in this sum which contain k functions $S^{(+)}$ and $n-k$ functions $S^{(-)}$. On relabelling the variables of integration $z_1,\ldots,z_k$ in the formula for $\psi^{(n)}$, we can rewrite the sum of terms of this type as

$$\sum_{\text{perm}\, i}\varepsilon_P\chi_F[y_{i_0}|f,g]S^{(+)}(y_{i_1}-z_1)\ldots S^{(+)}(y_{i_k}-z_k)$$
$$\times S^{(-)}(y_{i_{k+1}}-z_{k+1})\ldots S^{(-)}(y_{i_n}-z_n), \tag{79}$$

where $y_{i_0}, y_{i_1},\ldots,y_{i_n}$ are permutations of the set of $n+1$ space-time variables $x, z_1,\ldots,z_n$. Both the propagators $S^{(\pm)}$ and the function χ_F can be decomposed into complete systems of functions with positive and negative frequencies, $\chi_m^{(+)}$ and $\chi_m^{(-)}$. To fix our attention, we shall assume that the function χ_F decomposes into a series of functions $\chi_m^{(+)}$,

$$\chi_F[x|f,g] = \sum_m C_m\chi_m^{(+)}(x),$$

$$S^{(+)}(x-y) = i\sum_m \chi_m^{(+)}(x)\overline{\chi_m^{(+)}}(y).$$

Substitution of this series into formula (79) yields:

$$\sum_m \sum_{m_1} \dots \sum_{m_n} C_m \sum_{\text{perm}\, i} \varepsilon_{\mathrm{P}} \chi^{(+)}_m(y_{i_0})\, \chi^{(+)}_{m_1}(y_{i_1}) \dots \chi^{(+)}_{m_k}(y_{i_k}) \dots,$$

where the part indicated by dots is not essential for our considerations. In the foregoing formula the product of the negaton wave functions $\chi^{(+)}_{m_i}$ is fully antisymmetrized. In like manner it can be shown that all products of the positon wave functions $\chi^{(-)}_m$ also undergo antisymmetrization.

Thus, in its decomposition into a perturbation series the function $C[\mathscr{A}]\psi_{\mathrm{F}}$ contains contributions only from the properly antisymmetrized n-electron functions.

A by-product of the foregoing discussion on the antisymmetrization of wave functions is a formula for the amplitude $C[\mathscr{A}]$ in the form of a perturbation expansion. Let us now examine the properties of this expression. The zeroth-order term is equal to one and the first-order term is equal to zero since

$$C^{(1)}[\mathscr{A}] = e\int \mathrm{d}z\, S_{\mathrm{F}}(z, z)\gamma(z)\cdot\mathscr{A}(z) = e\int \mathrm{d}z\, \mathrm{Tr}\{S_{\mathrm{F}}(z-z)\gamma^\mu\}\mathscr{A}_\mu(z) = 0.$$

It will now be proved that just as the first-order contribution to the amplitude $C[\mathscr{A}]$, all other contributions in the odd orders of perturbation theory are zero. To prove this we shall use an important new representation of the amplitude $C[\mathscr{A}]$. This representation is obtained by means of the following identity,

$$\frac{\partial}{\partial e} C^{(n+1)}[\mathscr{A}] = \int \mathrm{d}z\, C^{(n)}[z, z|\mathscr{A}]\, \gamma(z)\cdot\mathscr{A}(z),$$

which follows from formulae (74). On summing these identities with respect to n from 0 to ∞, we get a differential equation for $C[\mathscr{A}]$,

$$\frac{\partial}{\partial e} C[\mathscr{A}] = C[\mathscr{A}]\int \mathrm{d}z\, \mathrm{Tr}\{K_{\mathrm{F}}[z, z|\mathscr{A}]\gamma^\mu\}\mathscr{A}_\mu(z).$$

The solution of this equation which satisfies the condition $C[\mathscr{A}]|_{e=0} = 1$ is given by the formula

$$\ln C[\mathscr{A}] = \int_0^e \mathrm{d}e \int \mathrm{d}z\, \mathrm{Tr}\{K_{\mathrm{F}}[z, z|\mathscr{A}]\gamma^\mu\}\mathscr{A}_\mu(z). \tag{80}$$

Substitution of the perturbation expansion of the propagator K_{F} into this formula and integration with respect to e yield†

$$\ln C[\mathscr{A}] = -\sum_{n=2}^{\infty} \frac{(-e)^n}{n} \int \mathrm{d}z_1 \dots \mathrm{d}z_n$$
$$\times\, \mathrm{Tr}\{\gamma\cdot\mathscr{A}(z_1)\, S_{\mathrm{F}}(z_1-z_2)\gamma\cdot\mathscr{A}(z_2)\, S_{\mathrm{F}}(z_2-z_3) \dots \gamma\cdot\mathscr{A}(z_n)\, S_{\mathrm{F}}(z_n-z_1)\}.$$

† By virtue of the arguments presented earlier, we discard the terms $\mathrm{Tr}\{\gamma^\mu S_{\mathrm{F}}(0)\}$.

The invariance of the trace under cyclic permutations and under transposition as well as properties (49) of the propagators S_F lead to the relation

$$\operatorname{Tr}\{\gamma^{\mu_1} S_F(z_1-z_2)\gamma^{\mu_2} S_F(z_2-z_3) \dots \gamma^{\mu_n} S_F(z_n-z_1)\}$$
$$= (-1)^n \{\gamma^{\mu_1} S_F(z_1-z_n)\gamma^{\mu_n} S_F(z_n-z_{n-1}) \dots \gamma^{\mu_2} S_F(z_2-z_1)\}.$$

The deduction from this is that in the series under consideration all terms with odd values of n are equal to zero.†

Finally, therefore, the vacuum-to-vacuum amplitude may be rewritten as

$$C[\mathscr{A}] = \exp\Bigg[-\sum_{n=1}^{\infty} \frac{e^{2n}}{2n} \int dz_1 \dots dz_n$$
$$\times \operatorname{Tr}\{\gamma\cdot\mathscr{A}(z_1)\, S_F(z_1-z_2) \dots \gamma\cdot\mathscr{A}(z_{2n})\, S_F(z_{2n}-z_1)\}\Bigg]. \tag{81}$$

Next, applying relations (46) successively to propagators S_F, we prove the amplitude $C[\mathscr{A}]$ to be invariant with respect to the proper Lorentz transformations, space reflection, and time reversal. Each time, through a suitable change of the variables of integration, we convert the integrals in formula (81) to the original form.

In order to prove this amplitude to be gauge-invariant, we employ the straightforward criterion for such invariance which we gave in Section 13. Accordingly, let us calculate the following expression,

$$i\partial_\mu \frac{\delta}{\delta\mathscr{A}_\mu(x)} \ln C[\mathscr{A}] = \sum_{n=1}^{\infty} e^{2n} \int dz_2 \dots dz_{2n}$$
$$\times [\operatorname{Tr}\{-i\gamma\cdot\partial_x S_F(x-z_2)\gamma\cdot\mathscr{A}(z_2) \dots S_F(z_{2n}-x)\}$$
$$-\operatorname{Tr}\{S_F(x-z_2)\gamma\cdot\mathscr{A}(z_2) \dots S_F(z_{2n}-x) i\gamma\cdot\overleftarrow{\partial}_x\}].$$

By virtue of the equations satisfied by the propagators S_F, the two terms in the brackets cancel each other. Thus, finally we get the result

$$\partial_\mu \frac{\delta}{\delta\mathscr{A}_\mu(x)} C[\mathscr{A}] = 0, \tag{82}$$

which means that the amplitude $C[\mathscr{A}]$ is invariant under gauge transformations of the potential.

† The fact that these terms vanish also follows as a special case from Furry's theorem, which we discuss in Section 22.

The amplitude $C[\mathscr{A}]$ is also invariant under change of sign of the potential, $\mathscr{A}_\mu \to -\mathscr{A}_\mu$, since only terms of even order appear in expansion (81). This is a manifestation of the invariance of the theory with respect to charge conjugation.

18. ELECTRON FIELD OPERATORS

In the preceding section we demonstrated that the relativistic theory of the electron based on the Dirac equation is, in actual fact, the theory of many particles. With the exception of the static case, the electromagnetic field always creates pairs, transforming zero-electron (vacuum) or one-electron states into many-electron states.

In the previous section we investigated processes in which a system of many electrons occurred only in intermediate states. Now we shall examine processes in which systems of many electrons will also occur in asymptotic states.

Creation and Annihilation Operators

The most convenient method of describing systems consisting of many indistinguishable particles, especially for systems with a variable number of particles, is the method based on creation and annihilation operators and on field operators. We have already seen this in the non-relativistic theory of systems of indistinguishable particles and in the quantum theory of the electromagnetic field. The relativistic quantum mechanics of many electrons, which we shall take up in this section, testifies perhaps even more eloquently to the usefulness of the field operator method.

As in the discussion in non-relativistic theory, we shall begin by introducing the occupation-number representation and the creation and annihilation operators associated with it.

The fundamental definitions will first be given for the simplest case, one without the external electromagnetic field. Let us define the *creation and annihilation operators for negatons and positons*, denoting them by $a_i^\dagger$ and a_i, and $b_i^\dagger$ and b_i, respectively. In exactly the same way as in the non-relativistic theory, we shall associate with them operator-valued distributions $a^\dagger(\mathbf{p}, r)$ and $a(\mathbf{p}, r)$ and $b^\dagger(\mathbf{p}, r)$ and $b(\mathbf{p}, r)$. The definition of creation and annihilation operators will contain one small difference of a technical nature and we shall therefore write out explicitly the formulae defining these operators:

$$a_k^\dagger \Psi_{n_1 \dots n_k \dots ; m_1 \dots m_k \dots} = \begin{cases} \nu(n_1 \dots n_k) \Psi_{n_1 \dots n_k+1 \dots ; m_1 \dots m_k \dots}, & \text{when } n_k = 0, \\ 0, & \text{when } n_k = 1, \end{cases} \tag{1a}$$

$$a_k \Psi_{n_1 \dots n_k \dots ; m_1 \dots m_k \dots} = \begin{cases} 0, & \text{when } n_k = 0, \\ \nu(n_1 \dots n_k) \Psi_{n_1 \dots n_k-1 \dots ; m_1 \dots m_k \dots}, & \text{when } n_k = 1, \end{cases} \tag{1b}$$

$$b_k^\dagger \Psi_{n_1 \dots n_k \dots ; m_1 \dots m_k \dots} = \begin{cases} \nu(n_1 \dots n_k \dots) \nu(m_1 \dots m_k) \Psi_{n_1 \dots n_k \dots ; m_1 \dots m_k+1 \dots}, & \text{when } m_k = 0, \\ 0, & \text{when } m_k = 1, \end{cases} \tag{1c}$$

$$b_k \Psi_{n_1 \dots n_k \dots ; m_1 \dots m_k \dots} = \begin{cases} 0, & \text{when } m_k = 0, \\ \nu(n_1 \dots n_k \dots) \nu(m_1 \dots m_k) \Psi_{n_1 \dots n_k ; m_1 \dots m_k-1 \dots}, & \text{when } m_k = 1. \end{cases} \tag{1d}$$

In these formulae the symbol $\Psi_{n_1 \dots n ; m_1 \dots m_k \dots}$ denotes the vector of the state containing $n_1, \dots, n_k, \dots$ negatons in states described by the profiles $f_1(\mathbf{p}, r), \dots, f_k(\mathbf{p}, r), \dots$ and $m_1, \dots, m_k, \dots$ positons in states described by the profiles $g_1(\mathbf{p}, r), \dots, g_k(\mathbf{p}, r), \dots$. Without loss of generality we may assume that we use the same set of profiles for describing negatons and positons (i.e. that $f_i = g_i$ for all i's). This assumption will henceforth be made. The vector Ω, describing the state containing neither negatons ($n_k = 0$, $k = 1, 2, \dots$) nor positons ($m_k = 0, k = 1, 2, \dots$), is the *vacuum state vector.*

The minor modification mentioned earlier consists in introducing the factors $\nu(n_1 \dots n_k \dots)$,

$$\nu(n_1 \dots n_k \dots) \equiv (-1)^{\sum_{k=1}^{\infty} n_k}, \tag{2}$$

in the definitions of the positon creation and annihilation operators. We have done this in order to get simple anticommutation relations not only between the operators of negaton creation and annihilation and of positon creation and annihilation separately, but also between operators for negatons and positons. These are the relations

$$[a_i, a_j^\dagger]_+ = \delta_{ij} = [b_i, b_j^\dagger]_+, \tag{3a}$$

$$[a_i, a_j]_+ = 0 = [b_i, b_j]_+, \tag{3b}$$

$$[a_i, b_j]_+ = 0 = [a_i, b_j^\dagger]_+. \tag{3c}$$

Had we not introduced the factor (2), the operators for negatons would commute with the operators for positons, and this would complicate the description considerably.

Creation and annihilation operators are related to appropriate operator-valued distributions by the following formulae:

$$a_i = a[g_i], \quad a_i^\dagger = a^\dagger[g_i], \tag{4a}$$

$$b_i = b[g_i], \quad b_i^\dagger = b^\dagger[g_i], \tag{4b}$$

$$a^\dagger[g] = \sum_r \int d\Gamma(m) g(\mathbf{p}, r) a^\dagger(\mathbf{p}, r), \tag{5a}$$

$$a[g] = \sum_r \int d\Gamma(m) g^*(\mathbf{p}, r) a(\mathbf{p}, r), \tag{5b}$$

$$b^\dagger[g] = \sum_r \int d\Gamma(m) g(\mathbf{p}, r) b^\dagger(\mathbf{p}, r), \tag{5c}$$

$$b[g] = \sum_r \int d\Gamma(m) g^*(\mathbf{p}, r) b(\mathbf{p}, r). \tag{5d}$$

The commutation relations for annihilation and creation operators lead to the following commutation relations for operator-valued distributions:

$$[a(\mathbf{p}, r), a^\dagger(\mathbf{q}, s)]_+ = \delta_{rs}\delta_\Gamma(\mathbf{p}, \mathbf{q}) = [b(\mathbf{p}, r), b^\dagger(\mathbf{q}, s)]_+. \tag{6}$$

All the other anticommutators are equal to zero.

Wave functions in the momentum representation of the system of negatons and positons may be expressed in terms of the operators $a(\mathbf{p}, r)$ and $b(\mathbf{q}, s)$ as matrix elements of the following form,

$$f(\mathbf{p}_1 r_1, \ldots, \mathbf{p}_n r_n; \mathbf{q}_1 s_1, \ldots, \mathbf{q}_m s_m)$$

$$= \frac{1}{\sqrt{n!}} \frac{1}{\sqrt{m!}} (\Omega | b(\mathbf{q}_m, s_m) \ldots b(\mathbf{q}_1, s_1) a(\mathbf{p}_n, r_n) \ldots a(\mathbf{p}_1, r_1) \Psi). \quad (7)$$

This formula can be proved first for Fock basis vectors:

$$(\Omega | b(\mathbf{q}_m, s_m) \ldots b(\mathbf{q}_1, s_1) a(\mathbf{p}_n, r_n) \ldots a(\mathbf{p}_1, r_1) a^\dagger_{i_1} \ldots a^\dagger_{i_n} b^\dagger_{j_1} \ldots b^\dagger_{j_m} \Omega)$$

$$= \sum_{\text{perm}\, k_i} \varepsilon_P g_{i_1}(\mathbf{p}_{k_1}, r_{k_1}) \ldots g_{i_n}(\mathbf{p}_{k_n}, r_{k_n}) \sum_{\text{perm}\, l_j} \varepsilon_{P'} g_{j_1}(\mathbf{q}_{l_1}, s_{l_1}) \ldots g_{j_m}(\mathbf{q}_{l_m}, s_{l_m}),$$

and then extended by linearity to the entire space.

Field Operators

At present, we shall define the electron field operators patterned after field operators in non-relativistic theory and in the quantum theory of the electromagnetic field. Since we are concerned with two kinds of particles, we shall introduce two *field operators*: one for *negatons*, $\psi_N(x)$, and another for *positons*, $\psi_P(x)$. In constructing these operators we shall strive for an explicit relativistic invariance. It will be assumed that $\psi_N(x)$ and $\psi_P(x)$ have the transformation properties of a bispinor field. The transformation properties of the operator-valued distributions $a(\mathbf{p}, r)$ and $b(\mathbf{p}, r)$ are the same as those of wave-packet profiles for the solutions of the Dirac equation (cf. formulae (16.65)). This results, for instance, from relation (7) between these operators and the wave functions. The condition of explicit covariance implies that the field operators $\psi_N(x)$ and $\psi_P(x)$ should be of the form,†

$$\psi_N(x) = \sum_r \int d\Gamma(m) \chi^{(+)}(x; \mathbf{p}, r) a(\mathbf{p}, r), \quad (8a)$$

$$\psi_P(x) = \sum_r \int d\Gamma(m) \chi^{(+)}(x; \mathbf{p}, r) b(\mathbf{p}, r). \quad (8b)$$

† Up to a non-essential multiplicative constant.

On the basis of the condition of explicit covariance, we shall also construct the field operators $\psi_N^+(x)$ and $\psi_P^+(x)$, built up out of the creation operators:

$$\psi_N^+(x) = \sum_r \int d\Gamma(m)\, \chi^{(-)}(x, \mathbf{p}, r)\, a^\dagger(\mathbf{p}, r), \tag{9a}$$

$$\psi_P^+(x) = \sum_r \int d\Gamma(m)\, \chi^{(-)}(x; \mathbf{p}, r)\, b^\dagger(\mathbf{p}, r). \tag{9b}$$

All four bispinor field operators satisfy the free Dirac equation.

The field operators ψ_N and ψ_P^+ are used to construct the electron field operator

$$\psi(x) \equiv \psi_N(x) + \psi_P^+(x)$$
$$= \sum_r \int d\Gamma(m)\, [\chi^{(+)}(x; \mathbf{p}, r)\, a(\mathbf{p}, r) + \chi^{(-)}(x; \mathbf{p}, r)\, b^\dagger(\mathbf{p}, r)]. \tag{10}$$

The field operator so constructed is reminiscent of the structure of the solution of the Dirac equation (cf. formula (16.43)), but on the other hand it is patterned after the electromagnetic field operators (as the Fourier transform of the annihilation *and* creation operators).

Field operators at two different points in space-time, $\psi(x)$ and $\bar{\psi}(y)$, satisfy commutation relations of the form:

$$[\psi(x), \bar{\psi}(y)]_+ = \frac{1}{i} S(x-y), \tag{11a}$$

$$[\psi(x), \psi(y)]_+ = 0 = [\bar{\psi}(x), \bar{\psi}(y)]_+, \tag{11b}$$

where

$$S(x) \equiv S^{(+)}(x) + S^{(-)}(x) = (m + i\gamma \cdot \partial)\Delta(x). \tag{12}$$

It follows from these formulae that at the same instant of time the field operators ψ and $\psi^\dagger$ anticommute to δ, just as do the field operators for fermions in non-relativistic theory,

$$[\psi_a(\mathbf{x}, t), \psi_b^\dagger(\mathbf{y}, t)]_+ = \delta_{ab}\, \delta(\mathbf{x}-\mathbf{y}). \tag{13}$$

Field operators for negatons and positons separately do not satisfy such simple commutation relations at the same instant of time. The anticommutator of these operators at different points is the function $S^{(+)}(x-y)$. This anticommutator may, therefore, be written in terms of the Bessel function K_1 of the distance $|\mathbf{x}-\mathbf{y}|$ (cf. Appendix E).

The field operator $\psi(x)$ satisfies the free Dirac equation

$$(-i\gamma \cdot \partial + m)\,\psi(x) = 0. \tag{14}$$

Relations inverse to (10), expressing annihilation and creation operators in terms of field operators, follow from the properties of the solutions to the free Dirac equation, $\chi^{(\pm)}$, discussed in Section 16),

$$a[f] = \int \mathrm{d}\sigma_\mu \overline{\chi}^{(+)}[x|f]\gamma^\mu \psi(x), \tag{15a}$$

$$a^\dagger[f] = \int \mathrm{d}\sigma_\mu \overline{\psi}(x)\,\gamma^\mu\,\chi^{(+)}[x|f], \tag{15b}$$

$$b[f] = \int \mathrm{d}\sigma_\mu \overline{\psi}(x)\gamma^\mu \chi^{(-)}[x|f], \tag{15c}$$

$$b^\dagger[f] = \int \mathrm{d}\sigma_\mu \overline{\chi}^{(-)}[x|f]\gamma^\mu \psi(x). \tag{15d}$$

Operators representing physical quantities will be constructed from the field operators as bilinear forms. The *operator of electric current density* is of the form

$$j_\mu(x) \equiv e\!:\!\overline{\psi}(x)\gamma^\mu\psi(x)\!:, \tag{16}$$

where the symbol : : denotes the normal product of the operators.†
This operator is a Hermitian operator-valued distribution, i.e.

$$(\Psi|\,j^\mu(x)\Phi)^* = (\Phi|\,j^\mu(x)\Psi). \tag{17}$$

It satisfies the continuity equation

$$\partial_\mu j^\mu(x) = 0, \tag{18}$$

whereas the charge operator Q constructed from it is equal to the difference between the operators of the number of negatons and positons, multiplied by e,

$$\begin{aligned} Q &\equiv \int \mathrm{d}\sigma_\mu j^\mu(x) = \int \mathrm{d}^3x j^0(\mathbf{x}, t) \\ &= e \sum_r \int \mathrm{d}\Gamma(m)\,[a^\dagger(\mathbf{p}, r)\,a(\mathbf{p}, r) - b^\dagger(\mathbf{p}, r)\,b(\mathbf{p}, r)]. \end{aligned} \tag{19}$$

The current operator consists of four components,

$$j^\mu = \overline{\psi}_{\mathrm{N}}\gamma^\mu\psi_{\mathrm{N}} + \overline{\psi}{}^+_{\mathrm{P}}\gamma^\mu\psi_{\mathrm{N}} + \,:\!\overline{\psi}{}^+_{\mathrm{P}}\gamma^\mu\psi^+_{\mathrm{P}}\!:\, + \overline{\psi}_{\mathrm{N}}\gamma^\mu\psi^+_{\mathrm{P}},$$

† Cf. Appendix B.

which can be associated with the four elementary processes shown in Fig. 17.1.

The commutator of two current operators is:

$$[j^\mu(x), j^\nu(y)] = \frac{1}{i}\overline{\psi}(x)\gamma^\mu S(x-y)\gamma^\nu\psi(y)$$

$$-\frac{1}{i}\overline{\psi}(y)\gamma^\nu S(y-x)\gamma^\mu\psi(x). \tag{20a}$$

Since the function $S(x)$ vanishes for the spatial vectors x, the current operators commute for spatially separated points,

$$[j^\mu(x), j^\nu(y)] = 0, \quad \text{when } (x-y)^2 < 0. \tag{20b}$$

This property, called *locality*, is not possessed separately by the four components of the operator j^μ. Locality is required of all operators representing x-dependent physical quantities directly subject to measurements.

The current operator may also be written in the form of a commutator:

$$j^\mu(x) = \frac{e}{2}[\overline{\psi}(x)\gamma^\mu\psi(x) - \mathrm{Tr}\{\psi(x)\overline{\psi}(x)\gamma^\mu\}]$$

$$= -\frac{e}{2}\mathrm{Tr}\{[\psi(x), \overline{\psi}(x)\gamma^\mu]\}, \tag{21}$$

or in an equivalent form containing the chronological product,

$$j^\mu(x) = \frac{e}{2} \lim_{\substack{\eta\to 0 \\ \eta\cdot\eta>0,\ \eta^0>0}} T\big(\overline{\psi}(x+\eta)\gamma^\mu\psi(x) + \overline{\psi}(x-\eta)\gamma^\mu\psi(x)\big). \tag{22}$$

The *operators of the components of the energy–momentum tensor* $T^{\mu\nu}(x)$ may also be expressed in terms of the field operators,

$$T^{\mu\nu}(x) = \frac{1}{2} : [\overline{\psi}(x)\gamma^\mu i\overleftrightarrow{\partial^\nu}\psi(x) + \overline{\psi}(x)\gamma^\nu i\overleftrightarrow{\partial^\mu}\psi(x)] :. \tag{23}$$

This tensor satisfies the continuity equation, whereas by integrating the appropriate components over three-dimensional space, we obtain ten generators of the Poincaré group.

Electrons in an External Electromagnetic Field

In the presence of the electromagnetic field, the construction of creation and annihilation operators and of the field operators will undergo modifications. First of all, there no longer exists a universal vacuum state, i.e. one in which no particles are present at any moment. If the electromagnetic field vanishes beyond the region of space-time bounded in time,† two vacuum states can nevertheless be introduced: one in which there are no particles in the remote past, and another in which there are no particles in the far future. The corresponding state vectors will be denoted by Ω^{in} and Ω^{out}. The system in the Ω^{in} state is characterized by the absence of particles in the remote past, when the electromagnetic field has not yet started to act, but we find pairs of negatons and positons when we examine this state in the future or in the region where the electromagnetic field does act. A similar picture is associated with the Ω^{out} state.

Just as the concept of vacuum state, so the concept of n-particle state in the presence of an external electromagnetic field loses its absolute character. It is possible, however, to consider IN and OUT states characterized by the content of negatons and positons in the past and in the future. One-particle states of negatons and positons in the past and in the future will be described by means of the same set of wave packets with profiles $g_k(\mathbf{p}, r)$. We distinguish two Fock bases formed by the vectors:

$$\Psi^{\text{in}}_{n_1 \ldots n_k \ldots;\, m_1 \ldots m_k \ldots} = \Psi^{\text{in}}_{\{n_k;\, m_k\}} \tag{24a}$$

and

$$\Psi^{\text{out}}_{n_1 \ldots n_k \ldots;\, m_1 \ldots m_k \ldots} = \Psi^{\text{out}}_{\{n_k;\, m_k\}}. \tag{24b}$$

States described by these vectors are characterized by the occupation numbers $n_1, \ldots, n_k, \ldots$ for negatons and $m_1, \ldots, m_k, \ldots$ for positons, respectively, in the remote past and in the far future.

Proceeding from these two bases, we define two systems (IN and OUT) of negaton creation and annihilation operators, $a_k^{\dagger\,\text{in}\atop\text{out}}$ and $a_k^{\text{in}\atop\text{out}}$, respectively, and positon creation and annihilation operators, $b_k^{\dagger\,\text{in}\atop\text{out}}$ and $b_k^{\text{in}\atop\text{out}}$,

† The electromagnetic field may also go over into a static field in the remote past and in the far future.

respectively. The formulae defining these operators are of the same form as formulae (1), except that we add the superscripts IN and OUT everywhere.

The products of the operators $a_k^{\dagger in}$ and $b_k^{\dagger in}$, acting on the state vector Ω^{in}, create state vectors (24a). Similarly, the state vectors (24b) are obtained by operating on the vector Ω^{out} with products of the operators $a_k^{\dagger out}$ and $b_k^{\dagger out}$.

IN and OUT operator-valued distributions can be associated with the IN and OUT creation and annihilation operators in the same way as in the field-free case; then IN and OUT field operators of current, charge, components of the energy–momentum tensor, and Poincaré group generators can also be associated with the latter.

Now we shall go on to describe various processes in which negatons and positons take part in the electromagnetic field and to consider the transition amplitudes describing these processes. The *transition amplitudes* can be built up out of the *fundamental amplitudes* $A\{n_k, m_k; n'_k, m'_k\}$ defined as scalar products of the IN and OUT basis vectors,

$$A\{n_k, m_k; n'_k, m'_k\} \equiv (\Psi^{out}_{\{n_k, m_k\}} | \Psi^{in}_{\{n'_k, m'_k\}}). \tag{25}$$

Here $A\{n_k, m_k; n'_k, m'_k\}$ is the amplitude of the probability of finding in the far future a system of negatons with occupation numbers $\{n_k\}$ and positons with occupation numbers $\{m_k\}$ if in the remote past the system consists of negatons with occupation numbers $\{n'_k\}$ and positons with occupation numbers $\{m'_k\}$.

In special cases, for scattering processes when the initial and final states are ones of at most one negaton and one positon, these amplitudes should be equal to the scattering amplitudes $S^{\pm\pm}$ investigated in the preceding section. The proper form of these four transition amplitudes is guaranteed by the following relations between the IN and OUT creation and annihilation operators,

$$a^{out}(\mathbf{p}, r) = \sum_s \int d\Gamma(m) [S_Q^{++}(\mathbf{p}, r; \mathbf{q}, s)\, a^{in}(\mathbf{q}, s)$$
$$+ S_Q^{+-}(\mathbf{p}, r; \mathbf{q}, s)\, b^{\dagger out}(\mathbf{q}, s)], \tag{26a}$$

$$b^{\dagger in}(\mathbf{p}, r) = \sum_s d\Gamma(m) [S_Q^{--}(\mathbf{p}, r; \mathbf{q}, s)\, b^{\dagger out}(\mathbf{q}, s)$$
$$+ S_Q^{-+}(\mathbf{p}, r; \mathbf{q}, s)\, a^{in}(\mathbf{q}, s)]. \tag{26b}$$

For, if we take account of these relations and the commutation relations for the IN and OUT creation and annihilation operators, we get

$$(a^{\dagger\text{out}}[f]\,\Omega^{\text{out}}|a^{\dagger\text{in}}[g]\,\Omega^{\text{in}}) = (\Omega^{\text{out}}|\Omega^{\text{in}})\,S_{\text{Q}}^{++}[f,g], \tag{27a}$$

$$(\Omega^{\text{out}}|b^{\dagger\text{in}}[f]\,a^{\dagger\text{in}}[g]\Omega^{\text{in}}) = (\Omega^{\text{out}}|\Omega^{\text{in}})\,S_{\text{Q}}^{-+}[f,g], \tag{27b}$$

$$(b^{\dagger\text{out}}[g]\,\Omega^{\text{out}}|b^{\dagger\text{in}}[f]\,\Omega^{\text{in}}) = (\Omega^{\text{out}}|\Omega^{\text{in}})\,S_{\text{Q}}^{--}[f,g], \tag{27c}$$

$$(a^{\dagger\text{out}}[f]\,b^{\dagger\text{out}}[g]\,\Omega^{\text{out}}|\Omega^{\text{in}}) = (\Omega^{\text{out}}|\Omega^{\text{in}})\,S_{\text{Q}}^{+-}[f,g]. \tag{27d}$$

To give an example we shall prove formula (27b). When the operator $b^{\dagger\text{in}}$, written as the right-hand side of eq. (26b), is substituted into this formula, we obtain two terms. The first does not make any contribution to the amplitude since the expression $(\Omega^{\text{out}}|b^{\dagger\text{out}}\ldots$ is equal to zero. The second term contains the operator $a^{\text{in}}(\mathbf{q}, s)$ which, after commutation with the creation operators, also yields zero. All that remains is the term resulting from commutation, which has exactly the form of the right-hand side of eq. (27b).

Since we are considering systems of mutually non-interacting electrons, any transition amplitude may be expressed in terms of the vacuum-to-vacuum transition amplitude and four one-particle quasi-amplitudes $S_{\text{Q}}^{\pm\pm}$ (cf. the discussion in Section 5 concerning mutually non-interacting particles). The general transition amplitude may be written in terms of the scalar products of the OUT and IN basis vectors,

$$(a_{i_1}^{\dagger\text{out}}\ldots a_{i_n}^{\dagger\text{out}}\,b_{j_1}^{\dagger\text{out}}\ldots b_{j_m}^{\dagger\text{out}}\,\Omega^{\text{out}}|b_{j_1'}^{\dagger\text{in}}\ldots b_{j_k'}^{\dagger\text{in}}\,a_{i_1'}^{\dagger\text{in}}\ldots a_{i_l'}^{\dagger\text{in}}\,\Omega^{\text{in}}).$$

On the other hand, by using relations (26) between the IN and OUT operators, we can rewrite this scalar product as

$$(\Omega^{\text{out}}|\Omega^{\text{in}})\sum S_{\text{Q}}[\cdot\,,\cdot]\ldots S_{\text{Q}}[\cdot\,,\cdot].$$

The arguments of the transition quasi-amplitudes occurring in this formula are not written out since we shall not be using this formula. The point was merely to demonstrate that in the given case of mutually non-interacting electrons all transition amplitudes can be expressed in terms of the vacuum-to-vacuum amplitude and one-electron quasi-amplitudes.

The S Operator

In order to show that the theory we have constructed for a system of electrons in an external electromagnetic field is correct, i.e. that it satisfies the principle of conservation of probability and the causality condition, we shall construct the S operator and examine its properties. The S operator is defined† through its matrix elements in the basis of IN vectors,

$$(\Psi^{\text{in}}_{\{n_k;\, m_k\}}|\mathsf{S}[\mathscr{A}]\Psi^{\text{in}}_{\{n'_k;\, m'_k\}}) \equiv (\Psi^{\text{out}}_{\{n_k;\, m_k\}}|\Psi^{\text{in}}_{\{n'_k;\, m'_k\}}). \tag{28}$$

From this definition and from the completeness of the set of vectors $\Psi^{\text{in}}_{\{n_k;\, m_k\}}$ and $\Psi^{\text{out}}_{\{n_k;\, m_k\}}$, it follows that the operator $\mathsf{S}[\mathscr{A}]$ must satisfy the following relations for all i's:

$$a_i^{\text{in}}\,\mathsf{S}[\mathscr{A}] = \mathsf{S}[\mathscr{A}]\,a_i^{\text{out}}, \tag{29a}$$

$$b_i^{\text{in}}\,\mathsf{S}[\mathscr{A}] = \mathsf{S}[\mathscr{A}]\,b_i^{\text{out}}, \tag{29b}$$

$$a_i^{\dagger\text{in}}\,\mathsf{S}[\mathscr{A}] = \mathsf{S}[\mathscr{A}]\,a_i^{\dagger\text{out}}, \tag{29c}$$

$$b_i^{\dagger\text{in}}\,\mathsf{S}[\mathscr{A}] = \mathsf{S}[\mathscr{A}]\,b_i^{\dagger\text{out}}, \tag{29d}$$

$$\mathsf{S}[\mathscr{A}]\Omega^{\text{out}} = \Omega^{\text{in}}. \tag{29e}$$

Since the relations between the IN and OUT operators are given by formulae (26), and the vacuum-to-vacuum amplitude $(\Omega^{\text{in}}|\mathsf{S}[\mathscr{A}]\Omega^{\text{in}})$ by formula (17.81), eqs. (29) uniquely determine the operator $\mathsf{S}[\mathscr{A}]$.

The operator $\mathsf{S}[\mathscr{A}]$ with the required properties will now be shown to be given by the formula

$$\mathsf{S}[\mathscr{A}] = T\exp\left[-i\int d^4x\, j_\mu^{\text{in}}(x)\,\mathscr{A}^\mu(x)\right]. \tag{30}$$

The operator defined by formula (30) is *unitary*, since the operator of the energy density of the interaction of electrons with an external electromagnetic field,

$$\mathscr{H}^{\text{int}}(x) \equiv j_\mu^{\text{in}}(x)\,\mathscr{A}^\mu(x), \tag{31}$$

is a Hermitian operator.

The operator $\mathsf{S}[\mathscr{A}]$ also satisfies the *causality condition* in the relativistic form (cf. formulae (5.56) and (13.35)),

$$\frac{\delta}{\delta\mathscr{A}_\mu(x)}\left(\mathsf{S}^\dagger[\mathscr{A}]\frac{\delta\mathsf{S}[\mathscr{A}]}{\delta\mathscr{A}_\nu(y)}\right) = 0, \quad \text{when } x \gtrsim y, \tag{32}$$

since the density of the interaction Hamiltonian depends locally on the field $\mathscr{A}_\mu(x)$.

† The same symbol $\mathsf{S}[\mathscr{A}]$ as in Section 13 is used here since this is the same physical quantity, though in a different theory.

Next we shall prove that the operator defined by formula (30) obeys the relations (29). The proof is in two parts. In the first part, we shall prove that relations (29a)–(29d) are satisfied.

For this proof it is convenient to introduce the *operator of the electron field in an external electromagnetic field*,

$$\psi[x|\mathscr{A}] \equiv \mathsf{S}^\dagger[\mathscr{A}]\, T(\psi^{\mathrm{in}}(x)\,\mathsf{S}[\mathscr{A}]). \tag{33}$$

It follows from the unitarity of the operator $\mathsf{S}[\mathscr{A}]$ and the properties of the chronological product that in the remote past the operator $\psi[x|\mathscr{A}]$ coincides with the operator $\psi^{\mathrm{in}}(x)$,

$$\begin{aligned}\psi[x|\mathscr{A}] &= \psi^{\mathrm{in}}(x)\\ &= \sum_r \int \mathrm{d}\Gamma(m)\,[\chi^{(+)}(x;\mathbf{p},r)\,a^{\mathrm{in}}(\mathbf{p},r)+\chi^{(-)}(x;\mathbf{p},r)\,b^{\dagger\mathrm{in}}(\mathbf{p},r)],\\ &\text{when } t < T_1.\end{aligned} \tag{34}$$

This operator obeys the Dirac equation with electromagnetic potential,

$$[-i\gamma\cdot\partial+m+e\gamma\cdot\mathscr{A}(x)]\,\psi[x|\mathscr{A}] = 0. \tag{35}$$

The proof of this equation is based on the properties of the time derivatives of the chronological products (cf. Appendix B) and on the relation

$$\begin{aligned}&\gamma^0\delta(t-t')\,[\psi^{\mathrm{in}}(x), j_\mu^{\mathrm{in}}(y)\mathscr{A}^\mu(y)]\\ &\quad = e\gamma\cdot\mathscr{A}(x)\,\psi^{\mathrm{in}}(x)\,\delta^{(4)}(x-y),\end{aligned}$$

which results from the commutation relations for the operators ψ^{in} and $\bar{\psi}^{\mathrm{in}}$.

Equation (35) and the initial condition (34) determine the solution, hence also the form of the operator $\psi[x|\mathscr{A}]$ in the far future. The operator coefficients in the parts with positive and negative frequencies of this operator in the far future will be denoted by the symbols $\tilde{a}^{\mathrm{out}}$ and $\tilde{b}^{\dagger\mathrm{out}}$. The operator $\psi[x|\mathscr{A}]$ may be written in terms of the Feynman propagator K_{F} by formula (17.38) in which functions have been replaced by the field operators,

$$\psi[x|\mathscr{A}] = \int \mathrm{d}^4y\, K_{\mathrm{F}}[x, y|\mathscr{A}]\, \overleftarrow{D}_y\, \psi_{\mathrm{F}}(y), \tag{36}$$

where

$$\psi_F(x) \equiv \sum_r \int d\Gamma(m)[\chi^{(+)}(x; \mathbf{p}, r)\, a^{in}(\mathbf{p}, r) + \chi^{(-)}(x; \mathbf{p}, r)\, \tilde{b}^{\dagger out}(\mathbf{p}, r)]. \tag{37}$$

The form of the operator $\psi_F(x)$ has been read from the integral equation

$$\psi[x|\mathscr{A}] = \psi_F(x) - e\int d^4y S_F(x-y)\, \gamma\cdot\mathscr{A}(y)\, \psi[y|\mathscr{A}] \tag{38}$$

and from the asymptotic conditions satisfied by the propagator S_F and the operator $\psi[x|\mathscr{A}]$.

From the representation (36) of the operator $\psi[x|\mathscr{A}]$ we get

$$\tilde{a}^{out}[f] = \int_{\sigma_F} d\sigma_\mu \int d^4y \bar{\chi}^{(+)}[x|f]\, K_F[x, y|\mathscr{A}]\, \overleftarrow{D}_y \psi_F(y)$$
$$= a^{in}[f] - i\int d^4x d^4y \bar{\chi}^{(+)}[x|f]\, D_x K_F[x, y|\mathscr{A}]\, \overleftarrow{D}_y \psi_F(y). \tag{39}$$

In similar fashion we express the operator $b^{\dagger in}$ in terms of $\tilde{b}^{\dagger out}$ and $\psi_F(x)$. When these formulae are stripped of the wave-packet profiles we obtain relations between $\tilde{a}^{out}(\mathbf{p}, r)$, $\tilde{b}^{\dagger out}(\mathbf{p}, r)$, $a^{in}(\mathbf{p}, r)$, and $b^{\dagger in}(\mathbf{p}, r)$ which are the same as the relations (26) between a^{out}, $b^{\dagger out}$, a^{in}, and $b^{\dagger in}$. The operators $\tilde{a}^{out}$ and $\tilde{b}^{\dagger out}$ thus are equal to the operators a^{out} and $b^{\dagger out}$.

The operators of the electron field in an external electromagnetic field, $\psi[x|\mathscr{A}]$, therefore goes over in the far future into the field operator $\psi^{out}(x)$,

$$\psi[x|\mathscr{A}] = \psi^{out}(x)$$
$$= \sum_r \int d\Gamma(m)\, [\chi^{(+)}(x; \mathbf{p}, r) a^{out}(\mathbf{p}, r) + \chi^{(-)}(x; \mathbf{p}, r)\, b^{\dagger out}(\mathbf{p}, r)], \tag{40}$$
$$\text{when} \quad t > T_2.$$

Let us go back to the definition (33) of the operator $\psi[x|\mathscr{A}]$ and compare the asymptotic properties of both sides of the equality in the far future. We then find that the operator $\mathsf{S}[\mathscr{A}]$ defined by formula (30) satisfies equalities (29a)–(29d).

In the second part of the proof we shall demonstrate that the vacuum-to-vacuum transition amplitude determined by operator (30) is equal to the amplitude $C[\mathscr{A}]$ given by formula (17.81).

For this purpose, let us calculate the functional derivative of the amplitude $(\Omega^{\text{in}}|\mathsf{S}[\mathscr{A}]\Omega^{\text{in}})$ with respect to $\mathscr{A}_\mu(x)$:

$$\frac{\delta(\Omega^{\text{in}}|\mathsf{S}[\mathscr{A}]\Omega^{\text{in}})}{\delta\mathscr{A}_\mu(x)} = -i\big(\Omega^{\text{in}}|T(\mathsf{S}[\mathscr{A}]\,j^{\text{in}\mu}(x))\,\Omega^{\text{in}}\big)$$

$$= \frac{-ie}{2} \lim_{\substack{\eta\to 0\\ \eta\cdot\eta>0,\ \eta^0>0}} \big(\Omega^{\text{in}}|T(\mathsf{S}[\mathscr{A}]\,(\bar\psi^{\text{in}}(x+\eta)\gamma^\mu\psi^{\text{in}}(x)$$

$$+\bar\psi^{\text{in}}(x-\eta)\,\gamma^\mu\psi^{\text{in}}(x)))\,\Omega^{\text{in}}\big). \tag{41}$$

The expectation value of the chronological product of the operators

$$\big(\Omega^{\text{in}}|T(\mathsf{S}[\mathscr{A}]\,\psi^{\text{in}}(x)\,\bar\psi^{\text{in}}(y))\,\Omega^{\text{in}}\big) \tag{42}$$

satisfies the differential equation

$$[-i\gamma\cdot\partial_x+m+e\gamma\cdot\mathscr{A}(x)]\,\big(\Omega^{\text{in}}|T(\mathsf{S}[\mathscr{A}]\,\psi^{\text{in}}(x)\,\bar\psi^{\text{in}}(y))\,\Omega^{\text{in}}\big)$$

$$= -i(\Omega^{\text{in}}|\mathsf{S}[\mathscr{A}]\,\Omega^{\text{in}})\,\delta^{(4)}(x-y), \tag{43}$$

and, as a function in both variables x and y, obeys the Feynman asymptotic conditions. Accordingly, the equality

$$\big(\Omega^{\text{in}}|T(\mathsf{S}[\mathscr{A}]\,\psi^{\text{in}}(x)\,\bar\psi^{\text{in}}(y))\,\Omega^{\text{in}}\big)$$

$$= -i(\Omega^{\text{in}}|\mathsf{S}[\mathscr{A}]\,\Omega^{\text{in}})\,K_F[x,y|\mathscr{A}] \tag{44}$$

holds. The functional derivative of the vacuum-to-vacuum transition amplitude thus is

$$\frac{\delta(\Omega^{\text{in}}|\mathsf{S}[\mathscr{A}]\,\Omega^{\text{in}})}{\delta\mathscr{A}_\mu(x)} = (\Omega^{\text{in}}|\mathsf{S}[\mathscr{A}]\,\Omega^{\text{in}})\,e\,\mathrm{Tr}\{\gamma^\mu K_F[x,x|\mathscr{A}]\}. \tag{45}$$

In view of the fact that $\mathscr{A}_\mu(x)$ and e always appear in the product $e\mathscr{A}_\mu(x)$, we can use the equality

$$e\frac{\partial}{\partial e} = \int \mathrm{d}^4x\,\mathscr{A}_\mu(x)\frac{\delta}{\delta\mathscr{A}_\mu(x)}\,.$$

Thus, we get the same differential equation for the amplitude $(\Omega^{\text{in}}|\mathsf{S}[\mathscr{A}]\Omega^{\text{in}})$ as we did for $C[\mathscr{A}]$ (cf. formula (17.80)). The initial condition is also identical

$$(\Omega^{\text{in}}|\mathsf{S}[\mathscr{A}]\,\Omega^{\text{in}})|_{e=0} = 1,$$

whereby

$$(\Omega^{\text{in}}|\mathsf{S}[\mathscr{A}]\,\Omega^{\text{in}}) = C[\mathscr{A}]. \tag{46}$$

This final result completes the proof that the operator $\mathsf{S}[\mathscr{A}]$ given by formula (30) is correct.

Propagators

The matrix element of the chronological product of field operators (42), considered in the proof above, has turned out to be equal, up to the factor i, to the propagator $C[x, y|\mathscr{A}]$ which was introduced in the previous section. Many-electron propagators patterned after that propagator will now be constructed:

$$C[x_1 \ldots x_n, y_n \ldots y_1|\mathscr{A}] \equiv i^n\left(\Omega^{\text{in}}|T(\mathsf{S}[\mathscr{A}]\psi^{\text{in}}(x_1) \ldots \psi^{\text{in}}(x_n)\overline{\psi}^{\text{in}}(y_n) \ldots \overline{\psi}^{\text{in}}(y_1))\Omega^{\text{in}}\right). \tag{47}$$

These propagators may also be rewritten as

$$C[x_1 \ldots x_n, y_n \ldots y_1|\mathscr{A}] = i^n\left(\Omega^{\text{out}}|T(\psi[x_1|\mathscr{A}] \ldots \psi[x_n|\mathscr{A}]\overline{\psi}[y_n|\mathscr{A}] \ldots \overline{\psi}[y_1|\mathscr{A}])\,\Omega^{\text{in}}\right). \tag{48}$$

The proof is analogous to that for formula (5.65) in the non-relativistic theory of many particles.

By formula (30), the propagator $C[x_1 \ldots, y_n \ldots|\mathscr{A}]$ may also be written in the form of the following series:

$$C[x_1 \ldots x_n, y_n \ldots y_1|\mathscr{A}] = i^n \sum_{k=0}^{\infty} \frac{(-i)^k}{k!} \int d^4 z_1 \ldots d^4 z_k$$
$$\times \mathscr{A}_{\mu_1}(z_1) \ldots \mathscr{A}_{\mu_k}(z_k)\left(\Omega^{\text{in}}|T(\psi^{\text{in}}(x_1) \ldots \psi^{\text{in}}(x_n)\right.$$
$$\left.\times \overline{\psi}^{\text{in}}(y_n) \ldots \overline{\psi}^{\text{in}}(y_1)\, j^{\text{in}\mu_1}(z_1) \ldots j^{\text{in}\mu_k}(z_k))\,\Omega^{\text{in}}\right). \tag{49}$$

Since, as is shown in Appendix H, the equality

$$i^n\left(\Omega^{\text{in}}|T(\psi^{\text{in}}(x_1) \ldots \psi^{\text{in}}(x_n)\overline{\psi}^{\text{in}}(y_n) \ldots \overline{\psi}^{\text{in}}(y_1))\,\Omega^{\text{in}}\right) = S_F(x_1 \ldots x_n, y_1 \ldots y_n) \tag{50}$$

holds, this series may be written in the form:†

$$C[x_1 \ldots x_n, y_n \ldots y_1|\mathscr{A}] = \sum_{k=0}^{\infty} \frac{e^k}{k!} \int dz_1 \ldots dz_k$$
$$\times \gamma(z_1)\cdot\mathscr{A}(z_1) \ldots \gamma(z_k)\cdot\mathscr{A}(z_k)\, S_F(x_1 \ldots x_n z_1 \ldots z_k, y_1 \ldots y_n z_1 \ldots z_k). \tag{51}$$

† Here we shall employ the simplified notation used at the end of Section 17.

In the special case, for $n = 1$, this formula coincides with the formula obtained for the propagator $C[x, y|\mathscr{A}]$ in Section 17 by the Fredholm method.

Now let us express the propagator $C[x_1 \ldots x_n, y_n \ldots y_1|\mathscr{A}]$ in terms of the amplitude $C[\mathscr{A}]$ and the products of the propagators K_F. It will be shown that

$$C[x_1 \ldots x_n, y_n \ldots y_1|\mathscr{A}] = C[\mathscr{A}]\, K_F[x_1 \ldots x_n, y_1 \ldots y_n|\mathscr{A}], \quad (52)$$

where

$$K_F[x_1 \ldots x_n, y_1 \ldots y_n|\mathscr{A}] \equiv \text{Det}\begin{vmatrix} K_F[x_1, y_1|\mathscr{A}] & \ldots & K_F[x_1, y_n|\mathscr{A}] \\ \vdots & & \vdots \\ K_F[x_n, y_1|\mathscr{A}] & \ldots & K_F[x_n, y_n|\mathscr{A}] \end{vmatrix}. \quad (53)$$

To prove formula (52) we shall make use of the property of "splitting up" by the functional differentiation (17.32), which property is possessed by the propagator $K_F[x, y|\mathscr{A}]$. By virtue of this property, we have

$$\frac{\delta}{\delta \mathscr{A}_\mu(z)} K_F[x_1 \ldots x_n, y_1 \ldots y_n|\mathscr{A}]$$

$$= -e \sum_{\text{perm}\, i} \varepsilon_P \sum_{k=1}^{n} K_F[x_1, y_{i_1}|\mathscr{A}] \ldots K_F[x_k, z|\mathscr{A}]\, \gamma^\mu K_F[z, y_{i_k}|\mathscr{A}] \ldots K_F[x_n, y_{i_n}|\mathscr{A}]$$

$$= e\gamma^\mu(z)\, K_F[x_1 \ldots x_n z, y_1 \ldots y_n z|\mathscr{A}]$$

$$- e\gamma^\mu(z) K_F[z, z|\mathscr{A}] K_F[x_1 \ldots x_n, y_1 \ldots y_n|\mathscr{A}].$$

Application of formula (45) for the functional derivative of the amplitude $C[\mathscr{A}]$ yields

$$\frac{\delta}{\delta \mathscr{A}_\mu(z)} (C[\mathscr{A}] K_F[x_1 \ldots x_n, y_1 \ldots y_n|\mathscr{A}])$$

$$= C[\mathscr{A}] e\gamma^\mu(z)\, K_F[x_1 \ldots x_n z, y_1 \ldots y_n z|\mathscr{A}].$$

Knowing the functional derivatives, we go on to build the following Volterra's series:

$$C[\mathscr{A} + \tilde{\mathscr{A}}]\, K_F[x_1 \ldots x_n, y_1 \ldots y_n|\mathscr{A} + \tilde{\mathscr{A}}]$$

$$= \exp\left(\int dz\, \mathscr{A}_\mu(z) \frac{\delta}{\delta \tilde{\mathscr{A}}_\mu(z)}\right) C[\tilde{\mathscr{A}}] K_F[x_1 \ldots x_n, y_1 \ldots y_n|\tilde{\mathscr{A}}]$$

$$= C[\tilde{\mathscr{A}}] \sum_{k=0}^{\infty} \frac{e^k}{k!} \int dz_1 \ldots dz_k\, \gamma(z_1)\cdot\mathscr{A}(z_1) \ldots \gamma(z_k)\cdot\mathscr{A}(z_k)$$

$$\times K_F[x_1 \ldots x_n z_1 \ldots z_k, y_1 \ldots y_n z_1 \ldots z_k|\tilde{\mathscr{A}}].$$

On putting $\tilde{\mathscr{A}}_\mu = 0$, we get the right-hand side of formula (51) for the propagator $C[x_1 \ldots x_n, y_n \ldots y_1|\mathscr{A}]$.

In conclusion, we shall give the relation between the operator $\mathsf{S}[\mathscr{A}]$ and the propagators. We shall begin the deduction of this relation by writing out formulae which are a natural modification of the *reduction formulae* (5.61) obtained in non-relativistic theory:

$$[a^{\mathrm{in}}[g], \mathsf{S}T(\mathcal{O}_1(x_1) \dots \mathcal{O}_n(x_n))]_{\mp}$$
$$= i\int \mathrm{d}^4x\, \overline{\chi}^{(+)}[x|g]\, D_x \mathsf{S}T(\psi[x|\mathscr{A}]\, \mathcal{O}_1(x_1) \dots \mathcal{O}_n(x_n)), \tag{54a}$$
$$[\mathsf{S}T(\mathcal{O}_1(x_1) \dots \mathcal{O}_n(x_n)), a^{\dagger\mathrm{in}}[g]]_{\mp}$$
$$= i\int \mathrm{d}^4x\, \mathsf{S}T(\mathcal{O}_1(x_1) \dots \mathcal{O}_n(x_n)\, \overline{\psi}[x|\mathscr{A}])\, \overleftarrow{D}_x \chi^{(+)}[x|g], \tag{54b}$$
$$[\mathsf{S}T(\mathcal{O}_1(x_1) \dots \mathcal{O}_n(x_n)), b^{\mathrm{in}}[g]]_{\mp}$$
$$= i\int \mathrm{d}^4x\, \mathsf{S}T(\mathcal{O}_1(x_1) \dots \mathcal{O}_n(x_n)\, \overline{\psi}[x|\mathscr{A}])\, \overleftarrow{D}_x \chi^{(-)}[x|g], \tag{54c}$$
$$[b^{\dagger\mathrm{in}}[g], \mathsf{S}T(\mathcal{O}_1(x_1) \dots \mathcal{O}_n(x_n))]_{\mp}$$
$$= i\int \mathrm{d}^4x\, \overline{\chi}^{(-)}[x|g]\, D_x \mathsf{S}T(\psi[x|\mathscr{A}]\, \mathcal{O}_1(x_1) \dots \mathcal{O}_n(x_n)). \tag{54d}$$

The proof of these formulae is left to the reader.

Our subsequent procedure is also patterned after the relevant considerations in Section 5. In this way we get a formula for the operator $\mathsf{S}[\mathscr{A}]$ in the form of a sum of normal products:

$$\mathsf{S}[\mathscr{A}] = \sum_{n=0}^{\infty} \frac{i^n}{(n!)^2} \int \mathrm{d}x_1 \dots \mathrm{d}x_n \mathrm{d}y_1 \dots \mathrm{d}y_n$$
$$\times :\overline{\psi}^{\mathrm{in}}(x_n) \dots \overline{\psi}^{\mathrm{in}}(x_1) D_{x_1} \dots D_{x_n} C[x_1 \dots x_n, y_n \dots y_1 | \mathscr{A}]\, \overleftarrow{D}_{y_1} \dots \overleftarrow{D}_{y_n}$$
$$\times \psi^{\mathrm{in}}(y_1) \dots \psi^{\mathrm{in}}(y_n):. \tag{55}$$

By utilizing the representation (52) of the propagators C in terms of the products of the propagators K_{F} we can substantially simplify the formula for the operator $\mathsf{S}[\mathscr{A}]$. Since the normal products of the field operators are antisymmetrical, we can collect groups of $n!$ equal terms and sum up the series to an exponential function,

$$\mathsf{S}[\mathscr{A}] = C[\mathscr{A}] : \exp\left(i\int \mathrm{d}x\mathrm{d}y\, \overline{\psi}^{\mathrm{in}}(x)\, D_x K_{\mathrm{F}}[x, y|\mathscr{A}]\, \overleftarrow{D}_y \psi^{\mathrm{in}}(y)\right):. \tag{56}$$

This formula is a summary of our considerations on the probability amplitudes of processes occurring in the electromagnetic field when the mutual interaction of the electrons is not taken into account. All the amplitudes can then be expressed in terms of two quantities: the vacuum-to-vacuum transition amplitude and the Feynman one-electron propagator.

The theory of mutually non-interacting electrons may be generalized to the case when, in addition to an external field $\mathscr{A}_\mu(x)$, there is an external current $\mathscr{I}^\mu(x)$ (independent of the field $\mathscr{A}_\mu$). Just as in Section 13, we take the existence of the external current into account by multiplying the S operator by a numerical phase factor $\exp(-i\int d^4x\, \mathscr{I}^\mu(x)\mathscr{A}_\mu(x))$.

The vacuum-to-vacuum amplitude in the theory with external current thus is given by the formula

$$(\Omega^{\text{out}}|\Omega^{\text{in}}) = (\Omega^{\text{in}}|\mathsf{S}[\mathscr{A}, \mathscr{I}]\,\Omega^{\text{in}}) = C[\mathscr{A}]\exp\left(-i\int d^4x\, \mathscr{I}^\mu(x)\mathscr{A}_\mu(x)\right). \tag{57}$$

This amplitude will be denoted by the symbol

$$C[\mathscr{A}, \mathscr{I}] \equiv C[\mathscr{A}]\exp\left(-i\int d^4x\, \mathscr{I}^\mu(x)\mathscr{A}_\mu(x)\right). \tag{58}$$

It follows from formula (56) that the remaining transition amplitudes depend on the external current only through the amplitude $C[\mathscr{A}, \mathscr{I}]$.

This generalization will be used in the next chapter.

CHAPTER 6

THE FORMULATION OF QUANTUM ELECTRODYNAMICS

19. THE GENERAL POSTULATES OF QUANTUM ELECTRODYNAMICS

QUANTUM electrodynamics is the theory of mutually interacting electrons and the quantized electromagnetic field, and their interaction with an external electromagnetic field and external currents. We shall construct it by combining two elements already familiar to us: the quantum theory of many electrons not interacting with each other and the quantum theory of the electromagnetic field interacting with an external current. We shall also make use of the results obtained in Chapter 2 concerning a system of mutually interacting particles.

Quantum electrodynamics is an example of relativistic quantum field theory.# However, it is much more complicated than other theories of interacting fields because of its gauge invariance and the related vanishing of photon mass. The complications arising from gauge invariance were discussed in Section 8, using a simple example of classical theory which is invariant under gauge transformation of electromagnetic potentials and the phase of the complex field. We showed there that a canonical formalism which is compatible with the postulates of relativistic invariance for gauge-dependent quantities cannot be constructed. In that theory, gauge-dependent objects were auxiliary quantities and did not represent any measurable physical quantities.

Quantum field theory which is based on a set of precisely formulated postulates is called axiomatic field theory. The formulation of axiomatic field theory was first given by A. S. Wightman (1956) (cf. also the monographs of Streater and Wightman, and of Bogolyubov, Logunov, and Todorov).

The vanishing of the photon mass brings about the infrared catastrophe which we discussed in Section 15 with the simple example of a field interacting with given sources.

Quantum electrodynamics fits within the general scheme of quantum theories presented in Chapter 1. We shall discuss the general principles of this theory, casting them in the form of separate postulates. At first we shall restrict ourselves to the simplest case when neither external fields nor external currents appear. We can then make full use of the symmetry under Poincaré transformations.

0. Quantum electrodynamics is a quantum theory. The postulates I-V of quantum theory formulated in Chapter 1 hold in it.

1. *Relativistic invariance.* The unitary representation $\mathsf{U}(a, \Lambda)$ of the Poincaré group acts in the Hilbert space of the state vectors of the system. The generators of this representation, P_μ and $M_{\mu\nu}$, are interpreted as operators of the energy, momentum, and four-dimensional angular momentum.

2. *Existence of a vacuum vector.* We postulate the existence of exactly one vector Ω, invariant under all transformations $\mathsf{U}(a, \Lambda)$,

$$\mathsf{U}(a, \Lambda)\Omega = \Omega.$$

The state represented by the vector Ω is called the *vacuum state.*

3. The *spectral condition.* We assume that the spectrum of the energy operator P_0 lies on the positive semi-axis and that the only eigenvector of this operator is the vector of the vacuum state Ω. The vacuum state thus is at the same time the ground state of the system.

It follows from the spectral condition that the spectrum of the operators P_μ lies on and inside the upper cone in four-dimensional momentum space.

4. Field operators. Physical quantities of a field character are assigned operator-valued distributions which we shall call the *operators of physical fields.*† A privileged role among these operators is played by the operators of the components of the energy-momentum tensor $T^{\mu\nu}(x)$ and

† Readers interested in a mathematically more precise formulation of this postulate and the next one are referred to the monographs mentioned in the preceding footnote.

of the current density vector $j^{\mu}(x)$. These operators satisfy the continuity equations

$$\partial_{\mu} T^{\mu\nu}(x) = 0,$$
$$\partial_{\mu} j^{\mu}(x) = 0.$$

The generators of the Poincaré transformations are obtained by integrating over all space the appropriate components of the operator $T^{\mu\nu}(x)$ or the components multiplied by coordinates. The total charge operator, on the other hand, is obtained by integrating the charge density operator $j^{0}(x)$ over all space.

The operators of *physical fields* represent only those quantities which are invariant under gauge transformations. Apart from the components of the operators $T^{\mu\nu}(x)$ and $j^{\mu}(x)$, other quantities of this kind are, for example, the components of the electromagnetic field intensities to which we assign the operators $f_{\mu\nu}(x)$. Thus, the operators of physical fields do not include electromagnetic potentials or the electron field. In the theory of electrons not interacting with each other, the potential and electron field undergo the following changes under gauge transformations,

$$\mathscr{A}_{\mu}(x) \to \mathscr{A}_{\mu}(x) + \partial_{\mu} \Lambda(x),$$
$$\psi[x|\mathscr{A}] \to \mathrm{e}^{-ie\Lambda(x)} \psi[x|\mathscr{A}],$$
$$\bar{\psi}[x|\mathscr{A}] \to \mathrm{e}^{ie\Lambda(x)} \bar{\psi}[x|\mathscr{A}].$$

This means in particular that in quantum electrodynamics, field operators transforming according to relativistic transformation laws do not include operators creating and annihilating individual electrons. This is a manifestation of a basic law of conservation in nature, the law of charge conservation.

Under the action of the unitary transformations $\mathsf{U}(a, \Lambda)$, physical field operators transform in accordance with the tensor character of the physical quantities they represent (cf. formula (10.3)).

5. *Locality.* The physical field operators $F_1(x)$ and $F_2(y)$ commute at spatially separated points,

$$[F_1(x), F_2(y)] = 0, \quad \text{when} \quad (x-y)^2 < 0.$$

This postulate means that measurements made in spatially separated regions are compatible; physical fields are local.

In the presence of external fields and currents, the foregoing postulates undergo modification which follows from the considerations of earlier sections. If the external field and currents vanish in the remote past and in the far future, as we most frequently assume, the theory of mutually interacting electrons in an external field has two different relativistic structures in a single Hilbert space. There exist two different vacuum vectors Ω^{in} and Ω^{out}, and two different Poincaré group representations, $\mathsf{U}^{\text{in}}(a, \Lambda)$ and $\mathsf{U}^{\text{out}}(a, \Lambda)$, with which are associated two sets of generators, P_μ^{in}, $M_{\mu\nu}^{\text{in}}$ and P_μ^{out}, $M_{\mu\nu}^{\text{out}}$. The field operators depend on the external fields and interpolate between two structures, in the past and in the future.

The Fundamental Dynamical Postulate

The general postulates of quantum electrodynamics discussed in the preceding part of this section are still not sufficient for determining the probabilities of physical processes. For these were only postulates of a general nature, concerning every relativistic quantum field theory. What is still lacking is a postulate that would determine the dynamics of electromagnetic processes. The starting point for the formulation of such a postulate will be the theory of mutually non-interacting electrons moving in an external electromagnetic field $\mathscr{A}_\mu(x)$ in the presence of an external current $\mathscr{I}^\mu(x)$.

The *fundamental dynamical postulate* is in the form of a relation between the matrix elements (between the IN and OUT vacuum states) of the chronological products of physical field operators in the theory of mutually non-interacting electrons and in the complete quantum electrodynamics.

Let us consider a system of local physical fields, such as the intensity of the electromagnetic field, current density, energy and momentum density, etc. Operators (operator-valued distributions) representing these physical quantities in the theory of mutually non-interacting electrons will be denoted by $F_1^{(0)}[x|\mathscr{A}, \mathscr{I}], \ldots, F_n^{(0)}[x|\mathscr{A}, \mathscr{I}]$, whereas the corresponding operators in quantum electrodynamics will be denoted by $F_1[x|\mathscr{A}, \mathscr{I}], \ldots, F_n[x|\mathscr{A}, \mathscr{I}]$. We postulate the existence of the following relation between the matrix elements in the IN and OUT vacuum states of the chronological products of the operators† $F_i^{(0)}[x]$ and $F_i[x]$,

$$(\Omega^{\text{out}}|T(F_1[x_1]\,F_2[x_2]\,\dots\,F_n[x_n])\,\Omega^{\text{in}})$$

$$= \exp\left(\frac{1}{2i}\int\frac{\delta}{\delta\mathscr{A}}D^{\text{F}}\frac{\delta}{\delta\mathscr{A}}\right)(\Omega^{\text{out}}_{(0)}|T(F_1^{(0)}[x_1]F_2^{(0)}[x_2]\,\dots\,F_n^{(0)}[x_n])\,\Omega^{\text{in}}_{(0)}), \quad (1)$$

where the operation in the exponent is defined as in Section 13,

$$\int\frac{\delta}{\delta\mathscr{A}}D^{\text{F}}\frac{\delta}{\delta\mathscr{A}} \equiv \int d^4z\,d^4z'\frac{\delta}{\delta\mathscr{A}_\mu(z)}\left(-g_{\mu\nu}D_{\text{F}}(z-z')\right)\frac{\delta}{\delta\mathscr{A}_\nu(z')}. \quad (2)$$

In what sense does this postulate determine the field operators F_i in quantum electrodynamics?# To find the answer to this question we shall consider a particular operator, $F[x]$, and we shall choose two sets of operators $F_i[x_i]$ and $F_j'[x_j']$ so that for the operators of the first group the points x_i be later than x whereas for the operators of the second group the points x_j' be earlier than x. The matrix element appearing on the left-hand side of formula (1) can then be written as:

$$(\{T(F_1[x_1]\,\dots\,F_k[x_k])\}^\dagger\Omega^{\text{out}}|F[x]\,T(F_1'[x_1']\,\dots\,F_l'[x_l'])\,\Omega^{\text{in}}).$$

On the basis of the fundamental dynamical postulate, therefore, it is possible to determine all the matrix elements of every physical field operator $F[x]$ between (generalized) state vectors of the form:

$$\{T(F_1[x_1]\,\dots\,F_k[x_k])\}^\dagger\,\Omega^{\text{out}}, \qquad (x_1, \dots, x_k) \gtrsim x, \quad (3a)$$

$$T(F_1'[x_1']\,\dots\,F_l'[x_l'])\,\Omega^{\text{in}}, \qquad (x_1', \dots, x_l') \lesssim x. \quad (3b)$$

The fundamental dynamical postulate thus determines physical field operators in the Hilbert space spanned by vectors of type (3a) or (3b).

The electromagnetic field is the simplest, and at the same time the most basic, physical field. In the theory of mutually non-interacting electrons this field is described by a set of numerical functions $\partial_{[\mu}\mathscr{A}_{\nu]}(x)$. In the complete quantum electrodynamics the electromagnetic field

† To simplify the notation, we disregard the dependence on $\mathscr{A}$ and $\mathscr{I}$.

The complete answer to this question has been obtained in the relativistic quantum theory of non-interacting fields. It has been shown that the vacuum-state matrix elements of the products of the current density operators fully determine the field operators (cf. the papers by J. Langerholc and B. Schroer (1967)).

is described by the field operator (operator-valued distribution) $f_{\mu\nu}(x)$. By virtue of the fundamental dynamical postulate, the vacuum matrix elements of the chronological products of the operators $f_{\mu\nu}$ are given by the formula

$$\left(\Omega^{\text{out}}|T\left(f_{\mu_1\nu_1}(x_1)\dots f_{\mu_k\nu_k}(x_k)\right)\Omega^{\text{in}}\right)$$
$$= \exp\left(\frac{1}{2i}\int\frac{\delta}{\delta\mathscr{A}}D^{\mathrm{F}}\frac{\delta}{\delta\mathscr{A}}\right)\partial_{[\mu_1}\mathscr{A}_{\nu_1]}(x_1)\dots\partial_{[\mu_k}\mathscr{A}_{\nu_k]}(x_k)\,(\Omega^{\text{out}}_{(0)}|\Omega^{\text{in}}_{(0)}). \tag{4}$$

We shall begin the analysis of quantum electrodynamics by investigating the properties of these elements with the aid of the perturbation theory.

20. PERTURBATION THEORY AND FEYNMAN DIAGRAMS

The fundamental computational method in quantum electrodynamics consists of formal perturbation-series expansions in powers of the elementary charge. Such formal expansions constitute a convenient starting point for obtaining various relations which can be given a meaning, irrespective of the convergence of the perturbation expansions.

The elementary charge e of the electron plays a double role in quantum electrodynamics. Firstly, it specifies the coupling of electrons with an external field and, secondly, it specifies the coupling of electrons with the quantum electromagnetic field. This is a manifestation of the principle of the universality of electromagnetic interactions, which was mentioned in Section 13. In the first case the charge e always appears multiplied by the potential of the external field $\mathscr{A}$. Expansion with respect to $e\mathscr{A}$ is justified only when the external field is sufficiently weak. In the second case, the charge e occurs without the potential $\mathscr{A}$. By perturbation theory we shall mean expansion in powers of the charge e appearing in the second role. The power of the parameter e appearing in front of expressions obtained by expansion into a perturbation series determines the order of the perturbation theory. We expect that on account of the smallness of the dimensionless constant associated with this expansion (fine-structure constant $\alpha = e^2/4\pi\hbar c \simeq 1/137$), the sum of the first several terms of the expansion will be a good approximation of the full theory. This will be a good approximation even when, as is believed at present, the perturbation expansion will be only an asymptotic (divergent!) series. Comparison of theoretical results obtained by the perturbation theory method with experimental data fully confirms this expectation.

Where do the coefficients e (without the potential $\mathscr{A}$) in the formulae for the vacuum matrix elements in quantum electrodynamics come from? Their source is found in the functional derivatives with respect to $\mathscr{A}$. By virtue of formulae (17.32) and (18.45) each functional derivative acting on K_F or $C[\mathscr{A}]$ introduces exactly one factor e.

For convenience we shall introduce quantities which play the part of potentials for vacuum matrix elements (19.4) which we shall henceforth call many-photon propagators (cf. formula (13.58)),

$$T_{\mu_1 \ldots \mu_k}[z_1 \ldots z_k | \mathscr{A}, \mathscr{I}]$$

$$\equiv \exp\left(\frac{1}{2i}\int \frac{\delta}{\delta\mathscr{A}} D^F \frac{\delta}{\delta\mathscr{A}}\right) \mathscr{A}_{\mu_1}(z_1) \ldots \mathscr{A}_{\mu_k}(z_k)\,(\Omega^{\text{out}}_{(0)}|\Omega^{\text{in}}_{(0)}). \tag{1}$$

The matrix elements (19.4) of the operators of the electromagnetic field are obtained by antisymmetric differentiation of the propagator $T_{\mu_1 \ldots \mu_k}$ with respect to all arguments z_i.

The perturbation series expansion of a many-photon propagator will be determined by expanding the operation $\exp\left(\frac{1}{2i}\int \frac{\delta}{\delta\mathscr{A}} D^F \frac{\delta}{\delta\mathscr{A}}\right)$ into a power series,

$$T[z_1 \ldots z_k | \mathscr{A}, \mathscr{I}]$$

$$= \sum_{m=0}^{\infty} \frac{(-i)^m}{2^m m!} \int dw_1 \ldots dw_{2m} D^F(w_1 - w_2) \ldots D^F(w_{2m-1} - w_{2m})$$

$$\times \frac{\delta}{\delta\mathscr{A}(w_1)} \cdots \frac{\delta}{\delta\mathscr{A}(w_{2m})} \mathscr{A}(z_1) \ldots \mathscr{A}(z_k)\, C[\mathscr{A}, \mathscr{I}]. \tag{2}$$

To simplify the notation, in this formula we have omitted the vector indices just as we did from the beginning with bispinor indices. This will henceforth be done wherever it does not cause confusion. The photon propagator $D^F_{\mu\nu}$ with omitted vector indices is denoted by the symbol D^F, in contrast to the scalar propagator D_F which does not possess vector indices at all.

To calculate a many-photon propagator in any order of perturbation theory it is sufficient in principle to use the familiar formulae for functional differentiation of the vacuum-to-vacuum amplitude $C[\mathscr{A}, \mathscr{J}]$ and one-electron propagators K_F appearing as a result of the differentiation of $C[\mathscr{A}, \mathscr{J}]$. In view of the large number of similar terms making up the full propagator in the given order of perturbation theory these calculations are, however, very tedious. A systematic and at the same time illustrative classification of all the expressions obtained in perturbation theory is made possible by the Feynman diagram method to which we shall devote the rest of this section.

The Feynman diagram method was created† precisely for describing perturbation expansions in quantum electrodynamics. In Section 5 this method was applied on a limited scale to non-relativistic quantum mechanics. In quantum electrodynamics the fundamental graphical elements will be associated with the propagators $K_F[x, y|\mathscr{A}]$ and $D^F_{\mu\nu}(z-z')$, matrices $e\gamma^\mu$, external potentials $\mathscr{A}_\mu(z)$ and currents $\mathscr{J}^\mu(z)$, the quantities out of which many-photon propagators are constructed. The way in which these elements are combined into complete diagrams is dictated by the structure of the perturbation expansion of the propagators.

The General Principles of Constructing Feynman Diagrams

All the functional differentiations in formula (2) may be performed by using three basic relations:

$$\frac{\delta}{\delta \mathscr{A}_\mu(w)} \mathscr{A}_\nu(z) = \delta^\mu_\nu \delta^{(4)}(z-w), \tag{3a}$$

$$\frac{\delta}{\delta \mathscr{A}_\mu(w)} C[\mathscr{A}, \mathscr{J}] = \left(-i\mathscr{J}^\mu(w) + e\operatorname{Tr}\{\gamma^\mu K_F[w, w|\mathscr{A}]\}\right) C[\mathscr{A}, \mathscr{J}], \tag{3b}$$

$$\frac{\delta}{\delta \mathscr{A}_\mu(w)} K_F[x, y|\mathscr{A}] = -eK_F[x, w|\mathscr{A}]\gamma^\mu K_F[w, y|\mathscr{A}]. \tag{3c}$$

† R. P. Feynman (1949b).

For illustration let us calculate the second functional derivative of the expression $\mathcal{A}(z_1)\mathcal{A}(z_2)C[\mathcal{A},\mathcal{J}]$,

$$\frac{\delta^2}{\delta\mathcal{A}(w_1)\,\delta\mathcal{A}(w_2)}\,\mathcal{A}(z_1)\,\mathcal{A}(z_2)\,C[\mathcal{A},\mathcal{J}]$$

$$= C[\mathcal{A},\mathcal{J}]\{\delta(z_1-w_1)\,\delta(z_2-w_2)+\delta(z_1-w_2)\,\delta(z_2-w_1)$$

$$+\big(\mathcal{A}(z_1)\,\delta(z_2-w_2)+\mathcal{A}(z_2)\,\delta(z_1-w_2)\big)\big(-i\mathcal{J}(w_1)+e\gamma(w_1)K_F[w_1,w_1|\mathcal{A}]\big)$$

$$+\big(\mathcal{A}(z_1)\,\delta(z_2-w_1)+\mathcal{A}(z_2)\,\delta(z_1-w_1)\big)\big(-i\mathcal{J}(w_2)+e\gamma(w_2)K_F[w_2,w_2|\mathcal{A}]\big)$$

$$+\mathcal{A}(z_1)\,\mathcal{A}(z_2)\,\big(-i\mathcal{J}(w_1)+e\gamma(w_1)K_F[w_1,w_1|\mathcal{A}]\big)$$

$$\times\big(-i\mathcal{J}(w_2)+e\gamma(w_2)K_F[w_2,w_2|\mathcal{A}]\big)$$

$$-e^2\mathcal{A}(z_1)\,\mathcal{A}(z_2)\gamma(w_1)\,\gamma(w_2)K_F[w_1,w_2|\,\mathcal{A}]\,K_F[w_2,w_1|\mathcal{A}]\}.$$

Here we have made use of the abbreviated notations explained in Section 17. The product of the Kronecker symbol and the four-dimensional Dirac δ-function $\delta^{\nu}_{\mu}\,\delta^{(4)}(z-w)$ has also been abbreviated to $\delta(z-w)$.

Analytical expression	Graphical element
$-iK_F[x,y\|\mathcal{A}]$	
$-ie\gamma(w)$	
$+ie\gamma(w)K_F[x,w\|\mathcal{A}]\,K_F[w,y\|\mathcal{A}]$	
$e\gamma(w)K_F[w,w\|\mathcal{A}]$	
$-i\mathcal{J}(w)$	
$\mathcal{A}(z)$	
$\delta(z-w)$	
$-iD^F(w_i-w_j)$	

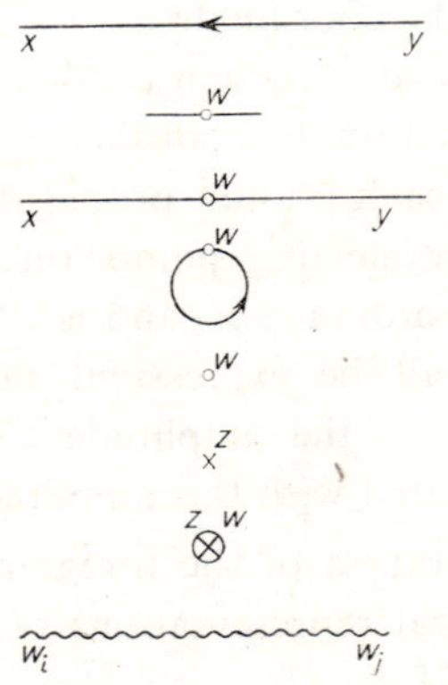

The number of terms increases rapidly with the multiplicity of the derivatives. The diagram method helps to put into order the expressions obtained by differentiation.

This method is based on the following rules of association (collected together in the table on page 319).

1. With each electron propagator $-i\,K_F[x_i, y_j|\mathscr{A}]$ we associate a segment of line which we shall call an *electron line*. The ends of the electron line will be denoted by the coordinates x_i and y_j, whereas the line itself will be provided with an arrow directed from the point y_j to the point x_i. If the same argument appears in two propagators, the end and the beginning of the corresponding electron lines are joined.

   If both arguments of the same propagator are identical, we connect the end and the beginning of the corresponding electron line, thus forming a closed loop. Each closed loop introduces an extra minus sign.
2. With each matrix $-ie\gamma(w_i)$ we associate a small circle labelled with the coordinate w_i, drawn at the point where the corresponding two electron lines meet or at the point where the corresponding electron line closes to form a loop.
3. With each external current $-i\mathscr{J}(w_i)$ we associate a small circle labelled with the coordinate w_i (not lying on electron lines).
4. With each external potential $\mathscr{A}(z_i)$ we associate a cross labelled with the coordinate z_i.
5. With each function $\delta(z_i - w_j)$ we associate a small cross in a circle, denoted by the coordinates z_i and w_j.
6. With each photon propagator $-iD^F(w_i - w_j)$ we associate a wavy line connecting points on the diagram which are labelled with the coordinates w_i and w_j. We shall call it a *photon line*.
7. Since all the expressions making up the propagator are proportional to the amplitude $C[\mathscr{A}, \mathscr{J}]$, no graphical element will be associated with this amplitude.

As illustration of the foregoing association rules, in Fig. 20.1 we give graphical representations of the third functional derivative of the amplitude $C[\mathscr{A}, \mathscr{J}]$.

These rules are sufficient for associating diagrams not only with functional derivatives of the expression $\mathscr{A}(z_1) \dots \mathscr{A}(z_k)C[\mathscr{A}, \mathscr{J}]$, but also with all functional derivatives of more general expressions constructed out of products of the one-electron propagators K_F and

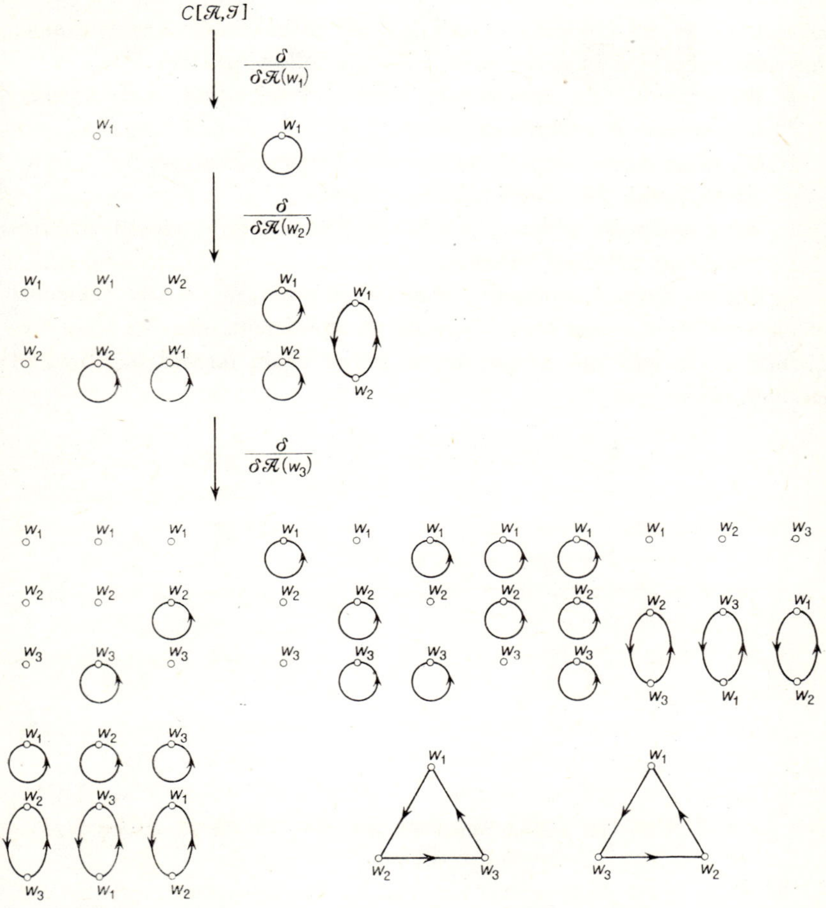

FIG. 20.1. The third functional derivative of the amplitude $C[\mathscr{A}, \mathscr{J}]$.

the amplitude $C[\mathscr{A}, \mathscr{J}]$. Expressions of this type, which we shall subsequently need, are of the form:

$$\mathscr{A}(z_1) \dots \mathscr{A}(z_k)\, C[\mathscr{A}, \mathscr{J}](-i)^n K_F[x_1, y_1|\mathscr{A}] \dots K_F[x_n, y_n|\mathscr{A}]. \quad (4)$$

The graphical representation of this expression is a set of k crosses and n electron lines.

It follows from the rules for the differentation of products that the set of diagrams representing the i-th derivative with respect to

$\mathscr{A}(w_1), \ldots, \mathscr{A}(w_i)$ contains exactly all the *topologically non-equivalent diagrams*† obtained in accordance with the following rules:

a. We draw all the crosses and lines corresponding to the expression subject to differentiation.
b. On these crosses and lines, and also beyond them, we put i circles labelled with the coordinates $w_1, \ldots, w_i$.
c. We connect an arbitrary number of free circles to closed electron lines in an arbitrary manner.

It follows from formula (3c) that the orientations of two segments of electron lines joined by a circle are in agreement; thus we shall frequently draw only one arrow for an entire chain (closed or open) of electron lines.

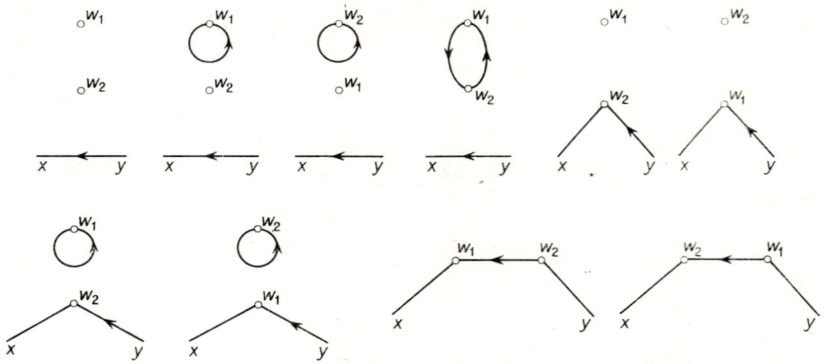

FIG. 20.2. The second functional derivative of the expression $-iC[\mathscr{A}, \mathscr{I}]K_F[x, y|\mathscr{A}]$.

These rules are illustrated in Fig. 20.2 wherein we give the construction of diagrams representing the second functional derivative of the expression $-iC[\mathscr{A}, \mathscr{I}]K_F[x, y|\mathscr{A}]$.

† Topologically non-equivalent diagrams are ones which could not be transformed into each other by rotations and bends, if models of these diagrams were to be made of a plastic material, with the vertices and ends of the lines on these models labelled and with the directions of the arrows marked. In other words, topologically non-equivalent diagrams cannot be transformed into each other through a deformation of the lines.

On the basis of the diagrams drawn in accordance with the rules a, b, c, we can write out the analytical expressions for the functional derivative of any order of product (4). To this end it is sufficient to apply association rules 1–7 and an *additional rule concerning the sign*:

The analytical expression constructed in accordance with rules 1-7 is multiplied by $(-1)^L$, where L is the number of closed chains of electron lines in the diagram.

For example, the following analytical expressions correspond to the diagrams in Fig. 20.3:

$$-(-i)^6 e^3 \mathrm{Tr}\{\gamma^{\mu_1} K_F[w_1, w_2|\mathscr{A}]\gamma^{\mu_2} K_F[w_2, w_3|\mathscr{A}]\gamma^{\mu_3} K_F[w_3, w_1|\mathscr{A}]\} C[\mathscr{A}, \mathscr{I}],$$

$$(-i)^5 e^2 K_F[x, w_1|\mathscr{A}]\, \gamma^{\mu_1} K_F[w_1, w_2|\mathscr{A}]\, \gamma^{\mu_2} K_F[w_2, y|\mathscr{A}]\, C[\mathscr{A}, \mathscr{I}].$$

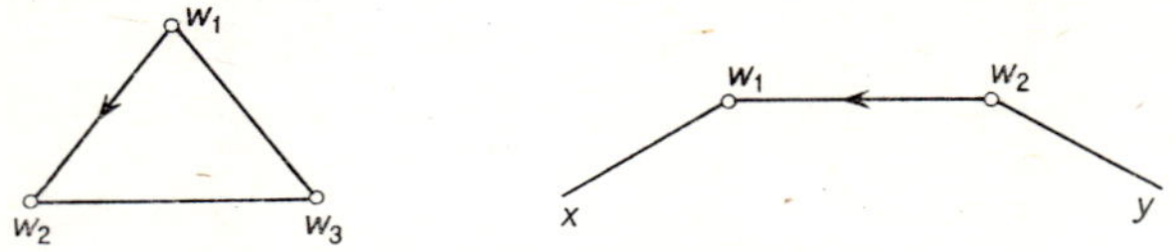

FIG. 20.3. Examples of diagrams.

In the series expansion (2) of the many-photon propagator the result of the functional differentiation is multiplied by the product of the photon propagators and is integrated with respect to all the variables $w_1, \ldots, w_{2m}$. In diagrams these operations are represented by connecting circles w_1 with w_2, then w_3 with w_4, etc. by wavy lines (photon lines). The photon lines may begin and end at free circles, at circles with crosses, and at circles lying on electron lines.

To simplify the diagrams we shall not draw the crosses with circles at the ends of photon lines, but leave only the coordinate labels of these ends. The junction of a photon line and two electron lines, or one closed electron loop is called a *vertex of a diagram*. Since one factor $-ie\gamma(w)$ is associated with each vertex, the number of vertices in a diagram determines the order of the perturbation theory for the expression represented by the given diagram. Small circles representing an external current will be called *external vertices of the diagram*. Photon lines connecting the vertices of a diagram are called *internal photon lines*. The other photon lines will be referred to as *external lines*. One or both ends of external lines are labelled with coordinates z_i. Each contribution to a propagator $T[z_1 \ldots z_k]$ is a p-fold integral with respect to the coordinates $w_1, \ldots, w_p$ labelling the vertices of the corresponding diagram. The remaining integrations may be performed thanks to the functions $\delta(z_i - w_j)$.

In the diagrams representing the final expressions composing the full propagator we can drop the labelling of the vertices with the variables w_i, for they are always variables of integration. The analytical expressions obtained by integration with respect to the variables labelling the vertices are identical if the corresponding diagrams differ from each other only as to the labelling of the vertices. In Fig. 20.4, as an illustration,

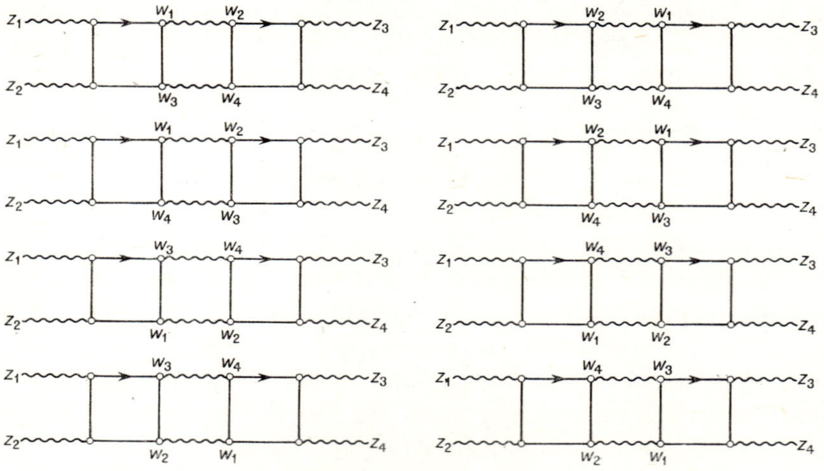

FIG. 20.4. Eight diagrams yielding the same result on integration with respect to the variables w_1, w_2, w_3, and w_4.

we give eight diagrams representing integrand expressions which yield identical results on integration. In order to evaluate the contribution to the propagator from all the expressions represented by these eight diagrams, it is sufficient to calculate the contribution from one of them and multiply the result by 8. All the essential information contained in the diagrams in Fig. 20.4 can be given by drawing the single diagram in Fig. 20.5 in which only the ends of the external lines are labelled.

Diagrams without labelled vertices will be called *Feynman diagrams*. Each Feynman diagram represents the set of all topologically non-equivalent diagrams, differing as to labelling of the vertices. The number

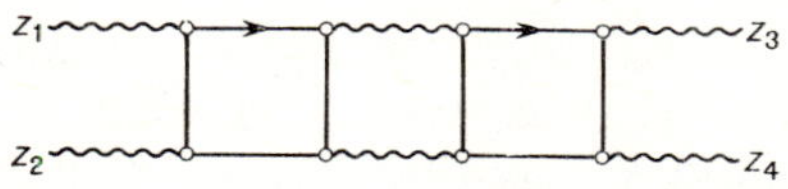

FIG. 20.5. Feynman diagram replacing the diagrams in Fig. 20.4.

of diagrams of this set multiplied by $(2^k k!)^{-1}$, where k is the number of internal photon lines, will be called the *combinatorial factor* of the given Feynman diagram. The combinatorial factor of the diagram in Fig. 20.5 is equal to 1 and those of the diagrams in Fig. 20.6 are: $\frac{1}{2}$ (diagram a), $\frac{1}{6}$ (diagram b), and 1 (diagrams c and d).

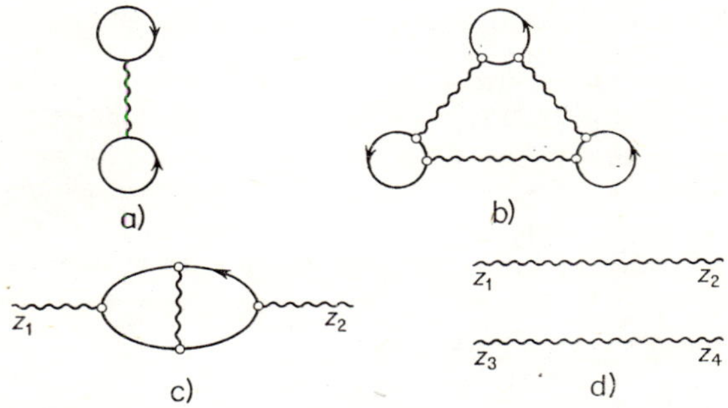

FIG. 20.6. Feynman diagrams.

Calculation of the combinatorial factors is facilitated by two lemmas given below.

Lemma 1. The combinatorial factor of a diagram having n internal photon lines is equal to $(2^n n!)^{-1} W$, where W is the number of topologically non-equivalent ways in which labelled internal photon lines with orientations marked can be arranged on a given diagram.

The proof of this lemma consists in the observation that the aforementioned equality of integrals has its source in the symmetry of the propagator $D^F_{\mu\nu}(w_1 - w_2)$ under the interchanges $w_1 \leftrightarrow w_2$, $\mu \leftrightarrow \nu$, and the symmetry of the products of the propagators D^F with respect to a change in the order of the propagators.

The combinatorial factor of any diagram can be expressed in terms of the combinatorial factors of the connected parts of the diagram. A *connected part of a diagram* is one which cannot be divided into two parts without cutting any line. A diagram which consists of only one connected part is called a *connected diagram*. Each diagram is divided into a vacuum part and a linked part. The *vacuum part* of the diagram is a system of lines not connected with any external lines of the diagram. The *linked part* of the diagram is the system of all its connected parts joined to external lines. The linked part is what remains when the vacuum part is separated. Both the vacuum part and the linked part of the diagram may be disconnected.

Lemma 2. The combinatorial factor of a whole diagram is the product of the combinatorial factor of the vacuum part and the combinatorial factors of all the connected parts constituting the linked part.

Suppose that there are n internal photon lines. The combinatorial factor of the entire diagram is equal to $(1/2)^n(1/n!)\,W$, where W is the number of ways in which the photon lines can be deployed in the diagram. If the vacuum part and the linked parts, respectively, contain $n_0, n_1, \ldots, n_k$ photon lines, then all the lines may be classified into groups occurring in the corresponding parts of the diagrams in $n!/(n_0!\, n_1! \ldots n_k!)$ ways. In turn suppose that within each of these parts the photon lines can be deployed in $W_0, W_1, \ldots, W_k$ non-equivalent ways. We then obtain the following equality:

$$\left(\frac{1}{2}\right)^n \frac{1}{n!} W = \left(\frac{1}{2}\right)^{n_0} \frac{1}{n_0!} W_0 \left(\frac{1}{2}\right)^{n_1} \frac{1}{n_1!} W_1 \ldots \left(\frac{1}{2}\right)^{n_k} \frac{1}{n_k!} W_k, \tag{5}$$

which is the essence of Lemma 2.

The combinatorial factor of the vacuum part is not in general equal to the product of the combinatorial factors of its connected parts. This follows from the fact that the vacuum part may contain several identical connected parts (cf. Fig. 20.1). An interchange of photon lines between such identical parts does not lead to a non-equivalent set of photon lines in the sense of Lemma 1.

Let us consider a general vacuum diagram $V(k_1, k_2, \ldots)$ containing k_1 copies of the connected part C_1, k_2 copies of the connected part C_2, etc. In dividing the photon lines into groups, as mentioned in Lemma 2, it is necessary to take account of the indistinguishability of the identical connected parts. As a result, the combinatorial factor of the diagram $V(k_1, k_2, \ldots)$ is equal to the product of the combinatorial factors of the connected diagrams divided by the factor $k_1!\,k_2!\,\ldots$. For vacuum diagrams the counterpart of formula (5) thus takes on the form:

$$\left(\frac{1}{2}\right)^n \frac{1}{n!} W = \frac{1}{k_1!}\left[\left(\frac{1}{2}\right)^{n_1} \frac{1}{n_1!} W_1\right]^{k_1} \frac{1}{k_2!}\left[\left(\frac{1}{2}\right)^{n_2} \frac{1}{n_2!} W_2\right]^{k_2} \cdots \tag{6}$$

Analysis of the Connectedness of Propagators

We shall now carry out an analysis of the connectedness of diagrams and the associated decomposition of propagators into simpler parts. In the next section we shall demonstrate the existence of a relation between the probability amplitudes of photon processes and propagators. A similar relation occurred in the simplified quantum theory of the electromagnetic field interacting only with an external current (cf. Section 13). The decomposition of propagators into connected parts thus will determine the decomposition of the corresponding transition amplitudes into connected parts.

In the simplified theories of electromagnetic interactions, described in Chapters 4 and 5, each transition amplitude for any system of particles could be expressed in terms of the vacuum-to-vacuum transition amplitude and one-particle amplitudes. The one-particle amplitudes were, respectively, the amplitudes of the emission and absorption of a single photon, or the negaton and positon scattering amplitudes and the creation and annihilation amplitudes of a single pair. In the theory now under discussion, the transition amplitudes are much more complicated in structure. An arbitrary transition amplitude cannot be expressed in terms of the transition amplitudes for a smaller number of particles. It will be possible only to separate from it the part constructed from simpler amplitudes, but a part which is not further decomposable always remains.

The first step in the analysis of the connectedness of propagators consists in isolating the vacuum parts out of the propagators. For this purpose we shall divide each diagram representing a contribution to a propagator into a vacuum part and a linked part.

The set of all diagrams representing a given propagator may be divided into subsets of diagrams in such a way that all diagrams possessing the same linked part belong to a given subset. Diagrams belonging to a given subset thus differ only as to their vacuum parts. The simplest diagrams of a given subset are presented in Fig. 20.7. The

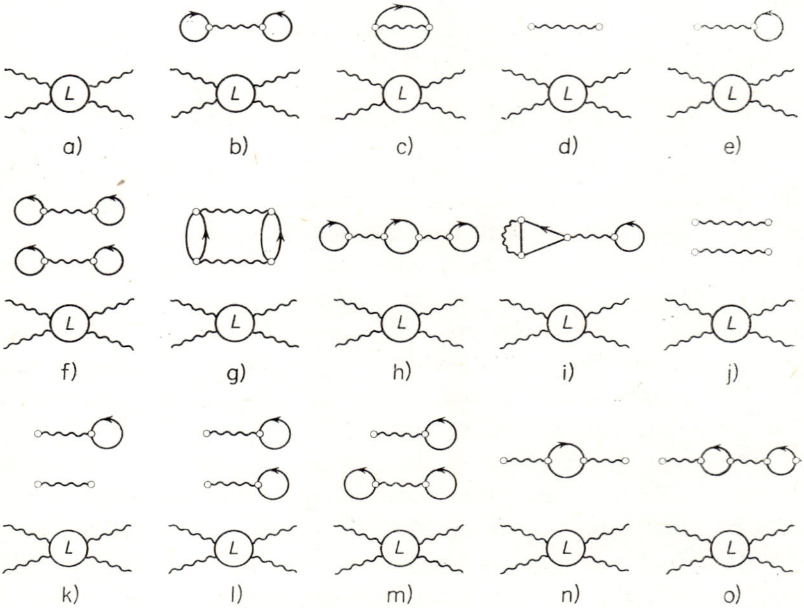

FIG. 20.7. The simplest diagrams possessing the same linked part.

part of the diagram consisting of a circle labelled with an L (for linked) and with symbolically marked external lines represents the common linked part. If the expressions corresponding to the vacuum parts of diagrams (a), (b), (c), (d), etc., are denoted by the symbols† V_0, V_1,

† The factor $C[\mathscr{A}, \mathscr{I}]$ which is not schown in the diagrams is incorporated into the vacuum part of the propagator. We thus have:

$$V_0 = C[\mathscr{A}, \mathscr{I}], \quad V_1 = -\frac{ie^2}{2}\int \mathrm{Tr}\{\gamma K_F\} D^F \mathrm{Tr}\{\gamma K_F\} C[\mathscr{A}, \mathscr{I}],$$

$$V_2 \frac{ie^2}{2}\int \mathrm{Tr}\{\gamma K_F \gamma K_F\} D^F C[\mathscr{A}, \mathscr{I}], \quad V_3 = \frac{i}{2}\int \mathscr{I} D^F \mathscr{I} C[\mathscr{A}, \mathscr{I}], \text{ etc.}$$

V_2, V_3, etc., and the expressions associated with the linked part by the symbol L, the sum of the expressions corresponding to the diagrams of a given subset may be written in the form

$$V_0 L + V_1 L + V_2 L + \ldots = \left(\sum_{i=0}^{\infty} V_i\right) L,$$

since by virtue of Lemma 2, the combinatorial factor of each diagram is the product of the combinatorial factors of the vacuum part and the linked part. The sum of all contributions to the propagator may thus be rewritten as the product

$$T[z_1 \ldots z_k | \mathscr{A}, \mathscr{I}] = V[\mathscr{A}, \mathscr{I}] T^L[z_1 \ldots z_k | \mathscr{A}, \mathscr{I}], \tag{7}$$

where $V[\mathscr{A}, \mathscr{I}]$ is the sum of the contributions from all the vacuum parts,

$$V[\mathscr{A}, \mathscr{I}] = \sum_{i=0}^{\infty} V_i,$$

whereas T^L contains all expressions corresponding only to linked diagrams. The expressions T^L will sometimes be called the *proper propagator* or simply *propagator*.

The sum of the contributions from all vacuum parts is a universal quantity which is independent of the propagator studied. It may be expressed by the formula

$$V[\mathscr{A}, \mathscr{I}] = \exp\left(\frac{1}{2i} \int \frac{\delta}{\delta \mathscr{A}} D^{\mathrm{F}} \frac{\delta}{\delta \mathscr{A}}\right) C[\mathscr{A}, \mathscr{I}]. \tag{8}$$

This is the vacuum-to-vacuum transition amplitude in full quantum electrodynamics.

After separating the vacuum part, we can proceed with the division by decomposing the proper propagators into connected parts.

In the general case, diagrams representing contributions to the proper propagator† $T^L[z_1 \ldots z_k]$ are not connected. These diagrams do not contain vacuum parts, it is true, but not all the external lines are joined with each other by internal lines. However, the following important rule does hold. Each connected part of a disconnected

† To simplify the notation, we omit the symbols $\mathscr{A}$ and $\mathscr{I}$.

diagram representing a contribution to the propagator $T[z_1 \dots z_k]$ and having l external photon lines is one of the connected diagrams representing a contribution to the propagator $T[z_1 \dots z_l]$. The converse statement is also true; each connected diagram corresponding to a propagator $T[z_1 \dots z_l]$ occurs as the connected part of disconnected diagrams corresponding to the propagator $T[z_1 \dots z_k]$ where $k > l$. The sum of the expressions corresponding to all the connected diagrams of a given propagator will be denoted by the symbol $T^C[z_1 \dots z_k]$ (C for connected) and called the *connected part of the propagator*. Each propagator is decomposable into connected parts in the following manner,

$$T^L[z_1 \dots z_k] = T^C[z_1 \dots z_k] + \sum T^C[z_{l_1} \dots] T^C[z_{l_2} \dots] \dots T^C[z_{l_k} \dots]. \tag{9}$$

The summation in this formula runs over all partitions of variables $z_1 \dots z_k$ into groups. Since we showed earlier that the same kind of diagrams correspond to both sides of equality (9), in order to prove this equality it is sufficient merely to note that by Lemma 2 the combinatorial factors are also the same.

Now we shall give examples to illustrate the decomposition of several propagators into connected parts. The simplest of them is the propagator $T^L[z]$, to which diagrams with a single external photon line correspond. We shall denote it here by the symbol $\mathfrak{A}[z]$. This propagator, which (as we shall demonstrate later) describes processes of photon emission and absorption caused by an external current and/or an external electromagnetic field, is represented solely by connected diagrams. The simplest of them are shown in Fig. 20.8. The decomposition of the propagator $\mathfrak{A}[z]$ into connected parts thus is of the form

$$\mathfrak{A}[z] \equiv T^L[z] = T^C[z] = (V[\mathscr{I}, \mathscr{A}])^{-1} T[z]. \tag{10}$$

The next propagator, $T^L[z_1 \; z_2]$, to which diagrams with two external lines correspond, describes processes in which two photons participate (independent emission and/or absorption of both photons under the influence of external currents and fields and the correlated emission of one photon and/or the absorption of the other). The

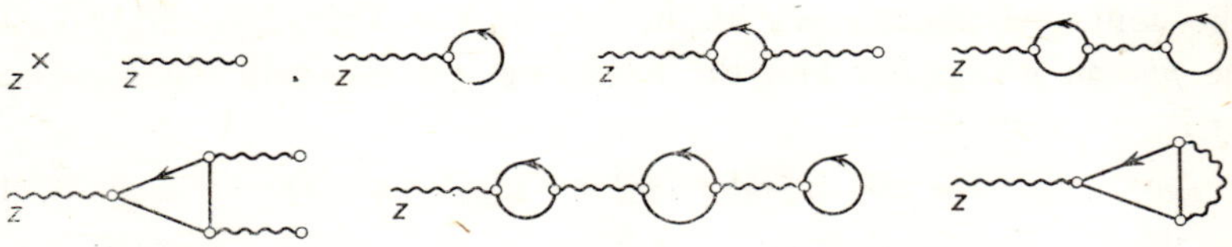

FIG. 20.8. The simplest diagrams representing the propagator $\mathfrak{A}[z]$.

simplest diagrams corresponding to the propagator $T^L [z_1 z_2]$ are shown in Fig. 20.9.

Some diagrams representing the propagator $T^L[z_1 z_2]$ are disconnected (for example, the diagrams in the upper part of Fig. 20.9), their connected parts corresponding to the propagator $T^L[z]$. The decomposition of

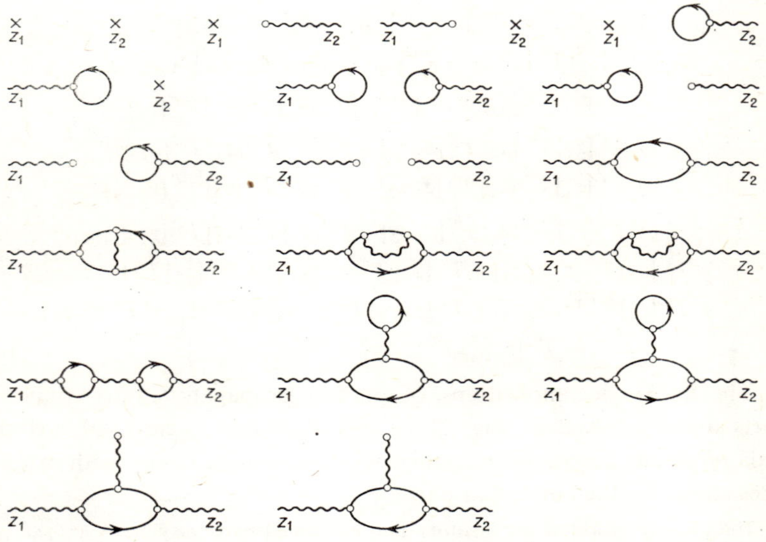

FIG. 20.9. The simplest diagrams representing the propagator $T^L[z_1 z_2]$.

the propagator $T^L[z_1 z_2]$ into connected parts thus is expressed by the formula

$$T^L[z_1 z_2] = T^C[z_1] T^C[z_2] + T^C[z_1 z_2]. \tag{11}$$

The connected component of the propagator $T^L[z_1 z_2]$ will be called the *photon propagator* and the following notation will be introduced for it,

$$-i\mathscr{G}_{\mu\nu}[z_1 z_2|\mathscr{A}, \mathscr{I}] \equiv T^C_{\mu\nu}[z_1 z_2|\mathscr{A}, \mathscr{I}]. \tag{12}$$

This object is a generalization of the free photon propagator and goes over into $D^{\mathrm{F}}_{\mu\nu}(z_1 - z_2)$ when we switch off the electromagnetic interaction of electrons with the quantum electromagnetic field (i.e. in the zeroth-order perturbation theory),

$$\mathscr{G}^{(0)}_{\mu\nu}[z_1 z_2] = D^{\mathrm{F}}_{\mu\nu}(z_1 - z_2). \tag{13}$$

The next propagators in turn, $T^L[z_1\, z_2\, z_3]$ and $T^L[z_1\, z_2\, z_3\, z_4]$, describe three- and four-photon processes. Their decomposition into connected parts is given by the formulae

$$\begin{aligned} T^L[z_1\, z_2\, z_3] = {} & T^C[z_1]T^C[z_2]T^C[z_3] + T^C[z_1]T^C[z_2\, z_3] \\ & + T^C[z_2]T^C[z_1\, z_3] + T^C[z_3]T^C[z_1\, z_2] + T^C[z_1\, z_2\, z_3], \end{aligned} \tag{14a}$$

$$\begin{aligned} T^L[z_1\, z_2\, z_3\, z_4] = {} & T^C[z_1]T^C[z_2]T^C[z_3]T^C[z_4] \\ & + T^C[z_1]T^C[z_2]T^C[z_3 z_4] + T^C[z_1]T^C[z_3]T^C[z_2 z_4] \\ & + T^C[z_1]T^C[z_4]T^C[z_2 z_3] + T^C[z_2]T^C[z_3]T^C[z_1 z_4] \\ & + T^C[z_2]T^C[z_4]T^C[z_1 z_3] + T^C[z_3]T^C[z_4]T^C[z_1 z_2] \\ & + T^C[z_1]T^C[z_2 z_3 z_4] + T^C[z_2]T^C[z_1 z_3 z_4] + T^C[z_3]T^C[z_1 z_2 z_4] \\ & + T^C[z_4]T^C[z_1 z_2 z_3] + T^C[z_1 z_2]T^C[z_3 z_4] + T^C[z_1 z_3]T^C[z_2 z_4] \\ & + T^C[z_1 z_4]T^C[z_2 z_3] + T^C[z_1 z_2 z_3 z_4]. \end{aligned} \tag{14b}$$

The above decompositions of photon propagators into connected parts are illustrated in Fig. 20.10 where double circles with external ends represent proper propagators, whereas shaded circles with external lines stand for the connected parts.

The decomposition of propagators into connected parts may be described concisely by means of the concept of a generating functional for photon propagators. Such a generating functional for the propagators $T[z_1 \ldots z_k]$ is the vacuum-to-vacuum transition amplitude treated as a functional of the external current. For on expanding the exponential function $\exp(-i\int \tilde{\mathscr{I}}\mathscr{A})$ into a series, we obtain the formula

$$V[\mathscr{A}, \mathscr{I} + \tilde{\mathscr{I}}] = \sum_{k=0}^{\infty} \frac{(-i)^k}{k!} \int \mathrm{d}z_1 \ldots \mathrm{d}z_k\, T[z_1 \ldots z_k|\mathscr{A}, \mathscr{I}]\tilde{\mathscr{I}} \ldots \tilde{\mathscr{I}}. \tag{15}$$

In order to determine the generating functional for the connected parts of the photon propagators we shall make an analysis of the connectedness of the vacuum diagrams. In the general case, vacuum diagrams are disconnected but they can all

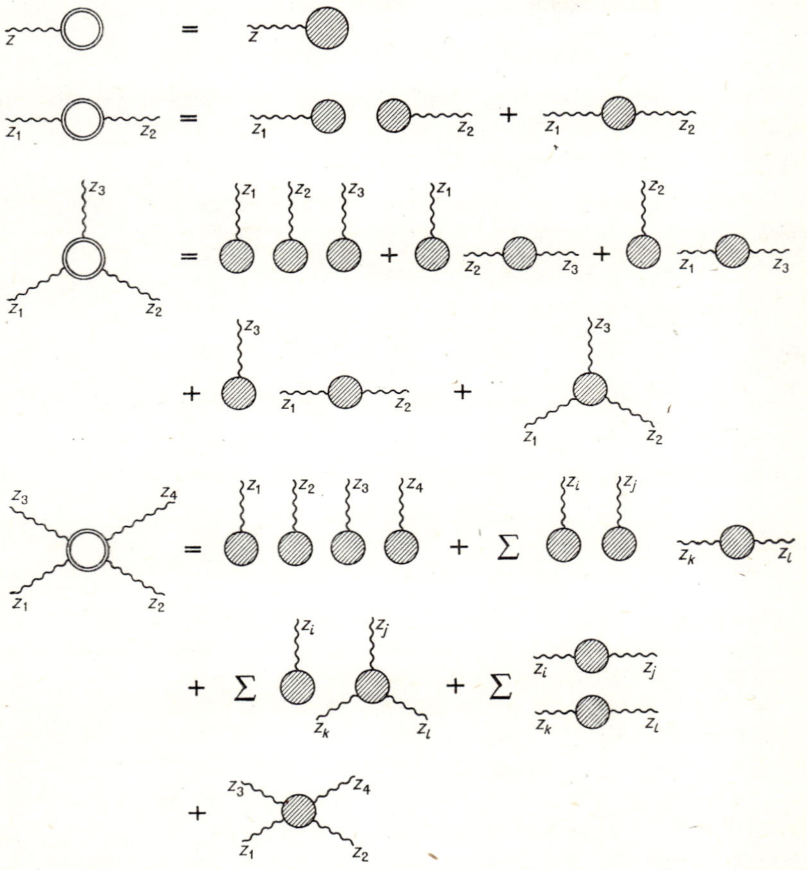

FIG. 20.10. The decomposition of many-photon propagators into connected parts.

be built out of universal connected vacuum diagrams. Let us use V_1^C, V_2^C, V_3^C, etc., to denote the contributions from the connected vacuum diagrams. A disconnected vacuum diagram in which the connected part V_1^C is contained k_1 times, the connected part V_2^C is contained k_2 times, etc., by virtue of the generalized Lemma 2 (formula (6)) makes the following contribution to the vacuum-to-vacuum amplitude:

$$C[\mathscr{A},\mathscr{I}]\frac{1}{k_1!}(V_1^C)^{k_1}\frac{1}{k_2!}(V_2^C)^{k_2}\ldots .$$

The sum of all vacuum contributions may then be rewritten as an exponential function:

$$V[\mathscr{A},\mathscr{I}] = C[\mathscr{A},\mathscr{I}]\exp(W_1[\mathscr{A},\mathscr{I}]),$$

where $W_1[\mathscr{A},\mathscr{I}]$ is the sum of the contributions from all connected vacuum diagrams (omitting the factor $C[\mathscr{A},\mathscr{I}]$),

$$W_1[\mathscr{A},\mathscr{I}] = V_1^C + V_2^C + \ldots .$$

In the preceding chapter, in similar fashion we presented the amplitude $C[\mathscr{A},\mathscr{I}]$ which, by virtue of formulae (17.81) and (18.58), is of the form

$$C[\mathscr{A},\mathscr{I}] = \exp(W_0[\mathscr{A},\mathscr{I}]),$$

where

$$W_0[\mathscr{A},\mathscr{I}] = -i\int dz\,\mathscr{I}(z)\cdot\mathscr{A}(z)$$

$$-\sum_{n=1}^{\infty}\frac{e^{2n}}{2n}\int dz_1\ldots dz_n \operatorname{Tr}\{\gamma\cdot\mathscr{A}(z_1)S_F(z_1-z_2)\ldots\gamma\cdot\mathscr{A}(z_{2n})S_F(z_{2n}-z_1)\}.$$

Finally, therefore, the vacuum-to-vacuum transition amplitude is written as

$$V[\mathscr{A},\mathscr{I}] = \exp(W[\mathscr{A},\mathscr{I}]),$$

where

$$W[\mathscr{A},\mathscr{I}] = W_0[\mathscr{A},\mathscr{I}] + W_1[\mathscr{A},\mathscr{I}].$$

The functional $W[\mathscr{A},\mathscr{I}]$ contains contributions from all the connected vacuum diagrams (described by W_1) and contributions from all vacuum diagrams in the theory of mutually non-interacting electrons (described by W_0).

The functional $W[\mathscr{A},\mathscr{I}]$ is the generating functional for the connected parts of the propagators, i.e.

$$W[\mathscr{A},\mathscr{I}+\tilde{\mathscr{I}}] = W[\mathscr{A},\mathscr{I}]$$

$$+\sum_{k=1}^{\infty}\frac{(-i)^k}{k!}\int dz_1\ldots dz_k\, T^C[z_1\ldots z_k|\mathscr{A},\mathscr{I}]\,\tilde{\mathscr{I}}(z_1)\ldots\tilde{\mathscr{I}}(z_k). \tag{16}$$

For the proof, we decompose the propagators $T[z_1\ldots z_k|\mathscr{A},\mathscr{I}]$ in formula (15) into the sum of the products of connected components (cf. formula (7)), and we then sum up the resulting series into an exponential function. This is done most easily by applying the procedure of successively separating the factors:

$$\exp\left(-i\int T^C[z]\cdot\tilde{\mathscr{I}}(z)\right),\quad \exp\left(-\frac{1}{2}\int T^C[z_1z_2]\tilde{\mathscr{I}}(z_1)\tilde{\mathscr{I}}(z_2)\right),\ \text{etc.}$$

Our analysis of the propagators with the aim of separating from them parts with a relatively simple structure may be continued once the connected components have been separated. Namely, we shall separate strongly connected parts from the connected components.

A diagram is said to be *weakly connected* if by breaking one internal line we can divide it into two diagrams, each of which contains some external lines of the original diagram. Otherwise, the diagram is said to be *strongly connected.* Examples of weakly and strongly connected diagrams are given in Fig. 20.11.

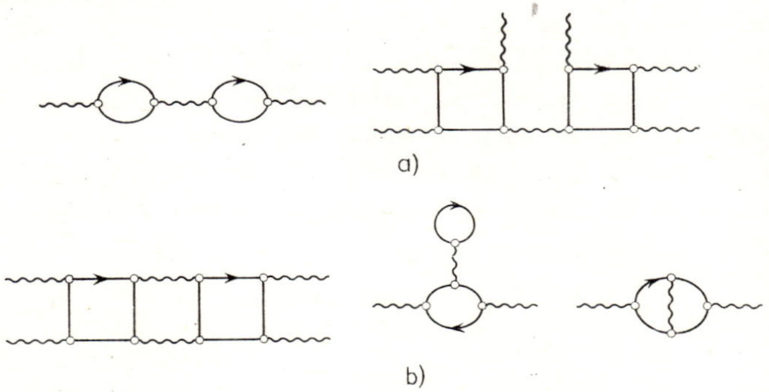

FIG. 20.11. Examples of (a) weakly and (b) strongly connected diagrams.

The sum of all expressions corresponding to strongly connected diagrams (without external lines), representing a given propagator, is called the *strongly connected part* of the propagator and is denoted by the symbol $\Delta^{\mu_1 \ldots \mu_k}[z_1 \ldots z_k]$.

For the simplest propagator T_μ $[z]$ we have:

$$T_\mu[z] = \mathscr{A}_\mu(z) - i \int \mathscr{G}_{\mu\nu}[z z'] \, \Delta^\nu[z']. \tag{17}$$

Let us examine the decomposition of the photon propagator into strongly connected parts. Among the diagrams in Fig. 20.9, all connected diagrams, with the exception of the last one, are at the same time strongly connected. Strongly connected diagrams representing the photon propagator are called *photon self-energy diagrams.* All weakly connected diagrams representing the photon propagator can be built out of the photon self-energy diagrams if the ends of the external lines are connected in pairs into one chain. Each series connection of photon self-energy diagrams yields one of the diagrams representing the photon propagator. This construction may be visualized graphically, as we have done in Fig. 20.12, where the cross-hatched circle with external lines

FIG. 20.12. The decomposition of a photon propagator into strongly connected parts.

attached represents the set of all diagrams of the photon self-energy. The analytical expression which is the sum of all the expressions corresponding to the diagrams obtained from photon self-energy diagrams by detaching external lines is called the *photon self-energy function* and is denoted by the symbol $i\,\Pi^{\lambda\varrho}[z_1 z_2|\mathscr{A}\mathscr{I}]$. The symbolical equality shown in Fig. 20.12 may be expressed in terms of the photon self-energy function as follows:

$$\begin{aligned}
-i\mathscr{G}_{\mu\nu}[z_1 z_2] &= -iD^{\mathrm{F}}_{\mu\nu}(z_1-z_2)\\
&+(-i)^2 i\int \mathrm{d}z_1'\,\mathrm{d}z_2'\,D^{\mathrm{F}}_{\mu\lambda}(z_1-z_1')\Pi^{\lambda\varrho}[z_1'z_2']D^{\mathrm{F}}_{\varrho\nu}(z_2'-z_2)\\
&+(-i)^3\, i^2\int \mathrm{d}z_1'\,\mathrm{d}z_2'\,\mathrm{d}z_3'\,\mathrm{d}z_4'\,D^{\mathrm{F}}_{\mu\lambda}(z_1-z_1')\Pi^{\lambda\varrho}[z_1'z_2']\,D^{\mathrm{F}}_{\varrho\varkappa}(z_2'-z_3')\\
&\times\Pi^{\varkappa\sigma}[z_3'z_4']\,D^{\mathrm{F}}_{\sigma\nu}(z_4'-z_2)+\ \ldots . \qquad (18)
\end{aligned}$$

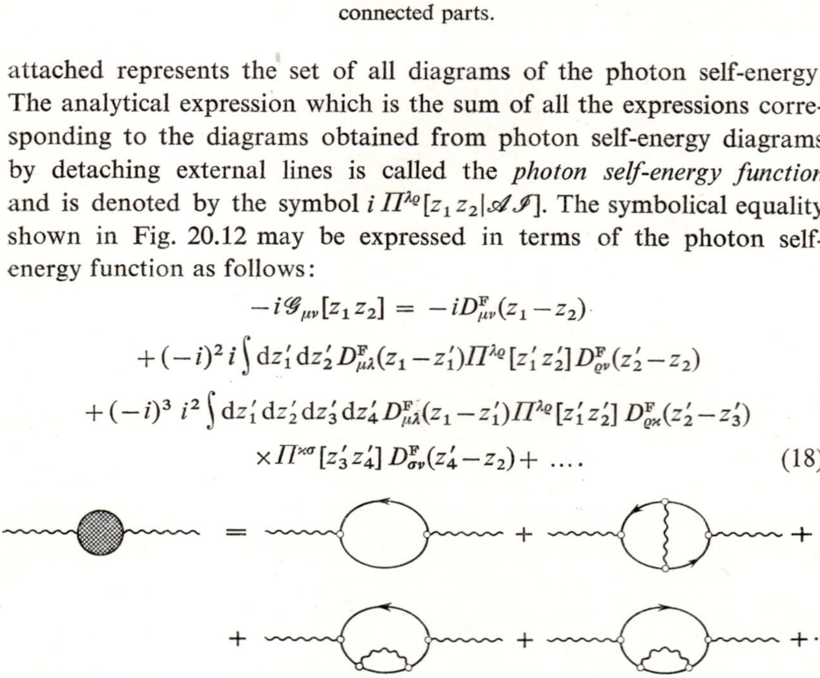

FIG. 20.13. The simplest diagrams of photon self-energy.

In Fig. 20.13 we give the four simplest diagrams of the photon self-energy. Corresponding to them are the following analytical expressions for the photon self-energy function:

$$-ie^2\,\mathrm{Tr}\{\gamma^{\lambda}K_{\mathrm{F}}[z_1 z_2|\mathscr{A}]\,\gamma^{\varrho}K_{\mathrm{F}}[z_2 z_1|\mathscr{A}]\},$$

$$e^4\int \mathrm{d}z_1'\,\mathrm{d}z_2'\,D^{\mathrm{F}}_{\mu\nu}(z_1'-z_2')$$

$$\times \operatorname{Tr}\{\gamma^{\lambda} K_F[z_1 z_1'|\mathscr{A}]\,\gamma^{\mu} K_F[z_1' z_2|\mathscr{A}]\,\gamma^{\varrho} K_F[z_2 z_2'|\mathscr{A}]\,\gamma^{\nu} K_F[z_2' z_1|\mathscr{A}]\},$$

$$e^4 \int dz_1'\, dz_2'\, D^F_{\mu\nu}(z_1'-z_2')$$

$$\times \operatorname{Tr}\{\gamma^{\lambda} K_F[z_1 z_1'|\mathscr{A}]\,\gamma^{\mu} K_F[z_1' z_2'|\mathscr{A}]\,\gamma^{\nu} K_F[z_2' z_2|\mathscr{A}]\,\gamma^{\varrho} K_F[z_2 z_1|\mathscr{A}]\},$$

$$e^4 \int dz_1'\, dz_2'\, D^F_{\mu\nu}(z_1'-z_2')$$

$$\times \operatorname{Tr}\{\gamma^{\lambda} K_F[z_1 z_2|\mathscr{A}]\,\gamma^{\varrho} K_F[z_2 z_1'|\mathscr{A}]\,\gamma^{\mu} K_F[z_1' z_2'|\mathscr{A}]\,\gamma^{\nu} K_F[z_2' z_1|\mathscr{A}]\}.$$

The decomposition of the propagator $T^C[z_1 z_2 z_3 z_4]$ into strongly connected parts is of the form

$$T^C_{\mu\nu\lambda\varrho}[z_1 z_2 z_3 z_4] = (-i)^4 \int dw_1 \dots dw_4$$
$$\times \mathscr{G}_{\mu\alpha}[z_1 w_1]\,\mathscr{G}_{\nu\beta}[z_2 w_2]\,\mathscr{G}_{\lambda\gamma}[z_3 w_3]\,\mathscr{G}_{\varrho\delta}[z_4 w_4]$$
$$\times \Big\{\Delta^{\alpha\beta\gamma\delta}[w_1 w_2 w_3 w_4]$$
$$- i \int dw_5\, dw_6\, \Delta^{\alpha\beta\sigma}[w_1 w_2 w_5]\,\mathscr{G}_{\sigma\tau}[w_5 w_6]\,\Delta^{\tau\gamma\delta}[w_6 w_3 w_4]$$
$$- i \int dw_5\, dw_6\, \Delta^{\alpha\gamma\sigma}[w_1 w_3 w_5]\,\mathscr{G}_{\sigma\tau}[w_5 w_6]\,\Delta^{\tau\beta\delta}[w_6 w_2 w_4]$$
$$- i \int dw_5\, dw_6\, \Delta^{\alpha\delta\sigma}[w_1 w_4 w_5]\,\mathscr{G}_{\sigma\tau}[w_5 w_6]\,\Delta^{\tau\beta\gamma}[w_6 w_2 w_3]\Big\}. \qquad (19)$$

This decomposition is illustrated in Fig. 20.14 in which the connected parts are indicated by shaded circles, in keeping with the convention adopted earlier. The strongly connected parts of the propagator have been marked by cross-hatched circles.

In analysing connected diagrams representing many-photon propagators, we shall employ the concept of *truncated diagram*. This is the name we give a diagram from which diagrams corresponding to photon propagators cannot be separated by breaking one internal line. Each strongly connected diagram is at the same time truncated, but the converse is not true. In Fig. 20.15 as an example we give two truncated diagrams and two that are not truncated. The first truncated diagram is weakly connected whereas the other is strongly connected.

The sum of analytical expressions corresponding to all diagrams obtained by detaching external lines from the truncated diagrams representing the connected part of a given propagator is called the *truncated part of the propagator*. This is a generalization of the concept of photon self-energy and is denoted by $T_T^{\mu_1\dots\mu_k}[z_1 \dots z_k|\mathscr{A}, \mathscr{I}]$.

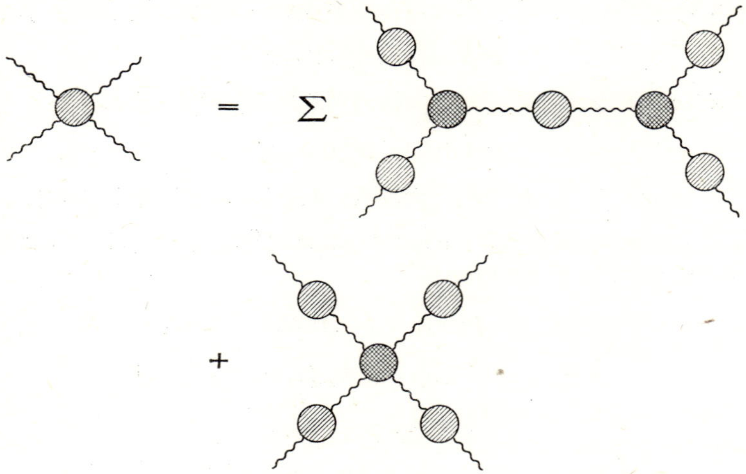

FIG. 20.14. The decomposition of the propagator $T^C[z_1 z_2 z_3 z_4]$ into strongly connected parts.

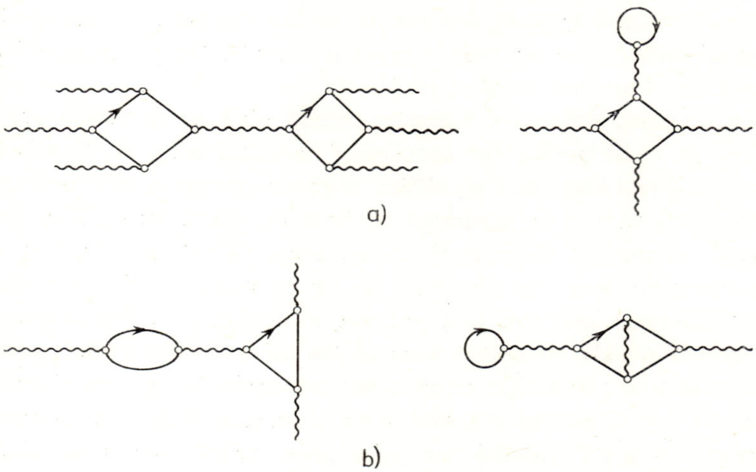

FIG. 20.15. Examples of (a) truncated diagrams and (b) untruncated diagrams.

It follows from this definition that the truncated part and the connected part of a propagator are related by the formula

$$T^C[z_1 \dots z_k] = (-i)^k \int \mathrm{d}z_1' \dots \mathrm{d}z_k' \mathcal{G}[z_1 z_1'] \dots \mathcal{G}[z_k z_k'] \, T_T[z_1' \dots z_k']. \quad (20)$$

The concepts of truncated diagrams and truncated parts are not applicable to the photon propagator $\mathcal{G}_{\mu\nu}$.

Only for the propagators $\mathfrak{A}[z]$ and $T[z_1 z_2 z_3]$ is each truncated diagram at the same time a strongly connected diagram,

$$T_T^\mu[z] = \Delta^\mu[z],$$

$$T_T^{\mu\nu\lambda}[z_1 z_2 z_3] = \Delta^{\mu\nu\lambda}[z_1 z_2 z_3].$$

The entire discussion of this section has been carried out on the basis of the perturbation expansion of the propagators. Finally, the relations between propagators and their connected, strongly connected, and truncated parts can, however, be adopted as definitions of these quantities without recourse to perturbation theory. For example, formulae (9), (10), (11), and (14) may be solved for T^C and regarded as a definition of the connected part of the propagator, regardless of whether expansion in e is justified or not.

The relations for the three simplest propagators, solved for T^C, are of the form:

$$T^C[z] = T^L[z], \quad (21a)$$

$$T^C[z_1 z_2] = T^L[z_1 z_2] - T^L[z_1] T^L[z_2], \quad (21b)$$

$$T^C[z_1 z_2 z_3] = T^L[z_1 z_2 z_3] - T^L[z_1 z_2] T^L[z_3]$$
$$- T^L[z_2 z_3] T^L[z_1] - T^L[z_1 z_3] T^L[z_2] + 2T^L[z_1] T^L[z_2] T^L[z_3]. \quad (21c)$$

21. PHOTON PROCESSES

In this section we shall consider such scattering processes in which the initial state and final state contain a given number of photons. We shall call them *photon processes*. The probability amplitudes of photon processes (abbreviated to transition amplitudes) are defined in the same way as in the quantum theory of mutually non-interacting photons. To this end we choose two sets of state vectors $\Psi^{\text{in}}_{\{n_k\}}$ and

$\Psi^{\text{out}}_{\{n_k\}}$ representing states with the following physical properties. The vector $\Psi^{\text{in}}_{\{n_k\}}$ ($\Psi^{\text{out}}_{\{n_k\}}$) describes the state which in the limit, when $t \to -\infty$ ($t \to +\infty$), is characterized by the photon occupation numbers $n_1 \ldots n_k \ldots$ The passage to the limit $t \to \pm\infty$ is taken to mean the same thing here as in the non-relativistic theory discussed in Section 5.

The Relation between Transition Amplitudes and Propagators

Transition amplitudes are scalar products of the vectors Ψ^{out} and Ψ^{in}. Now let us examine the relations between these products and many-photon propagators.

For this purpose we shall separate from the operator $f_{\mu\nu}(x)$ the parts $f^{\text{in}}_{\mu\nu}(x)$ and $f^{\text{out}}_{\mu\nu}(x)$ which create and annihilate IN and OUT photons. In the simplified version of electrodynamics, this could be done very easily; it was sufficient merely to consider the field operators $f_{\mu\nu}(x)$ in the past and in the future beyond the region in which the external current $\mathscr{I}^{\mu}(x)$ was present (cf. formulae (12.2)). For in the past and in the future the field operator $f_{\mu\nu}(x)$ satisfies the free Maxwell equations and it could be written as a linear combination of the creation and annihilation operators.

In the complete quantum electrodynamics such a straightforward procedure is inapplicable since the field operator $f_{\mu\nu}(x)$ does not satisfy the free Maxwell equations even in the region free of external fields and currents, for in addition there always is the current of charged particles, described by the operator j^{μ}. Only a particular fragment of the operator $f_{\mu\nu}(x)$ creates and annihilates single IN and OUT photons. The other parts have a more complicated structure, being built out of the products of IN and OUT creation and annihilation operators.

The relativistic invariance of the theory imposes limitations on the functions which are the coefficients in such an expansion. It is easiest to determine the form of the *linear* part in the creation and annihilation operators and, in addition, this is the part we need most. First of all, let us consider the field operator in the past. The representation of the Poincaré group U^{in} operates in the region free of external fields and external currents. It follows from the transformation laws of the field operators that the linear part in the IN creation and annihilation operators is proportional to the operator $f^{\text{in}}_{\mu\nu}(x)$,

$$f_{\mu\nu}(x) = N f^{\text{in}}_{\mu\nu}(x) + (\text{terms non-linear in } c^{\text{in}} \text{ and } c^{\dagger\text{in}}). \tag{1}$$

The proportionality factor is a real number since the field operators are Hermitian. The square of the coefficient N is called the *renormalization constant of the photon propagator*. We shall denote it by the traditional# symbol Z_3,

$$Z_3 = N^2.$$

The relation between the renormalization constant Z_3 and the photon propagator will be determined subsequently in this section in terms of the Källén–Lehmann representation of the propagator.

Formula (1) may be rewritten as

$$f^{\text{in}}_{\mu\nu}(x) = Z_3^{-\frac{1}{2}} f_{\mu\nu}(x) + (\text{terms non-linear in } c^{\text{in}} \text{ and } c^{\dagger\text{in}}). \tag{2}$$

The annihilation operators† $c^{\text{in}}[f]$ are related to the field operators $f^{\text{in}}_{\mu\nu}$ by the formula‡ (cf. formula (11.18a)):

$$\begin{aligned} c^{\text{in}}[f] &= (\varphi_{\text{T}}[f] \,|\, \hat{f}^{\text{in}}) \\ &= Z_3^{-\frac{1}{2}} (\varphi_{\text{T}}[f] \,|\, \hat{f})_{t=\text{const}} + (\text{terms non-linear in } c^{\text{in}} \text{ and } c^{\dagger\text{in}}). \end{aligned} \tag{3}$$

We deduce from this formula that the *asymptotic conditions* in quantum electrodynamics must be modified in comparison with non-relativistic quantum mechanics; the renormalization constant Z_3 will appear in the asymptotic conditions:

$$c^{\text{in}}[f] = Z_3^{-\frac{1}{2}} \lim_{t \to -\infty} (\varphi_{\text{T}}[f] \,|\, \hat{f})_t, \tag{4a}$$

$$c^{\dagger\text{in}}[f] = Z_3^{-\frac{1}{2}} \lim_{t \to -\infty} (\hat{f} \,|\, \varphi_{\text{T}}[f])_t. \tag{4b}$$

This notation was introduced in the first article (F. J. Dyson (1949)) on the theory of renormalization in quantum electrodynamics.

† The formulae for the creation operators are obtained by taking the Hermitian conjugates of the formulae for the annihilation operators.

‡ The scalar product defined by formula (11.17) may be calculated for any two tensor fields. If one of these fields does not satisfy the free Maxwell equations, the scalar product so defined is time-dependent.

In order to obtain formulae for the OUT operators the transition to the limit $t \to -\infty$ in formula (4) should be replaced with the transition to the limit $t \to +\infty$. In view of the invariance of electrodynamics with respect to time reversal, the same renormalization constant Z_3 appears in the asymptotic conditions for the IN and OUT operators.

Using the asymptotic conditions, we shall express the transition amplitudes in terms of the propagators. Let us do this first of all for the simplest case: the emission amplitude of a single photon. We then have

$$(c^{\dagger\text{out}}[f]\Omega^{\text{out}}|\Omega^{\text{in}}) = Z_3^{-\frac{1}{2}} \lim_{t\to\infty} (\varphi_{\text{T}}[f]|\mathscr{T}[\mathscr{A}, \mathscr{I}]), \tag{5}$$

where $\mathscr{T}_{\mu\nu}[z|\mathscr{A}, \mathscr{I}]$ is an antisymmetrical tensor built out of the derivatives of the propagator $T_\mu[z|\mathscr{A}, \mathscr{I}]$,

$$\mathscr{T}_{\mu\nu}[z|\mathscr{A}, \mathscr{I}] = \partial_{[\mu} T_{\nu]}[z|\mathscr{A}, \mathscr{I}]. \tag{6}$$

Since one factor in this scalar product has been expressed in terms of the potential vector (propagator), it will be convenient also to express the other factor, the tensor wave function $\varphi_{\mu\nu}$, in terms of the potential vector φ_μ (vector wave function). For we can then use the form (13.12d) of the scalar product, and we get:

$$(c^{\dagger\text{out}}[f]\,\Omega^{\text{out}}|\Omega^{\text{in}}) = Z_3^{-\frac{1}{2}} (\varphi_V[f]|T)_{\sigma\to+\infty}$$

$$= iZ_3^{-\frac{1}{2}} \lim_{\sigma\to+\infty} \int d\sigma_\mu\{\partial^{[\mu}\varphi^{\nu]*}[z|f]\,T_\nu[z] - \varphi_\nu^*[z|f]\,\partial^{[\mu}T^{\nu]}[z]\}. \tag{7}$$

In similar fashion we obtain the formula for the photon absorption amplitude,

$$(\Omega^{\text{out}}|c^{\dagger\text{in}}[f]\Omega^{\text{in}}) = Z_3^{-\frac{1}{2}} (T^*|\varphi_V[f])_{\sigma\to-\infty}$$

$$= iZ_3^{-\frac{1}{2}} \lim_{\sigma\to-\infty} \int d\sigma_\mu\{\partial^{[\mu}T^{\nu]*}[z]\,\varphi_\nu[z|f] - T_\nu[z]\,\partial^{[\mu}\varphi^{\nu]}[z|f]\}. \tag{8}$$

These formulae differ from their counterparts in the quantum theory of the electromagnetic field interacting solely with external currents (cf. formulae (13.62)) in that the renormalization constant Z_3 and the transition to the limit $t \to \pm\infty$ appear in them. Previously, such a transition to the limit was not necessary since in the region free of external currents the propagator $T[z]$ satisfied the free Maxwell equations and, in consequence, in this region the scalar product did not depend on the choice of the hypersurface σ. Now, to put it figuratively, though not very precisely, the propagator $T[z]$ does not describe free propagation until the limit, when $z^0 \to \pm\infty$, is reached.

The formulae for photon emission and absorption amplitudes may be simplified by expressing the scalar products of the propagators and vector wave functions of the photons as the results of the operation of certain linear functionals on the propagators. Let us for this purpose define two differential operations,

$$\Phi^{*\mu\nu}[z|f] \equiv i\{\partial^{[\mu}\varphi^{\nu]*}[z|f] - \varphi^{\nu*}[z|f]\overleftarrow{\partial}^{\mu} + g^{\mu\nu}\varphi^*_\lambda[z|f]\overleftarrow{\partial}^{\lambda}\}, \tag{9a}$$

$$\Phi^{\nu\mu}[z|f] \equiv i\{\overleftarrow{\partial}^{\mu}\varphi^{\nu}[z|f] - g^{\mu\nu}\overleftarrow{\partial}^{\lambda}\varphi_\lambda[z|f] - \partial^{[\mu}\varphi^{\nu]}[z|f]\}. \tag{9b}$$

With the aid of these operations, we write the photon emission and define absorption amplitudes in the form:

$$(c^{\dagger\text{out}}[f]\Omega^{\text{out}}|\Omega^{\text{in}}) = Z_3^{-\frac{1}{2}} \lim_{\sigma\to+\infty} \int d\sigma_\mu \Phi^{*\mu\nu}[z|f]T_\nu[z], \tag{10a}$$

$$(\Omega^{\text{out}}|c^{\dagger\text{in}}[f]\Omega^{\text{in}}) = Z_3^{-\frac{1}{2}} \lim_{\sigma\to-\infty} \int d\sigma_\mu T_\nu[z]\Phi^{\nu\mu}[z|f]. \tag{10b}$$

These amplitudes may also be expressed in the form of integrals over time-space. The procedure which we shall employ for this purpose is highly reminiscent of the derivation of the reduction formulae (5.61) and (18.54). The integral over a hypersurface lying in the far future in formula (10a) will be transformed into an integral over a hypersurface lying in the remote past by the addition of an integral of four-dimensional divergence over all space-time. Next, we proceed in like manner with the integral over the hypersurface lying in the remote past, which figures in formula (10b). As a result, we arrive at:

$$\lim_{\sigma\to\infty} \int d\sigma_\mu \Phi^{*\mu\nu}[z|f]T_\nu[z] = \lim_{\sigma\to-\infty} \int d\sigma_\mu \Phi^{*\mu\nu}[z|f]T_\nu[z]$$
$$+ \int d^4z\,\partial_\mu\{\Phi^{*\mu\nu}[z|f]T_\nu[z]\}, \tag{11a}$$

$$\lim_{\sigma\to-\infty} \int d\sigma_\mu T_\nu[z]\Phi^{\nu\mu}[z|f] = \lim_{\sigma\to\infty} \int d\sigma_\mu T_\nu[z]\Phi_{\mu\nu}[z|f]$$
$$- \int d^4z\,\partial_\mu\{T_\nu[z]\Phi^{\nu\mu}[z|f]\}. \tag{11b}$$

By virtue of the asymptotic conditions, the integral over the hypersurface $\sigma \to -\infty$ vanishes in formula (11a) and so does the integral over the hypersurface $\sigma \to \infty$ in formula (11b), since we can identify them as, respectively, $Z_3^{-\frac{1}{2}}(\Omega^{\text{out}}|c^{\text{in}}[f]\,\Omega^{\text{in}})$ and $Z_3^{-\frac{1}{2}}(c^{\text{out}}[f]\,\Omega^{\text{out}}|\,\Omega^{\text{in}})$. On the other hand, by virtue of the properties of wave functions the four-dimensional divergences are reduced to the action of a simple differential operator on $T_\nu[z]$,

$$\partial_\mu\{\Phi^{*\mu\nu}[z|f]\,T_\nu[z]\} = i\varphi_\mu^*[z|f]\,M_z^{\mu\nu}\,T_\nu[z], \tag{12a}$$

$$\partial_\mu\{T_\nu[z]\,\Phi^{\nu\mu}[z|f]\} = -iT_\nu[z]\,\overleftarrow{M}_z^{\nu\mu}\,\varphi_\mu[z|f], \tag{12b}$$

where

$$M_z^{\mu\nu} \equiv -g^{\mu\nu}\Box_z + \partial_z^\mu\,\partial_z^\nu. \tag{13}$$

Finally, we get

$$(c^{\dagger\text{out}}[f]\Omega^{\text{out}}|\Omega^{\text{in}}) = iZ_3^{-\frac{1}{2}}\int d^4z\varphi_\mu^*[z|f]\,M_z^{\mu\nu}\,T_\nu[z], \tag{14a}$$

$$(\Omega^{\text{out}}|c^{\dagger\text{in}}[f]\,\Omega^{\text{in}}) = iZ_3^{-\frac{1}{2}}\int d^4z T_\nu[z]\,\overleftarrow{M}_z^{\nu\mu}\varphi_\mu[z|f]. \tag{14b}$$

Utilizing the properties of the differential operations defined by formulae (9), we express the general many-photon transition amplitude in terms of propagators,

$$(c^{\dagger\text{out}}[f_1]\,\dots\,c^{\dagger\text{out}}[f_l]\,\Omega^{\text{out}}|c^{\dagger\text{in}}[f_{l+1}]\,\dots\,c^{\dagger\text{in}}[f_k]\,\Omega^{\text{in}})$$

$$= Z_3^{-\frac{k}{2}}\lim\int_{\sigma_1\to\infty} d\sigma_{\mu_1}\dots\int_{\sigma_l\to\infty} d\sigma_{\mu_l}\int_{\sigma_{l+1}\to-\infty} d\sigma_{\mu_{l+1}}\dots\int_{\sigma_k\to-\infty} d\sigma_{\mu_k}$$

$$\times\,\Phi^{*\mu_1\nu_1}[z_1|f_1]\,\dots\,\Phi^{*\mu_l\nu_l}[z_l|f_l]\,T_{\nu_1\dots\nu_k}[z_1\,\dots\,z_k]$$

$$\times\,\Phi^{\nu_{l+1}\mu_{l+1}}[z_{l+1}|f_{l+1}]\,\dots\,\Phi^{\nu_k\mu_k}[z_k|f_k]. \tag{15}$$

This amplitude may also be written as multiple integrals over space-time by means of the same procedure which led to formulae (14) for photon emission and absorption amplitudes. Let us illustrate this procedure with the example of photon-photon transition amplitude.

This amplitude is given by the formula:

$$Z_3^{-1} \lim_{\sigma_1 \to \infty} \lim_{\sigma_2 \to -\infty} \int d\sigma_{\mu_1} \int d\sigma_{\mu_2} \Phi^{*\mu_1\nu_1}[z|_1 f_1] T_{\nu_1\nu_2}[z_1 z_2] \Phi^{\nu_2\mu_2}[z_2|f_2]$$

$$= Z_3^{-1} \lim_{\sigma_1 \to -\infty} \lim_{\sigma_2 \to -\infty} \int d\sigma_{\mu_1} \int d\sigma_{\mu_2} \Phi^{*\mu_1\nu_1}[z_1|f_1] T_{\nu_1\nu_2}[z_1 z_2] \Phi^{\nu_2\mu_2}[z_2|f_2]$$

$$+ iZ_3^{-1} \int d^4 z_1 \lim_{\sigma_2 \to -\infty} \int d\sigma_{\mu_2} \varphi^*_{\mu_1}[z_1|f_1] M^{\mu_1\nu_1}_{z_1} T_{\nu_1\nu_2}[z_1 z_2] \Phi^{\nu_2\mu_2}[z_2|f_2]. \quad (16)$$

By the asymptotic conditions, the first term on the right-hand side may be rewritten as:

$$(\Omega^{\text{out}}|c^{\text{in}}[f_1] c^{\dagger\text{in}}[f_2] \Omega^{\text{in}}) = (f_1|f_2)(\Omega^{\text{out}}|\Omega^{\text{in}}).$$

Next, let us convert the integral over the hypersurface σ_2 into an integral over space-time and an integral over a hypersurface lying in the far future. The latter integral makes no contribution to the transition amplitude. A rigorous proof of this assertion requires a detailed analysis of the various passages to the limit which appear in our considerations. Such proof will not be given here. An indirect proof that the transformations carried out are correct lies in the very form of the final result which is an exact analogue of the formulae arrived at in Chapter 5 (cf. formula (18.55)). For in the final account, we have

$$(c^{\dagger\text{out}}[f_1]\Omega^{\text{out}}|c^{\dagger\text{in}}[f_2]\Omega^{\text{in}}) = (f_1|f_2)(\Omega^{\text{out}}|\Omega^{\text{in}})$$

$$+ i^2 Z_3^{-1} \int d^4 z_1 \, d^4 z_2 \, \varphi^*_{\mu_1}[z_1|f_1] M^{\mu_1\nu_1}_{z_1} T_{\nu_1\nu_2}[z_1 z_2] \overleftarrow{M}^{\nu_2\mu_2}_{z_2} \varphi_{\mu_2}[z_2|f_2]. \quad (17)$$

The first term describes the transition of a photon from the initial to the final state without interaction, whereas the second describes the transition with interaction. Using the decomposition of the propagator $T[z_1 z_2]$ into connected parts, we get the decomposition of the transition amplitude into connected parts:

$$(c^{\dagger\text{out}}[f_1]\Omega^{\text{out}}|c^{\dagger\text{in}}[f_2]\Omega^{\text{in}}) = (\Omega^{\text{out}}|\Omega^{\text{in}})\Big\{(f_1|f_2)$$

$$+ \Big(iZ_3^{-\frac{1}{2}} \int d^4 z \, \varphi^*_\mu [z|f_1] M^{\mu\nu}_z \mathfrak{A}_\nu[z]\Big)\Big(iZ_3^{-\frac{1}{2}} \int d^4 z \, \mathfrak{A}_\lambda[z] \overleftarrow{M}^{\lambda\varrho}_z \varphi_\varrho[z|f_2]\Big)$$

$$+ iZ_3^{-1} \int d^4 z_1 d^4 z_2 \, \varphi^*_{\mu_1}[z_1|f_1] M^{\mu_1\nu_1}_{z_1} \mathcal{G}_{\nu_1\nu_2}[z_1 z_2] \overleftarrow{M}^{\nu_2\mu_2}_{z_2} \varphi_{\mu_2}[z_2|f_2]\Big\}. \quad (18)$$

The decomposition contains the product of the amplitudes of independent processes of emission and absorption as well as a term describing correlated two-photon processes. This latter term did not appear in the theory of mutually non-interacting photons. There was no correlation between acts of photon emission and absorption in that theory.

Generalization of the formulae (17) and (18) to the case of many-photon amplitudes is, in principle, simple but leads to complicated formulae. Just as in the theories considered earlier, so here, too, the amplitude of a transition with the participation of k photons is the sum of amplitudes describing a transition without interaction, one with the interaction of only one photon, one with two photons, etc. Of most interest is the last part of the k-photon amplitude, describing the interaction of all k photons. This amplitude, written in terms of integrals over space-time, is of the form:

$$i^k Z_3^{-k/2} \int \mathrm{d}z_1 \ldots \mathrm{d}z_k \varphi^*[z_1|f_1] \ldots \varphi^*[z_l|f_l]$$
$$\times M_{z_1} \ldots M_{z_l} T[z_1 \ldots z_k] \overleftarrow{M}_{z_{l+1}} \ldots \overleftarrow{M}_{z_k} \varphi[z_{i+1}|f_{i+1}] \ldots \varphi[z_k|f_k]. \quad (19)$$

This leads to the following formula for the S *operator in the subspace of photon states* (cf. formulae (5.60) and (5.63)),

$$\mathsf{S} = \sum_{n=0}^{\infty} \frac{1}{n!} \sum_{m=0}^{\infty} \frac{1}{m!} \sum_{\lambda_1 \ldots \lambda_n} \sum_{\lambda_1' \ldots \lambda_m'} \int \mathrm{d}\Gamma_1 \ldots \mathrm{d}\Gamma_n \mathrm{d}\Gamma_1' \ldots \mathrm{d}\Gamma_m'$$
$$\times c^{\dagger\text{in}}(\mathbf{k}_1, \lambda_1) \ldots c^{\dagger\text{in}}(\mathbf{k}_n, \lambda_n) \sigma^{(n,m)}(k_1 \lambda_1 \ldots k_n \lambda_n ; k_1' \lambda_1' \ldots k_n' \lambda_m')$$
$$\times c^{\text{in}}(\mathbf{k}_1', \lambda_1') \ldots c^{\text{in}}(\mathbf{k}_m', \lambda_m'), \quad (20)$$

where

$$\sigma^{(n,m)}(k_1 \lambda_1 \ldots k_n \lambda_n ; k_1' \lambda_1' \ldots k_m' \lambda_m')$$
$$= i^{n+m} Z_3^{-\frac{n+m}{2}} \varepsilon^*(\mathbf{k}_1 \lambda_1) \ldots \varepsilon^*(\mathbf{k}_n \lambda_n) \int \mathrm{d}z_1 \ldots \mathrm{d}z_n \mathrm{d}z_1' \ldots \mathrm{d}z_m'$$
$$\times \exp\Big(i \sum_{i=1}^{n} k_i \cdot z_i\Big) M_{z_1} \ldots M_{z_n} T[z_1 \ldots z_n z_1' \ldots z_m'] \overleftarrow{M}_{z_1'} \ldots \overleftarrow{M}_{z_m'}$$
$$\times \exp\Big(-i \sum_{j=1}^{m} k_j' \cdot z_j'\Big) \varepsilon(\mathbf{k}_1' \lambda_1') \ldots \varepsilon(\mathbf{k}_m' \lambda_m'). \quad (21)$$

In this formula we have used the short-hand notation:

$$\varepsilon_\mu(\mathbf{k}\,\lambda) = \begin{cases} \varepsilon_\mu(\mathbf{k}), & \lambda = +1, \\ -\varepsilon_\mu^*(\mathbf{k}), & \lambda = -1. \end{cases}$$

We prove representation (20) to be correct by demonstrating that the matrix elements of the S operator in the IN vector basis are equal to the transition amplitudes defined earlier.

The Källén–Lehmann Representation of the Photon Propagator

Let us consider the case when the external field and external currents are absent. The vacuum then is a stable state of the system ($\Omega^{\text{in}} = \Omega^{\text{out}} = \Omega$).

Now let us examine the vacuum-state expectation value of the product of two electromagnetic field operators,

$$\frac{1}{i}\mathscr{G}^{(+)}_{\mu\nu\lambda\varrho}(z_1\,z_2) \equiv (\Omega|f_{\mu\nu}(z_1)\,f_{\lambda\varrho}(z_2)\Omega). \tag{22}$$

In the absence of external fields and currents, the Poincaré group representation acts in the state-vector space. Components of the momentum operator P_μ are translation generators. Thus, the relation

$$f_{\mu\nu}(z) = \mathrm{e}^{iP\cdot(z-a)}f_{\mu\nu}(a)\mathrm{e}^{-iP\cdot(z-a)} \tag{23}$$

is satisfied. It hence follows that

$$\frac{1}{i}\mathscr{G}^{(+)}_{\mu\nu\lambda\varrho}(z_1\,z_2) = \frac{1}{i}\mathscr{G}^{(+)}_{\lambda\nu\lambda\varrho}(z_1-z_2) = (\Omega|\,f_{\mu\nu}(a)\mathrm{e}^{-iP\cdot(z_1-z_2)}f_{\lambda\varrho}(a)\Omega).$$

Between the field operators in this formula we insert the Dirac unit operator (cf. Section 4) built out of operators projecting onto generalized vectors which describe the asymptotic one-photon, two-photon, etc., states (we choose the IN vectors). Since we are using generalized vectors, we shall apply the Dirac notation:

$$1 = |\Omega\rangle\,\langle\Omega| + \sum_{\lambda_1}\int \mathrm{d}\Gamma_1\,\overset{\text{in}}{|\mathbf{k}_1\,\lambda_1\rangle}\,\overset{\text{in}}{\langle\mathbf{k}_1\,\lambda_1|}$$

$$+\sum_{\lambda_1,\lambda_2}\mathrm{d}\Gamma_1 d\Gamma_2\,\overset{\text{in}}{|\mathbf{k}_1\lambda_1,\mathbf{k}_2\,\lambda_2\rangle}\,\overset{\text{in}}{\langle\mathbf{k}_1\,\lambda_1,\mathbf{k}_2\,\lambda_2|} + \dots. \tag{24}$$

States in which not only photons are present may also appear in the sum on the right-hand side. On insertion of the unit operator, we get:

$$\frac{1}{i}\mathscr{G}^{(+)}_{\mu\nu\lambda\varrho}(z_1-z_2) = (\Omega|f_{\mu\nu}(a)\Omega)(\Omega|f_{\lambda\varrho}(a)\Omega)$$

$$+\sum_{\lambda_1}\int \mathrm{d}\Gamma_1\langle\Omega|f_{\mu\nu}(a)\,|\overset{\text{in}}{\mathbf{k}_1\,\lambda_1}\rangle\,\langle\overset{\text{in}}{\mathbf{k}_1\,\lambda_1}\,|f_{\lambda\varrho}(a)\,|\Omega\rangle\,\mathrm{e}^{-ik_1\cdot(z_1-z_2)}$$

$$+\sum_{\lambda_1,\lambda_2}\int \mathrm{d}\Gamma_1\,\mathrm{d}\Gamma_2\langle\Omega|f_{\mu\nu}(a)\,|\overset{\text{in}}{\mathbf{k}_1\,\lambda_1,\mathbf{k}_2\,\lambda_2}\rangle$$

$$\times\langle\overset{\text{in}}{\mathbf{k}_1\,\lambda_1,\mathbf{k}_2\,\lambda_2}\,|f_{\lambda\varrho}(a)\,|\Omega\rangle\mathrm{e}^{-i(k_1+k_2)\cdot(z_1-z_2)}+\ \ldots. \tag{25}$$

The first term in this sum vanishes because of the relativistic invariance of the theory. The remaining terms are written in the form:

$$\mathscr{G}^{(+)}_{\mu\nu\lambda\varrho}(z_1-z_2) = i\int\frac{\mathrm{d}^4k}{(2\pi)^4}\,\mathrm{e}^{-ik\cdot(z_1-z_2)}\varrho_{\mu\nu\lambda\varrho}(k),$$

where

$$\varrho_{\mu\nu\lambda\varrho}(k) = \sum_{\lambda_1}\int \mathrm{d}\Gamma_1(2\pi)^4\delta^{(4)}(k-k_1)\,\langle\Omega\,|f_{\mu\nu}(a)\,|\overset{\text{in}}{\mathbf{k}_1\,\lambda_1}\rangle\,\langle\overset{\text{in}}{\mathbf{k}_1\,\lambda_1}|f_{\lambda\varrho}(a)\,|\Omega\rangle$$

$$+\sum_{\lambda_1,\lambda_2}\int \mathrm{d}\Gamma_1\,\mathrm{d}\Gamma_2(2\pi)^4\delta^{(4)}(k-k_1-k_2)$$

$$\times\langle\Omega\,|f_{\mu\nu}(a)\,|\overset{\text{in}}{\mathbf{k}_1\,\lambda_1,\mathbf{k}_2\,\lambda_2}\rangle\,\langle\overset{\text{in}}{\mathbf{k}_1\,\lambda_1,\mathbf{k}_2\,\lambda_2}\,|f_{\lambda\varrho}(a)\,|\Omega\rangle+\ \ldots.$$

It follows from the translational invariance of the theory that the function $\varrho_{\mu\nu\lambda\varrho}(k)$ does not depend on a. From the invariance with respect to Lorentz transformations and from eq. (6.4b) which is satisfied by the operator $f_{\mu\nu}$ it follows that $\varrho_{\mu\nu\lambda\varrho}$ may be expressed in terms of one function of the variable k^2,

$$\varrho_{\mu\nu\lambda\varrho}(k) = k_{[\mu}g_{\nu][\lambda}k_{\varrho]}\theta(k_0)\varrho(k^2).$$

The function $\varrho(k^2)$, as follows from its definition, is a real non-negative function which vanishes when $k^2 < 0$.

From the decomposition (2) of the field operator it is evident that the contribution to $\varrho_{\mu\nu\lambda\varrho}(k)$ from the one-photon states is of the form

$$\varrho^{(1)}_{\mu\nu\lambda\varrho}(k) = k_{[\mu}g_{\nu][\lambda}k_{\varrho]}\theta(k_0)Z_3\,\delta(k^2).$$

The contribution from the other states is given by the formula

$$\varrho(k^2)-\varrho^{(1)}(k^2) = \int_{0^+}^{\infty}\mathrm{d}M^2\delta(k^2-M^2)\varrho(M^2),$$

where 0^+ in the lower limit of integration indicates that the integration does not include the point $k^2 = 0$.

As a result, the function $\mathscr{G}^{(+)}_{\mu\nu\lambda\varrho}$ may be rewritten as a linear combination of the propagators $D^{(+)}$ and $\Delta^{(+)}$,

$$\mathscr{G}^{(+)}_{\mu\nu\lambda\varrho}(z_1-z_2) = -\partial_{[\mu}g_{\nu][\lambda}\partial_{\varrho]}\Big\{ Z_3 D^{(+)}(z_1-z_2) + \int\limits_{0^+}^{\infty} \mathrm{d}M^2 \varrho(M^2)\Delta^{(+)}(z_1-z_2, M)\Big\}. \tag{26}$$

In similar fashion we can calculate the vacuum-state expectation value of the commutator of the field operator; this amounts to:

$$(\Omega\,|[f_{\mu\nu}(z_1), f_{\lambda\varrho}(z_2)]\,\Omega) = i\partial_{[\mu}g_{\nu][\lambda}\partial_{\varrho]}\Big\{Z_3 D(z_1-z_2) + \int\limits_{0^+}^{\infty} \mathrm{d}M^2 \varrho(M^2)\Delta(z_1-z_2, M)\Big\}. \tag{27}$$

The vacuum-state expectation value of the chronological product of two field operators is given by the formula

$$\big(\Omega\,|T\big(f_{\mu\nu}(z_1)f_{\lambda\varrho}(z_2)\big)\,\Omega\big)$$
$$= i\partial_{[\mu}g_{\nu][\lambda}\partial_{\varrho]}\Big\{Z_3 D_F(z_1-z_2) + \int\limits_{0^+}^{\infty} \mathrm{d}M^2 \varrho(M^2)\Delta_F(z_1-z_2, M)\Big\}. \tag{28}$$

In the case now under consideration, when there are no external fields and currents, the propagator $T_{\mu\nu}[z_1 z_2]$ is equal to its connected part since $\mathfrak{A}_\mu[z] = 0$,

$$T_{\mu\nu}[z_1 z_2|\mathscr{A}, \mathscr{I}]|_{\mathscr{A}=0=\mathscr{I}} = -i\mathscr{G}_{\mu\nu}[z_1 z_2|\mathscr{A}, \mathscr{I}]|_{\mathscr{A}=0=\mathscr{I}}. \tag{29}$$

As follows from previous considerations, the photon propagator is in this case a function of the difference of its arguments; we introduce the following notation for it:

$$\mathscr{G}_{\mu\nu}(z_1-z_2) \equiv \mathscr{G}_{\mu\nu}[z_1 z_2|\mathscr{A}, \mathscr{I}]|_{\mathscr{A}=0=\mathscr{I}}. \tag{30}$$

The invariance with respect to Lorentz transformations implies that the propagator $\mathscr{G}_{\mu\nu}(z)$ may be expressed in terms of two scalar functions,

$$\mathscr{G}_{\mu\nu}(z) = -g_{\mu\nu}\mathscr{G}(z) - \partial_\mu\partial_\nu\mathscr{G}_1(z). \tag{31}$$

Knowing the vacuum-state expectation value of the chronological product of two field operators, we can determine the propagator $\mathscr{G}_{\mu\nu}(z)$ to within the function $\mathscr{G}_1(z)$ in accordance with the formula:

$$\big(\Omega|T(f_{\mu\nu}(z_1)f_{\lambda\varrho}(z_2))\Omega\big) = -i\delta^{\alpha}_{[\mu}\delta^{\gamma}_{\nu]}\delta^{\beta}_{[\lambda}\delta^{\delta}_{\varrho]}\frac{\partial}{\partial z_1^{\alpha}}\frac{\partial}{\partial z_2^{\beta}}\mathscr{G}_{\gamma\delta}(z_1-z_2). \quad (32)$$

In this way we get

$$\mathscr{G}(z) = Z_3 D_F(z)+\int_{0+}^{\infty} \mathrm{d}M^2\varrho(M^2)\Delta_F(z, M). \quad (33)$$

The Fourier transform of the function $\mathscr{G}(z)$ thus is of the form

$$\tilde{\mathscr{G}}(k^2) = \frac{Z_3}{-k^2-i\varepsilon}+\int_{0+}^{\infty} \mathrm{d}M^2\frac{\varrho(M^2)}{M^2-k^2-i\varepsilon}. \quad (34)$$

Formulae (33) and (34) are known as the *Källén–Lehmann representation.*#

The decomposition of the photon propagator into the sum of the free photon propagator $D^F_{\mu\nu}$ and the integral of the propagators $g^{\mu\nu}\Delta_F$ with respect to M^2 has a figurative physical interpretation. It may be said that in the theory with interaction, perturbations of the electromagnetic field move in space-time from the point z_1 to the point z_2 as the combination of propagation with zero mass described by $Z_3 D^F$ and propagation with continuously distributed mass defined by the function $\varrho(M^2)$. The renormalization constant Z_3 determines the weight with which the component describing free propagation with zero mass enters into this superposition. In the theory *without* mutual interaction of photons with charges, only the free component appeared in the photon propagator and the constant Z_3 was equal to unity.

An additional argument for precisely such an interpretation of the representations obtained for the propagator is the *sum rule*,

$$1 = Z_3+\int_{0+}^{\infty} \mathrm{d}M^2\varrho(M^2), \quad (35)$$

This representation was given independently by G. Källén (1952) and J. Lehmann (1954).

from which it follows that, in accordance with the fundamental properties of the quantum theory, the quantities Z_3 and $\varrho(M^2)$ may be assigned the meaning of the probability of the various types of propagation occurring.

The sum rule (35) may be derived from the commutation relations for the operators **E** and **B** if we put $\mu = 0$, $\nu = k$, $\lambda = i$, and $\varrho = j$ in formula (27) and then make use of the properties (E.23) of the propagators D and Δ.

Renormalization of an External Current

The constant Z_3 has a lucid physical interpretation which we will find by studying the emission of radiation by weak currents. Such processes were investigated in detail in Section 12. At present, we shall compare the results obtained in quantum electrodynamics with those arrived at previously in the simplified theory.

To do this, we shall examine the photon emission amplitude in quantum electrodynamics, which we write as:

$$(c^{\dagger\text{out}}[f]\Omega^{\text{out}}|\Omega^{\text{in}}) = i\sum_{\lambda}\int \mathrm{d}\Gamma f(\mathbf{k},\lambda)\,\iota_{\text{QED}}(\mathbf{k},\lambda)(\Omega^{\text{out}}|\Omega^{\text{in}}), \tag{36}$$

where

$$\iota_{\text{QED}}(\mathbf{k},\lambda) = Z_3^{-\frac{1}{2}}\varepsilon_\mu^*(\mathbf{k},\lambda)\int \mathrm{d}^4z\, e^{ik\cdot z} M_z^{\mu\nu} T_\nu[z]. \tag{37}$$

For simplicity, we shall assume here that no external electromagnetic field is present. In the weak external current approximation it is sufficient to take account only of the term linear in $\mathcal{J}^\mu$. We then have†

$$T_\nu[z] \simeq \int \mathrm{d}^4z' \left.\frac{\delta T_\nu[z]}{\delta \mathcal{J}^\varrho(z')}\right|_{\mathcal{A}=0=\mathcal{J}} \mathcal{J}^\varrho(z')$$

$$= -i\int \mathrm{d}^4z'\, T_{\nu\varrho}[zz']|_{\mathcal{A}=0=\mathcal{J}}\mathcal{J}^\varrho(z') = -\int \mathrm{d}^4z'\,\mathcal{G}_{\nu\varrho}(z-z')\mathcal{J}^\varrho(z'). \tag{38}$$

Using the Källén–Lehmann representation for the photon propagator $\mathcal{G}_{\nu\varrho}(z-z')$ figuring in this formula, and putting the result into formula (37), we get:

$$\iota_{\text{QED}}(\mathbf{k},\lambda) = -\varepsilon_\mu^*(\mathbf{k},\lambda)Z_3^{\frac{1}{2}}\tilde{\mathcal{J}}^\mu(\mathbf{k},\omega). \tag{39}$$

† No term of zeroth order in $\mathcal{J}^\mu$ appears since $T_\mu[z|\mathcal{A},\mathcal{J}]|_{\mathcal{A}=0=\mathcal{J}} = 0$.

This differs from the result obtained in the approximate theory only by the factor $Z_3^{\frac{1}{2}}$. The probability amplitude of photon emission by a weak current thus is determined in quantum electrodynamics by the expression $Z_3^{\frac{1}{2}}\mathcal{J}^\mu$, and not by the current $\mathcal{J}^\mu$ itself. However, how can the intensity of external currents be determined in quantum theory? This can be done only by studying such effects as emission of photons, etc., caused by these currents. It follows from our considerations that in the process of photon emission, the effective current is equal to the product $Z_3^{\frac{1}{2}}\,\mathcal{J}^\mu$. This product will be called the *renormalized current*:

$$\mathcal{J}^\mu_{\text{ren}}(x) \equiv Z_3^{\frac{1}{2}}\mathcal{J}^\mu(x). \tag{40}$$

The renormalized current directly affects the radiation; this is an actually measurable quantity.

The current renormalization procedure described here has many features in common with the properties of classical electrodynamics described in the diagram of Section 8 (cf. the discussion on p. 108). The primary current $\mathcal{J}^\mu$ can not only generate radiation directly. It can also produce radiation indirectly by creating an electromagnetic field which affects the motion of charges (negaton-positon pairs). This motion of charges is in turn a source of radiation. It is the complete current generated by the primary current $\mathcal{J}^\mu$ which is the observable source of radiation:

motion of charges	←	electromagnetic field	←	primary current
↓				↓
radiation		+		radiation
		↓		
		total radiation		

22. ELECTRON-PHOTON PROCESSES

Compensating Current

Processes in which both photons and electrons appear in the initial and/or final states will be called *electron-photon processes*. The description of these processes will be preceded by a brief analysis of actual experiments with charged particles. We begin by obserwing that the natural state of all matter (at least, under terrestrial conditions) is the electrically neutral state. Charged particles thus are "produced" in experiments by separating them from electrically neutral agglomerations of matter. For example, electrons are produced by stripping them off individual atoms, metal, plasma, etc. The appearance of each free charge is always accompanied by the appearance of a hole (i.e. positive ion). The detection of a charged particle, in turn, results in the neutralization of its charge. Since both before and after the experiment the state of the entire system composed of the apparatus and the particles under study is locally (in macroscopic regions) an electrically neutral state, the following three stages can be distinguished in each experiment with charged particles. Separation of charges occurs in the first stage; a flow of currents takes place in the second; and, finally, in the third stage the separated charges recombine. It thus follows that each experiment with charged particles involves a flow of charge in the apparatus, compensating the flow of charge carried by the particles being studied. The macroscopic currents flowing in the apparatus will henceforth be referred to as *compensating currents*. The form of the compensating currents depends on the conditions of the experiment, but we expect this form to have only a negligible influence on the result of the experiment. The condition for the influence of the compensating current on the result of the experiment to be negligible is that this compensating current and the particles under study be separated by a considerable distance when the particles interact with each other (or with an external field).

Now let us go on to a description of processes with the participation of electrons. The appearance of each electron in the initial state and its disappearance in the final state is accompanied by a compensating current. The simplest model of the compensating current is that of a current described by numerical functions. A given compensating current of this kind will be denoted by the symbol $\mathscr{I}^{\mu}_{c}(x)$. The basic difference between the compensating current and the external current with which we have had to deal in our considerations thus far is that the compensating current does not satisfy the continuity equation,

$$\partial_{\mu}\mathscr{I}^{\mu}_{c}(x) \neq 0 .$$

The compensating current has sources at all points where the given electrons appear or disappear. In the general case, in addition to the compensating current there may also be an ordinary source-free external current $\mathscr{I}^{\mu}(x)$.

The role of the compensating current will be illustrated first of all with the simple example of the theory of mutually non-interacting electrons (developed in Chapter 5). For this purpose let us consider the propagation of electrons† from the points $y_1, \ldots, y_n$ to the points $x_1, \ldots, x_n$, which is accompanied by a flow of the external current $\mathscr{I}^{\mu}(z)$. Such a process is described by the propagator

$$C[x_1 \ldots x_n, y_n \ldots y_1|\mathscr{A}, \mathscr{I}_c+\mathscr{I}]$$
$$= C[\mathscr{A}, \mathscr{I}_c+\mathscr{I}]K_F[x_1 \ldots x_n, y_n \ldots y_1|\mathscr{A}]. \tag{1}$$

In the case under consideration, the compensating current has point sources with an intensity of e at all the points x_i at which electrons disappear, and point sources of intensity $-e$ at all points where electrons appear:

$$\partial_{\mu}\mathscr{I}^{\mu}_{c}(z) = e\sum_{i=1}^{n}\left(\delta^{(4)}(z-x_i)-\delta^{(4)}(z-y_i)\right). \tag{2}$$

Owing to the presence of the compensating current, expression (1) is invariant with respect to gauge transformations.

The action of the operation $\mathscr{B}(z)$,

$$\mathscr{B}(z) \equiv \partial_{\mu}\frac{\delta}{\delta\mathscr{A}_{\mu}(z)},$$

on the propagator $C[x_1 \ldots x_n, y_n \ldots y_1|\mathscr{A}, \mathscr{I}_c+\mathscr{I}]$ may be rewritten as follows by virtue of formulae (17.82) and (17.32) as well as equations (17.29) for the propagators K_F:

$$\mathscr{B}(z)C[x_1 \ldots x_n, y_n \ldots y_1|\mathscr{A}, \mathscr{I}_c+\mathscr{I}]$$
$$= i\left\{-\partial_{\mu}(\mathscr{I}^{\mu}(z)+\mathscr{I}^{\mu}_{c}(z))+e\sum_{i=1}^{n}[\delta(z-x_i)-\delta(z-y_i)]\right\}$$
$$\times C[x_1 \ldots x_n, y_n \ldots y_1|\mathscr{A}, \mathscr{I}_c+\mathscr{I}].$$

† This may be either propagation of negatons and positons, or the creation and annihilation of pairs.

Thus, by condition (2) we finally arrive at:

$$\mathscr{B}(z)C[x_1 \ldots x_n, y_n \ldots y_1|\mathscr{A}, \mathscr{I}_c+\mathscr{I}] = 0, \tag{3}$$

which, as we know, is the condition for the gauge invariance of the propagator.

The propagator $C[x_1 \ldots x_n, y_n \ldots y_1|\mathscr{A}, \mathscr{I}_c+\mathscr{I}]$ may also be rewritten in the form of a matrix element between the vacuum-state vectors Ω^{out} and Ω^{in} of the chronological product of the operators,

$$C[x_1 \ldots x_n, y_n \ldots y_1|\mathscr{A}, \mathscr{I}_c+\mathscr{I}]$$
$$= \left(\Omega^{\text{out}}|T\left\{\psi(x_1) \ldots \psi(x_n)\bar{\psi}(y_n) \ldots \bar{\psi}(y_1)\exp\left(-i\int (\mathscr{I}_c+\mathscr{I})\cdot\mathscr{A}\right)\right\}\Omega^{\text{in}}\right). \tag{4}$$

Electron-Photon Propagators

Let us now proceed to construct expressions analogous to propagators (4) in the complete quantum electrodynamics. Guiding ourselves by the analogy between many-photon and many-electron propagators in the simplified theories, we shall consider the following expressions in full quantum electrodynamics:

$$T[x_1 \ldots x_n, y_n \ldots y_1|\mathscr{A}, \mathscr{I}_c+\mathscr{I}]$$
$$\equiv (-i)^n \exp\left(\frac{1}{2i}\int \frac{\delta}{\delta\mathscr{A}} D^F \frac{\delta}{\delta\mathscr{A}}\right) C[x_1 \ldots x_n, y_n \ldots y_1|\mathscr{A}, \mathscr{I}_c+\mathscr{I}], \tag{5}$$

which we shall call *n-electron propagators.* They depend on both the field $\mathscr{A}$ and the current $\mathscr{I}$, and on the compensating current $\mathscr{I}_c$. These propagators are invariant with respect to gauge transformations. Accordingly, they may describe physical processes even though, as we have explained in Section 19, operators creating and annihilating individual electrons are not present among the operators of physical fields.

By differentiation with respect to $\mathscr{I}$, we build *electron-photon propagators* out of the n-electron propagator. The general propagator of this kind is defined by the formula

$$T[x_1 \ldots x_n, y_n \ldots y_1, z_1 \ldots z_k|\mathscr{A}, \mathscr{I}_c+\mathscr{I}]$$
$$\equiv (-i)^n \exp\left(\frac{1}{2i}\int \frac{\delta}{\delta\mathscr{A}} D^F \frac{\delta}{\delta\mathscr{A}}\right)$$
$$\times \mathscr{A}(z_1) \ldots \mathscr{A}(z_\kappa) C[x_1 \ldots x_n, y_n \ldots y_1|\mathscr{A}, \mathscr{I}_c+\mathscr{I}]. \tag{6}$$

The electron-photon propagator is not gauge-invariant but its gauge dependence is due solely to the potentials $\mathscr{A}(z_1), \ldots, \mathscr{A}(z_k)$, just as in the case of many-photon propagators. Subsequently in this section we shall give a physical meaning to electron-photon propagators by relating them to the probability amplitudes of electron-photon processes.

As in formula (19.4), so here, too, we may define gauge-invariant electron-photon propagators constructed out of antisymmetrized derivatives,

$$\begin{aligned} &\mathscr{T}[x_1 \ldots x_n, y_n \ldots y_1, z_1 \ldots z_k \,|\, \mathscr{A}, \mathscr{I}_c + \mathscr{I}] \\ &\quad \equiv (-i)^n \exp\left(\frac{1}{2i} \int \frac{\delta}{\delta \mathscr{A}} D^F \frac{\delta}{\delta \mathscr{A}}\right) \\ &\quad \times f(z_1) \ldots f(z_k) C[x_1 \ldots x_n, y_n \ldots y_1 | \mathscr{A}, \mathscr{I}_c + \mathscr{I}]. \end{aligned} \tag{7}$$

The Relation between the Compensating Current and the Gauge

We shall now examine the dependence of the propagators $\mathscr{T}[x \ldots, y \ldots, z \ldots | \mathscr{A}, \mathscr{I}_c + \mathscr{I}]$ on the choice of the compensating current. Let us take the simplest possible form of the compensating current, assuming it to be built of one universal vector function of two variables $a^\mu(z, x)$ satisfying the condition:

$$\frac{\partial}{\partial z^\mu} a^\mu(z, x) = \delta(z - x). \tag{8}$$

The compensating current for a system of n electrons appearing at the points y_i and disappearing at the points x_i will be the sum of $2n$ terms of the form

$$\mathscr{I}_c^\mu(z) = e \sum_{i=1}^{n} \left(a^\mu(z, x_i) - a^\mu(z, y_i)\right). \tag{9}$$

For such a compensating current, the dependence of the electron-photon propagators on the compensating current can be expressed in a simple manner. In doing this, we shall make use of the following auxiliary formulae:

$$\exp\left(\frac{1}{2i}\int\frac{\delta}{\delta\mathscr{A}}D^{\mathrm{F}}\frac{\delta}{\delta\mathscr{A}}\right)\exp\left(-i\int\mathscr{I}_{\mathrm{c}}\cdot\mathscr{A}\right)$$

$$=\exp\left(-i\int\mathscr{I}_{\mathrm{c}}\cdot\mathscr{A}\right)\exp\left(\frac{1}{2i}\int\left(\frac{\delta}{\delta\mathscr{A}}-i\mathscr{I}_{\mathrm{c}}\right)D^{\mathrm{F}}\left(\frac{\delta}{\delta\mathscr{A}}-i\mathscr{I}_{\mathrm{c}}\right)\right),$$

$$-i\mathscr{I}_{\mathrm{c}}^{\mu}(z)\,C[x_1\,\ldots\,x_n,y_n\,\ldots\,y_1|\mathscr{A}]$$

$$=-ie\int\mathrm{d}^4w\,a^{\mu}(z,w)\sum_{i=1}^{n}\left(\delta(w-x_i)-\delta(w-y_i)\right)C[x_1\,\ldots\,x_n,y_n\,\ldots\,y_1|\mathscr{A}]$$

$$=-\int\mathrm{d}^4w\,a^{\mu}(z,w)\,\mathscr{B}(w)\,C[x_1\,\ldots\,x_n,y_n\,\ldots\,y_1|\mathscr{A}]$$

$$=\int\mathrm{d}^4w\left(\partial_\lambda^{(w)}a^{\mu}(z,w)\right)\frac{\delta}{\delta\mathscr{A}_\lambda(w)}C[x_1\,\ldots\,x_n,y_n\,\ldots\,y_1|\mathscr{A}].$$

The final formula may be rewritten as:

$$\mathscr{T}[x_1\,\ldots\,x_n,y_n\,\ldots\,y_1,z_1\,\ldots\,z_k|\mathscr{A},\mathscr{I}_{\mathrm{c}}+\mathscr{I}]$$
$$=(-i)^n\exp\left(\frac{1}{2i}\int\frac{\delta}{\delta\mathscr{A}}D^{\mathrm{F}}[a]\frac{\delta}{\delta\mathscr{A}}\right)$$
$$\times f(z_1)\ldots f(z_k)\,C[x_1\,\ldots\,x_n,y_n\,\ldots\,y_1|\mathscr{A}[a],\mathscr{I}], \tag{10}$$

where

$$D^{\mathrm{F}}_{\mu\nu}[w,w'|a]\equiv\int\mathrm{d}^4z\,\mathrm{d}^4z'\left(\delta_\mu^\lambda\delta(w-z)+\partial_\mu^{(w)}a^\lambda(z,w)\right)$$
$$\times\left(-g_{\lambda\varrho}D_{\mathrm{F}}(z-z')\right)\left(\delta_\nu^\varrho\delta(w'-z')+\partial_\nu^{(w')}a^\varrho(z',w')\right), \tag{11}$$

while

$$\mathscr{A}_\mu[z|a]\equiv\mathscr{A}_\mu(z)+\partial_\mu\int\mathrm{d}^4w\,\mathscr{A}_\lambda(w)\,a^\lambda(w,z). \tag{12}$$

The right-hand side in formula (10) depends on the compensating current only through the potential $\mathscr{A}[a]$ and through the free propagator $D^{\mathrm{F}}[a]$.

The potential $\mathscr{A}[a]$ is physically equivalent to the initial potential $\mathscr{A}$, since it differs from it only by the gauge transformation. It follows from properties (9) of the function a^μ that the potential $\mathscr{A}[z|a]$ satisfies the following *gauge condition for the potential,*

$$\int d^4z \mathscr{A}_\mu[z|a] a^\mu(z, w) = 0. \tag{13}$$

We shall say that the potential $\mathscr{A}[a]$ is chosen in the gauge defined by a^μ.

Similarly, we shall say that the free photon propagator $D^F[a]$ has been chosen in the gauge defined by a^μ. This propagator satisfies the following *gauge conditions for the propagator*:

$$\int d^4w\, a^\mu(w, z) D^F_{\mu\nu}[w, w'|a] = 0 = \int d^4w' D^F_{\mu\nu}[w, w'|a] a^\nu(w', z). \tag{14}$$

The *radiation gauge of the potential* is obtained by choosing the vector a^μ in formula (13) in the form

$$a^\mu_{\text{rad}}(z, w) = (\partial^\mu - n^\mu(n\cdot\partial))(\Box - (n\cdot\partial)^2)^{-1}\delta(z-w). \tag{15}$$

The free photon propagator in the radiation gauge is the same as the expression $\mathscr{D}^F_{\mu\nu}(z-z', n)$ defined in Section 13.

The radiation gauge condition determines the potential in the class of bounded functions of the variables x, y, z, t. Another important gauge, the *Lorentz gauge,* does not possess this property since the Lorentz condition

$$\partial^\mu \mathscr{A}_\mu(z) = 0$$

still leaves freedom to make the gauge transformation,

$$\mathscr{A}_\mu \to \mathscr{A}_\mu + \partial_\mu \Lambda,$$

with the function Λ obeying the wave equation,

$$\Box \Lambda = 0.$$

Potentials satisfying the Lorentz condition may be made unique only through a choice of particular asymptotic conditions which are met by the potential. In view of the applications to the description of propagators, we shall choose the Feynman boundary conditions.† The Lorentz gauge with the Feynman asymptotic conditions is known as the *Landau gauge.* It is characterized by the choice of a^μ in the form

$$a^\mu_L(z, w) = \partial^\mu D_F(z-w).$$

The uniqueness of potentials in the Landau gauge stems from the fact that the condition

$$\int d^4z\, \partial_\mu \Lambda(z) \partial^\mu D_F(z-w) = 0$$

† The external potential field $\mathscr{A}_\mu(z)$ should be interpreted as the matrix element of the corresponding operator between the vacuum IN and OUT states (cf. eg. (20.17)). Therefore, the fact that $\mathscr{A}_\mu(z)$ takes on complex values required by Feynman asymptotic conditions is not inconsistent with the rest of the theory.

for the function Λ satisfying the wave equation requires that $\Lambda = 0$. To show this it is sufficient to integrate by parts and evaluate the boundary terms.

The photon propagator in the Landau gauge is of the form:

$$D^{\mathrm{F}}_{\mu\nu}(z-z') = (-g_{\mu\nu} + (\Box - i\varepsilon)^{-1}\partial_\mu\partial_\nu)\, D_{\mathrm{F}}(z-z'), \tag{16}$$

where the choice of the asymptotic conditions is indicated by the symbol $i\varepsilon$. The free photon propagator in the Landau gauge is relativistically invariant in form.

Another example of the gauge condition is obtained if we choose the function a^μ in the form

$$a^\mu(z, w) = \delta^\mu_0 \theta(z^0 - w^0)\, \delta(\mathbf{z}-\mathbf{w}).$$

A potential in this gauge has its component $\mathscr{A}_0$ identically equal to zero.

In all three cases, the functions a^μ, and along with them the free photon propagators as well, were functions of the difference of coordinates, $z-z'$. Such a choice of gauge is most convenient since it does not violate the invariance of the theory under translations in space and in time.

Now let us return to formula (10) for the electron-photon propagator. On the basis of the foregoing considerations, this formula may be interpreted as follows. The gauge-invariant propagator $\mathscr{T}$, with compensating current $\mathscr{I}_c$ determined by the function a^μ, is equal to the gauge-dependent propagator (without compensating current) calculated in the gauge determined by a^μ. Both the potential $\mathscr{A}_\mu[a]$ and the free photon propagator $D^{\mathrm{F}}_{\mu\nu}[a]$ should be chosen in this gauge.

The aforementioned relation gives a physical meaning to the gauge-dependent propagators.# We shall henceforth make use of just such propagators, since they are simpler to handle in calculations especially in the absence of external fields and external currents. For after performing the differentiation with respect to $\mathscr{A}$ in the absence of external fields and currents, we substitute the free propagators S_{F} for the propagators K_{F}. The entire dependence on the compensating current then is contained only in the free photon propagators $D^{\mathrm{F}}_{\mu\nu}[a]$.

Feynman Diagrams

The methods described in Section 20 can be used to obtain perturbation expansions of electron-photon propagators. The diagram technique is extended to electron-photon propagators without any additional assumptions. The only difference is that wavy lines in diagrams repre-

---

This observation was first made by K. Johnson (1964).

senting contributions to electron-photon propagators (10) correspond to free photon propagators $D^{\mathrm{F}}_{\mu\nu}[z, z'|a]$. To simplify the notation we shall henceforth in general neglect the dependence on a^{μ} of the photon propagator D^{F}, and since we shall confine ourselves solely to compensating currents which do not violate translational invariance, we shall use the notations $D^{\mathrm{F}}_{\mu\nu}(z-z')$ for the free photon propagator.

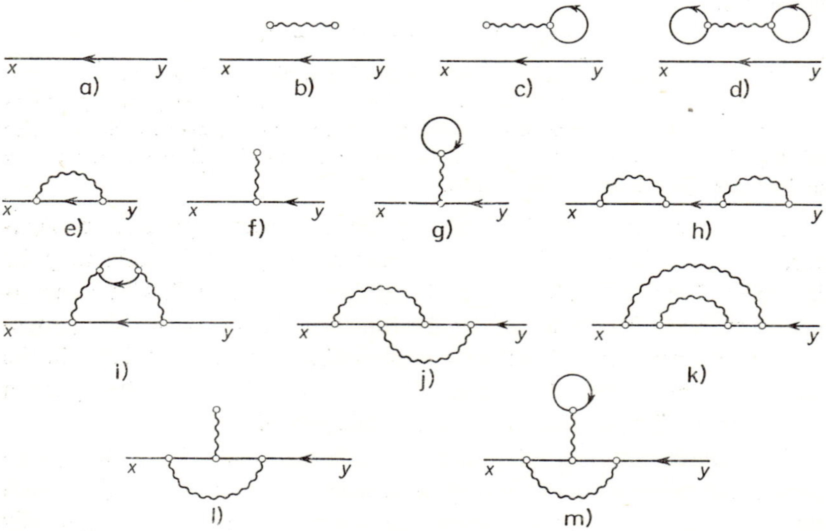

FIG. 22.1. Diagrams representing a one-electron propagator.

For illustration, in Fig. 22.1 we give the simplest diagrams representing a one-electron propagator.

The analysis of connectedness properties of the propagators in Section 20 can be applied to electron-photon propagators. First, we separate the vacuum part $V[\mathscr{A}, \mathscr{J}]$ from all the propagators, and in this way we get the linked parts.

The simplest example of the linked part is that of the one-electron propagator. We shall simply call it the *electron propagator* and introduce a special notation for it:

$$-iG[x, y|\mathscr{A}, \mathscr{J}] \equiv T^{L}[x, y|\mathscr{A}, \mathscr{J}]. \tag{17}$$

A similar notation is introduced for the linked part of the two-electron propagator:

$$(-i)^2 G[x_1 x_2, y_2 y_1|\mathscr{A}, \mathscr{I}] \equiv T^L[x_1 x_2, y_2 y_1|\mathscr{A}, \mathscr{I}], \tag{18}$$

and, in general, for the n-electron propagator:

$$(-i)^n G[x_1 \dots x_n, y_n \dots y_1|\mathscr{A}, \mathscr{I}] \equiv T^L[x_1 \dots x_n, y_n \dots y_1|\mathscr{A}, \mathscr{I}]. \tag{19}$$

The electron propagator is also the connected part of the one-electron propagator,

$$T^L[x, y|\mathscr{A}, \mathscr{I}] = T^C[x, y|\mathscr{A}, \mathscr{I}], \tag{20}$$

since an electron line entering each diagram must be joined in a continuous manner with an electron line leaving the diagram (because of charge conservation).

The linked part of the two-electron propagator, however, contains unconnected parts. The analogue of formula (20.9) for the two-electron propagator is of the form

$$\begin{aligned} T^L[x_1 x_2, y_2 y_1|\mathscr{A}, \mathscr{I}] &= T^C[x_1 x_2, y_2 y_1|\mathscr{A}, \mathscr{I}] \\ &+ T^C[x_1, y_1|\mathscr{A}, \mathscr{I}]T^C[x_2, y_2|\mathscr{A}, \mathscr{I}] \\ &- T^C[x_1, y_2|\mathscr{A}, \mathscr{I}]T^C[x_2, y_1|\mathscr{A}, \mathscr{I}]. \end{aligned} \tag{21}$$

This decomposition may also be expressed in terms of the propagators G,

$$\begin{aligned} G[x_1 x_2, y_2 y_1] = G^C[x_1 x_2, y_2 y_1] &+ G[x_1, y_1]G[x_2, y_2] \\ &- G[x_2, y_2]G[x_2, y_1]. \end{aligned} \tag{22}$$

To illustrate the foregoing considerations, in Fig. 22.2 we give the simplest diagrams representing the linked part of the two-electron propagator. Since antisymmetrization with respect to the electron coordinates x_i and y_i occurs in the formula for $C[x_1 \dots x_n, y_n \dots y_1|\mathscr{A}, \mathscr{I}]$, diagrams (a) and (b), (c) and (d), etc., make contributions to the propagator with opposite signs. The general prescription for determining the sign of the contribution from an arbitrary diagram to the electron-photon propagator is:

The analytical expression built in accordance with rules 1–7 of Section 20 is multiplied by $(-1)^L \varepsilon_p$, where L is the number of closed loops of electron lines, while ε_p is equal to 1 or -1, depending on whether the numeration of the ends of the electron lines is obtained from the numeration of the origins of these lines by even or odd permutation.

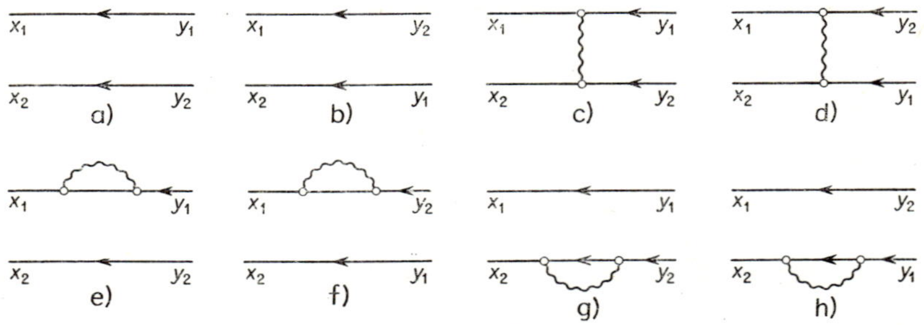

FIG. 22.2. The linked part of the two-electron propagator.

The concepts of strongly connected diagrams and strongly connected parts for electron-photon propagators are introduced in the same way as for photon propagators. The sum of all expressions corresponding to strongly connected diagrams, without external lines, representing a given electron-photon propagator will be denoted by the symbol $T_{\mathrm{S}}^{\mu_1 \dots \mu_k}[x_1 \dots x_n, y_n \dots y_1, z_1 \dots z_k|\mathscr{A}, \mathscr{I}]$. To avoid confusion, in the drawings we shall indicate the external lines of strongly connected diagrams (as in Section 20). Traditional symbols will also be used to denote the strongly connected parts of some chosen propagators. For example, in accordance with the considerations of Section 20 we shall denote the strongly connected parts of photon propagators with the symbols $\Delta^{\mu_1 \dots \mu_k}[z_1 \dots z_k|\mathscr{A}, \mathscr{I}]$.

Strongly connected diagrams for the electron propagator are called the *electron self-energy diagrams* (cf. Fig. 22.3). All weakly connected diagrams representing the electron propagator can be built out of the electron self-energy diagrams by connecting the ends of the corresponding external lines. Each series connection of electron self-energy diagrams is one of the diagrams representing the electron propagator. This property has been shown graphically in Fig. 22.4 where the cross-hatched circles (together with the attached lines) represent the set of all self-energy diagrams.

The analytical expression which is the sum of the contributions from all the diagrams obtained from the electron self-energy diagrams by detaching the external lines is referred to as the *self-energy function*

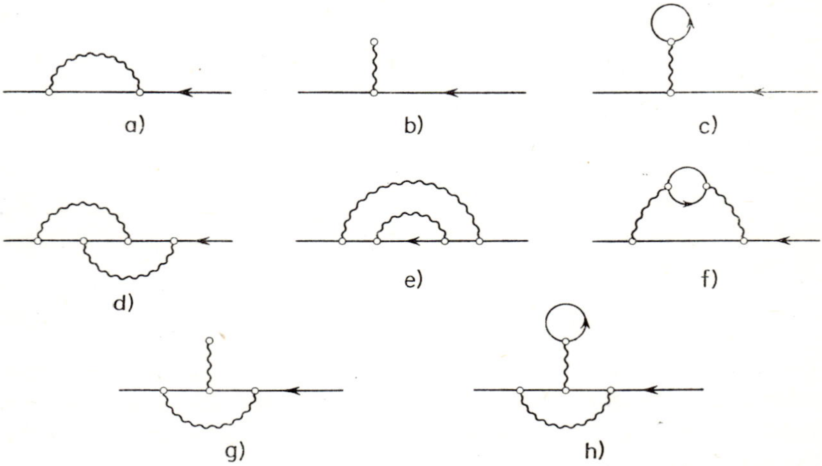

FIG. 22.3. Electron self-energy diagrams.

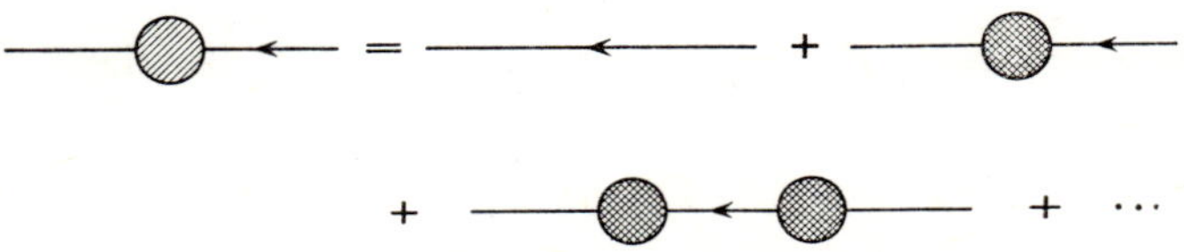

FIG. 22.4. The decomposition of the electron propagator into strongly connected parts.

of the electron. It will be denoted by the symbol $i\Sigma[x, y|\mathscr{A}, \mathscr{I}]$. The symbolical equality shown in Fig. 22.4 may be expressed in terms of the electron self-energy function as follows:

$$-iG[x, y|\mathscr{A}, \mathscr{I}] = -iK_{\mathrm{F}}[x, y|\mathscr{A}]$$
$$-i\int \mathrm{d}x'\,\mathrm{d}y'\,K_{\mathrm{F}}[x, x'|\mathscr{A}]\,\Sigma\,[x', y'|\mathscr{A}, \mathscr{I}]\,K_{\mathrm{F}}[y', y|\mathscr{A}] + \ldots. \tag{23}$$

The electron self-energy function, just as the electron propagator, is a matrix with two bispinor indices; the multiplication in formula (23) is to be understood as the matrix multiplication of Σ by two propagators K_{F}.

Below we give the analytical expression corresponding to the simplest electron self-energy diagrams (cf. Fig. 22.3):

$$\Sigma[x, y|\mathscr{A}, \mathscr{I}] = -ie^2\gamma^\mu K_F[x, y|\mathscr{A}]\gamma^\nu D^F_{\mu\nu}(x-y)$$

$$-e^2\delta(x-y)\gamma^\mu \int dz\, D^F_{\mu\nu}(x-z)\mathscr{I}^\nu(z)$$

$$+ie^2\delta(x-y)\gamma^\mu \int dz\, D^F_{\mu\nu}(x-z)\,\mathrm{Tr}\,\{\gamma^\nu K_F[z, z|\mathscr{A}]\} + \ldots .$$

The next electron-photon propagator for which we shall carry out an analysis of the connectedness is the propagator $T[x, y, z|\mathscr{A}, \mathscr{I}]$. The decomposition of this propagator into connected parts is given by the formula

$$T^L[x, y, z] = T^C[x, y, z] + T^C[x, y]T^C[z]. \tag{24}$$

For the strongly connected part of the propagator $T^C[x, y, z]$ we introduce a new symbol $\Gamma^\mu[x, y, z|\mathscr{A}, \mathscr{I}]$ and a special name: *vertex function*. In the definition of this function one finds it convenient to separate the factor $-ie$,

$$-ie\Gamma^\mu[x, y, z|\mathscr{A}, \mathscr{I}] \equiv T^\mu_S[x, y, z]. \tag{25}$$

The connected part of the propagator $T^\mu[x, y, z]$ is related to the vertex function by the formula

$$T^C_\mu[x, y, z]$$
$$= (-i)^3(-ie)\int d\xi\, d\eta\, d\zeta\, \mathscr{G}_{\mu\nu}[z\zeta]G[x, \xi]\Gamma^\nu[\xi, \eta, \zeta]\, G[\eta, y]. \tag{26}$$

Figure 22.5 shows the simplest diagrams representing the vertex function.

Similar notation is introduced for the strongly connected parts of the other propagators describing processes with the participation of one electron (for $k > 1$):

$$e^k\Gamma^{\mu_1 \ldots \mu_k}[x, y, z_1 \ldots z_k|\mathscr{A}, \mathscr{I}]$$
$$\equiv T^{\mu_1 \ldots \mu_k}_S[x, y, z_1 \ldots z_k|\mathscr{A}, \mathscr{I}]. \tag{27}$$

The strongly connected part of the propagator $T[x_1 x_2, y_2 y_1|\mathscr{A}, \mathscr{I}]$ is denoted by the symbol $E[x_1 x_2, y_2 y_1|\mathscr{A}, \mathscr{I}]$,

$$E[x_1 x_2, y_2 y_1|\mathscr{A}, \mathscr{I}] = T_S[x_1 x_2, y_2 y_1|\mathscr{A}, \mathscr{I}]. \tag{28}$$

The concept of truncated diagrams and truncated propagators may also be applied in a natural manner to electron-photon propagators.

A truncated diagram is a connected diagram from which we cannot separate any diagrams representing electron or photon propagators

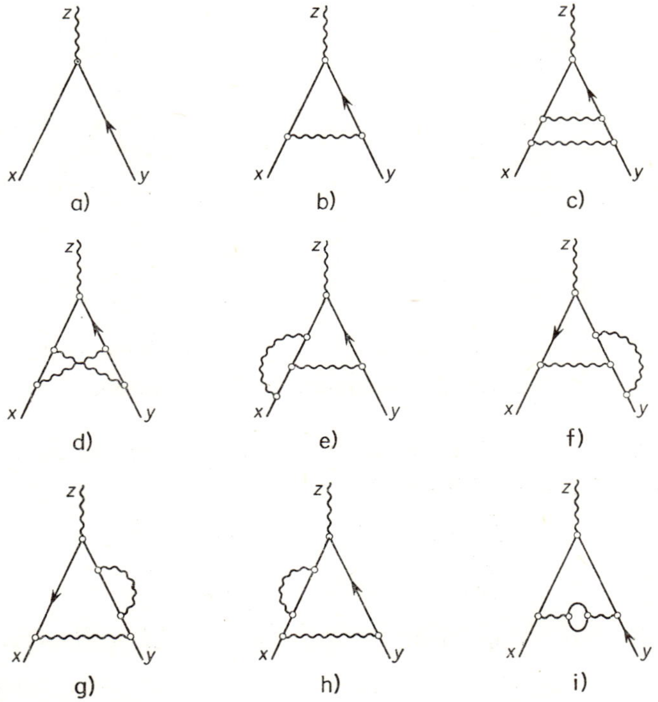

FIG. 22.5. Diagrams representing the vertex function.

by breaking only one internal electron or photon line. Each strongly connected diagram is truncated. In Fig. 22.6 we have two non-truncated diagrams and the corresponding truncated ones obtained by detaching the parts representing the photon and electron propagators. Once again in order to avoid confusion, we shall put the external lines on truncated diagrams in the figures.

The sum of the analytical expressions corresponding to all truncated diagrams, after detachment of the external lines, representing the connected part of the propagator will be called the *truncated part of the propagator* and will be denoted by the symbol $T_{\mathrm{T}}^{\mu_1 \ldots \mu_k}[x_1 \ldots x_n, y_n \ldots y_1, z_1 \ldots z_k|\mathscr{A}, \mathscr{I}]$. The notion of truncated diagrams is not applicable to the electron and photon propagators. From the definition of

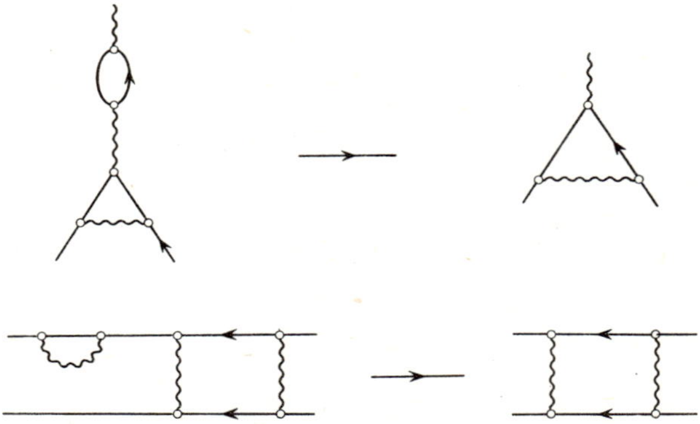

FIG. 22.6. The formation of truncated diagrams.

the truncated part it follows that the connected part and the truncated part of a propagator are related by the formula

$$T^{C}_{\mu_1 \ldots \mu_k}[x_1 \ldots x_n, y_n \ldots y_1, z_1 \ldots z_k]$$
$$= \int \mathrm{d}\zeta_1 \ldots \mathrm{d}\zeta_k \mathrm{d}\xi_1 \ldots \mathrm{d}\xi_n \mathrm{d}\eta_1 \ldots \mathrm{d}\eta_n$$
$$\times (-i)^{2n+k} \mathscr{G}_{\mu_1\nu_1}[z_1 \zeta_1] \ldots \mathscr{G}_{\mu_k\nu_k}[z_k \zeta_k] G[x_1, \xi_1] \ldots G[x_n, \xi_n]$$
$$\times T_{\mathrm{T}}^{\nu_1 \ldots \nu_k}[\xi_1 \ldots \xi_n, \eta_n \ldots \eta_1, \zeta_1 \ldots \zeta_k] G[\eta_n, y_n] \ldots G[\eta_1, y_1]. \quad (29)$$

To conclude this analysis, we shall give the relations between the truncated and strongly connected parts of the propagators $T[x, y, z_1 z_2]$ and $T[x_1 x_2, y_2 y_1]$:

$$T_{\mathrm{T}}^{\mu\nu}[x, y, z_1 z_2] = e^2 \Gamma^{\mu\nu}[x, y, z_1 z_2]$$
$$+ie^2 \int \mathrm{d}\xi \mathrm{d}\eta \Gamma^{\mu}[x, \xi, z_1] G[\xi, \eta] \Gamma^{\nu}[\eta, y, z_2]$$
$$+ie^2 \int \mathrm{d}\xi \mathrm{d}\eta \Gamma^{\nu}[x, \xi, z_2] G[\xi, \eta] \Gamma^{\mu}[\eta, y, z_1]$$
$$-e \int \mathrm{d}\zeta_1 \mathrm{d}\zeta_2 \Delta^{\mu\nu\lambda}[z_1 z_2 \zeta_1] \mathscr{G}_{\lambda\varrho}[\zeta_1 \zeta_2] \Gamma^{\varrho}[x, y, \zeta_2], \quad (30)$$

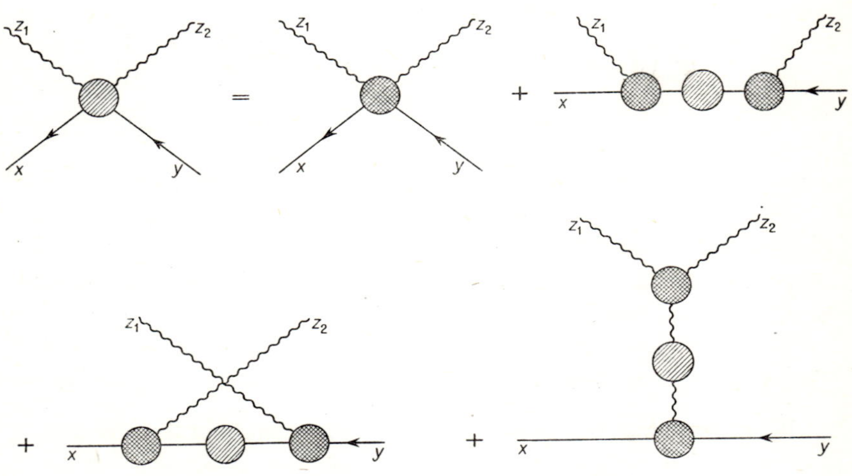

FIG. 22.7. The decomposition of the propagator $T[x, y, z_1 z_2]$ into strongly connected parts.

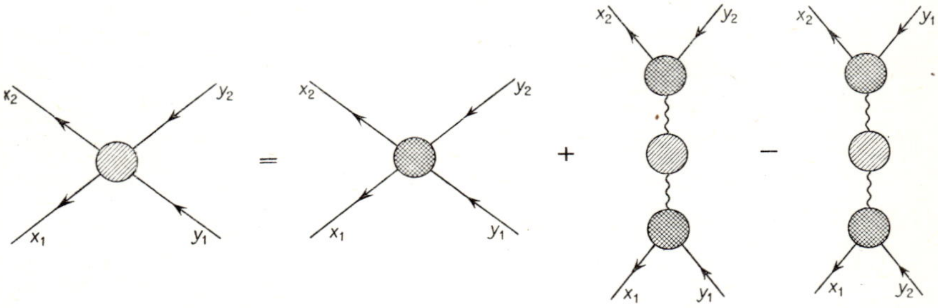

FIG. 22.8. The decomposition of the propagator $T[x_1 x_2, y_2 y_1]$ into strongly connected parts.

$$T_{\mathrm{T}}[x_1\, x_2, y_2 y_1] = E[x_1 x_2, y_2 y_1]$$

$$+ie^2 \int \mathrm{d}w_1\, \mathrm{d}w_2 \Gamma^{\mu}[x_1, y_1, w_1] \mathscr{G}_{\mu\nu}[w_1 w_2] \Gamma^{\nu}[x_2, y_2, w_2]$$

$$-ie^2 \int \mathrm{d}w_1 dw_2 \Gamma^{\mu}[x_1, y_2, w_1] \mathscr{G}_{\mu\nu}[w_1 w_2] \Gamma^{\nu}[x_2, y_1, w_2]. \quad (31)$$

These relations are shown graphically in Figs. 22.7 and 22.8.

Just as for photon propagators, the relations finally obtained between the propagators and their connected, strongly connected, and truncated parts may be taken as definitions of these quantities without recourse to perturbation theory.

The Relation between Propagators and Transition Amplitudes

In determining the relations between electron-photon propagators and transition amplitudes for electron-photon processes, we shall make use of the now familiar relations between the photon propagators and the transition amplitudes of photon processes. In Section 21 we showed that in the theory with interaction perturbations of the electromagnetic field described by the field operators $f_{\mu\nu}(x)$ propagate in space-time not only as individual photons but also as all possible aggregates composed of many photons and/or other particles interacting with photons. The one-photon component of the perturbation could be separated by means of an operation consisting of integration with the photon wave function over the hypersurface σ, the transition to the limit $\sigma \to \pm\infty$, and division by the square root of the renormalization constant Z_3.

We shall assume that a similar operation applied to the electron-photon propagators yields transition amplitudes for processes with the participation of electrons. This operation consists of integration of the propagator with the negaton or positon wave function over the hypersurface σ, the transition to the limit $\sigma \to \pm\infty$, and division by the square root of the *renormalization constant* Z_2.†

† This constant is denoted by the traditional symbol introduced by Dyson in his paper on renormalization (F. J. Dyson (1949)). The value of the renormalization constant Z_2 and the renormalization constant Z_3 is determined in the next chapter in the lowest order of perturbation theory.

This is a natural generalization of the results obtained in the non-relativistic theory of interacting particles, in the relativistic theory of non-interacting electrons, and in the theory of mutually interacting photons, which have been discussed in detail in preceding sections.

In the simplest case of the scattering of a single negaton, the formula for the transition amplitude takes the form (cf. formula (17.61a)):

$$S^{++}[f, g] = Z_2^{-1} \lim_{\substack{\sigma_x \to \infty \\ \sigma_y \to -\infty}} \int d\sigma_\mu \int d\sigma_\nu \overline{\chi^{(+)}}[x|f]\gamma^\mu T[x, y|\mathscr{A}, \mathscr{I}]\gamma^\nu \chi^{(+)}[y|g]. \tag{32}$$

We shall not write out the formulae for other transition amplitudes in the form of integrals over hypersurfaces since in subsequent applications we shall use more convenient formulae containing integrals over space-time (cf. formulae (17.62) and (21.19)).

When calculating transition amplitudes for processes with the participation of electrons we run into additional difficulties which did not occur in any of the three aforementioned simpler theories. These difficulties are caused by the vanishing photon rest mass resulting in the infrared catastrophe which was discussed in Section 15.

As mentioned earlier (cf. footnote on p. 220), initially we shall assume that the rest mass of photons is non-zero. Thanks to this, we shall be able to postpone the discussion of the infrared catastrophe until we have all the other elements of the theory at our disposal.

The definition of electron-photon propagators $\mathscr{T}[x_1 \ldots x_n, y_n \ldots y_1, z_1 \ldots z_k|\mathscr{A}, \mathscr{I}+\mathscr{I}_c]$ in the theory with photon rest mass μ can be introduced without difficulty. It is sufficient to replace the free propagator of the photon with zero mass,

$$D^{\mathrm{F}}_{\mu\nu}(z-z') = -g_{\mu\nu} D_{\mathrm{F}}(z-z'),$$

in the definition (7) by the propagator $\Delta^{\mathrm{F}}_{\mu\nu}$,

$$\Delta^{\mathrm{F}}_{\mu\nu}(z-z') = (-g_{\mu\nu} - \mu^{-2}\partial_\mu \partial_\nu)\Delta_{\mathrm{F}}(z-z'),$$

which describes free propagation with mass μ.

Since the propagator $\mathscr{T}$ is invariant under the gauge transformations, the term containing derivatives of the function Δ_{F} in the propagator $\Delta^{F}_{\mu\nu}$ may be omitted and the effective propagator obtained in the form:

$$-g_{\mu\nu}\Delta_{F}(z-z').$$

The propagators $\mathscr{T}$ in the theory with mass μ are thus obtained from propagators in the theory with zero rest mass by replacing the scalar propagator D_F with the scalar propagator with mass Δ_F. When the compensating current has been eliminated in the theory with mass, propagators $\Delta^{F}[a]$ dependent on the form of the compensating current appear in the formulae for the electron-photon propagators:

$$\Delta^{F}_{\mu\nu}[w, w'|a] = \int d^4z\, d^4z' \big(\delta^{\lambda}_{\mu}\delta(w-z)+\partial^{(w)}_{\mu}a^{\lambda}(z, w)\big) \times \big(-g_{\lambda\varrho}\Delta_{F}(z-z')\big)\big(\delta^{\varrho}_{\nu}\delta(w'-z')+\partial^{(w')}_{\nu}a^{\varrho}(z', w')\big). \quad (33)$$

To avoid further complication of the notation, we shall henceforth denote propagators in the theory with zero mass and in the theory with mass μ by the symbol $D^{F}_{\mu\nu}(z-z')$. The context will always make it evident which propagator is involved.

The introduction of a finite rest mass for the photon makes it possible to define in the state-vector space two (IN and OUT) Fock bases consisting of vectors which asymptotically describe free electrons and photons. For, owing to the finite photon mass, the interaction between electrons becomes an interaction of finite range. Instead of Coloumb interaction, Yukawa-type interaction appears.

In the operation on IN basis vectors, we define the creation and annihilation operators of asymptotically free electrons and photons which we use in turn to build the field operators $\psi^{in}(x)$, $\bar{\psi}^{in}(x)$, and $A^{in}_{\mu}(x)$ (cf. formulae (18.10) and (14.32)). These operators satisfy Dirac's free equations and Proca's free equations, respectively, in all space. The OUT operators can be built in similar manner.

A unitary operator mapping the OUT basis into the IN basis is the S operator in the complete quantum electrodynamics with finite photon mass. In view of the infrared catastrophe, this operator does not have a limit when $\mu \to 0$.

On the basis of results obtained hitherto (cf. formulae (5.68), (18.55) and (21.20)), we postulate the following relation between the full S operator in quantum electrodynamics and electron-photon propagators,

$$\mathrm{S}[\mathscr{A}, \mathscr{I}] = \sum_{n=0}^{\infty}\sum_{k=0}^{\infty} Z_2^{-n} Z_3^{-k/2} \frac{i^{2n}}{(n!)^2}\frac{i^k}{k!}\int \mathrm{d}x_1 \ldots \mathrm{d}x_n \mathrm{d}y_1 \ldots \mathrm{d}y_n \mathrm{d}z_1 \ldots \mathrm{d}z_k$$

$$\times :\overline{\psi}^{\mathrm{in}}(x_n) \ldots \overline{\psi}^{\mathrm{in}}(x_1) A^{\mathrm{in}}_{\mu_1}(z_1) \ldots A^{\mathrm{in}}_{\mu_k}(z_k) D_{x_1} \ldots D_{x_n}$$

$$\times K^{\mu_1\nu_1}_{z_1} \ldots K^{\mu_k\nu_k}_{z_k} T_{\nu_1\ldots\nu_k}[x_1 \ldots x_n, y_n \ldots y_1, z_1 \ldots z_k|\mathscr{A}, \mathscr{I}]$$

$$\times \overleftarrow{D}_{y_1} \ldots \overleftarrow{D}_{y_n} \psi^{\mathrm{in}}(y_1) \ldots \psi^{\mathrm{in}}(y_n): \tag{34}$$

Since the photon wave functions φ_μ now satisfy the Proca equation instead of the Maxwell equations, Proca operations $K_z^{\mu\nu}$ defined by formula (14.19) appear in place of the operations $M_z^{\mu\nu}$.

This formula for the S operator may be "derived" by ignoring all complications associated with the compensating currents and the gauge, treating quantum electrodynamics just like the quantum theory of interacting fields: a bispinor field described by the operators $\psi(x)$ and $\overline{\psi}(x)$ and a vector field described by the operators A_μ.

Formula (34) for the S operator then follows from the *reduction formulae*,

$$[a_i^{\mathrm{in}}, \mathrm{S}T(\mathcal{O}_1(x_1) \ldots \mathcal{O}_n(x_n))]_\mp$$

$$= iZ_2^{-1/2}\int \mathrm{d}^4x \overline{\chi}^{(+)}[x|g_i] D_x \mathrm{S}T(\psi(x)\mathcal{O}_1(x_1) \ldots \mathcal{O}_n(x_n)), \tag{35a}$$

$$[\mathrm{S}T(\mathcal{O}_1(x_1) \ldots \mathcal{O}_n(x_n)), a_i^{\dagger\mathrm{in}}]_\mp$$

$$= iZ_2^{-1/2}\int \mathrm{d}^4x \mathrm{S}T(\mathcal{O}_1(x_1) \ldots \mathcal{O}_n(x_n)\overline{\psi}(x)) \overleftarrow{D}_x \chi^{(+)}[x|g_i], \tag{35b}$$

$$[\mathrm{S}T(\mathcal{O}_1(x_1) \ldots \mathcal{O}_n(x_n)), b_i^{\mathrm{in}}]_\mp$$

$$= iZ_2^{-1/2}\int \mathrm{d}^4x \mathrm{S}T(\mathcal{O}_1(x_1) \ldots \mathcal{O}_n(x_n)\overline{\psi}(x)) \overleftarrow{D}_x \chi^{(-)}[x|g_i], \tag{35c}$$

$$[b_i^{\dagger\mathrm{in}}, \mathrm{S}T(\mathcal{O}_1(x_1) \ldots \mathcal{O}_n(x_n))]_\mp$$

$$= iZ_2^{-1/2}\int \mathrm{d}^4x \overline{\chi}^{(-)}[x|g_i] D_x \mathrm{S}T(\psi(x)\mathcal{O}_1(x_1) \ldots \mathcal{O}_n(x_n)), \tag{35d}$$

$$[c_k^{\mathrm{in}}, \mathrm{S}T(\mathcal{O}_1(x_1) \ldots \mathcal{O}_n(x_n))]_-$$

$$= iZ_3^{-1/2}\int \mathrm{d}^4z \varphi_\mu^*[z|f_k] K_z^{\mu\nu} \mathrm{S}T(A_\nu(z)\mathcal{O}_1(x_1) \ldots \mathcal{O}_n(x_n)), \tag{35e}$$

$$[\mathrm{S}T(\mathcal{O}_1(x_1) \ldots \mathcal{O}_n(x_n)), c_k^{\dagger\mathrm{in}}]_-$$

$$= iZ_3^{-1/2}\int \mathrm{d}^4z \varphi_\mu[z|f_k] K_z^{\mu\nu} \mathrm{S}T(\mathcal{O}_1(x_1) \ldots \mathcal{O}_n(x_n) A_\nu(z)), \tag{35f}$$

(in formulae (35a)–(35d) the commutators refer to the case when an even number of operators ψ and $\bar{\psi}$ occurs among the operators $\mathcal{O}_i(x_i)$, and the anticommutators to the case when this number is odd). The reduction formulae in turn emerge from the following asymptotic conditions for the field operators:

$$\lim_{\sigma\to\mp\infty} \int d\sigma_\mu \bar{\chi}^{(+)}[x|g_i]\gamma^\mu \psi(x) = Z_2^{1/2} a_i^{\substack{\text{in}\\ \text{out}}}, \tag{36a}$$

$$\lim_{\sigma\to\mp\infty} \int d\sigma_\mu \bar{\chi}^{(-)}[x|g_i]\gamma^\mu \psi(x) = Z_2^{1/2} b_i^{\dagger\substack{\text{in}\\ \text{out}}}, \tag{36b}$$

$$\lim_{\sigma\to\mp\infty} i \int d\sigma_\mu (\varphi_\lambda^*[x|f_k] f^{\lambda\mu}(x) - \varphi^{\lambda\mu *}[x|f_k] A_\lambda(x)) = Z_3^{1/2} c_k^{\substack{\text{in}\\ \text{out}}}, \tag{36c}$$

and from the conditions obtained by Hermitian conjugation of the foregoing formulae.

The decomposition of the propagators into connected parts allows the operator $\mathsf{S}[\mathcal{A}, \mathcal{J}]$ to be written in the form:

$$\mathsf{S}[\mathcal{A}, \mathcal{J}] = V[\mathcal{A}, \mathcal{J}]:\exp\left\{\sum_{n=0}^{\infty}\sum_{k=0}^{\infty} Z_2^{-n} Z_3^{-k/2} \frac{i^n}{(n!)^2} \frac{i^k}{k!}\right.$$

$$\times \int dx_1 \ldots dx_n dy_1 \ldots dy_n dz_1 \ldots dz_k \bar{\psi}^{\text{in}}(x_n) \ldots \bar{\psi}^{\text{in}}(x_1) A_{\mu_1}^{\text{in}}(z_1) \ldots A_{\mu_k}^{\text{in}}(z_k)$$

$$\times D_{x_1} \ldots D_{x_n} K_{z_1}^{\mu_1\nu_1} \ldots K_{z_k}^{\mu_k\nu_k} T^C_{\nu_1\ldots\nu_k}[x_1 \ldots x_n, y_n \ldots y_1, z_1 \ldots z_k|\mathcal{A}] \overleftarrow{D}_{y_1} \ldots \overleftarrow{D}_{y_n}$$

$$\left. \times \psi^{\text{in}}(y_1) \ldots \psi^{\text{in}}(y_n)\right\}:. \tag{37}$$

When we take only the first two terms in the exponent into account, we get:

$$\mathsf{S}[\mathcal{A}, \mathcal{J}] \simeq V[\mathcal{A}, \mathcal{J}]:\exp\left\{iZ_3^{-1/2}\int d^4z\, A_\mu^{\text{in}}(z) K_z^{\mu\nu} \mathfrak{A}_\nu[z|\mathcal{A}, \mathcal{J}]\right\}:$$

$$:\exp\left\{iZ_2^{-1}\int d^4x d^4y \bar{\psi}^{\text{in}}(x) D_x G[x, y|\mathcal{A}, \mathcal{J}] \overleftarrow{D}_y \psi^{\text{in}}(y)\right\}:. \tag{38}$$

The S operator given by this approximate formula describes a system of interacting particles such that each particle interacts separately with the effective field generated by all the other particles, this field being treated as an external one. The interactions of the individual particles are not correlated with each other. This is the quantum electrodynamic analogue of the Hartree-Fock approximation. The first exponential function in formula (38) describes the interactions in which only photons participate. By discarding the second exponential function, we would obtain a formula analogous to the one for the S operator in the theory of a system of photons interacting with given sources. The role of the external current $\mathscr{I}^{\mu}(x)$ is played by the expression

$$\mathscr{I}^{\mu}[z|\mathscr{A}, \mathscr{I}] \equiv -Z_3^{-1/2} K_z^{\mu\nu} \mathfrak{A}_{\nu}[z|\mathscr{A}, \mathscr{I}], \tag{39}$$

which we interpreted in the preceding section as a current induced by the external field and the external current. This current satisfies the continuity equation but is in general a complex quantity. The occurrence of an imaginary part is associated with the creation of real negaton-positon pairs and real photons by the external field. For slowly varying† fields (and only then can formula (38) be regarded as a good approximation) the imaginary part of the current $\mathscr{I}^{\mu}[z|\mathscr{A}, \mathscr{I}]$ is negligible compared with the real part.

The second exponential function in formula (38) describes processes in which only electrons take part. The transition amplitudes calculated by using this part of the S operator are products of the one-electron amplitudes in which the one-electron transition amplitudes $S[i, \lambda; j, \lambda']$ are given by

$$S[i, \lambda; j, \lambda'] = \delta_{ij}\delta_{\lambda\lambda'}$$
$$+ i\lambda Z_2^{-1} \int d^4x d^4y \bar{\chi}[x|i, \lambda] D_x G[x, y|\mathscr{A}, \mathscr{I}] \overleftarrow{D}_y \chi[y|j, \lambda']. \tag{40}$$

In the complete quantum electrodynamics, in which the S operator is given by formula (37), processes consisting in the independent emission and absorption of photons and the scattering of non-interacting electrons in the external field, described by means of expression (38), appear in conjunction with all other, more complex processes. In formula (37), each successive term in the sum figuring in the exponent describes processes in which the entire group of particles appear indi-

† We use this name for fields which vary slowly over space and time in comparison with the lengths and times characteristic of the electron, i.e. 3.8×10^{-11} cm and 1.3×10^{-21} sec.

visibly. These further terms are said to describe the *many-particle correlations*. For example, the expression

$$\int dx\,dy\,dz\,\bar{\psi}^{\rm in}(x)\,A_\mu^{\rm in}(z)\,D_x K_z^{\mu\nu} T_\nu^C[x, y, z|\mathscr{A}, \mathscr{I}]\,\overleftarrow{D}_y\psi^{\rm in}(y) \tag{41}$$

in the exponent in formula (37) describes the correlation between two electrons and one photon. If this term were taken into account in the approximate formula for the S operator, the formula would cease to describe the independent motion of photons and electrons and would also describe their mutual interaction, which is detected by observing the correlation between the measurements performed on electrons and on photons.

As follows from formula (37), each interaction of a system of many particles, however, takes place not only through the fully correlated interaction of all particles simultaneously, but also through partially correlated interactions.# Such partially correlated interactions consist in the interaction of groups of particles selected from the system of particles under consideration. In other words, in addition to a connected part (fully correlated), each scattering amplitude also contains an unconnected part composed of expressions describing the mutually independent interactions of the subsystems of particles. One such partial amplitude, coming from the part of the S operator given by formula (38), describes the independent emission and absorption of each photon and the independent scattering in the external field of each electron.

Feynman Diagrams in Momentum Representation

Let us now examine the propagators $T_{\mu_1 \dots \mu_k}(x_1 \dots x_n, y_n \dots y_1, z_1 \dots z_k)$ for a system of interacting electrons and photons with mass in the absence of an external field and an external current. We shall make use of the Feynman diagram technique to calculate the successive terms in the expansion of the propagator with respect to the parameter e. After performing all the functional differentiations,

An analysis of the topics discussed here, by means of the methods of quantum field theory, was given by E. Freese (1955). Within the framework of the general theory of the S operator, these problems were investigated by E. H. Wichmann and J. H. Crichton (1963).

we replace the functions $K_F[x, y|\mathscr{A}]$ with the functions $S_F(x-y)$. The results of Section 17 may be used to verify that when this has been done, all the expressions corresponding to diagrams which contain a closed electron line with only one vertex are equal to zero. For, corresponding to such a closed line is the expression $\lim_{\varepsilon\to 0} \mathrm{Tr}\{\gamma^\mu(S_F(\varepsilon) + S_F(-\varepsilon))\}$, which is equal to zero. A more general statement, known as *Furry's theorem*,# also holds. This theorem states that the contributions to a propagator from Feynman diagrams which contain a closed electron line with an odd number of vertices are equal to zero.

To prove this theorem, let us consider an arbitrary diagram containing a closed electron line with k labelled vertices. Corresponding to each such diagram there always is another which differs only as to the orientation of the closed line (cf. Fig. 22.9).

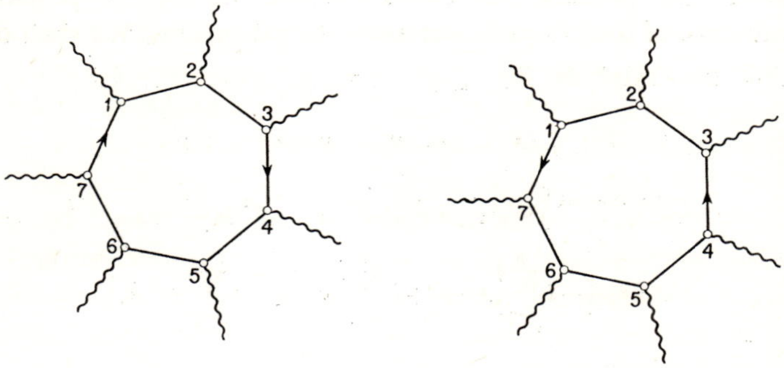

FIG. 22.9. Diagrams in the momentum representation, differing as to the orientation of the closed electron line.

If the number of vertices of the closed line is odd, then by the properties of the traces of products of the functions S_F and matrices γ^μ as discussed in Section 17, the analytical expressions corresponding to these two diagrams added together yield zero. In particular, it follows from Furry's theorem that all photon propagators with an odd number of coordinates $T_{\mu_1 \ldots \mu_{2k+1}}(z_1 \ldots z_{2k+1})$ are, in the absence of an external field and an external current, also equal to zero. This result may also be obtained directly from the definition of propagators, by making use of the invariance of the theory with respect to charge conjugation. This invariance implies the existence of an

W. H. Furry (1937).

operator C which changes particles into antiparticles and vice versa, and changes the sign of the electromagnetic field,

$$\mathsf{C}^{-1}f_{\mu\nu}(x)\mathsf{C} = -f_{\mu\nu}(x).$$

Since in the absence of external fields and currents the vacuum in theory which is invariant with respect to charge conjugation remains invariant under the action of the operator C, we have

$$\big(\Omega|T(f_{\mu_1\nu_1}(z_1)\dots f_{\mu_k\nu_k}(z_k))\,\Omega\big) = \big(\Omega|\mathsf{C}^{-1}T(f_{\mu_1\nu_1}(z_1)\dots f_{\mu_k\nu_k}(z_k))\,\mathsf{C}\Omega\big)$$

$$= (-1)^k\big(\Omega\,|\,T(f_{\mu_1\nu_1}(z_1)\dots f_{\mu_k\nu_k}(z_k))\,\Omega\big).$$

It follows from Furry's theorem that of the eight diagrams in Fig. 22.3, only four give a non-zero contribution to the propagator $G(x, y)$ and of the eighteen diagrams in Fig. 20.9, only five make non-zero contributions to the propagator $T_{\mu\nu}(z_1 z_2)$.

Now let us examine the Fourier transforms of the propagators in the absence of the external field and external current. We shall denote them by the symbols $\tilde{T}_{\mu_1\dots\mu_k}(p_1 \dots p_n, q_n \dots q_1, k_1 \dots k_k)$,

$$T_{\mu_1\dots\mu_k}(x_1 \dots x_n, y_n \dots y_1, z_1 \dots z_k)$$

$$= (2\pi)^{-8n-4k}\int d^4p_1 \dots d^4p_n d^4q_1 \dots d^4q_n d^4k_1 \dots d^4k_k$$

$$\times \exp\Big(-i\sum_{i=1}^{n} x_i\cdot p_i + i\sum_{i=1}^{n} y_i\cdot q_i - i\sum_{i=1}^{k} z_i\cdot k_i\Big)$$

$$\times\, \tilde{T}_{\mu_1\dots\mu_k}(p_1 \dots p_n, q_n \dots q_1, k_1 \dots k_k), \tag{42a}$$

$$\tilde{T}_{\mu_1\dots\mu_k}(p_1 \dots p_n, q_n \dots q_1, k_1 \dots k_k)$$

$$= \int d^4x_1 \dots d^4x_n d^4y_1 \dots d^4y_n d^4z_1 \dots d^4z_k$$

$$\times \exp\Big(i\sum_{i=1}^{n} x_i\cdot p_i - i\sum_{i=1}^{n} y_i\cdot q_i + i\sum_{i=1}^{k} z_i\cdot k_i\Big)$$

$$\times\, T_{\mu_1\dots\mu_k}(x_1 \dots x_n, y_n \dots y_1, z_1 \dots z_k). \tag{42b}$$

The propagators in the successive orders of the expansion with respect to e are built out of the functions $S_F(x)$, $D^{\mathrm{F}}_{\mu\nu}(x)$, and matrices γ^μ. By formulae (17.22a), (17.25), and (33) we obtain the following form of the

Fourier transforms of the functions $S_F(x)$ and $D^F_{\mu\nu}(x)$. These transforms are denoted by the symbols $\tilde{S}_F(p)$ and $\tilde{D}^F{}_{\mu\nu}(k)$,

$$S_F(x) = (2\pi)^{-4}\int d^4p\, e^{-ip\cdot x}\tilde{S}_F(p), \tag{43a}$$

$$\tilde{S}_F(p) = \frac{1}{m-\gamma\cdot p-i\varepsilon} = \frac{m+\gamma\cdot p}{m^2-p^2-i\varepsilon}, \tag{43b}$$

$$D^F_{\mu\nu}(x) = (2\pi)^{-4}\int d^4k\, e^{-ik\cdot x}\tilde{D}^F_{\mu\nu}(k), \tag{43c}$$

$$\tilde{D}^F_{\mu\nu}(k) = \frac{d_{\mu\nu}(k)}{\mu^2-k^2-i\varepsilon}, \tag{43d}$$

where

$$d_{\mu\nu}(k) = -\big(\delta^\lambda_\mu + ik_\mu\tilde{a}^\lambda(k)\big)g_{\lambda\varrho}\big(\delta^\varrho_\nu + ik_\nu\tilde{a}^\varrho(k)\big). \tag{43e}$$

The Fourier transform of the propagator in the k-th order of the perturbation expansion is obtained when in formula (42b) we substitute the propagator expressed in terms of the functions S_F and $D^F_{\mu\nu}$ and perform the integration with respect to the space-time variables. As a result of the integration with respect to these variables, for each vertex we obtain the function $(2\pi)^4\delta^{(4)}$ whose argument is the sum of the respective momenta. Each Fourier transform of the propagator in any order of perturbation theory is the sum of integrals over space-time of the products of the functions $\tilde{S}_F$, $\tilde{D}^F_{\mu\nu}$, and the matrices γ^μ. A *Feynman diagram in momentum representation* is associated with each term in this sum. These diagrams have the same structure as the Feynman diagrams in coordinate representation, the only difference being in the method of labelling the lines. All the external lines, and at times the internal lines as well, are labelled with momenta which are arguments of the relevant functions $\tilde{S}_F$ and $\tilde{D}^F_{\mu\nu}$. The rules establishing the association of the graphical elements with the analytical expressions follow from the corresponding rules for the coordinate representation; they have been listed in the table on page 378.

In each analytical expression corresponding to the connected part of a diagram with n vertices we can (because the Dirac $\delta^{(4)}$ functions appear) integrate with respect to $n-1$ momenta. One more function $\delta^{(4)}(\sum p_i - \sum q_i + \sum k_i)$ remains in each such expression; $\sum q_i$ denotes the sum of the momenta of all electron lines entering the diagram, $\sum p_i$ stands for the sum of the momenta of the outgoing electron lines, and $\sum k_i$ the sum of the momenta of the external photon lines.

Analytical expression	Graphical element
$-i(2\pi)^{-4}\tilde{S}_F(p)$	
$-i(2\pi)^{-4}\tilde{D}^F_{\mu\nu}(k)$	
$-ie(2\pi)^4\gamma^{\mu}$	○

The Fourier transform of the propagator can be reproduced from the Feynman diagrams in the momentum representation by using the following rules:

1. On the diagram we denote the momenta of all lines so that at each vertex the sum of the incoming momenta be equal to the sum of the outgoing momenta. In accordance with the definitions (42) and (43) that have been adopted for the Fourier transforms, the momenta of the electron lines are directed in agreement with the arrows, and the momenta of the external photon lines are directed out from the diagram. The directions of the momenta of the internal photon lines are fixed arbitrarily for each line.
2. To each chain of electron lines there corresponds a matrix product of the functions $-i(2\pi)^{-4}\tilde{S}_F$ and the matrices $-ie(2\pi)^4\gamma$, arranged so that the right-to-left order corresponds to motion in accordance with the arrow on the diagram. The trace of such a matrix product over the bispinor indices and the additional factor -1 corresponds to each closed loop.
3. To each photon line there corresponds a function $-i(2\pi)^{-4}\tilde{D}^F_{\mu\nu}$ whose vector indices are contracted with the indices of the matrix, just as in the coordinate representation.
4. We integrate with respect to all momentum variables which are not momenta of external lines.

5. The result is multiplied by the factor $(2\pi)^{4N}$, where N is the number of external lines, and by $\delta^{(4)}$, whose argument is the difference of incoming and outgoing momenta.
6. If the relative permutation of the momenta p_i and q_j labelling the ends of the chains of electron lines is odd, the analytical expression is further multiplied by a factor of -1.

To illustrate the foregoing considerations, let us examine the propagator $\tilde{T}(p_1 p_2, q_2 q_1)$ in the second-order expansion with respect to e. This propagator is used to calculate the scattering amplitude of two electrons with initial momenta q_1 and q_2 and final momenta p_1 and p_2. The propagator in question is represented, in this order, by the diagrams

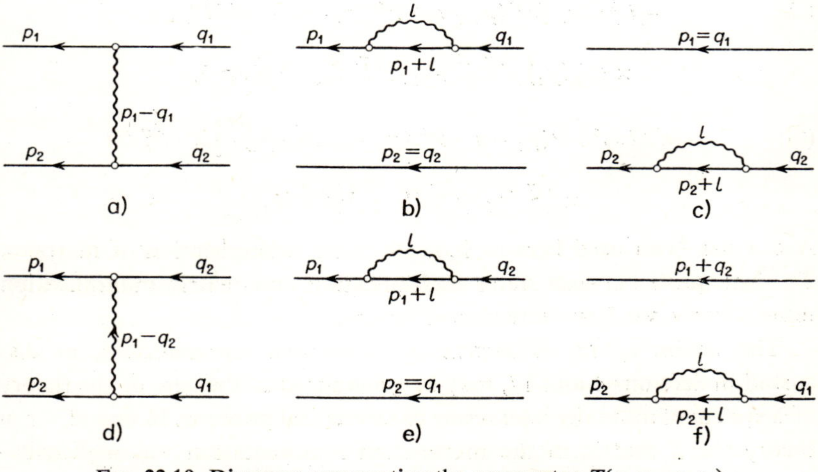

FIG. 22.10. Diagrams representing the propagator $T(p_1 p_2, q_2 q_1)$.

in Fig. 22.10. Only (a) and (d) are connected diagrams. The propagator $\tilde{T}(p_1 p_2, q_2 q_1)$ in the second-order expansion in e is the sum of the following six expressions corresponding to the six diagrams in Fig. 22.10:

(a): $$ie^2(2\pi)^4\delta^{(4)}(p_1+p_2-q_1-q_2)\{\tilde{S}_F(p_1)\gamma^\mu\tilde{S}_F(q_1)\}$$
$$\cdot\{\tilde{S}_F(p_2)\gamma^\nu\tilde{S}_F(q_2)\}\tilde{D}^F_{\mu\nu}(p_1-q_1),$$

(b): $$ie^2(2\pi)^4\delta^{(4)}(p_1-q_1)\,\delta^{(4)}(p_2-q_2)\int \mathrm{d}^4l\,\tilde{D}^{\mathrm{F}}_{\mu\nu}(l)$$
$$\times\{\tilde{S}_{\mathrm{F}}(p_1)\gamma^\nu\tilde{S}_{\mathrm{F}}(l+p_1)\gamma^\mu\tilde{S}_{\mathrm{F}}(q_1)\}\cdot\tilde{S}_{\mathrm{F}}(p_2),$$

(c): $$ie^2(2\pi)^4\delta^{(4)}(p_1-q_1)\,\delta^{(4)}(p_2-q_2)\,\tilde{S}_{\mathrm{F}}(p_1)\cdot\int \mathrm{d}^4l\,\tilde{D}^{\mathrm{F}}_{\mu\nu}(l),$$
$$\times\{\tilde{S}_{\mathrm{F}}(p_2)\gamma^\nu\tilde{S}_{\mathrm{F}}(l+p_2)\gamma^\mu\tilde{S}_{\mathrm{F}}(q_2)\}$$

(d): $$-ie^2(2\pi)^4\delta^{(4)}(p_1+p_2-q_1-q_2)\{\tilde{S}_{\mathrm{F}}(p_1)\gamma^\mu\tilde{S}_{\mathrm{F}}(q_2)\}$$
$$\cdot\{\tilde{S}_{\mathrm{F}}(p_2)\gamma^\nu\tilde{S}_{\mathrm{F}}(q_1)\}\,\tilde{D}^{\mathrm{F}}_{\mu\nu}(p_1-q_2),$$

(e): $$-ie^2(2\pi)^4\delta^{(4)}(p_1-q_2)\,\delta^{(4)}(p_2-q_1)\int \mathrm{d}^4l\,\tilde{D}^{\mathrm{F}}_{\mu\nu}(l)$$
$$\times\{\tilde{S}_{\mathrm{F}}(p_1)\gamma^\nu\tilde{S}_{\mathrm{F}}(l+p_1)\gamma^\mu\tilde{S}_{\mathrm{F}}(q_2)\}\cdot\tilde{S}_{\mathrm{F}}(p_2),$$

(f): $$-ie^2(2\pi)^4\delta^{(4)}(p_1-q_2)\,\delta^{(4)}(p_2-q_1)\,\tilde{S}_{\mathrm{F}}(p_1)\cdot\int \mathrm{d}^4l\,\tilde{D}^{\mathrm{F}}_{\mu\nu}(l)$$
$$\times\{\tilde{S}_{\mathrm{F}}(p_2)\gamma^\nu\tilde{S}_{\mathrm{F}}(l+p_2)\gamma^\mu\tilde{S}_{\mathrm{F}}(q_1)\}.$$

A dot has been used here to indicate outer multiplication of matrices. To distinguish between outer multiplication and matrix multiplication more clearly we have introduced braces.

The *notion of the S matrix in momentum representation*, as discussed in Sections 4 and 17, may be generalized to the case of the theory of a system of mutually interacting electrons and photons. In one-electron theory the S matrix in the momentum representation was a distribution $S(\mathbf{p}, r, \lambda\colon \mathbf{q}, s, \lambda')$. In the theory now under discussion, each transition amplitude is a linear functional of all wave-packet profiles of particles in the initial state and an antilinear functional of all wave-packet profiles of particles in the final state. The S matrix in momentum representation is the name we give to a distribution $S(\mathbf{p}_1, s_1, \lambda_1, \ldots; \mathbf{q}_n, s'_n, \lambda'_n, \ldots; \mathbf{k}_1, r_1, \varkappa_1, \ldots)$, such that on multiplication by the wave-packet profiles and integration over the appropriate hypersurfaces we get the transition amplitude. The Fourier transforms of the propagators are related directly to the S matrix. It follows from the representation (34) of the S operator that the S matrix in the momentum representation is the sum of the expressions constructed from the Fourier transform of the propagators in the following manner:

$$i^{2n}(-i)^k Z_2^{-n} Z_3^{-k/2} \bar{u}(\mathbf{p}_1, s_1, \lambda_1) \dots \bar{u}(\mathbf{p}_n, s_n, \lambda_n) \varepsilon^{\mu_1}(\mathbf{k}_1, r_1, \varkappa_1) \dots \varepsilon^{\mu_k}(\mathbf{k}_k, r_k, \varkappa_k)$$
$$\times [(m-\lambda_1 \gamma \cdot p_1) \dots (m-\lambda_n \gamma \cdot p_n)(\mu^2 - k_1^2) \dots (\mu^2 - k_k^2)$$
$$\times \tilde{T}_{\mu_1 \dots \mu_k}(\lambda_1 p_1 \dots \lambda_n p_n, \lambda_n' q_n \dots \lambda_1' q_1, x_1 k_1 \dots, \varkappa_k k_k)$$
$$\times (m-\lambda_1' \gamma \cdot q_1) \dots (m-\lambda_n' \gamma \cdot q_n)] u(\mathbf{q}_1, s_1', \lambda_1') \dots u(\mathbf{q}_n, s_n', \lambda_n'). \quad (44)$$

In this formula we have introduced the notations:

$$u(\mathbf{p}, s, \lambda) \equiv \begin{cases} u(\mathbf{p}, s), & \lambda = +1, \\ v(\mathbf{p}, s), & \lambda = -1, \end{cases} \quad (45a)$$

$$\varepsilon_\mu(\mathbf{k}, r, \varkappa) \equiv \begin{cases} \varepsilon_\mu^*(\mathbf{k}, r), & \varkappa = +1, \\ \varepsilon_\mu(\mathbf{k}, r), & \varkappa = -1, \end{cases} \quad (45b)$$

and made use of the relations:

$$\varepsilon_\mu(\mathbf{k})[-g^{\mu\nu}(\mu^2 - k^2) - k^\mu k^\nu] = -\varepsilon^\nu(\mathbf{k})(\mu^2 - k^2). \quad (46)$$

CHAPTER 7

RENORMALIZATION THEORY

23. THE NECESSITY FOR RENORMALIZATION

THE method of computing propagators by expanding them into a power series in e, as described in earlier sections, leads to fundamental difficulties. For it turns out that some integrals appearing in the successive orders of perturbation theory for propagators are divergent. In this chapter we shall give a formulation of perturbation theory such that these difficulties are circumvented. It will be shown that consideration of divergent integrals may be avoided if we expand into perturbation series not the propagators themselves, but the products of the propagators and the reciprocals of the renormalization constants Z_2 and Z_3 out of which the transition amplitudes are built (cf. formula (22.34)), perform the expansion with respect to the observable electron charge e_{obs} (this will be shown to differ from the e which we have used hitherto), and take account of the change caused in the rest masses of particles by the interaction.

These problems will be illustrated first of all with the example of the three simplest propagators in the absence of external fields and external currents.

THE ELECTRON PROPAGATOR

To begin with, let us calculate the Fourier transform $\tilde{G}(p)$ of the electron propagator up to terms of the order e^2. To simplify the calculations, we shall employ the Landau gauge for the free photon propagator, i.e. we choose the Fourier transform of the vector a^μ describing the compensating current in the form

$$\tilde{a}^{\mu}(k) = \frac{-ik^{\mu}}{-k^2 - i\varepsilon}.$$

With the compensating current so chosen the electron-photon propagators in all orders of perturbation theory are covariant under Poincaré transformations.

Using the method of Feynman diagrams in the momentum representation, we obtain the following expression:

$$\tilde{G}^{(0)}(p) + \tilde{G}^{(2)}(p) = \tilde{S}_{\mathrm{F}}(p)$$
$$-ie^2 \tilde{S}_{\mathrm{F}}(p) \int \frac{\mathrm{d}^4 k}{(2\pi)^4} \gamma^{\mu} \tilde{S}_{\mathrm{F}}(p+k) \gamma^{\nu} \tilde{D}^{\mathrm{F}}_{\mu\nu}(k) \tilde{S}_{\mathrm{F}}(p). \tag{1}$$

On substitution of expressions (22.43) for the Fourier transforms of the propagators $\tilde{S}_{\mathrm{F}}$ and $\tilde{D}^{\mathrm{F}}_{\mu\nu}$, the function of the electron self-energy in the momentum representation, $\tilde{\Sigma}(p)$, in the second-order perturbation theory takes on the form

$$\tilde{\Sigma}^{(2)}(p) = -ie^2 \int \frac{\mathrm{d}^4 k}{(2\pi)^4} \gamma^{\mu} \frac{1}{m - \hat{p} - \hat{k} - i\varepsilon} \gamma^{\nu} \frac{1}{\mu^2 - k^2 - i\varepsilon} \left(-g_{\mu\nu} - \frac{k_{\mu} k_{\nu}}{-k^2 - i\varepsilon} \right) \tag{2}$$

where we have introduced the notation

$$\hat{p} = \gamma \cdot p, \quad \hat{k} = \gamma \cdot k.$$

The integral above is divergent since the integrand tends too slowly to zero (as k^{-3}), when the components of the vector tend to infinity.† A finite part can, however, be separated in unique manner from this integral. It further turns out that the divergent parts do not affect any observable properties of electrons and photons.

Note, first of all, that because of the relativistic covariance the function of the electron self-energy depends on the components of the vector p only through the expression $\hat{p} = \gamma \cdot p$ (we recall that $p^2 = (\hat{p})^2$). This will be confirmed by direct calculation further on. The self-energy

† This problem will be explained in more detail subsequently in this section.

function may thus be expressed in terms of a single function Σ of one variable ξ:

$$\tilde{\Sigma}^{(2)}(p) = \Sigma^{(2)}(\hat{p}) = \Sigma^{(2)}(\xi)\Big|_{\xi=\hat{p}}.$$

Two-fold differentiation of both sides of equality (2) with respect to the components of the momentum p yields a convergent integral on the right-hand side (the integrand tends to zero as k^{-5}, when the components of the vector k tend to infinity). It thus follows that if the function $\Sigma^{(2)}(\hat{p})$ is written in the form

$$\Sigma^{(2)}(\hat{p}) = -A^{(2)}+B^{(2)}(m-\hat{p})+(m-\hat{p})^2\Sigma_{\mathrm{R}}(\hat{p}), \tag{3}$$

where $A^{(2)}$ and $B^{(2)}$ are constants, then divergent integrals appear only in those constants, whereas the rest of the self-energy function is regular. The constants $A^{(2)}$ and $B^{(2)}$ may be written as

$$A^{(2)} = -\Sigma^{(2)}(\hat{p})\Big|_{\hat{p}=m},$$

$$B^{(2)} = -\frac{\partial\Sigma^{(2)}(\hat{p})}{\partial\hat{p}}\Bigg|_{\hat{p}=m}.$$

Let us investigate how the electron propagator depends on the constants $A^{(2)}$ and $B^{(2)}$. By formula (1) we get

$$\begin{aligned}\tilde{G}^{(0)}(p)+\tilde{G}^{(2)}(p) &= \frac{1}{m-\hat{p}}+\frac{1}{m-\hat{p}}\tilde{\Sigma}^{(2)}(p)\frac{1}{m-\hat{p}}\\ &= \frac{1}{m-\hat{p}}-\frac{1}{m-\hat{p}}A^{(2)}\frac{1}{m-\hat{p}}+\frac{B^{(2)}}{m-\hat{p}}+\Sigma_{\mathrm{R}}^{(2)}(\hat{p})\\ &\simeq (1+B^{(2)})\frac{1+(m+A^{(2)}-\hat{p})\Sigma_{\mathrm{R}}^{(2)}(\hat{p})}{m+A^{(2)}-\hat{p}}.\end{aligned} \tag{4}$$

The foregoing equality is approximate, being valid only in the second-order perturbation theory. From this formula it is seen that the electron mass (pole of the propagator) is displaced by $\delta m = A^{(2)}$, while the normalization of the propagator at the pole changes by a factor of $1+B^{(2)}$ relative to the free propagator $\tilde{S}_{\mathrm{F}}(p)$. Both of these changes have a lucid physical interpretation. The change in mass, i.e. the so-called *mass renormalization*, results from the interaction of the electron with the electromagnetic field, for, the self-field of the charge makes its contribution to the rest mass of the charged particle. Such an effect takes place in both classical electrodynamics (cf. Section 8) and quantum electrodynamics. For a point particle in linear electrodynamics (based on Maxwell's equations) the mass correction is infinite, it is true, but this correction does not figure in expressions which have a direct physical meaning. Before the results of theory are compared with experiment, all the quantities need to be expressed in terms of the observable electron mass m_{obs} ($m_{obs} = 0.511$ Mev). The magnitude of m_{obs} differs by δm from m which characterizes an electron not interacting with the electromagnetic field:

$$m_{obs} = m + \delta m.$$

In second-order perturbation theory, the electron propagator written in terms of m_{obs} is of the form

$$\tilde{G}^{(0)}(p)+\tilde{G}^{(2)}(p) \simeq (1+B^{(2)})\,\frac{1+(m_{obs}-\hat{p})\Sigma_{R}^{(2)}(\hat{p})}{m_{obs}-\hat{p}}.$$

Replacement of the mass m in the function $\Sigma_{R}^{(2)}$ by the mass m_{obs} leads to changes of a higher order since the mass correction δm is at least of the order e^2. In consequence, the mass correction does not appear in the electron propagator if the latter is expressed in terms of the observable mass.

The constant $B^{(2)}$ affects only the normalization of the propagator. In describing scattering effects, one finds it more convenient to use *renormalized propagators*. Normalization of the renormalized electron propagator $\tilde{G}_{ren}(p)$ is defined in the same way as in the case of the free propagator, i.e.:

$$(m_{obs}-\hat{p})\,\tilde{G}_{ren}(p)\Big|_{\hat{p}=m} = 1. \tag{5}$$

The renormalized electron propagator is given to within second-order terms by the formula

$$(\tilde{G}^{(0)}(p)+\tilde{G}^{(2)}(p))_{ren} \simeq \frac{1+(m_{obs}-\hat{p})\Sigma_{R}^{(2)}(\hat{p})}{m_{obs}-\hat{p}}. \tag{6}$$

Now let us evaluate this expression, showing by direct calculation that the expression for the renormalized propagator contains no divergent integrals.

To avoid operating with divergent integrals in the intermediate stages of calculation, we shall employ *regularization*.† We choose a method based on the appropriate choice of the space dimension N. Not until the final formulae will we put $N = 4$.

Integrals with respect to the components of the momentum vector occur at every step in perturbation theory calculations in quantum electrodynamics and in other quantum field theories. Several computational methods have been worked out to facilitate evaluation of such integrals. Two of them are discussed in Appendix L.

Formula (2) is generalized to N dimensions as follows:

$$\tilde{\Sigma}^{(2)}(p) = -\frac{ie^2m^{2\beta}}{(4\pi)^\beta}\int\frac{d^Nk}{(2\pi)^N}$$

$$\times\gamma^\mu\frac{1}{m-\hat{p}-\hat{k}-i\varepsilon}\gamma^\nu\frac{1}{\mu^2-k^2-i\varepsilon}\left(-g_{\mu\nu}-\frac{k_\mu k_\nu}{-k^2-i\varepsilon}\right) \tag{7}$$

where the parameter β,

$$\beta = 2-\frac{N}{2},$$

measures the deviation from the true dimension $N = 4$, whereas the factor $\left(\frac{m^2}{4\pi}\right)^\beta$ has been introduced for convenience in calculations.‡

Before we proceed to evaluate the integral in formula (7) for $\tilde{\Sigma}^{(2)}(p)$, let us carry out some transformations in the integrand:

$$\gamma^\mu\frac{1}{m-\hat{p}-\hat{k}-i\varepsilon}\gamma^\nu g_{\mu\nu} = \frac{m\gamma^\mu\gamma_\mu+\gamma^\mu(\hat{p}+\hat{k})\gamma_\mu}{m^2-(p+k)^2-i\varepsilon} = \frac{Nm-(N-2)(\hat{p}+\hat{k})}{m^2-(p+k)^2-i\varepsilon},$$

$$\hat{k}\frac{1}{m-\hat{p}-\hat{k}-i\varepsilon}\hat{k} = [(m-\hat{p})-(m-\hat{p}-\hat{k})]\frac{1}{m-\hat{p}-\hat{k}-i\varepsilon}[(m-\hat{p})-(m-\hat{p}-\hat{k})]$$

$$= -\hat{k}-(m-\hat{p})+(m-\hat{p})\frac{m+\hat{p}+\hat{k}}{m^2-(p+k)^2-i\varepsilon}(m-\hat{p}).$$

† Various methods of regularization are discussed in Appendix K.

‡ Among other things, this is how we preserve the dimensionless nature (in units of $\hbar$ and c) of the electric charge.

The term $\hat{k}$ in the last line makes no contribution to the integral on account of the antisymmetry of the integrand under the change $k \to -k$. Thus, we get

$$\tilde{\Sigma}^{(2)}(p) = ie^2 \frac{m^{2\beta}}{(4\pi)^\beta} \int \frac{d^N k}{(2\pi)^N} \left[\frac{Nm-(N-2)(\hat{p}+\hat{k})}{(m^2-(p+k)^2-i\varepsilon)(\mu^2-k^2-i\varepsilon)} \right.$$

$$\left. - \frac{m-\hat{p}}{(\mu^2-k^2-i\varepsilon)(-k^2-i\varepsilon)} + \frac{(m-\hat{p})(m+\hat{p}+\hat{k})(m-\hat{p})}{(m^2-(p+k)^2-i\varepsilon)(\mu^2-k^2-i\varepsilon)(-k^2-i\varepsilon)} \right].$$

To calculate these integrals, we shall use the *Feynman–Schwinger method* which consists in combining denominators (cf. Appendix L):

$$\frac{1}{m^2-(p+k)^2-i\varepsilon} \frac{1}{\mu^2-k^2-i\varepsilon} = \int_0^1 du \frac{1}{[m^2u+\mu^2(1-u)-k^2-2p\cdot ku-p^2u-i\varepsilon]^2},$$

$$\frac{1}{\mu^2-k^2-i\varepsilon} \frac{1}{-k^2-i\varepsilon} = \int_0^1 du \frac{1}{[\mu^2u-k^2-i\varepsilon]^2},$$

$$\frac{1}{m^2-(p+k)^2-i\varepsilon} \frac{1}{\mu^2-k^2-i\varepsilon} \frac{1}{-k^2-i\varepsilon} = \frac{1}{m^2-(p+k)^2-i\varepsilon} \frac{1}{\mu^2} \left[\frac{1}{-k^2-i\varepsilon} \right.$$

$$\left. - \frac{1}{\mu^2-k^2-i\varepsilon} \right] = \frac{1}{\mu^2} \int_0^1 du \left\{ \frac{1}{[m^2u-k^2-2p\cdot ku-p^2u-i\varepsilon]^2} \right.$$

$$\left. - \frac{1}{[m^2u+\mu^2(1-u)-2p\cdot ku-p^2u-i\varepsilon]^2} \right\}.$$

In these integrals we displace the origin of the coordinate system in the space of momenta k,

$$k_\mu \to k_\mu - up_\mu,$$

and discard the antisymmetrical parts of the integrands. On changing the order of integration, we then get the expression:

$$\tilde{\Sigma}^{(2)}(p) = ie^2 \frac{m^{2\beta}}{(4\pi)^\beta} \int_0^1 du \int \frac{d^N k}{(2\pi)^N} \left\{ \frac{Nm-(N-2)\hat{p}(1-u)}{[m^2u+\mu^2(1-u)-k^2-p^2u(1-u)-i\varepsilon]^2} \right.$$

$$- \frac{m-\hat{p}}{[\mu^2u-k^2-i\varepsilon]^2} + \frac{(m-\hat{p})(m+\hat{p}(1+u))(m-\hat{p})}{\mu^2}$$

$$\left. \times \left[\frac{1}{[m^2u-k^2-p^2u(1-u)-i\varepsilon]^2} - \frac{1}{[m^2u+\mu^2(1-u)-k^2-p^2u(1-u)-i\varepsilon]^2} \right] \right\}.$$

Thus, in our calculations we have integrals of the type:

$$I = \left(\frac{m^2}{4\pi} \right)^\beta \int \frac{d^N k}{(2\pi)^N} \frac{1}{(D-k^2-i\varepsilon)^2}.$$

Before integrating, we rotate the contour of integration (here, the real axis) through an angle of $\pi/2$ around the origin of the system in the plane of the complex variable k_0. This operation is called *Wick's rotation.* Regardless of the value of D, on rotating the contour we do not encounter any singularities (ε is positive!), whereas the integral vanishes over the large half-circle. Figure 23.1 shows the position of the poles

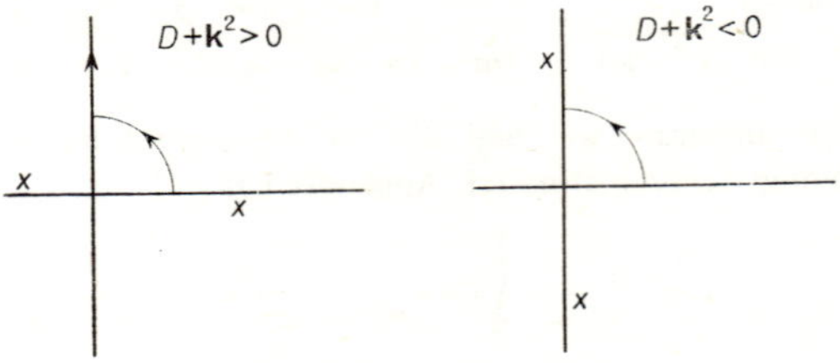

FIG. 23.1. The position of the poles of the integrands.

of the integrands in the cases when $D+\mathbf{k}^2$ is positive and when it is negative. Once the contour has been rotated, we replace k_0 by a new variable of integration k_N,

$$k_0 = ik_N.$$

While the variable k_N runs over the real axis from $-\infty$ to ∞, the variable k_0 runs over the imaginary axis. As a result of this change of variables, we get the integrals

$$I = i\left(\frac{m^2}{4\pi}\right)^\beta \int \frac{d^N k}{(2\pi)^N} \frac{1}{(D+\varkappa^2-i\varepsilon)^2},$$

where $\varkappa^2$ is the squared length of the vector k in space with the Euclidean metric,

$$\varkappa^2 = k_1^2+k_2^2+ \ldots +k_N^2.$$

This integral is calculated in spherical coordinates by using the formula for the surface S_N of a sphere in N-dimensional space,

$$S_N = \frac{2\pi^{N/2}}{\Gamma(N/2)},$$

and the formula†

$$\int_0^\infty \frac{d\varkappa\, \varkappa^{N-1}}{(D+\varkappa^2-i\varepsilon)^2} = \frac{\Gamma(N/2)\Gamma(2-N/2)}{2\Gamma(2)} (D-i\varepsilon)^{\frac{N}{2}-2}.$$

In this way we arrive at

$$I = \frac{i}{16\pi^2}\Gamma(\beta)\left(\frac{m^2}{D-i\varepsilon}\right)^\beta.$$

---

† We recall that $\int_0^\infty dt \dfrac{t^{2x-1}}{(1+t^2)^{x+y}} = \dfrac{1}{2}\dfrac{\Gamma(x)\Gamma(y)}{\Gamma(x+y)}.$

The electron self-energy function in the second order of perturbation theory may thus be written as

$$\tilde{\Sigma}^{(2)}(p) = -\frac{e^2}{16\pi^2}\Gamma(\beta)m^{2\beta}\int_0^1 du\left\{\frac{(4-2\beta)m-(2-2\beta)\hat{p}(1-u)}{[m^2u+\mu^2(1-u)-p^2u(1-u)-i\varepsilon]^\beta}\right.$$

$$-\frac{m-\hat{p}}{[\mu^2u-i\varepsilon]^\beta}+\frac{(m-\hat{p})^2\left(m+\hat{p}(1-u)\right)}{\mu^2}$$

$$\left.\times\left[\frac{1}{[m^2u-p^2u(1-u)-i\varepsilon]^\beta}-\frac{1}{[m^2u+\mu^2(1-u)-p^2u(1-u)-i\varepsilon]^\beta}\right]\right\}. \quad (8)$$

The behaviour of the function $\Gamma(\beta)$ in the neighbourhood of the point $\beta = 0$ is given by the formula

$$\Gamma(\beta) = \frac{1}{\beta}-C+\left(\frac{\pi^2}{12}+\frac{C^2}{2}\right)\beta+\ \dots$$

where $C = 0.577\dots$ is Euler's constant.

It thus follows that the point $\beta = 0$ ($N = 4$) is a characteristic, and simultaneously a singular, point of the theory. At this point the function $\tilde{\Sigma}^{(2)}(p)$ possesses a pole as a function of the variable N. The occurrence of this pole is a manifestation of the divergence of integral (2) in the regularization method we have adopted.

The residue, however, is independent of $\hat{p}$,

$$\beta\tilde{\Sigma}^{(2)}(p)\Big|_{\beta=0} = -\frac{3me^2}{16\pi^2},$$

and in consequence that pole affects only the mass correction.

From the general discussion given earlier, the residue might be expected to contain both a constant term and one linear in $\hat{p}$. A term linear in $\hat{p}$ does not appear in our expression for $\tilde{\Sigma}^{(2)}(p)$ by chance, in a sense, as a result of the choice of a particular form for the free photon propagator. In other gauges (for other compensating currents) the coefficient in front of $\hat{p}$ is also infinite when $\beta = 0$.

We shall now calculate explicitly the function of the electron self-energy when $\beta \to 0$. The three parts of this function which appear on the right-hand side of formula (8) will be treated separately. The most awkward is the first part. The integral figuring there, to within terms tending to zero when $\beta \to 0$, may be written in the form:†

$$\Gamma(\beta)\int_0^1 du\Big[(4-2\beta)m-(2-2\beta)\hat{p}(1-u)$$

$$+(-4m+2\hat{p}(1-u))\beta\ln\frac{m^2u+\mu^2(1-u)-p^2u(1-u)}{m^2}\Big]$$

$$\simeq \frac{4m-\hat{p}}{\beta}+(6-4C)m-\hat{p}(1-C)$$

$$+2m\left[\frac{m^2-\mu^2-p^2}{p^2}\ln\frac{\mu^2}{m^2}+\frac{\sqrt{\Delta}}{p^2}\ln\frac{m^2+\mu^2-p^2+\sqrt{\Delta}}{m^2+\mu^2-p^2-\sqrt{\Delta}}\right]$$

$$-\hat{p}\Big[\frac{m^2-\mu^2}{p^2}+\frac{m^4+p^4-2\mu^2m^2-2\mu^2p^2+\mu^4}{2p^4}\ln\frac{\mu^2}{m^2}$$

$$+\frac{m^2+p^2-\mu^2}{2p^4}\sqrt{\Delta}\ln\frac{m^2+\mu^2-p^2+\sqrt{\Delta}}{m^2+\mu^2-p^2-\sqrt{\Delta}}\Big], \tag{9}$$

where

$$\Delta = m^4+p^4+\mu^4-2m^2\mu^2-2p^2\mu^2-2m^2p^2.$$

Substitution of the expression $p^2+i\varepsilon$ for p^2 leads to the proper branch of the logarithm function. In the limit, when $\mu^2 \to 0$, we get

$$\frac{4m-\hat{p}}{\beta}+(6-4C)m-\hat{p}(1-C)+4m\frac{m^2-p^2}{p^2}\ln\frac{m^2-p^2}{m^2}$$

$$-\hat{p}\left(\frac{m^2}{p^2}+\frac{m^4-p^4}{p^4}\ln\frac{m^2-p^2}{m^2}\right).$$

The foregoing expression does not have a derivative with respect to $\hat{p}$ at the point $\hat{p} = m$. Accordingly, in the theory with zero photon mass, the electron propagator cannot be normalized to unity at the point $\hat{p} = m$. This is one more manifestation of the infrared catastrophe.

† Use is made here of the expansion

$$x^\beta = e^{\beta\ln x} = 1+\beta\ln x+\ \dots.$$

In the intermediate stages of the calculation we shall omit $-i\varepsilon$.

The function of the electron self-energy in the theory with photon mass μ possesses a derivative at the point $\hat{p} = m$ and the part $\Sigma_{\mathrm{R}}(\hat{p})$ can be separated from this function in accordance with formula (3). The exact expression for $\Sigma_{\mathrm{R}}(\hat{p})$ is quite complicated. Here we shall give only the approximate formula in which we discard terms of the order μ^2/m^2, μ^2/p^2, and $\mu^2/(m^2-p^2)$, assuming that they are small in comparison with unity.

Above we have calculated the contribution from the first part of the integrand in formula (8) to the function $\tilde{\Sigma}^{(2)}(p)$. The second part of the integrand is linear in $\hat{p}$; thus it does not make any contribution to Σ_{R}. The third part is proportional to $(m-\hat{p})^2$ and, consequently, its value and that of its derivative at the point $\hat{p} = m$ are equal to zero. When we add the two parts, the result is

$$(m-\hat{p})^2\Sigma_{\mathrm{R}}^{(2)}(\hat{p}) \simeq -\frac{e^2}{8\pi^2}\left\{(m-\hat{p})\left(1-2\ln\frac{\mu^2}{m^2}\right)\right.$$

$$-(m-\hat{p})\frac{m\hat{p}}{p^2}+\left(4m\frac{m^2-p^2}{p^2}-\hat{p}\frac{m^4-p^4}{p^4}\right)\ln\frac{m^2-p^2-i\varepsilon}{m^2}$$

$$-\frac{(m-\hat{p})^2}{m^2-p^2}\left[(m+\hat{p})\ln\frac{\mu^2}{m^2}+\left(m\frac{m^2+p^2}{p^2}\right.\right.$$

$$\left.\left.\left.+\hat{p}\frac{m^4+p^4}{p^4}\right)\ln\frac{m^2-p^2-i\varepsilon}{m^2}\right]\right\}$$

$$= -\frac{e^2}{16\pi^2}(m-\hat{p})\left[1-3\ln\frac{\mu^2}{m^2}-\frac{m\hat{p}}{p^2}+3m\frac{m+\hat{p}}{p^2}\ln\frac{m^2-p^2-i\varepsilon}{m^2}\right]. \tag{10}$$

The expression $\Sigma_{\mathrm{R}}^{(2)}$ can also be evaluated in such a way that there is no need to regularize the divergent integrals. To this end we make use of the fact that differentiation with respect to the components of the vector p improves the convergence of integral (2). In doing this, we find it convenient to use the identity

$$\int_m^{\xi}d\xi\int_m^{\xi}d\xi\,\frac{\partial^2\Sigma^{(2)}(\xi)}{\partial\xi^2} = \Sigma^{(2)}(\xi)-\Sigma^{(2)}(m)$$

$$+(m-\xi)\left.\frac{\partial\Sigma^{(2)}(\xi)}{\partial\xi}\right|_{\xi=m} = (m-\xi)^2\Sigma_{\mathrm{R}}^{(2)}(\xi).$$

The second derivative of the function $\Sigma^{(2)}(\xi)$ with respect to ξ is expressible in terms of the derivatives of the self-energy functions $\tilde{\Sigma}^{(2)}(p)$ for $\hat{p} = \xi$ according to the formula

$$\frac{\partial^2 \Sigma(\xi)}{\partial \xi^2} = \frac{1}{p^2} p_\lambda p_\varrho \frac{\partial^2}{\partial p_\lambda \partial p_\varrho} \tilde{\Sigma}^{(2)}(p)_{\hat{p}} \Big|_{\hat{p}=\xi}.$$

The proof of this formula is based on the previously discussed possibility of rewriting the function $\tilde{\Sigma}(p)$ in the form

$$\tilde{\Sigma}(p) = \alpha(p^2) + \hat{p}\beta(p^2) = [\alpha(\xi^2) + \xi\beta(\xi^2)] \Big|_{\xi=\hat{p}}.$$

On carrying out two-fold differentiation of the function $\tilde{\Sigma}^{(2)}(p)$, given by formula (2), with respect to the components of the four-vector p, we get

$$\frac{\partial^2 \Sigma^{(2)}(\xi)}{\partial \xi^2}\Bigg|_{\xi=\hat{p}} = -\frac{1}{p^2} 2ie^2 \int \frac{d^4k}{(2\pi)^4} \gamma^\mu \frac{1}{m-\hat{p}-\hat{k}} \hat{p} \frac{1}{m-\hat{p}-\hat{k}}$$

$$\times \hat{p} \frac{1}{m-\hat{p}-\hat{k}} \gamma^\nu \frac{1}{\mu^2-k^2} \left(-g_{\mu\nu} - \frac{k_\mu k_\nu}{k^2}\right). \quad (11)$$

To calculate the derivatives we made use of the relation

$$\frac{\partial}{\partial p_\lambda} \frac{1}{m-\hat{p}} = \frac{1}{m-\hat{p}} \gamma^\lambda \frac{1}{m-\hat{p}},$$

which can be obtained by differentiating the identity

$$\frac{1}{m-\hat{p}}(m-\hat{p}) \equiv 1$$

with respect to p_λ.

The integrand in formula (11) tends sufficiently fast to zero (as k^{-5}) when $k \to \infty$ to ensure convergence of the integral.

This second method of evaluating the expression $\Sigma_R(\hat{p})$ is, however, less convenient for it requires the complicated integral (11) to be calculated. Another shortcoming is that the starting point for further calculations is the expression (2) for $\tilde{\Sigma}^{(2)}(p)$ which contains a divergent integral.

In the considered case of the electron propagator, no divergent integrals appear in the second-order perturbation theory if the renormalized propagator is calculated as a function of the observed mass of the electron. It turns out that in all higher orders of perturbation theory, and not only in the second order, the renormalized electron propagator $Z_2^{-1}\tilde{G}(p)$ is a finite quantity provided that we perform renormalization of mass and renormalization of charge, i.e. that we express this propagator in terms of the observable mass m_{obs} and the observable charge e_{obs}. The renormalization constant of the electron propagator Z_2 is chosen so as to ensure that the propagator is normalized in the same way as in the case of the free propagator,

$$[(m-\hat{p})Z_2^{-1}\tilde{G}(p)]\Big|_{\hat{p}=m} = 1. \tag{12}$$

The Photon Propagator

When we use the Feynman diagram method, to within terms of the order e^2 we get the following expression for the photon propagator:

$$\tilde{\mathscr{G}}^{(0)}_{\mu\nu}(k)+\tilde{\mathscr{G}}^{(2)}_{\mu\nu}(k)$$

$$= \tilde{D}^{\mathrm{F}}_{\mu\nu}(k)+ie^2\tilde{D}^{\mathrm{F}}_{\mu\lambda}(k)\int\frac{\mathrm{d}^4p}{(2\pi)^4}\,\mathrm{Tr}\{\gamma^\lambda\tilde{S}_{\mathrm{F}}(p+k)\gamma^\varrho\tilde{S}_{\mathrm{F}}(p)\}\tilde{D}^{\mathrm{F}}_{\varrho\nu}(k). \tag{13}$$

On substitution of the Fourier transforms of the free electron propagators, the Fourier transform of the photon self-energy function in the second-order perturbation theory assumes the form

$$\tilde{\Pi}^{\lambda\varrho(2)}(k) = ie^2\int\frac{\mathrm{d}^4p}{(2\pi)^4}\,\mathrm{Tr}\left\{\gamma^\lambda\frac{1}{m-\hat{p}-\hat{k}-i\varepsilon}\gamma^\varrho\frac{1}{m-\hat{p}-i\varepsilon}\right\}. \tag{14}$$

This integral is divergent, just like the one representing the electron self-energy function. Three-fold differentiation with respect to the components of the vector k, however, yields a convergent integral. It thus follows that the divergent part is a polynomial function of the second order in the coordinates of the vector k.

On the other hand, as we shall subsequently show by direct calculation, the photon self-energy function can be expressed in terms of a single scalar function of k^2,

$$\tilde{\Pi}^{\lambda\varrho(2)}(k) = (-g^{\lambda\varrho}k^2+k^\lambda k^\varrho)\Pi^{(2)}(k^2). \tag{15}$$

This is the result of the invariance of the theory under gauge transformations. It implies that if the function $\Pi^{(2)}(k^2)$ is written as

$$\Pi^{(2)}(k^2) = C^{(2)}-k^2\Pi^{(2)}_{\mathrm{R}}(k^2),$$

a divergent integral will appear only in the constant $C^{(2)}$ whereas the remaining part of the photon self-energy function is regular.

Let us investigate how the photon propagator depends on the constant $C^{(2)}$. We shall continue to use the Landau gauge:

$$\tilde{\mathscr{G}}^{(0)}_{\mu\nu}(k)+\tilde{\mathscr{G}}^{(2)}_{\mu\nu}(k) = \left(-g_{\mu\nu}+\frac{k_\mu k_\nu}{k^2}\right)\frac{1}{-k^2-i\varepsilon}$$

$$+\left(-g_{\mu\lambda}+\frac{k_\mu k_\lambda}{k^2}\right)\frac{1}{-k^2-i\varepsilon}\tilde{\Pi}^{\lambda\varrho(2)}(k)\left(-g_{\varrho\nu}+\frac{k_\varrho k_\nu}{k^2}\right)\frac{1}{-k^2-i\varepsilon}$$

$$\simeq (1+C^{(2)})\left(-g_{\mu\nu}+\frac{k_\mu k_\nu}{k^2}\right)\frac{1-k^2\Pi_{\mathrm{R}}(k^2)}{-k^2-i\varepsilon}. \tag{16}$$

The latter equality holds only in the second-order perturbation theory. Since no infrared catastrophe occurs in purely photon propagators, we have set $\mu^2 = 0$. From formula (16) we deduce that the constant $C^{(2)}$ affects only the normalization of the photon propagator. It follows from the discussion in Section 21 that scattering effects are described most conveniently by using the *renormalized photon propagator* $\tilde{\mathscr{G}}^{\mathrm{ren}}_{\mu\nu}(k)$, normalized in the same way as the free propagator, i.e.

$$\tilde{\mathscr{G}}^{\mathrm{ren}}_{\mu\nu}(k) = Z_3^{-1}\tilde{\mathscr{G}}_{\mu\nu}(k) = \left(-g_{\mu\nu}+\frac{k_\mu k_\nu}{k^2}\right)\tilde{\mathscr{G}}_{\mathrm{ren}}(k^2),$$

$$-k^2\tilde{\mathscr{G}}_{\mathrm{ren}}(k^2)\Big|_{k^2=0} = 1.$$

The renormalized photon propagator in the second-order perturbation theory is given by the formula:

$$\tilde{\mathscr{G}}^{(0)}_{\mathrm{ren}}(k^2)+\tilde{\mathscr{G}}^{(2)}_{\mathrm{ren}}(k^2) = \frac{1-k^2\Pi^{(2)}_{\mathrm{R}}(k^2)}{-k^2-i\varepsilon}. \tag{17}$$

Now let us calculate this expression by employing regularization in the intermediate stages.

In N-dimensional space, on applying the Feynman–Schwinger technique of combining denominators and shifting the variable of integration, we get

$$\tilde{\Pi}^{\lambda\varrho(2)}(k) = ie^2\left(\frac{m^2}{4\pi}\right)^{\beta}\int_0^1 \mathrm{d}u\int\frac{\mathrm{d}^N p}{(2\pi)^N}$$

$$\times\frac{\mathrm{Tr}\{\gamma^\lambda(m+\hat{p}-u\hat{k})\gamma^\varrho(m+\hat{p}+(1-u)\hat{k})\}}{[m^2-p^2-k^2u(1-u)-i\varepsilon]^2}. \tag{18}$$

The traces of the products of matrices γ can be calculated independently of the number of space-time dimensions only on the basis of the commutation relations for those matrices. In this way we get

$$\mathrm{Tr}\{\gamma^\mu\gamma^\nu\} = \mathrm{Tr}\{I\}g^{\mu\nu},$$

$$\mathrm{Tr}\{\gamma^\mu\gamma^\nu\gamma^\lambda\gamma^\varrho\} = \mathrm{Tr}\{I\}(g^{\mu\nu}g^{\lambda\varrho}-g^{\mu\lambda}g^{\nu\varrho}+g^{\mu\varrho}g^{\nu\lambda}).$$

The dependence of the traces of the unit matrix in bispinor space on the dimension of space-time will not be specified. In the final formulae, of course, we shall put

$$\mathrm{Tr}\{I\} = 4.$$

Any (regular) dependence of this expression on the number of dimensions of space-time has an effect only on the finite additive constants in the constants $A^{(2)}$, $B^{(2)}$, $C^{(2)}$.

Integration with respect to the four-vector p in formula (18) is performed in the same manner as before, being preceded by Wick's rotation:

$$\tilde{\Pi}^{\lambda\varrho(2)}(k) = -e^2\left(\frac{m^2}{4\pi}\right)^\beta \frac{S_N}{(2\pi)^N}\mathrm{Tr}\{I\}$$

$$\times\int_0^1 du\int_0^\infty d\varkappa\,\frac{\varkappa^{N-1}}{[m^2+\varkappa^2-k^2u(1-u)-i\varepsilon]^2}$$

$$\times[g^{\lambda\varrho}(m^2+\varkappa^2(1-2/N)+k^2u(1-u))-2u(1-u)k^\lambda k^\varrho]$$

$$= -\frac{e^2}{2\pi^2}\Gamma(\beta)\frac{1}{4}\mathrm{Tr}\{I\}(g^{\lambda\varrho}k^2-k^\lambda k^\varrho)\int_0^1 du\,\frac{u(1-u)m^{2\beta}}{[m^2-k^2u(1-u)-i\varepsilon]^\beta}.$$

Use has been made here of the formula given in the footnote on p. 385 and the following relation for integrals, with respect to angular variables, of the products of two components of a vector in the Euclidean space:

$$\int d\Omega\,\varkappa_\mu\varkappa_\nu = \frac{S_N}{N}\delta_{\mu\nu}\varkappa^2.$$

With an accuracy to within terms tending to zero when $\beta\to 0$, we obtain

$$\tilde{\Pi}^{\lambda\varrho(2)}(k) \simeq (g^{\lambda\varrho}k^2-k^\lambda k^\varrho)\left(C^{(2)}-k^2\Pi_{\mathrm{R}}^{(2)}(k^2)\right),$$

where

$$C^{(2)} = -\frac{e^2}{12\pi^2}\left(\frac{1}{\beta} - C\right),$$

$$-k^2\Pi_R^{(2)}(k^2) = \frac{e^2}{2\pi^2}\int_0^1 du\, u(1-u)\ln\frac{m^2-k^2u(1-u)-i\varepsilon}{m^2}$$

$$= \frac{e^2}{12\pi^2}\left[-\frac{5}{3} - \frac{4m^2}{k^2}\right.$$

$$\left.+\left(1+\frac{2m^2}{k^2}\right)\sqrt{1-\frac{4m^2}{k^2}}\ln\frac{1+\sqrt{1-4m^2/k^2}}{1-\sqrt{1-4m^2/k^2}}\right]. \tag{19}$$

In accordance with our preliminary analysis, only the constant $C^{(2)}$ contains a singular term when $\beta \to 0$.

The special form of the photon self-energy function described by formula (15) can be used to calculate the function $\Pi_R^{(2)}(k^2)$ as well, without resorting to regularization. With this in view, we shall make use of the formula

$$\frac{\partial \Pi(k^2)}{\partial k^2} = -\frac{1}{18k^4}(-g_{\alpha\beta}k^2+k_\alpha k_\beta)k_\varrho \frac{\partial^3}{\partial k^\lambda \partial k_\alpha \partial k_\beta}(-g^{\lambda\varrho}k^2+k^\lambda k^\varrho)\Pi(k^2),$$

which is satisfied by every function of the variable k^2. The third derivative of the photon self-energy function, in the second order of perturbation theory, may be rewritten as the integral

$$\frac{\partial^3 \tilde{\Pi}^{\lambda\varrho(2)}(k)}{\partial k_\mu \partial k_\nu \partial k} = ie^2\int \frac{d^4p}{(2\pi)^4}$$

$$\times \operatorname{Tr}\left\{\gamma^\lambda \frac{1}{m-\hat{p}-\hat{k}}\gamma^{(\mu}\frac{1}{m-\hat{p}-\hat{k}}\gamma^\nu \frac{1}{m-\hat{p}-\hat{k}}\gamma^{\sigma)}\frac{1}{m-\hat{p}-\hat{k}}\gamma^\varrho \frac{1}{m-\hat{p}}\right\},$$

which is convergent since the integrand tends to zero as p^{-5}, when $p \to \infty$.

This method, however, has the same drawbacks as the analogous procedure applied in the case of the electron propagator.

The Vertex Function

In the absence of an external field, the vertex function $\Gamma^\mu(x, y, z)$ is a function only of the difference of its arguments. We shall take $x-z$ and $z-y$ as independent variables and examine the Fourier transform $\tilde{\Gamma}^{\mu}(p, q)$ of the vertex function with respect to these variables:

$$\Gamma^{\mu}(x, y, z) = \int \frac{d^4p}{(2\pi)^4} \frac{d^4q}{(2\pi)^4} e^{-ip\cdot(x-z)-iq\cdot(z-y)}\tilde{\Gamma}^{\mu}(p, q).$$

In the zeroth and second order of perturbation theory, the Feynman diagram technique yields the expression

$$\tilde{\Gamma}^{\mu(0)}(p, q)+\tilde{\Gamma}^{\mu(2)}(p, q) = \gamma^{\mu}-ie^2\int \frac{d^4k}{(2\pi)^4} \gamma^{\lambda}\tilde{S}_F(p+k)\gamma^{\mu}\tilde{S}_F(q+k)\gamma^{\varrho}\tilde{D}_{\lambda\varrho}(k)$$

$$= \gamma^{\mu}-ie^2\int \frac{d^4k}{(2\pi)^4}\gamma^{\lambda}\frac{1}{m-\hat{p}-\hat{k}-i\varepsilon}\gamma^{\mu}\frac{1}{m-\hat{q}-\hat{k}-i\varepsilon}\gamma^{\varrho}$$

$$\times\left(-g_{\lambda\varrho}+\frac{k_{\lambda}k_{\varrho}}{k^2+i\varepsilon}\right)\frac{1}{\mu^2-k^2-i\varepsilon}. \tag{20}$$

The integral figuring in this formula is divergent. However, it is sufficient to differentiate this integral once with respect to the components of the vector p or the components of the vector q in order for an additional factor $1/(m-\hat{p}-\hat{k})$ or $1/(m-\hat{q}-\hat{k})$ to make its appearance and, hence, to get a convergent integral.

In second-order perturbation theory we can separate the regular part $\Lambda_R^{\mu(2)}$ from the vertex function, in accordance with the formula

$$\tilde{\Gamma}^{\mu(0)}(p, q)+\tilde{\Gamma}^{\mu(2)}(p, q) = \gamma^{\mu}+D^{(2)}\gamma^{\mu}+\Lambda_R^{\mu(2)}(p, q), \tag{21}$$

where the singular part, when $\beta \to 0$, is contained only in the constant $D^{(2)}$, whereas the regular part $\Lambda_R^{\mu(2)}(p, q)$ is so defined as to satisfy the condition

$$\Lambda_R^{\mu(2)}(p, p)\Big|_{\hat{p}=m} = 0.$$

The *renormalized vertex function* is defined by means of the normalization condition

$$\tilde{\Gamma}^{\mu}_{\text{ren}}(p, p)\Big|_{\hat{p}=m} = \gamma^{\mu}. \tag{22}$$

In second-order perturbation theory, the right-hand side of eq. (21) can be written as the product of the renormalization constant $1+D^{(2)}$ and the renormalized vertex function,

$$\tilde{\Gamma}^{\mu(0)}(p, q)+\tilde{\Gamma}^{\mu(2)}(p, q) = (1+D^{(2)})\tilde{\Gamma}^{\mu(2)}_{\text{ren}}(p, q),$$

where

$$\tilde{\Gamma}^{\mu(2)}_{\text{ren}}(p, q) = \gamma^{\mu}+\Lambda^{\mu(2)}_{\text{R}}(p, q).$$

We can use the regularization method to calculate the vertex function and to separate the regular part from it. The constant $D^{(2)}$ can be related to the constant $B^{(2)}$ which was introduced during investigation of the self-energy function of the electron. On comparing formulae (2) and (20), we arrive at the relation

$$\tilde{\Gamma}^{\mu(2)}(p, p) = \frac{\partial}{\partial p_{\mu}}\tilde{\Sigma}^{(2)}(p)$$

from which it follows that

$$D^{(2)} = -B^{(2)}.$$

Thus, the renormalization of the electron propagator and the renormalization of the vertex function are interrelated in second-order perturbation theory.

This is associated with the gauge invariance, which implies *Ward's identity*:#

$$\tilde{\Gamma}^{\mu}(p, p) = -\frac{\partial}{\partial p_{\mu}}\tilde{G}^{-1}(p). \tag{23}$$

As we shall demonstrate subsequently, this identity is satisfied in all orders of perturbation theory.

It follows from Ward's identity that the renormalization constant of the electron propagator Z_2 is at the same time the renormalization constant for the vertex function,

$$\tilde{\Gamma}^{\mu}_{\text{ren}}(p, q) = Z_2\tilde{\Gamma}^{\mu}(p, q). \tag{24a}$$

This identity was discovered by J. C. Ward (1950).

For we then have the relation

$$\tilde{\Gamma}^{\mu}_{\text{ren}}(p, p)\Big|_{\hat{p}=m} = \gamma^{\mu}, \tag{24b}$$

which establishes the proper normalization of the vertex function.

Charge Renormalization

It will now be shown that not only the mass but also the charge of the electron experiences change as a result of interaction with the quantum electromagnetic field. The observable electron charge e_{obs} differs from the parameter e, with respect to which we have expanded, just as the observable electron mass m_{obs} differs from the parameter figuring in the electron propagator in zeroth-order perturbation theory.

We shall demonstrate that the electron charge e_{obs} is related to the parameter e by

$$e_{\text{obs}} = Z_3^{1/2} e. \tag{25}$$

To prove this relation, let us consider low-energy scattering of two electrons, for small value of the momentum transfer t between electrons. The initial and final momenta of the electrons (q_1 and q_2, and p_1 and p_2) then satisfy the inequalities:

$$s - 4m^2 = (q_1 + q_2)^2 - 4m^2 \ll m^2, \tag{26a}$$

$$-t = -(q_1 - p_1)^2 = -(q_2 - p_2)^2 \ll m^2. \tag{26b}$$

In the classical picture of this process the small momentum transfer means that the electrons fly past each other at a large distance and their momenta suffer little change as a result of the scattering. When inequalities (26) are satisfied, the non-relativistic approximation may be used. The differential cross-section for such scattering is given in classical mechanics (and also in quantum mechanics) by the *Rutherford formula.* In the centre-of-mass system, when the indistinguishability of the electrons is taken into account we get the formula

$$\frac{d\sigma}{d\Omega} = \left(\frac{e^2}{16\pi m v^2}\right)^2 \left[\frac{1}{\sin^4\left(\frac{1}{2}\theta\right)} + \frac{1}{\cos^4\left(\frac{1}{2}\theta\right)} - \frac{1}{\sin^2\left(\frac{1}{2}\theta\right)\cos^2\left(\frac{1}{2}\theta\right)}\right]. \tag{27}$$

In quantum electrodynamics the probability amplitude for this scattering process is an infinite power series in the parameter e. With the assumption of inequalities (26), however, a dominant part may be separated from this amplitude. Note first of all that in every order of perturbation theory diagrams representing processes in which electrons exchange only one photon between them make a greater contribution in the limit (when $t \to 0$ and $s \to 4m^2$) than do the other diagrams. Transition amplitudes corresponding to such diagrams contain a large factor $1/t$. Half of such truncated diagrams in the second and in the fourth orders of perturbation theory are shown in Fig. 23.2. The other

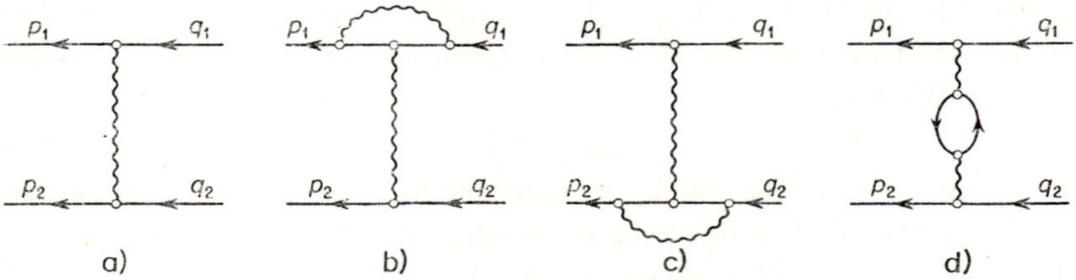

FIG. 23.2. Scattering of negatons in the second and fourth orders of perturbation theory.

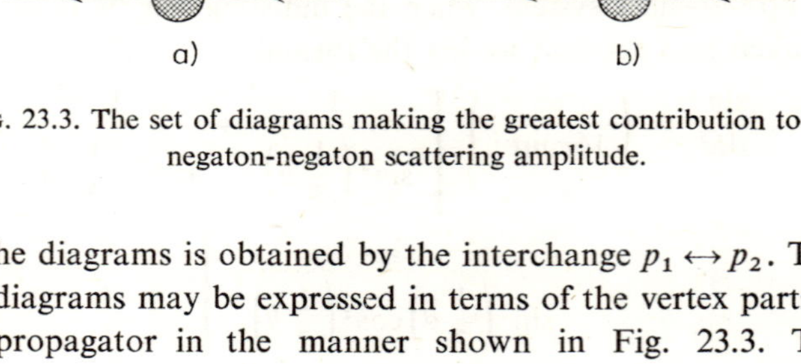

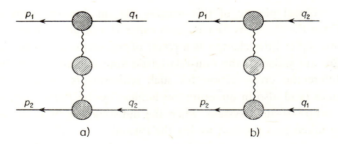

FIG. 23.3. The set of diagrams making the greatest contribution to the negaton-negaton scattering amplitude.

half of the diagrams is obtained by the interchange $p_1 \leftrightarrow p_2$. The set of all such diagrams may be expressed in terms of the vertex parts and the photon propagator in the manner shown in Fig. 23.3. The contribution to the scattering amplitude from the expressions corresponding to these diagrams is of the following analytical form:

$$e^2 Z_2^2 [\{\bar{u}(\mathbf{p}_1, r_1)\tilde{\Gamma}^{\mu}(p_1, q_1)u(\mathbf{q}_1, s_1)\}\tilde{\mathscr{G}}_{\mu\nu}(p_1 - q_1)$$
$$\times \{\bar{u}(\mathbf{p}_2, r_2)\tilde{\Gamma}^{\nu}(p_2, q_2)u(\mathbf{q}_2, s_2)\} - \{\bar{u}(\mathbf{p}_1, r_1)\tilde{\Gamma}^{\mu}(p_1, q_2)u(\mathbf{q}_2, s_2)\}$$
$$\times \tilde{\mathscr{G}}_{\mu\nu}(p_1 - q_2)\{\bar{u}(\mathbf{p}_2, r_2)\tilde{\Gamma}^{\nu}(p_2, q_1)u(\mathbf{q}_1, s_1)\}]. \tag{28}$$

In order to determine the limit of this expression when $p_1 \to q_1$, $p_2 \to q_2$, and $p_1 \to p_2$, we make use of the formulae (24), from which it follows that

$$\tilde{\Gamma}^{\mu}(p, p)\Big|_{\hat{p}=m} = Z_2^{-1}\gamma^{\mu},$$

and of the spectral representation of the photon propagator, from which it follows that

$$\tilde{\mathscr{G}}^{\text{tr}}_{\mu\nu}(k)_{k\to 0} \to \frac{Z_3}{k^2}\left(g_{\mu\nu} - \frac{k_\mu k_\nu}{k^2}\right).$$

The symbol $\tilde{\mathscr{G}}^{\text{tr}}_{\mu\nu}(k)$ denotes the transverse part of the photon propagator. The longitudinal part (i.e. the part proportional to $k_\mu k_\nu$) makes no contribution to the amplitude (28).

The principal contribution to the negaton-negaton scattering amplitude for momentum values satisfying conditions (26) may be written as

$$Z_3 e^2 \left[\frac{\{\bar{u}(\mathbf{p}_1, r_1)\gamma^{\mu}u(\mathbf{q}_1, s_1)\}\{\bar{u}(\mathbf{p}_2, r_2)\gamma_{\mu}u(\mathbf{q}_2, s_2)\}}{(p_2 - q_1)^2}\right.$$
$$\left. - \frac{\{\bar{u}(\mathbf{p}_1, r_1)\gamma^{\mu}u(\mathbf{q}_2, s_2)\}\{\bar{u}(\mathbf{p}_2, r_2)\gamma_{\mu}u(\mathbf{q}_1, s_1)\}}{(p_1 - q_2)^2}\right]. \tag{29}$$

This formula is of the same form as that for the electron scattering amplitude in second-order perturbation theory, except that $Z_3 e^2$ appears here instead of e^2. In the next chapter we shall show that from this expression for the transition amplitude it is possible to obtain in the non-relativistic approximation the Rutherford formula for the differential cross section, differing from formula (27) in that the role of the charge e is now played by $\sqrt{Z_3}e$.

Relation (25) between e and e_{obs} can thus be justified by reference to the principle of correspondence between quantum electrodynamics and classical mechanics (or non-relativistic quantum mechanics). Formula (25) can also be obtained by applying the correspondence principle to other low-energy processes, e.g. to the scattering of long-wavelength photons by free electrons. We shall return to this topic in Chapter 8.

24. EQUATIONS FOR RENORMALIZED PROPAGATORS#

In the preceding section, with the aid of straightforward examples we showed that in quantum electrodynamics divergent integrals appear in the expressions for propagators calculated by the Feynman diagram technique. These divergences occur, however, only in auxiliary quantities, devoid of any direct physical meaning. From these quantities we can construct expressions representing measurable physical quantities which, as it turns out, are free of divergences.

The mathematical procedure by which all quantities which are not directly observable are expressed in terms of quantities accessible to measurement is called *renormalization*. An example of renormalization was met in the preceding section in second-order perturbation theory. In the present section, we shall describe renormalization in the general case for arbitrary propagators in all orders of perturbation theory. It should be emphasized that although renormalization does eliminate all divergent expressions in quantum electrodynamics, its physical justification by no means rests in the occurrence of divergent integrals at intermediate stages in the formulation of the theory. Renormalization is applicable to all physical theories in which an interaction alters the properties of a system relative to the original theoretical picture. The need to renormalize crops up in nearly all quantum theories of systems of many particles. As a result of renormalization, for example, the mass of electrons in the solid state is changed† and the forces of the mutual

The renormalization method presented in this chapter was given by I. Białynicki-Birula (1965b). A renormalization method similar to ours has been worked out by E. S. Fradkin (1955b and 1965).

† The effective mass at times differs significantly from the mass of free electrons.

interaction of electrons in plasma are modified (screening of the Coulomb potential). It may be said that the occurrence of divergent integrals merely provided a drastic reminder of the need to recast the theory in a way so that its predictions concern the relationships between observables.

Our considerations will deal with electron-photon propagators $T[x_1 \ldots x_n, y_n \ldots y_1, z_1 \ldots z_k | \mathscr{A}, \mathscr{I}+\mathscr{I}_c]$ for a system with an external field $\mathscr{A}$, external current $\mathscr{I}$, and compensating current $\mathscr{I}_c$.

In order to achieve complete equivalence with the more traditional formulation, the dependence on the compensating current will be replaced by dependence on the gauge of the free photon propagator, in accordance with the procedure described in Section 22. The external potential of the field will, however, be treated as an arbitrary quantity, unrestricted by any gauge conditions.

Renormalized electron-photon propagator is the name we give to the expression

$$T_{\text{ren}}[x_1 \ldots x_n, y_n \ldots y_1, z_1 \ldots z_k] = Z_2^{-n} Z_3^{-k/2}\, T[x_1 \ldots x_n, y_n \ldots y_1, z_1 \ldots z_k], \tag{1}$$

which is treated as a function of the observable electron mass m_{obs} and the observable charge e_{obs}, as well as a functional of the observable electromagnetic field and the observable current.

Set of Equations for Propagators

Equations, which have solutions free of divergent integrals in all orders of perturbation theory, will now be derived for renormalized propagators. For our starting point we take the equations for unrenormalized propagators which emerge from the definition of these propagators from the following identity for functional derivatives,

$$\exp\left(\frac{1}{2i}\int \frac{\delta}{\delta\mathscr{A}} D^{\mathrm{F}} \frac{\delta}{\delta\mathscr{A}}\right) \mathscr{A}_\mu(z) \exp\left(-\frac{1}{2i}\int \frac{\delta}{\delta\mathscr{A}} D^{\mathrm{F}} \frac{\delta}{\delta\mathscr{A}}\right)$$

$$\equiv \mathscr{A}_\mu(z) - i\int D^{\mathrm{F}} \frac{\delta}{\delta\mathscr{A}},$$

and from the properties (17.29), (17.32), and (18.45) of the propagators K_F and the functional $C[\mathscr{A}, \mathscr{I}]$. These are the quantities out of which the complete propagators are constructed in quantum electrodynamics.

Below is the complete set of equations for electron-photon propagators:

$$\left\{D_{x_i}+e\gamma^{\mu}(x_i)\mathscr{A}_{\mu}(x_i)-ie\gamma^{\mu}(x_i)\int d^4z D^{F}_{\mu\nu}(x_i-z)\frac{\delta}{\delta\mathscr{A}_{\nu}(z)}\right\}$$
$$\times T_{\mu_1\ldots\mu_k}[x_1\ldots x_n, y_n\ldots y_1, z_1\ldots z_k|\mathscr{A},\mathscr{I}]$$
$$= -i\sum_{j=1}^{n}(-1)^{j-i}\delta(x_i-y_j)$$
$$T_{\mu_1\ldots\mu_k}[x_1\ldots\check{x}_i\ldots x_n, y_n\ldots\check{y}_j\ldots y_1, z_1\ldots z_k|\mathscr{A},\mathscr{I}], \tag{2}$$

$$T_{\mu_1\ldots\mu_k}[x_1\ldots x_n, y_n\ldots y_1, z_1\ldots z_k|\mathscr{A},\mathscr{I}]$$
$$\times\left\{\overleftarrow{D}_{y_i}+e\gamma^{\mu}(y_i)\mathscr{A}_{\mu}(y_i)-ie\gamma^{\mu}(y_i)\int d^4z D^{F}_{\mu\nu}(y_i-z)\frac{\overleftarrow{\delta}}{\delta\mathscr{A}_{\nu}(z)}\right\}$$
$$= -i\sum_{j=1}^{n}(-1)^{j-i}\delta(x_j-y_i)$$
$$T_{\mu_1\ldots\mu_k}[x_1\ldots\check{x}_j\ldots x_n, y_n\ldots\check{y}_i\ldots y_1, z_1\ldots z_k|\mathscr{A},\mathscr{I}], \tag{3}$$

$$\left\{i\frac{\delta}{\delta\mathscr{A}_{\mu}(z)}-\mathscr{I}^{\mu}(z)\right\}T_{\mu_1\ldots\mu_k}[x_1\ldots x_n, y_n\ldots y_1, z_1\ldots z_k|\mathscr{A},\mathscr{I}]$$
$$= e\gamma^{\mu}(z)\,T_{\mu_1\ldots\mu_k}[x_1\ldots x_n z, zy_n\ldots y_1, z_1\ldots z_k|\mathscr{A},\mathscr{I}]$$
$$+i\sum_{l=1}^{k}\delta^{\mu}_{\mu_l}\,\delta(z-z_l)\,T_{\mu_1\ldots\check{\mu}_l\ldots\mu_k}[x_1\ldots x_n, y_n\ldots y_1, z_1\ldots\check{z}_l\ldots z_k|\mathscr{A},\mathscr{I}], \tag{4}$$

$$\left\{i\frac{\delta}{\delta\mathscr{I}^{\mu}(z)}-\mathscr{A}_{\mu}(z)+\int d^4z' D^{F}_{\mu\nu}(z-z')\mathscr{I}^{\nu}(z')\right\}$$
$$\times T_{\mu_1\ldots\mu_k}[x_1\ldots x_n, y_n\ldots y_1, z_1\ldots z_k|\mathscr{A},\mathscr{I}]$$
$$= e\int d^4z' D^{F}_{\mu\nu}(z-z')\gamma^{\nu}(z')\,T_{\mu_1\ldots\mu_k}[x_1\ldots x_n z', z'y_n\ldots y_1, z_1\ldots z_k|\mathscr{A},\mathscr{I}]$$
$$-i\sum_{l=1}^{k}D^{F}_{\mu\mu_l}(z-z_l)\,T_{\mu_1\ldots\check{\mu}_l\ldots\mu_k}[x_1\ldots x_n, y_n\ldots y_1, z_1\ldots\check{z}_l\ldots z_k|\mathscr{A},\mathscr{I}]. \tag{5}$$

In addition to these equations, the following algebraic relations (not containing derivatives) between the propagators also hold:

$$\begin{aligned}
T_{\mu\mu_1\ldots\mu_k}&[x_1 \ldots x_n, y_n \ldots y_1, zz_1 \ldots z_k|\mathscr{A}, \mathscr{I}] \\
&= \left\{\mathscr{A}_\mu(z) - \int d^4z' D^F_{\mu\nu}(z-z')\mathscr{I}^\nu(z')\right\} \\
&\times T_{\mu_1\ldots\mu_k}[x_1 \ldots x_n, y_n \ldots y_1, z_1 \ldots z_k|\mathscr{A}, \mathscr{I}] \\
+e\int d^4z' D^F_{\mu\nu}(z-z')\gamma^\nu(z')\, &T_{\mu_1\ldots\mu_k}[x_1 \ldots x_n z', z' y_n \ldots y_1, z_1 \ldots z_k|\mathscr{A}, \mathscr{I}] \\
-\sum_{l=1}^{k} D^F_{\mu\mu_l}(z-z_l)\, &T_{\mu_1\ldots\check{\mu}_l\ldots\mu_k}[x_1 \ldots x_n, y_n \ldots y_1, z_1 \ldots \check{z}_l \ldots z_k|\mathscr{A}, \mathscr{I}]. \qquad (6)
\end{aligned}$$

These relations can be derived just like eqs. (2)–(5) from the definition of propagators. It follows from relations (6) that the propagators for an arbitrary value of the parameter k may be constructed out of propagators with a smaller value of k but with larger values of n.

Equations (2)–(6) are sufficient for determining the propagators by using perturbation theory, if eqs. (2) and (3) are supplemented with the Feynman asymptotic conditions.† This problem will not be tackled at present since the purpose of our discussion here is to derive equations for renormalized propagators possessing a direct physical meaning. After obtaining such equations, we shall demonstrate that renormalized propagators can be determined from them.

Elimination of the Field $\mathscr{A}_\mu(z)$

The first step on the road to obtaining equations for the renormalized propagators consists in replacing the external field $\mathscr{A}_\mu(z)$, hitherto figuring as an "independent variable", by the complete field $\mathfrak{A}_\mu(z)$ composed of both the external field and the field produced in vacuum by the motion of charges. The physical properties of the field $\mathfrak{A}_\mu$ will be discussed in Section 28; for the time being, the replacement of $\mathscr{A}_\mu$ by $\mathfrak{A}_\mu$ will be treated simply as a change of variables.

The complete field $\mathfrak{A}_\mu(z)$ will be defined as the linked part of the propagator $T_\mu[z|\mathscr{A}, \mathscr{I}]$ introduced in Section 20.

---

† The reader is advised to verify this fact by direct calculation for the simplest propagators in the lowest orders of perturbation theory.

To eliminate the field $\mathscr{A}$ and the derivatives with respect to $\mathscr{A}$, we shall use the following formulae for the derivatives of the field $\mathfrak{A}$ with respect to $\mathscr{J}$ and $\mathscr{A}$ and for the derivatives of the field $\mathscr{A}$ with respect to $\mathfrak{A}$:

$$\left(\frac{\delta\mathfrak{A}_\mu(z_1)}{\delta\mathscr{J}^\nu(z_2)}\right)_{\mathscr{A}} = -\mathscr{G}_{\mu\nu}[z_1 z_2], \tag{7a}$$

$$\left(\frac{\delta\mathfrak{A}_\mu(z_1)}{\delta\mathscr{A}_\nu(z_2)}\right)_{\mathscr{J}} = \delta_\mu^\nu\,\delta(z_1-z_2)+\int \mathrm{d}^4z\,\mathscr{G}_{\mu\lambda}[z_1 z]\Pi^{\lambda\nu}[zz_2], \tag{7b}$$

$$\left(\frac{\delta\mathscr{A}_\mu(z_1)}{\delta\mathfrak{A}_\nu(z_2)}\right)_{\mathscr{J}} = \delta_\mu^\nu\,\delta(z_1-z_2)-\int \mathrm{d}^4z D^{\mathrm{F}}_{\mu\lambda}(z_1-z)\Pi^{\lambda\nu}[zz_2], \tag{7c}$$

where $\Pi^{\lambda\nu}$ is the photon self-energy function introduced in Section 20. To indicate what derivatives are meant in the foregoing formulae, we have used notation similar to that employed in thermodynamics for derivatives calculated with one of the variables fixed.

Formula (7a) follows from the definition of the photon propagator as the linked part of the propagator $T_{\mu\nu}[z_1 z_2]$ and from the formula

$$i^l \frac{\delta^l T_{\mu_1\dots\mu_k}[x_1\dots x_n, y_n\dots y_1, z_1\dots z_k|\mathscr{A},\mathscr{J}]}{\delta\mathscr{J}^{\nu_1}(w_1)\dots\delta\mathscr{J}^{\nu_l}(w_l)}$$
$$= T_{\mu_1\dots\mu_k\nu_1\dots\nu_l}[x_1\dots x_n, y_n\dots y_1, z_1\dots z_k w_1\dots w_l|\mathscr{A},\mathscr{J}]$$

for $n = 0$. For we have

$$\frac{\delta\mathfrak{A}_\mu(z_1)}{\delta\mathscr{J}^\nu(z_2)} = \frac{\delta}{\delta\mathscr{J}^\nu(z_2)} V^{-1} i \frac{\delta V}{\delta\mathscr{J}^\mu(z_1)} = -iV^{-1}T_{\mu\nu}[z_1 z_2]$$
$$+iV^{-2}T_\mu[z_1]T_\nu[z_2] = -iT^L_{\mu\nu}[z_1 z_2]+iT^L_\mu[z_1]T^L_\nu[z_2]. \tag{8}$$

The proofs of formulae (7b) and (7c) are somewhat more complicated. In these proofs we make use of the relation

$$\int \mathrm{d}^4z\,D^{\mathrm{F}}_{\mu\lambda}(z_1-z)\frac{\delta\mathfrak{A}_\nu(z_2)}{\delta\mathscr{A}_\lambda(z)} = \mathscr{G}_{\mu\nu}[z_1 z_2], \tag{9}$$

which follows from formula (7a) and from the equation

$$\left\{i\int \mathrm{d}^4z' D^{\mathrm{F}}_{\mu\lambda}(z-z')\frac{\delta}{\delta\mathscr{A}_\lambda(z')}+i\frac{\delta}{\delta\mathscr{J}^\mu(z)}-\mathscr{A}_\mu(z)\right\}$$
$$\times T_{\mu_1\dots\mu_k}[x_1\dots x_n, y_n\dots y_1, z_1\dots z_k|\mathscr{A},\mathscr{J}] = 0. \tag{10}$$

The latter equation is satisfied by all the propagators by virtue of eqs. (4) and (5). It furthermore follows from the above that

$$\int d^4z' D^F_{\mu\lambda}(z-z') \frac{\delta}{\delta \mathscr{A}_\lambda(z')} = \int d^4z' d^4z'' D^F_{\mu\lambda}(z-z') \frac{\delta \mathfrak{A}_\varrho(z'')}{\delta \mathscr{A}_\lambda(z')} \frac{\delta}{\delta \mathfrak{A}_\varrho(z'')}$$

$$= \int d^4z' \mathscr{G}_{\mu\lambda}[zz'] \frac{\delta}{\delta \mathfrak{A}_\lambda(z')}. \tag{11}$$

We shall now rewrite the definition of the field $\mathfrak{A}$ in the form

$$\mathfrak{A}_\mu(z) = \mathscr{A}_\mu(z) - \int d^4z' D^F_{\mu\lambda}(z-z') \mathscr{J}^\lambda[z'|\mathscr{A}, \mathscr{I}], \tag{12}$$

where we have introduced the complete current $\mathscr{J}^\lambda$ built out of the external current and the current induced in the vacuum by the external field:

$$\mathscr{J}^\lambda[z|\mathscr{A}, \mathscr{I}] = V^{-1} i \frac{\delta}{\delta \mathscr{A}_\lambda(z)} V = i \frac{\delta}{\delta \mathscr{A}_\lambda(z)} (\ln V) = \mathscr{I}^\lambda(z) + ie \operatorname{Tr}\{\gamma^\lambda G[z, z]\}. \tag{13}$$

On acting on both sides of this equation with the operation $D^F \dfrac{\delta}{\delta \mathscr{A}}$ and utilizing eqs (9), (11), and (12) we obtain

$$\mathscr{G}_{\mu\nu}[z_1 z_2] = D^F_{\mu\nu}(z_1 - z_2) - \int d^4z \int d^4z' D^F_{\mu\lambda}(z_1 - z) \left(\frac{\delta}{\delta \mathfrak{A}_\varrho(z')} \mathscr{J}^\lambda[z] \right) \mathscr{G}_{\varrho\nu}[z' z_2]. \tag{14}$$

This relation may be treated as an integral equation for the photon propagator. The iterative solution of this equation is of the form

$$\mathscr{G}_{\mu\nu}[z_1 z_2] = D^F_{\mu\nu}(z_1 - z_2) - \int d^4z \int d^4z' D^F_{\mu\lambda}(z_1 - z) \frac{\delta \mathscr{J}^\lambda[z]}{\delta \mathfrak{A}_\varrho(z')} D^F_{\varrho\nu}(z' - z_2) + \dots .$$

By comparison with eq. (20.18), we identify the derivative of $\mathscr{J}$ with respect to $\mathfrak{A}$ as the self-energy function of the photon,

$$\Pi^{\varrho\lambda}[z' z] = - \frac{\delta \mathscr{J}^\lambda[z]}{\delta \mathfrak{A}_\varrho(z')}. \tag{15}$$

In conclusion of this argumentation, we differentiate eq. (12) with respect to $\mathscr{A}$ and, on applying formulae (11) and (13), we arrive at

$$\frac{\delta \mathfrak{A}_\mu(z_1)}{\delta \mathscr{A}_\nu(z_2)} = \delta^\nu_\mu \delta(z_1 - z_2) - \int d^4z \mathscr{G}_{\mu\lambda}[z_1 z] \frac{\delta}{\delta \mathfrak{A}_\lambda(z)} \mathscr{J}^\nu[z_2]$$

$$= \delta^\nu_\mu \delta(z_1 - z_2) + \int d^4z \mathscr{G}_{\mu\lambda}[z_1 z] \Pi^{\lambda\nu}[z z_2].$$

The proof of formula (7c) proceeds in similar fashion:

$$\frac{\delta \mathscr{A}_\mu(z_1)}{\delta \mathfrak{A}_\nu(z_2)} = \frac{\delta}{\delta \mathfrak{A}_\nu(z_2)} \left(\mathfrak{A}_\mu(z_1) + \int d^4z\, D^{\mathrm{F}}_{\mu\lambda}(z_1 - z) \mathscr{J}^\lambda[z] \right)$$

$$= \delta^\nu_\mu \delta(z_1 - z_2) - \int d^4z\, \Pi^{\nu\lambda}[z_2 z] D^{\mathrm{F}}_{\lambda\mu}(z - z_1).$$

Use must still be made of the symmetry of the free photon propagator and the self-energy function with respect to the interchange of arguments.

The consistency of foımulae (7b) and (7c) is ensured by an identity which may be abbreviated to

$$(1 + \mathscr{G}\Pi)(1 - D^{\mathrm{F}}\Pi) = 1. \tag{16}$$

Instead of the full electron-photon propagators it is more convenient to investigate the linked parts of these propagators. These quantities, treated as functionals of the field $\mathfrak{A}$ and the current $\mathscr{J}$, are denoted by the same symbol $T^L_{\mu_1 \dots \mu_k}[x_1 \dots x_n, y_n \dots y_1, z_1 \dots z_k | \mathfrak{A}, \mathscr{J}]$, even though their functional dependences on $(\mathscr{A}, \mathscr{J})$ and $(\mathfrak{A}, \mathscr{J})$ differ. The functional dependence will, moreover, be frequently omitted, leaving only the brackets.

With the change of variables $(\mathscr{A}, \mathscr{J}) \to (\mathfrak{A}, \mathscr{J})$, the following rules for the derivatives hold:

$$\frac{\delta}{\delta \mathscr{A}_\mu(z)} \to \frac{\delta}{\delta \mathfrak{A}_\mu(z)} + \int d^4z' \int d^4z''\, \Pi^{\mu\nu}[zz'] \mathscr{G}_{\nu\lambda}[z'z''] \frac{\delta}{\delta \mathfrak{A}_\lambda(z'')}, \tag{17a}$$

$$\frac{\delta}{\delta \mathscr{J}^\mu(z)} \to \frac{\delta}{\delta \mathscr{J}^\mu(z)} - \int d^4z'\, \mathscr{G}_{\mu\nu}[zz'] \frac{\delta}{\delta \mathfrak{A}_\nu(z')}. \tag{17b}$$

The equations for the propagators T^L obtained from eqs. (2)–(6) when the field $\mathscr{A}$ is eliminated take on the form:

$$\left\{ D_{x_i} + e\gamma^\mu(x_i) \mathfrak{A}_\mu(x_i) - ie\gamma^\mu(x_i) \int d^4z\, \mathscr{G}_{\mu\nu}[x_i z | \mathfrak{A}, \mathscr{J}] \frac{\delta}{\delta \mathfrak{A}_\nu(z)} \right\}$$

$$\times T^L_{\mu_1 \dots \mu_k}[x_1 \dots x_n, y_n \dots y_1, z_1 \dots z_k | \mathfrak{A}, \mathscr{J}]$$

$$= -i \sum_{j=1}^{n} (-1)^{j-i} \delta(x_i - y_j)$$

$$\times T^L_{\mu_1 \dots \mu_k}[x_1 \dots \check{x}_i \dots x_n, y_n \dots \check{y}_j \dots y_1, z_1 \dots z_k | \mathscr{A}, \mathscr{J}], \tag{18}$$

$$T^L_{\mu_1\ldots\mu_k}[x_1 \ldots x_n, y_n \ldots y_1, z_1 \ldots z_k|\mathfrak{A}, \mathscr{I}]$$
$$\times\left\{\overleftarrow{D}_{y_i} + e\gamma^\mu(y_i)\mathfrak{A}_\mu(y_i) - ie\gamma^\mu(y_i)\int \mathrm{d}^4 z\, \frac{\overleftarrow{\delta}}{\delta\mathfrak{A}_\nu(z)}\,\mathscr{G}_{\nu\mu}[zy_i|\mathfrak{A}, \mathscr{I}]\right\}$$
$$= -i\sum_{j=1}^{n}(-1)^{j-i}\delta(x_j - y_i)$$
$$T^L_{\mu_1\ldots\mu_k}[x_1 \ldots \check{x}_j \ldots x_n, y_n \ldots \check{y}_i \ldots y_1, z_1 \ldots z_k|\mathfrak{A}, \mathscr{I}], \tag{19}$$

$$i\frac{\delta}{\delta\mathfrak{A}_\mu(z)}\, T^L_{\mu_1\ldots\mu_k}[x_1 \ldots x_n, y_n \ldots y_1, z_1 \ldots z_k|\mathfrak{A}, \mathscr{I}]$$
$$= \int \mathrm{d}^4 z' \left\{\delta^\mu_\nu\, \delta(z - z') - \int \mathrm{d}^4 z''\, \Pi^{\mu\lambda}[zz'']\, D^{\mathrm{F}}_{\lambda\nu}(z'' - z')\right\}$$
$$\times\Big\{- ie\gamma^\nu(z')\, G[z', z'|\mathfrak{A}, \mathscr{I}]\, T^L_{\mu_1\ldots\mu_k}[x_1 \ldots x_n, y_n \ldots y_1, z_1 \ldots z_k|\mathfrak{A}, \mathscr{I}]$$
$$- e\gamma^\nu(z')\, T^L_{\mu_1\ldots\mu_k}[x_1 \ldots x_n z', z' y_n \ldots y_1, z_1 \ldots z_k|\mathfrak{A}, \mathscr{I}]$$
$$+ i\sum_{l=1}^{k}\delta^\nu_{\mu_l}\, \delta(z' - z_l)\, T^L_{\mu_1\ldots\check{\mu}_l\ldots\mu_k}[x_1 \ldots x_n, y_n \ldots y_1, z_1 \ldots \check{z}_l \ldots z_k|\mathfrak{A}, \mathscr{I}]\Big\}, \tag{20}$$

$$\Big\{i\frac{\delta}{\delta\mathscr{I}^\mu(z)} - i\int \mathrm{d}^4 z'\, \mathscr{G}_{\mu\nu}[zz']\, \frac{\delta}{\delta\mathfrak{A}_\nu(z')}$$
$$+ ie\int \mathrm{d}^4 z'\, D^{\mathrm{F}}_{\mu\nu}(z - z')\, \gamma^\nu(z')\, G[z', z']\Big\}$$
$$\times\, T^L_{\mu_1\ldots\mu_k}[x_1 \ldots x_n, y_n \ldots y_1, z_1 \ldots z_k|\mathfrak{A}, \mathscr{I}]$$
$$= e\int \mathrm{d}^4 z'\, D^{\mathrm{F}}_{\mu\nu}(z - z')\, \gamma^\nu(z')\, T^L_{\mu_1\ldots\mu_k}[x_1 \ldots x_n z', z' y_n \ldots y_1, z_1 \ldots z_k|\mathfrak{A}, \mathscr{I}]$$
$$- i\sum_{l=1}^{k} D^{\mathrm{F}}_{\mu\mu_l}(z - z_l)\, T^L_{\mu_1\ldots\check{\mu}_l\ldots\mu_k}[x_1 \ldots x_n, y_n \ldots y_1, z_1 \ldots \check{z}_l \ldots z_k|\mathfrak{A}, \mathscr{I}]. \tag{21}$$

On elimination of the field $\mathscr{A}$, relations (6) become:

$$T^L_{\mu\mu_1\ldots\mu_k}[x_1 \ldots x_n, y_n \ldots y_1, zz_1 \ldots z_k|\mathfrak{A}, \mathscr{I}]$$
$$= \left\{\mathfrak{A}_\mu(z) + ie\int \mathrm{d}^4 z'\, D^{\mathrm{F}}_{\mu\nu}(z - z')\, \gamma^\nu(z')\, G[z', z'|\mathfrak{A}, \mathscr{I}]\right\}$$
$$\times\, T^L_{\mu_1\ldots\mu_k}[x_1, \ldots x_n, y_n \ldots y_1, z_1 \ldots z_k|\mathfrak{A}, \mathscr{I}]$$
$$+ \int \mathrm{d}^4 z'\, D^{\mathrm{F}}_{\mu\nu}(z - z')\, \gamma^\nu(z')\, T^L_{\mu_1\ldots\mu_k}[x_1 \ldots x_n z', z' y_n \ldots y_1, z_1 \ldots z_k|\mathfrak{A}, \mathscr{I}]$$
$$- i\sum_{l=1}^{\mu} D^{\mathrm{F}}_{\mu\mu_l}(z - z_l)\, T^L_{\mu_1\ldots\check{\mu}_l\ldots\mu_k}[x_1 \ldots x_n, y_n \ldots y_1, z_1 \ldots \check{z}_l \ldots z_k|\mathfrak{A}, \mathscr{I}]. \tag{22}$$

Identity (16) has been used in deriving these equations.

Equations (18)–(22) for the linked propagators will be supplemented with equations for the vacuum-to-vacuum transition amplitude $V[\mathfrak{A}, \mathcal{J}]$ treated as the functional of $\mathfrak{A}$ and $\mathcal{J}$. Knowledge of this amplitude is necessary for constructing the transition amplitudes when the linked parts of the propagators are known. The equations for V are obtained from equations (4) and (5):

$$\Big\{i\frac{\delta}{\delta\mathfrak{A}_\mu(z)} - \mathcal{J}^\mu[z|\mathfrak{A}, \mathcal{J}] + \int d^4z' \int d^4z'' \Pi^{\mu\lambda}[zz'] D^{\mathrm{F}}_{\lambda\nu}(z'-z'')\mathcal{J}^\nu[z''|\mathfrak{A}, \mathcal{J}]\Big\} V[\mathfrak{A}, \mathcal{J}] = 0, \tag{23}$$

$$\Big\{i\frac{\delta}{\delta\mathcal{J}^\mu(z)} - i\int d^4z' \mathcal{G}_{\mu\nu}[zz'] \frac{\delta}{\delta\mathfrak{A}_\nu(z')} - \mathfrak{A}_\mu(z) + \int d^4z' D^{\mathrm{F}}_{\mu\nu}(z-z')\mathcal{J}^\nu[z'|\mathfrak{A}, \mathcal{A}]\Big\} V[\mathfrak{A}, \mathcal{J}] = 0. \tag{24}$$

Equations (18)–(24) no longer contain either the external field $\mathcal{A}$ itself or derivatives with respect to it. Solutions of these equations may thus be sought as the functionals $\mathfrak{A}$ and $\mathcal{J}$.

The transition from the variables $(\mathcal{A}, \mathcal{J})$ to $(\mathfrak{A}, \mathcal{J})$ is a close counterpart of the transition from the Lagrangian variables **E** and **B** to the canonical ones **D** and **B** described in Section 7.

Gauge Invariance and the Ward Identity

The Ward identity is a result of the invariance of the theory with respect to gauge transformations. The gauge invariance of $\mathcal{A}$-dependent quantities was expressed in Section 22 in terms of the operation $\mathcal{B}(z) = \partial_\mu(\delta/\delta\mathcal{A}_\mu(z))$. Since we are now treating all quantities as functionals of the complete field $\mathfrak{A}$, we should also express the operation $\mathcal{B}(z)$ in terms of differentiation with respect to that field. It turns out that these operations have the same form with both choices of variables,

$$\partial_\mu \frac{\delta}{\delta\mathcal{A}_\mu(z)} = \partial_\mu \frac{\delta}{\delta\mathfrak{A}_\mu(z)}, \tag{25}$$

This follows from the fact that under gauge transformations the fields $\mathscr{A}$ and $\mathfrak{A}$ experience the same change since the complete current $\mathscr{J}$ does not depend on the gauge of the potential:

$$\mathfrak{A}_\mu[z|\mathscr{A}+\partial\Lambda, \mathscr{J}] = \mathfrak{A}_\mu[z|\mathscr{A}, \mathscr{J}]+\partial_\mu\Lambda(z).$$

Relation (25) may also be obtained from eq. (5) and the equation

$$\partial_\mu \Pi^{\mu\nu}[zz'] = 0. \tag{26}$$

The latter is a result of the continuity equation for the complete current $\mathscr{J}$.

Next, we shall make use of eq. (20) which, on taking the divergence on either side and utilizing eqs. (18) and (19) as well as eq. (26), yields the *functional Ward identity*:

$$i\partial_\mu \frac{\delta}{\delta\mathfrak{A}_\mu(z)} T^L_{\mu_1\ldots\mu_k}[x_1 \ldots x_n, y_n \ldots y_1, z_1 \ldots z_k|\mathfrak{A}, \mathscr{J}]$$

$$= -e\sum_{i=1}^{n}[\delta(x_i-z)-\delta(z-y_i)]$$

$$\times\, T^L_{\mu_1\ldots\mu_k}[x_1 \ldots x_n, y_n \ldots y_1, z_1 \ldots z_k|\mathfrak{A}, \mathscr{J}]$$

$$+i\sum_{l=1}^{k}\partial_{\mu_l}\delta(z-z_l)\, T^L_{\mu_1\ldots\check{\mu}_l\ldots\mu_k}[x_1 \ldots x_n, y_n \ldots y_1, z_1 \ldots \check{z}_l \ldots z_k|\mathfrak{A}, \mathscr{J}]. \tag{27}$$

The gauge-dependence of the propagators might seem disquieting. The seeming contradiction vanishes, however, when we realize that gauge invariance holds only for propagators containing compensating current; these are equal to the propagators under consideration, it is true, but this is so only when the latter are calculated in a gauge defined by the functions a^μ which determine the compensating current. Propagators dependent on the gauge of the free photon propagator and potential, which we are using at present, are convenient quantities in terms of which quantum electrodynamics is most frequently formulated.

In the particular case, when $n = 1$, $k = 0$, the functional Ward identity yields the formula

$$i\partial_\mu \frac{\delta}{\delta\mathfrak{A}_\mu(z)} G[x, y] = -e[\delta(x-z)-\delta(z-y)]G[x, y]. \tag{28}$$

The functional derivative of the electron propagators with respect to $\mathfrak{A}$ can be expressed in terms of the vertex function by the formula

$$\frac{\delta}{\delta\mathfrak{A}_\lambda(\zeta)}G[x,y] = -eG[x,\xi]\Gamma^\lambda[\xi,\eta,\zeta]G[\eta,y]. \tag{29}$$

In the foregoing formula we have employed a convention which will be used frequently in what follows. The convention is that with respect to repeated coordinates we perform integration over all space-time, combined with summation with respect to the relevant bispinor or vector indices.

The validity of formula (29) is proved by multiplying both sides by $\mathscr{G}_{\mu\lambda}[z\zeta]$, integrating with respect to ζ, and rewriting the left-hand side as

$$\int d^4\zeta\mathscr{G}_{\mu\lambda}[z\zeta]\frac{\delta}{\delta\mathfrak{A}_\lambda(\zeta)}G[x,y] = \int d^4\zeta D_{\mu\lambda}(z-\zeta)\frac{\delta}{\delta\mathscr{A}_\lambda(\zeta)}G[x,y].$$

Next, we use the counterpart of formula (10) for the propagator T^L,

$$\left\{\int d^4z' D^F_{\mu\nu}(z-z')\frac{\delta}{\delta\mathscr{A}_\nu(z')}+\frac{\delta}{\delta\mathscr{I}^\mu(z)}\right\}$$
$$\times T^L_{\mu_1\ldots\mu_k}[x_1\ldots x_n, y_n\ldots y_1, z_1\ldots z_k|\mathscr{A},\mathscr{I}] = 0,$$

and apply formula (8) to arrive at

$$\int d^4\zeta D^F_{\mu\lambda}(z-\zeta)\frac{\delta}{\delta\mathscr{A}_\lambda(\zeta)}G[x,y] = -\frac{\delta}{\delta\mathscr{I}^\mu(z)}G[x,y]$$
$$= \mathfrak{A}_\mu(z)T^L[x,y]-T^L_\mu[x,y,z] = -T^C_\mu[x,y,z].$$

By formula (22.26) for the decomposition of the propagator $T^C_\mu[x,y,z]$, we thus get

$$\mathscr{G}_{\mu\lambda}[z\zeta]\frac{\delta}{\delta\mathfrak{A}^\wedge(\zeta)}G[x,y] = -T^C_\mu[x,y,z]$$
$$= -e\mathscr{G}_{\mu\lambda}[z\zeta]G[x,\xi]\Gamma^\lambda[\xi,\eta,\zeta]G[\eta,y].$$

"Dividing" both sides by the propagator $\mathscr{G}_{\mu\lambda}[z\zeta]$, we obtain formula (29).

The "division" by the photon propagator requires some elucidation. Just as the free photon propagator $D^F_{\mu\nu}$, this propagator—as the kernel of an integral operator with respect to the variables z_1 and z_2—corresponds to a projection operator, since it satisfies the gauge conditions

$$\int d^4z'\mathscr{G}_{\mu\nu}[zz']a^\nu(z') = 0 = \int d^4z' a^\mu(z')\mathscr{G}_{\mu\nu}[z'z].$$

These conditions may be obtained, for instance, by making use of the gauge conditions (22.14) for the free photon propagator and the iterative solution of equation (14). In all general considerations, when we are concerned only with those properties of the propagators which follow from the topological properties of the diagrams, the propagators $D^{\mathrm{F}}_{\mu\nu}$ and $\mathscr{G}_{\mu\nu}$ may be regarded as quantities possessing the integral inverse operators $(D^{\mathrm{F}}_{\mu\nu})^{-1}$ and $(\mathscr{G}_{\mu\nu})^{-1}$. The entire analysis of the connectedness of the diagrams and other general discussions can be carried out without any modifications for the case when the free photon propagator is not singular.

This was tacitly exploited once earlier when we identified the derivative of $\mathscr{J}$ with respect to $\mathfrak{A}$ with the self-energy function of the photon. Subsequently in our general considerations we shall frequently treat the photon propagator as a function having the inverse as an integral operator.

Equations (28) and (29) lead to the *generalized Ward identity*#

$$i\frac{\partial}{\partial z^{\mu}}\Gamma^{\mu}[x, y, z] = [\delta(x-z)-\delta(y-z)]G^{-1}[x, y], \tag{30}$$

which, in the absence of an external field and external current, may be rewritten as

$$(p-q)_{\mu}\tilde{\Gamma}^{\mu}(p, q) = -\tilde{G}^{-1}(p)+\tilde{G}^{-1}(q). \tag{31}$$

Differentiation of both sides with respect to p_{μ} and the substitution $p = q$ yield the Ward identity (23.23).

Symmetric Expressions for the Electron and Photon Propagators and the Vertex Function

The equations (18)–(22) for the electron-photon propagator $T^{L}_{\mu_1\ldots\mu_k}[x_1 \ldots x_n, y_n \ldots y_1, z_1 \ldots z_k]$ also contain the propagators $T^{L}_{\mu_1\ldots\mu_k}[x_1 \ldots x_{n-1}, y_{n-1}, z_1 \ldots z_k]$, $T^{L}_{\mu_1\ldots\mu_k}[x_1 \ldots x_{n+1}, y_{n+1} \ldots y_1, z_1 \ldots z_k]$ and $T^{L}_{\mu_1\ldots\mu_{k-1}}[x_1 \ldots x_n, y_n \ldots y_1, z_1 \ldots z_{k-1}]$, the photon propagator $\mathscr{G}$, and the electron propagator G.

The solution of the set of these equations should thus be started by tackling the equations for the propagators with the lowest values of n and k, and then successively solving the equations for increasingly complicated propagators.

This formula, which bears the name of the generalized Ward identity, was discovered independently by E. S. Fradkin (1955a), H. S. Green (1953), and Y. Takahashi (1957).

Let us begin with the equations which contain only the photon propagator and the electron propagator. The equations for the electron propagator are obtained by putting $n = 1$ and $k = 0$ in eqs. (18) and (19). In simplified notation they may be written as

$$\left\{K_{\mathrm{F}}^{-1}[x, \xi|\mathfrak{A}] - ie\gamma(x, \xi, \zeta)\mathscr{G}[\zeta\zeta']\frac{\delta}{\delta\mathfrak{A}(\zeta')}\right\}G[\xi, y] = \delta(x-y), \tag{32a}$$

$$G[x, \xi]\left\{K_{\mathrm{F}}^{-1}[\xi, y|\mathfrak{A}] - ie\gamma(y, \xi, \zeta)\cdot\frac{\overleftarrow{\delta}}{\delta\mathfrak{A}(\zeta')}\mathscr{G}[\zeta\zeta']\right\} = \delta(x-y), \tag{32b}$$

where the convention on integration holds and

$$\gamma^{\mu}(x, y, z) \equiv \gamma^{\mu}(z)\delta(x-z)\,\delta(z-y).$$

The function that is "inverse" to the propagator K_{F} satisfies the conditions

$$\int d^4\xi K_{\mathrm{F}}^{-1}[x, \xi|\mathfrak{A}]K_{\mathrm{F}}[\xi, y|\mathfrak{A}] = \delta(x-y),$$

$$\int d^4\xi K_{\mathrm{F}}[x, \xi|\mathfrak{A}]K_{\mathrm{F}}^{-1}[\xi, y|\mathfrak{A}] = \delta(x-y),$$

and has the following explicit representation:

$$K_{\mathrm{F}}^{-1}[x, y|\mathfrak{A}] = [D_x + e\gamma\cdot\mathfrak{A}(x)]\delta(x-y). \tag{33}$$

On being multiplied through by K_{F} and integrated, eq. (32a) for the electron propagator takes on the form

$$G[x, y] = K_{\mathrm{F}}[x, y|\mathfrak{A}]$$
$$+ieK_{\mathrm{F}}[x, \xi|\mathfrak{A}]\gamma(\xi, \eta, \zeta)\mathscr{G}[\zeta\zeta']\frac{\delta}{\delta\mathfrak{A}(\zeta')}G[\eta, y]. \tag{34}$$

A similar equation for the propagator $\mathscr{G}$ is obtained from eq. (14):

$$\mathscr{G}[z_1 z_2] = D^{\mathrm{F}}(z_1 - z_2)$$
$$-ieD^{\mathrm{F}}(z_1-\zeta)\gamma(\eta, \xi, \zeta)\left(\frac{\delta}{\delta\mathfrak{A}(\zeta')}G[\xi, \eta]\right)\mathscr{G}[\zeta' z_2]. \tag{35}$$

The expression on the right-hand side in eqs. (33) and (34) can, by formula (29), also be rewritten in terms of the vertex function:

$$G[x, y] = K_F[x, y|\mathfrak{A}]$$
$$-ie^2 K_F[x, \xi|\mathfrak{A}]\gamma(\xi, \eta, \zeta)\mathcal{G}[\zeta\zeta']G[\eta, \xi']\Gamma[\xi', \eta', \zeta']G[\eta', y], \quad (36)$$
$$\mathcal{G}[z_1 z_2] = D^F(z_1 - z_2)$$
$$+ie^2 D^F(z_1 - \zeta)\gamma(\eta, \xi, \zeta)G[\xi, \eta']\Gamma[\eta', \xi', \zeta']G[\xi', \eta]\mathcal{G}[\zeta' z_2]. \quad (37)$$

On the basis of these equations we make the identifications

$$\Sigma[x, y] = -ie^2\gamma(x, \eta, \zeta)\mathcal{G}[\zeta\zeta']G[\eta, \xi]\Gamma[\xi, y, \zeta'], \quad (38)$$
$$\Pi[zz'] = ie^2\gamma(\eta, \xi, z)G[\xi, \eta']\Gamma[\eta', \xi', z']G[\xi', \eta], \quad (39)$$

because the iterative solutions of the equations,

$$G[x, y] = K_F[x, y|\mathfrak{A}] + K_F[x, \xi|\mathfrak{A}]\Sigma[\xi, \eta]G[\eta, y], \quad (40)$$
$$\mathcal{G}[zz'] = D^F(z - z') + D^F(z - \zeta)\Pi[\zeta\zeta']\mathcal{G}[\zeta' z'], \quad (41)$$

obtained after (38) and (39) have been identified, coincide with formulae (22.23) and (20.18).

Equation (32b) for the electron propagator will be rewritten as

$$G[x, y] = K_F[x, y|\mathfrak{A}] + G[x, \xi]\Sigma[\xi, \eta]K_F[\eta, y|\mathfrak{A}]. \quad (42)$$

The iterative solution of this equation is the same as that of equation (40). An analogous "transposed" equation for the photon propagator can also be derived.

The next step towards getting equations for the renormalized propagators will be to obtain symmetric expressions for the self-energy functions Σ and Π. The expressions in formulae (38) and (39) are not symmetric with respect to the coordinates x and y, and z_1 and z_2. A dependence on one of these coordinates appears in the vertex function Γ, and a dependence on the other in the zeroth-order vertex function γ.

In order to get a symmetric expression for the self-energy function of the electron and thereby for the electron propagator, we eliminate the function $\gamma(x, y, z)$ from expression (38). To this end we define the kernel $\mathcal{K}[x, y, z_1 z_2]$,

$$e^2\mathcal{K}[x, y, z_1 z_2] \equiv ie^2\Gamma[x, \xi, z_2]G[\xi, \eta]\Gamma[\eta, y, z_1]$$
$$-e\Delta[z_1 z_2 \zeta_1]\mathcal{G}[\zeta_1 \zeta_2]\Gamma[x, y, \zeta_2] + e^2\Gamma[x, y, z_1 z_2] \quad (43)$$

and the resolvent $\mathscr{R}[x, y, z_1 z_2]$ satisfying an integral equation with the kernel $\mathscr{K}[x, y, z_1 z_2]$,

$$\mathscr{R}[x, y, z_1 z_2] = G[x, y]\mathscr{G}[z_1 z_2] + e^2\mathscr{R}[x, \xi, z_1 \zeta_1]\mathscr{K}[\xi, \eta, \zeta_1 \zeta_2]G[\eta, y]\mathscr{G}[\zeta_2 z_2]. \tag{44}$$

If the relations

$$\Delta[z_1 z_2 z_3] = -i\frac{\delta}{\delta\mathfrak{A}(z_1)}\mathscr{G}^{-1}[z_2 z_3], \tag{45}$$

$$e\Gamma[x, y, z] = \frac{\delta}{\delta\mathfrak{A}(z)}G^{-1}[x, y], \tag{46}$$

$$e\Gamma[x, y, z_1 z_2] = -i\frac{\delta}{\delta\mathfrak{A}(z_1)}\Gamma[x, y, z_2], \tag{47}$$

which follow† from the definitions of the functions Δ and Γ are used, the kernel $\mathscr{K}$ may be rewritten as

$$e\mathscr{G}[z_1 \zeta]G[x, \xi]\mathscr{K}[\xi, y, \zeta z_2] = -i\frac{\delta}{\delta\mathfrak{A}(z_2)}(\mathscr{G}[z_1 \zeta]G[x, \xi]\Gamma[\xi, y, \zeta]). \tag{48}$$

Equation (36) is transformed into

$$G^{-1}[x, y] = K_{\mathrm{F}}^{-1}[x, y|\mathfrak{A}] + ie^2\gamma(x, \xi, \zeta)\mathscr{G}[\zeta\zeta']G[\xi, \xi']\Gamma[\xi', y, \zeta'],$$

and this is then differentiated with respect to $\mathfrak{A}$. The result of the differentiation is expressed in terms of the kernel $\mathscr{K}$:

$$\gamma(x, y, z) = \Gamma[x, y, z] + e^2\gamma(x, \xi, \zeta_1)\mathscr{G}[\zeta_1 \zeta_2]G[\xi, \eta]\mathscr{K}[\eta, y, \zeta_2 z]. \tag{49}$$

The resulting relation may be treated as an integral equation for the function $\gamma(x, y, z)$. The solution of this equation is expressed with the aid of the resolvent $\mathscr{R}$ as follows:

$$\gamma(x, \xi, \zeta)\mathscr{G}[\zeta z]G[\xi, y] = \Gamma[x, \xi, \zeta]\mathscr{R}[\xi, y, \zeta z]. \tag{50}$$

---

† The proofs of these formulae are left to the reader. They are very similar to the proofs of other formulae discussed in this section.

On inserting the function γ, as expressed by this formula, into eq. (38), we arrive at the desired symmetric representation of the self-energy function,

$$\Sigma[x, y] = -ie^2\Gamma[x, \xi, \zeta_1]\mathscr{R}[\xi, \eta, \zeta_1\zeta_2]\Gamma[\eta, y, \zeta_2]. \tag{51}$$

Now we shall give a graphic representation of the electron self-energy function Σ. For this purpose we solve the integral equation for $\mathscr{R}$ by iteration considering only the first few terms of the series. The expressions so obtained for the function Σ are presented in Fig. 24.1

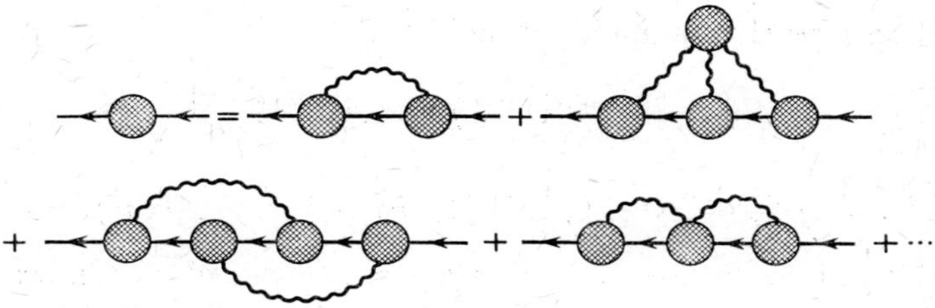

FIG. 24.1. Picture of the electron self-energy function.

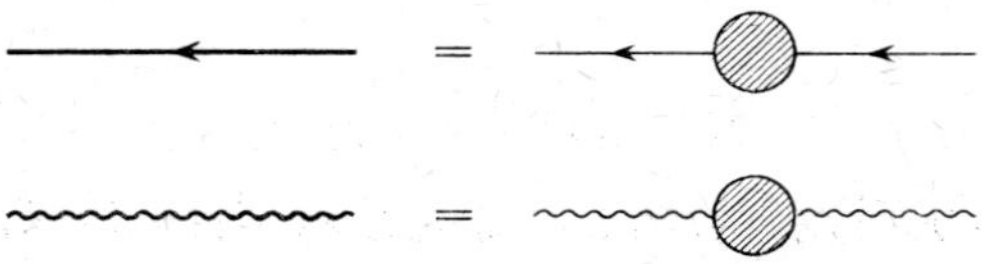

FIG. 24.2. Lines representing the complete electron and photon propagator.

(In this Figure, a heavy solid line has been used to indicate the full electron propagator $\mathscr{G}$, and a thick wavy line the full photon propagator $\mathscr{G}$. This is illustrated in Fig. 24.2.) Figure 24.1 shows that both vertices in the self-energy diagrams have been treated with complete symmetry.

Moreover, eqs. (49) and (50) lead to the following integral equation for the vertex function:

$$\Gamma[x, y, z] = \gamma(x, y, z) - e^2\Gamma[x, \xi, \zeta_1]\mathscr{R}[\xi, \eta, \zeta_1\zeta_2]\mathscr{K}[\eta, y, \zeta_2 z]. \tag{52}$$

To obtain a symmetric expression for the photon self-energy function, we define the kernel $K[x_1 x_2, y_2 y_1]$,

$$e^2 K[x_1 x_2, y_2 y_1] \equiv ie^2 \Gamma[x_1, y_2, \zeta_1] \mathscr{G}[\zeta_1 \zeta_2] \Gamma[x_2, y_1, \zeta_2] + E[x_1 x_2, y_2 y_1], \tag{53}$$

and the resolvent $R[x_1 x_2, y_2 y_1]$ of the integral equation with this kernel,

$$R[x_1 x_2, y_2 y_1] = G[x_1, y_2] G[x_2, y_1] + e^2 R[x_1 \xi_1, \eta_1 y_1] K[\eta_1 \eta_2, \xi_2 \xi_1] G[\xi_2, y_2] G[x_2, \eta_2]. \tag{54}$$

In addition, we shall utilize the condition

$$\mathscr{G}[z\zeta] \frac{\delta G[x, y]}{\delta \mathfrak{A}(\zeta)} = D^{\mathrm{F}}(z-\zeta) \frac{\delta G[x, y]}{\delta \mathscr{A}(\zeta)}$$

$$= eD^{\mathrm{F}}(z-\zeta) \gamma(\eta, \xi, \zeta) (G[x\xi, \eta y] - G[x, y] G[\xi, \eta]), \tag{55}$$

which is satisfied by the electron propagator and which follows from eqs. (11) and (4). When we use eq. (37) and the decomposition of the propagator $G[x_1 x_2, y_2 y_1]$ into strongly connected parts, then on further employing eq. (53) we get an integral equation for the function $\gamma(x, y, z)$:

$$\gamma(x, y, z) = \Gamma[x, y, z] + e^2 \gamma(\xi_1, \eta_1, z) G[\eta_1, \eta_2] K[\eta_2\, x, y \xi_2] G[\xi_2, \xi_1]. \tag{56}$$

The solution of this equation is expressed in terms of the resolvent R,

$$\gamma(\xi, \eta, z) G[\eta, y] G[x, \xi] = \Gamma[\xi, \eta, z] R[\eta x, y \xi]. \tag{57}$$

Insertion of the function γ as expressed by this formula into eq. (39) yields the desired symmetric representation of the self-energy function of the photon:

$$\Pi[z_1 z_2] = ie^2 \Gamma[\xi_1, \xi_2, z_1] R[\xi_2\, \eta_2, \eta_1\, \xi_1] \Gamma[\eta_1, \eta_2, z_2]. \tag{58}$$

The graphical representation of the self-energy function obtained from the first few terms of the iterative solution of the equation for R is given in Fig. 24.3.

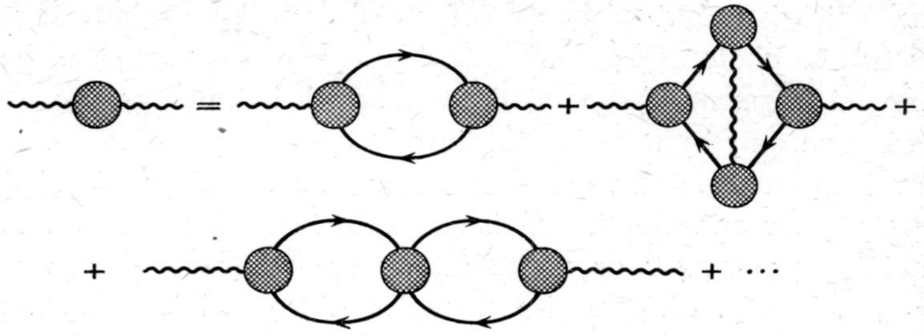

FIG. 24.3. Picture of the photon self-energy function.

An integral equation for the vertex function, which is different from eq. (52), also follows from eqs. (56) and (57):

$$\Gamma[x, y, z] = \gamma(x, y, z)$$
$$-e^2\Gamma[\xi_1, \eta_1, z]R[\eta_1\eta_2, \xi_2\xi_1]K[\xi_2 x, y\eta_2]. \qquad (59)$$

THE SKELETON STRUCTURE OF DIAGRAMS

The equations derived in this section for the electron and photon propagators and for the vertex function do not constitute a closed set of equations since they contain the functions $\Gamma[x, y, z_1 z_2]$, $\Delta[z_1 z_2 z_3]$, and $E[x_1 x_2, y_2 y_1]$, for which we do not yet have equations. Such equations could be derived by using the set of equations discussed in this section for all propagators. However, even more complicated functions would appear in these equations. To overcome this difficulty we shall proceed in a different manner and use the Feynman diagram technique for constructing the aforementioned functions from *functions of the first kind* as we shall call the functions $G[x, y]$, $\mathcal{G}[z_1 z_2]$, and $\Gamma[x, y, z]$. All other functions, which we shall call *functions of the second kind*, may be written uniquely in the form of infinite series, whose terms are integrals of the products of functions of the first kind.

To examine the structure of these expressions, we shall introduce the notion of *irreducible diagrams* and *skeleton diagrams*. With this in mind, let us consider any connected diagram which is not a diagram of a function of the first kind. Each connected fragment of such a diagram, which is linked to the rest of the diagram by only two electron lines or two photon lines, will be called a *propagator insert*, and each connected fragment of such a diagram, which is linked to the rest of the diagram only by two electron lines and one photon line and is not an ordinary vertex, will be called a *vertex insert*. A diagram which has no propagator or vertex inserts is called an *irreducible diagram*. Those containing propagator or vertex inserts are referred to as *reducible diagrams*. In Fig. 24.4 we give examples of irreducible and reducible

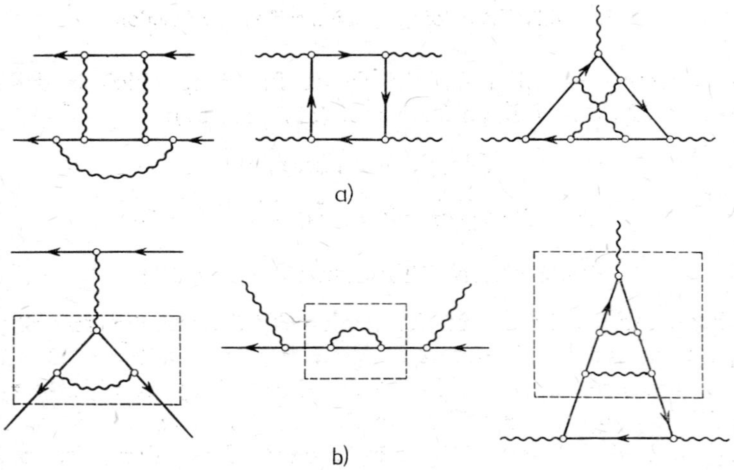

FIG. 24.4. Examples of (a) irreducible diagrams and (b) reducible diagrams.

diagrams. Propagator and vertex inserts have been marked on the reducible diagrams. On replacing all the propagator inserts by ordinary electron or photon lines, and vertex inserts by ordinary vertices, we obtain an irreducible diagram from each reducible diagram.

All connected diagrams corresponding to an arbitrary propagator of the second kind may be divided into classes of diagrams. One class comprises all diagrams which yield the same irreducible diagram when ali inserts are removed. The procedure may also be reversed. All diagrams may be obtained by drawing all possible inserts to all irreducible diagrams. The set of all diagrams associated with a given irreducible diagram can thus be represented by a diagram built out of heavy lines corresponding to the full electron and photon propagators, $G[x, y]$ and $\mathcal{G}[z_1 z_2]$, and of the vertices corresponding to the vertex function $\Gamma[x, y, z]$; such a diagram has the same structure as the irreducible diagram considered. A diagram representing the set of all reducible diagrams associated with a particular irreducible diagram is called a *skeleton diagram.* In Fig. 24.5 we have shown the simplest skeleton diagrams representing the strongly connected parts of the propagators $T[z_1 z_2 z_3]$, $T[x, y\, z_1 z_2]$, and $T[x_1 x_2, y_2 y_1]$. Just as in the case of ordinary diagrams, an analytical expression built out of the functions $G[x, y]$, $\mathcal{G}[z_1 z_2]$, and $\Gamma[x, y, z]$ corresponds to each skeleton diagram. Since skeleton diagrams with n vertices do not make any contribution to the propagator under consideration until the n-th order of perturbation theory, in each particular order of perturbation theory we have a finite number of skeleton diagrams.

Accordingly, in any order of perturbation theory the equations for functions of the first kind can be complemented to a closed set of equations by the addition of formulae for the functions $\Delta[z_1 z_2 z_3]$, $\Gamma[x, y, z_1 z_2]$ and $E[x_1 x_2, y_2 y_1]$, obtained with the aid of skeleton diagrams. Below we give these formulae in the lowest order of perturbation theory:

$$\Delta[z_1 z_2 z_3] = -e^3 \Gamma[\xi_1, \xi_2, z_1] G[\xi_2, \xi_3] \Gamma[\xi_3, \xi_4, z_2] G[\xi_4, \xi_5] \times \Gamma[\xi_5, \xi_6, z_3] G[\xi_6, \xi_1] + (z_1 \leftrightarrow z_2) + \ldots, \tag{60a}$$

$$\begin{aligned} e^2 \Gamma[x, y, z_1 z_2] = {} & e^4 \Gamma[x, \xi_1, \zeta_1] G[\xi_1, \xi_2] \Gamma[\xi_2, \xi_3, z_1] G[\xi_3, \xi_4] \\ & \times \Gamma[\xi_4, \xi_5, z_2] G[\xi_5, \xi_6] \Gamma[\xi_6, y, \zeta_2] \mathcal{G}[\zeta_1 \zeta_2] + (z_1 \leftrightarrow z_2) \\ & - e^4 \Gamma[x, y, \zeta_1] \mathcal{G}[\zeta_1 \zeta_2] \Gamma[\xi_1, \xi_2, \zeta_2] G[\xi_2, \xi_3] \Gamma[\xi_3, \xi_4, z_1] \\ & \times G[\xi_4, \xi_5] \Gamma[\xi_5, \xi_6, z_2] G[\xi_6, \xi_1] - (z_1 \leftrightarrow z_2) + \ldots, \end{aligned} \tag{60b}$$

$$\begin{aligned} E[x_1 x_2, y_2 y_1] = {} & e^4 \Gamma[x_1, \xi_1, \zeta_1] G[\xi_1, \xi_2] \Gamma[\xi_2, y_2, \zeta_2] (\mathcal{G}[\zeta_1 \zeta_3] \\ & \times \mathcal{G}[\zeta_2 \zeta_4] + \mathcal{G}[\zeta_1 \zeta_4] \mathcal{G}[\zeta_2 \zeta_3]) \Gamma[x_2, \eta_1, \zeta_3] G[\eta_1, \eta_2] \Gamma[\eta_2, y_2, \zeta_4] \\ & - (y_1 \leftrightarrow y_2) + \ldots . \end{aligned} \tag{60c}$$

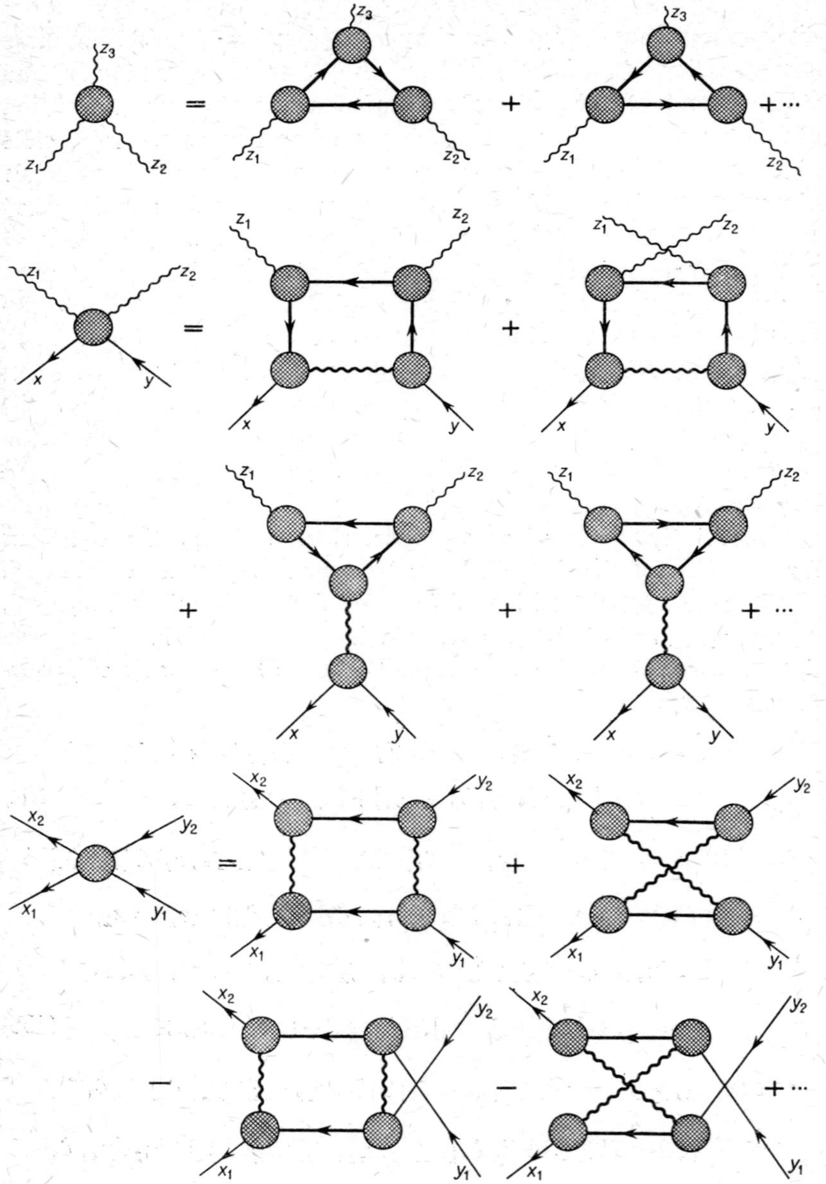

FIG. 24.5. Skeleton diagrams representing the strongly connected parts of the propagators $T[z_1 z_2 z_3]$, $T[x, y, z_1 z_2]$, and $T[x_1 x_2, y_2 y_1]$.

An arbitrary propagator of the second kind can thus be written, with arbitrary accuracy, as a sum

$$T[x_1 \dots x_n, y_n \dots y_1, z_1 \dots z_k]$$
$$= \sum_i \mathscr{S}_i[x_1 \dots x_n, y_n \dots y_1, z_1 \dots z_k | G, \mathscr{G}, \Gamma, e]. \tag{61}$$

Here, the symbol $\mathscr{S}_i[G, \mathscr{G}, \Gamma, e]$ denotes an expression, corresponding to a particular skeleton diagram, built out of functions of the first kind and the parameter e, whereas the summation is taken over all skeleton diagrams.

Renormalization

The formulae for the transition amplitudes contain propagators multiplied by renormalization constants (in second-order perturbation theory, exactly such products were expressed by means of convergent integrals). In view of this, in the equations derived in this section we shall eliminate all functions which have appeared hitherto, replacing them by so-called *renormalized functions* defined by the formulae:

$$T_{\text{ren}}[x_1 \dots x_n, y_n \dots y_1, z_1 \dots z_k] =$$
$$= Z_2^{-n} Z_3^{-k/2} T[x_1 \dots x_n, y_n \dots y_1, z_1 \dots z_k], \tag{62a}$$

$$\mathfrak{A}_{\text{ren}}(z) = Z_3^{-1/2} \mathfrak{A}(z), \tag{62b}$$

$$\mathscr{I}_{\text{ren}}(z) = Z_3^{1/2} \mathscr{I}(z), \tag{62c}$$

$$\Gamma_{\text{ren}}[x, y, z] = Z_2 \Gamma[x, y, z], \tag{62d}$$

$$\mathscr{K}_{\text{ren}}[x, y, z_1 z_2] = Z_2 \mathscr{K}[x, y, z_1 z_2], \tag{62e}$$

$$\mathscr{R}_{\text{ren}}[x, y, z_1 z_2] = Z_2^{-1} Z_3^{-1} \mathscr{R}[x, y, z_1 z_2], \tag{62f}$$

$$K_{\text{ren}}[x_1 x_2, y_2 y_1] = Z_2^2 Z_3^{-1} K[x_1 x_2, y_2 y_1], \tag{62g}$$

$$R_{\text{ren}}[x_1 x_2, y_2 y_1] = Z_2^{-2} R[x_1 x_2, y_2 y_1]. \tag{62h}$$

The parameter e will be expressed as a function of the electron charge e_{obs} by the formula

$$e_{\text{obs}}^2 = Z_3 e^2. \tag{62i}$$

The renormalized field has been introduced in order to replace the product $e\mathfrak{A}$ by an expression which does not contain either the parameter e or the renormalization constants, $e\mathfrak{A} = e_{\text{obs}} \mathfrak{A}_{\text{ren}}$. The renormalized current is given by formula (62c) since the current is proportional to the charge which is to be renormalized in accordance with formula (62i). All renormalized functions will be regarded as functions of the parameters e_{obs} and m_{obs}, as well as the functionals of the renormalized field $\mathfrak{A}_{\text{ren}}$ and the renormalized current $\mathscr{I}_{\text{ren}}$. The equations for the renormalized functions lead to the following complete set of integral equations for the renormalized functions:†

$$G_{\text{ren}}[x, y] = Z_2^{-1}K_{\text{F}}[x, y] + Z_2^{-1}K_{\text{F}}[x, \xi]\Sigma_{\text{ren}}[\xi, \eta]G_{\text{ren}}[\eta, y], \tag{63a}$$

$$\begin{aligned}\mathscr{G}_{\text{ren}}[z_1 z_2] &= Z_3^{-1}D^{\text{F}}(z_1 - z_2)\\ &+ Z_3^{-1}D^{\text{F}}(z_1 - \zeta_1)\Pi_{\text{ren}}[\zeta_1\zeta_2]\mathscr{G}_{\text{ren}}[\zeta_2 z_2],\end{aligned} \tag{63b}$$

$$\begin{aligned}\Sigma_{\text{ren}}[x, y] &= -ie_{\text{obs}}^2\Gamma_{\text{ren}}[x, \xi, \zeta_1]\mathscr{R}_{\text{ren}}[\xi, \eta, \zeta_1\zeta_2]\Gamma_{\text{ren}}[\eta, y, \zeta_2]\\ &+ Z_2\,\delta m\,\delta(x-y),\end{aligned} \tag{63c}$$

$$\Pi_{\text{ren}}[z_1 z_2] = ie_{\text{obs}}^2\Gamma_{\text{ren}}[\xi_1, \xi_2, z_1]R_{\text{ren}}[\xi_2\eta_2, \eta_1\xi_1]\Gamma_{\text{ren}}[\eta_1, \eta_2, z_2], \tag{63d}$$

$$\begin{aligned}\Gamma_{\text{ren}}[x, y, z] &= Z_2\gamma(x, y, z)\\ &- e_{\text{obs}}^2\Gamma_{\text{ren}}[x, \xi, \zeta_1]\mathscr{R}_{\text{ren}}[\xi, \eta, \zeta_1\zeta_2]\mathscr{K}_{\text{ren}}[\eta, y, \zeta_2 z],\end{aligned} \tag{63e}$$

$$\begin{aligned}\Gamma_{\text{ren}}[x, y, z] &= Z_2\gamma(x, y, z)\\ &- e_{\text{obs}}^2\Gamma_{\text{ren}}[\xi_1, \eta_1, z]R_{\text{ren}}[\eta_1\eta_2, \xi_2\xi_1]K_{\text{ren}}[\xi_2 x, y\eta_2],\end{aligned} \tag{63f}$$

$$\begin{aligned}\mathscr{R}_{\text{ren}}[x, y, z_1 z_2] &= G_{\text{ren}}[x, y]\mathscr{G}_{\text{ren}}[z_1 z_2]\\ &+ e_{\text{obs}}^2\mathscr{R}_{\text{ren}}[x, \xi, z_1\zeta_1]\mathscr{K}_{\text{ren}}[\xi, \eta, \zeta_1\zeta_2]G_{\text{ren}}[\eta, y]\mathscr{G}_{\text{ren}}[\zeta_2 z_1],\end{aligned} \tag{63g}$$

$$\begin{aligned}R_{\text{ren}}[x_1 x_2, y_2 y_1] &= G_{\text{ren}}[x_1, y_2]G_{\text{ren}}[x_2, y_1]\\ &+ e_{\text{obs}}^2 R_{\text{ren}}[x_1\xi_1, \eta_1 y_1]K_{\text{ren}}[\eta_1\eta_2, \xi_2\xi_1]G_{\text{ren}}[\xi_2, y_2]G_{\text{ren}}[x_2, \eta_2],\end{aligned} \tag{63h}$$

$$\begin{aligned}&T_{\text{ren}}[x_1 \ldots x_n, y_n \ldots y_1, z_1 \ldots z_k]\\ &= \sum_i \mathscr{S}_i[x_1 \ldots x_n, y_n \ldots y_1, z_1 \ldots z_k|G_{\text{ren}}, \mathscr{G}_{\text{ren}}, \Gamma_{\text{ren}}, e_{\text{obs}}].\end{aligned} \tag{63i}$$

To verify that eq. (63a) is equivalent to eq. (36), we can compare the expressions for G^{-1} obtained from the two equations. The photon mass μ will not be renormalized since in the final formulae we shall always set $\mu = 0$.

† The renormalized versions of eqs. (23) and (24) for the vacuum-to-vacuum amplitude will not be examined here because in applications, as a rule, use is made of the theory either without external fields and currents, or with a static field. The vacuum-to-vacuum amplitude is then an unobservable phase factor.

We adopt the convention that the connected, truncated, and strongly connected parts of renormalized propagators are defined just like the analogous parts of unrenormalized propagators, the difference being that all functions occurring in the defining formulae are replaced by renormalized functions. With this convention, we obtain the relations:

$$\Delta_{\text{ren}}[z_1 z_2 z_3] = Z_3^{3/2} \Delta[z_1 z_2 z_3], \tag{64a}$$

$$\Gamma_{\text{ren}}[x, y, z_1 z_2] = Z_2 \Gamma[x, y, z_1 z_2], \tag{64b}$$

$$E_{\text{ren}}[x_1 x_2, y_2 y_1] = Z_2^2 E[x_1 x_2, y_2 y_1]. \tag{64c}$$

Formulae for the renormalized kernels $\mathscr{K}_{\text{ren}}$ and K_{ren} follow from this:

$$\begin{aligned} e_{\text{obs}}^2 \mathscr{K}_{\text{ren}}[x, y, z_1 z_2] = {} & i e_{\text{obs}}^2 \Gamma_{\text{ren}}[x, \xi, z_2] G_{\text{ren}}[\xi, \eta] \Gamma_{\text{ren}}[\eta, y, z_1] \\ & - e_{\text{obs}} \Delta_{\text{ren}}[z_1 z_2 \zeta_1] \mathscr{G}_{\text{ren}}[\zeta_1 \zeta_2] \Gamma_{\text{ren}}[x, y, \zeta_2] \\ & + e_{\text{obs}}^2 \Gamma_{\text{ren}}[x, y, z_1 z_2], \end{aligned} \tag{65a}$$

$$\begin{aligned} & e_{\text{obs}}^2 K_{\text{ren}}[x_1 x_2, y_2 y_1] \\ & = i e_{\text{obs}}^2 \Gamma_{\text{ren}}[x_1, y_2, \zeta_1] \mathscr{G}_{\text{ren}}[\zeta_1 \zeta_2] \Gamma_{\text{ren}}[x_2, y_1, \zeta_2] \\ & \quad + E_{\text{ren}}[x_1 x_2, y_2 y_1]. \end{aligned} \tag{65b}$$

The set of eqs. (63) and (65) may be divided into two groups. The first group will comprise eqs. (63a)–(63f) for functions of the first kind; we shall call these *inhomogeneous equations.* The remaining equations will be classified in the second group and will be called *homogeneous equations.* The inhomogeneous equations contain the free functions K_F, D^F, and γ, the renormalization constants† Z_2, Z_3 and the mass correction δm. This is a characteristic feature which distinguishes inhomogeneous equations from the homogeneous equations constructed entirely from renormalized functions. Investigation of the relevant formulae immediately reveals that only renormalized functions appear in the homogeneous equations (63g), (63h), and (65). The fact that also in eq. (63i) only renormalized functions appear is substantiated by the following reasoning. The numbers of electron lines N_e, photon lines N_p, and vertices N_v in a skeleton diagram are related to the

† The renormalization constant Z_1 does not figure in the renormalization theory presented here. The symbol Z_1 is at times used to denote the renormalization constant of the vertex function. This constant is, however, equal to the constant Z_2 by the Ward identity.

numbers of coordinates n and k of an arbitrary propagator by the formulae

$$n = N_e - N_v, \tag{66a}$$

$$k = 2N_p - N_v. \tag{66b}$$

It thus follows that owing to the replacement of functions of the first kind and the parameter e by suitable renormalized quantities, the expressions corresponding to an arbitrary skeleton diagram of the propagator $T[x_1 \dots x_n, y_n \dots y_1, z_1 \dots z_k]$ are subject to multiplication by the appropriate factor $Z_2^{-n} Z_3^{-k/2}$.

The equations derived in this section for the renormalized functions may be solved by using perturbation theory with respect to the parameter e_{obs}. This is what we shall do in the next section.

25. RENORMALIZED PERTURBATION THEORY

Renormalized perturbation theory is a method by means of which we obtain solutions of equations for renormalized propagators (24.63) and (24.65) in the form of infinite power series in the parameter e_{obs}, or, to be more precise, in the dimensionless parameter α called the *fine-structure constant*: $\alpha = e_{\text{obs}}^2/4\pi\hbar c \simeq 1/137$. The successive terms in these series are well-defined mathematical expressions which contain no infinite integrals. Since the parameter α is much smaller than unity, one may hope that the first few terms in the expansion will constitute a good approximation.† This is fully confirmed by analysis of experimental data and comparison of such data with the results of calculations made by this method. We shall take up this point in the next chapter.

THE GENERAL PROPERTIES OF RENORMALIZED PERTURBATION THEORY

In order to obtain a perturbative solution of eqs. (24.63)–(24.65), all of the functions figuring in them and the constants Z_2, Z_3, and δm must be expanded in power series of e_{obs}^2. Odd powers of the charge e_{obs}

† The expansion in the parameter e did not possess this feature. From the property (21.35) of the constant Z_3 it follows that the parameter e is greater than e_{obs}. Moreover, it turns out that the constant Z_3^{-1} calculated by perturbation theory is infinite.

do not occur in these expansions if the whole product $e_{\text{obs}}\,\mathfrak{A}_{\text{ren}}$ is regarded as a function given beforehand and is not expanded with respect to the parameter e_{obs} appearing in this product.

The structure of the equations for renormalized propagators is such that if all the functions figuring in these equations are known in the $2n$-th order of perturbation theory with respect to e_{obs}, they can be determined in the $2(n+1)$-th order. The constants Z_2 and Z_3 are found in each order from the condition that the Fourier transforms of the renormalized electron and photon propagators have the residue equal to one at the pole in the absence of an external field. The mass correction δm is obtained from the assumption that the pole of the Fourier transform of the electron propagator lies at the point $\gamma \cdot p = m_{\text{obs}}$. To avoid having divergent integrals when determining Z_2, Z_3, and δm, one must introduce regularization at the intermediate stages of the calculation. Alternatively, one may determine functions of the first kind from functions of the second kind with the use of Ward identities (24.27).

The perturbation expansion of renormalized propagators with respect to e_{obs} lends itself to representation by means of Feynman diagrams. The equations for renormalized propagators are very similar in structure to those for unrenormalized propagators. The differences are that instead of the free electron and photon propagators, K_{F} and D^{F}, and the matrices γ, we have the same quantities multiplied by renormalization constants. The observable charge e_{obs} appears in place of the charge e. The equation for the electron propagator in renormalized theory differs from the initial equation by an additional term containing δm. Accordingly, additional terms describing the mass renormalization occur in perturbation theory. To illustrate this, let us write out—in abbreviated notation—three terms of the iterative solution of the equation for G_{ren}:

$$G_{\text{ren}} = Z_2^{-1}K_{\text{F}} + Z_2^{-1}K_{\text{F}}\Sigma_{\text{ren}}Z_2^{-1}K_{\text{F}} + Z_2^{-1}K_{\text{F}}\Sigma_{\text{ren}}Z_2^{-1}K_{\text{F}}\Sigma_{\text{ren}}Z_2^{-1}K_{\text{F}} + \ldots$$

In addition to terms which are analogous in structure to terms figuring in Σ, the self-energy function Σ_{ren} also contains the term $Z_2\,\delta m\,\delta(x-y)$. This term will be indicated on diagrams with a cross labelled with δm, on the electron line. The set of diagrams representing a renormalized electron propagator will then include additional diagrams describing the mass renormalization. All diagrams in renormalized perturbation theory are obtained by putting a cross with the label δm in every possible manner on the electron lines in diagrams corresponding to unrenormalized perturbation theory. On adding these extra diagrams, we can use the Feynman diagram technique in renormalized perturbation theory but we must make the following changes in the analytical expression corresponding to the diagrams:

$$K_F[x, y; m, e|\mathfrak{A}] \to Z_2^{-1} K_F[x, y; m_{obs}, e_{obs}|\mathfrak{A}_{ren}],$$

$$D^F_{\mu\nu}(z_1 - z_2) \to Z_3^{-1} D^F_{\mu\nu}(z_1 - z_2),$$

$$\gamma^\mu \to Z_2 \gamma^\mu,$$

$$e \to e_{obs},$$

$$\mathscr{I} \to \mathscr{I}_{ren}.$$

Analytical expression	Graphical element
$-iZ_2^{-1} S_F(x-y; m_{obs})$	
$-iZ_3^{-1} D^F_{\mu\nu}(z_1 - z_2)$	
$-ie_{obs} Z_2 \gamma^\mu$	
$-iZ_2 \delta m \delta(x-y)$	

The table lists all the analytical expressions corresponding to the basic elements of diagrams in renormalized theory without external fields and currents. The reader is advised to verify for himself that second-order renormalized perturbation theory yields renormalized electron and photon propagators in agreement with the expressions obtained in Section 23.

In perturbation theory the renormalization constants Z_2 and Z_3 as well as δm are in the form of power series in e_{obs}. Consequently, when the expression corresponding to a particular diagram is expanded into a series in e_{obs}, the resulting series contains arbitrarily high powers of that parameter. This question will not be dealt with in detail since, as will be demonstrated at the end of this section, the renormalization constants of electron and photon propagators may be eliminated completely from the formulae for the probabilities of all physical processes.

The fundamental property of renormalized perturbation theory is that the expressions for all renormalized propagators in every order are free of divergent integrals. On the other hand, since the S operator is expressible in terms of renormalized propagators, this method makes it possible to obtain results which lend themselves to comparison with experimental results. The proof that finite expressions are always obtained for propagators in all orders when renormalized perturbation theory is used is very complicated.# We shall not cite it here but shall confine ourselves merely to discussing the essential steps in it.

The proof is usually carried out with the simplifying assumption that no external field and external current exist. The first step in the proof consists in showing that the assumption of renormalized functions of the first kind being finite in every order of perturbation theory implies that the renormalized functions of the second kind built by the skeleton diagram method are also finite. For this purpose, the integrals representing these functions in the momentum representation are analysed for convergence. An essential role in this analysis is played by the Weinberg theorem from which it follows that in every order of perturbation theory the functions $\tilde{G}_{\text{ren}}(p)$, $\tilde{\mathscr{G}}_{\text{ren}}(k)$, and $\tilde{\Gamma}_{\text{ren}}(p, q)$ behave for large values of momenta just like the respective free functions, up to logarithmic factors. The second step in the proof consists in demonstrating that in every order of renormalized perturbation theory, a choice of constants Z_2, Z_3, and δm such that it guarantees the proper position of the poles and values of the propagator residues also ensures a finite value of the functions of the first kind.

Proof (incomplete) that renormalized perturbation theory yields finite expressions for the transition amplitudes in all orders was first given by F. J. Dyson (1949). Two supplemented versions of this proof were provided by J. C. Ward (1950) and A. Salam (1951 a, b). The detailed proof of the renormalizability of the theory (cf. the book by Bjorken and Drell) draws on a complicated theorem given by S. Weinberg (1960) about the asymptotic behaviour of the Fourier transforms of propagators. A simplified proof of this theorem was recently given by W. Zimmermann (1970).

The proof under discussion, however, has a gap since the aforementioned Weinberg theorem is applicable to integrals appearing in renormalization theory only after the path of integration with respect to each energy variable has been changed from the real axis to the imaginary axis. Justification for this change in the general case has never been given. Hepp (1966) obviated this difficulty and for the first time proved rigorously that in each order of perturbation theory renormalized propagators are well-defined distributions. Hepp's proof uses the renormalization method given in the book by Bogolyubov and Shirkov.

Renormalized Transition Probabilities

Let us now proceed to derive formulae for the transition probabilities for arbitrary processes in which electrons and photons participate.# For simplification, we shall concern ourselves only with the case with no external fields and currents.

On the basis of the considerations of Section 22, the following formula may be written for the transition amplitude A for any process as a function of the electron and photon momenta:

$$A(p; q; k) = (2\pi)^4 \delta^{(4)}(\Sigma p_i + \Sigma q_j + \Sigma k_l) M(p; q; k), \tag{1}$$

where

$$\begin{aligned} M(p; q; k) \equiv\ & i^{2n}(-i)^m Z_2^{-n} Z_3^{-m/2} \bar{u}(\mathbf{p}_1) \dots \bar{u}(\mathbf{p}_n) \varepsilon^{\mu_1}(\mathbf{k}_1) \dots \varepsilon^{\mu_m}(\mathbf{k}_m) \\ & \times [(m-\hat{p}_1) \dots (m-\hat{p}_n)(\mu^2-k_1^2) \dots (\mu^2-k_m^2) T^C_{\mu_1 \dots \mu_m}(p_1 \dots p_n, \\ & q_n \dots q_1, k_1 \dots k_m)(m-\hat{q}_n) \dots (m-\hat{q}_1)] u(\mathbf{q}_1) \dots u(\mathbf{q}_n). \end{aligned} \tag{2}$$

To simplify the formulae in comparison with formula (22.44) we have dropped the bispinor indices and made the assumption that positons do not participate in the process and that all photons occur in the final state. Our considerations may be generalized to all other cases by virtue of formula (22.44).

Note that only the connected part of the propagator T is involved in formula (2). The unconnected parts describe several processes which occur independently. Each of these processes may be described sepa-

This subsection is based on a paper by I. Białynicki-Birula (1970), to which the reader interested in the details omitted here is referred.

rately by using the connected part of the corresponding propagator. The connected part of the propagator is further expressed in terms of the truncated part by the formula

$$\tilde{T}^{C}_{\mu_1\ldots\mu_m}(p;q;k) = (-i)^{2n+m}\tilde{\mathscr{G}}_{\mu_1\nu_1}(k_1)\ldots\tilde{\mathscr{G}}_{\mu_m\nu_m}(k_m)\tilde{G}(p_1)\ldots\tilde{G}(p_n)$$
$$\times\, T_T^{\nu_1\ldots\nu_m}(p;q;k)\tilde{G}(q_n)\ldots\tilde{G}(q_1), \tag{3}$$

which we obtain from formula (22.29) by calculating the Fourier transforms by sides.

The electron and photon propagators have poles at the points $p^2 = m^2_{\text{obs}}$ and $k^2 = \mu^2$, respectively. The behaviour of these propagators (these are unrenormalized quantities) in the neighbourhood of these poles is given by the formulae:

$$\tilde{G}(p) = \frac{Z_2}{m_{\text{obs}}-\hat{p}-i\varepsilon} + (\text{finite terms when } p^2 = m^2_{\text{obs}}), \tag{4a}$$

$$\tilde{\mathscr{G}}_{\mu\nu}(k) = \left(-g_{\mu\nu}+\frac{k_\mu k_\nu}{k^2}\right)\frac{Z_3}{\mu^2-k^2-i\varepsilon} + (\text{finite terms when } k^2 = \mu^2). \tag{4b}$$

In order to get finite expressions for the transition amplitudes the parameter m figuring in the expressions $(m-\hat{p}_i)$ and $(m-\hat{q}_j)$ in formula (2) should be identified with the observable, true mass of the electron. The factors $(m_{\text{obs}}-\hat{p}_i)$ and $(m_{\text{obs}}-\hat{q}_j)$ cancel the poles in the electron propagators. A similar role was played by the factors $(m-\hat{p})$ and $(m-\hat{q})$ in formula (17.70) for the transition amplitude in the theory of mutually non-interacting electrons.†

On identifying the parameter m in formula (2) with the observable mass and on utilizing the representation of the propagator T^C in terms of the truncated propagator T_T, we arrive at the following formula for M:

$$M(p;q;k) = Z_2^n Z_3^{m/2}\bar{u}(\mathbf{p}_1)\ldots\bar{u}(\mathbf{p}_n)\varepsilon_{\mu_1}(\mathbf{k}_1)\ldots\varepsilon_{\mu_m}(\mathbf{k}_m)$$
$$\times\, \tilde{T}_T^{\mu_1\ldots\mu_m}(p;q;k)u(\mathbf{q}_n)\ldots u(\mathbf{q}_1). \tag{5}$$

We have made use of formulae (23.12) and (23.16b) describing the behaviour of the propagators of an electron and a photon on the mass

† Cf. the remarks following formula (17.70).

shell and the following corollary from the functional Ward identity:

$$k_{\mu_l}^{(l)} \tilde{T}_T^{\mu_1\dots\mu_l\dots\mu_m}(p_1 \dots p_n, q_n \dots q_1, k_1 \dots k_l \dots k_m)\Big|_{p_i^2=m^2=q_i^2} = 0. \tag{6}$$

The transition probability (more precisely, the density of transition probability in momentum space) $P(p; q; k)$ is defined by the formula

$$P(p; q; k) = (2\pi)^4 \delta^{(4)}\left(\sum p_i - \sum q_j + \sum k_l\right) |M|^2. \tag{7}$$

In the next chapter we shall give a detailed justification of this formula for the case of two-particle collisions. For the moment, we shall merely note that expression (7) may be regarded as the squared modulus of the transition amplitude per unit volume of space-time:

$$P(p; q; k) = \lim_{\substack{V\to\infty \\ T\to\infty}} \frac{|A(p; q; k)|^2}{VT},$$

since in a box of space-time of volume VT the following equinvalence holds:

$$(2\pi)^4\delta\left(\sum p\right)(2\pi)^4\delta\left(\sum p\right) = (2\pi)^4\delta\left(\sum p\right)\int_{TV} d^4x e^{-i\Sigma p\cdot x} = VT(2\pi)^4\delta\left(\sum p\right).$$

The squared modulus of the amplitude M may in turn be expressed in terms of the density matrices for electrons and photons, $\Lambda(p)$ and $\varrho_{\mu\nu}(k)$, in the initial and final states:

$$\Lambda(p, s) \equiv u(\mathbf{p}, s)\bar{u}(\mathbf{p}, s), \tag{8a}$$

$$\varrho_{\mu\nu}(k, r) \equiv \varepsilon_\mu(\mathbf{k}, r)\varepsilon_\nu^*(\mathbf{k}, r). \tag{8b}$$

In the simplest case, when polarization of electrons and photons is not observed, the density matrices are given by the formulae

$$\Lambda(p) = m + \hat{p}, \tag{9a}$$

$$\varrho_{\mu\nu}(k) = -g_{\mu\nu} + \mu^{-2}k_\mu k_\nu, \tag{9b}$$

in which summation has been performed over the spin states. We shall confine ourselves to this simple case since this simplifies the formulae substantially, whereas generalization entails no difficulties.

The transition probability for unpolarized particles takes on the form:

$$P(p;q;k) = (2\pi)^4\delta^{(4)}\Big(\sum p_i - \sum q_j + \sum k_l\Big) Z_2^{2n} Z_3^m$$
$$\times(-g_{\mu_1\nu_1}+\mu^{-2}k_{\mu_1}k_{\nu_1}) \dots (-g_{\mu_m\nu_m}+\mu^{-2}k_{\mu_m}k_{\nu_m})$$
$$\times \mathrm{Tr}\{(m+\hat{p}_1)\dots(m+\hat{p}_n)\tilde{T}_T^{\mu_1\dots\mu_m}(p;q;k)$$
$$\times (m+\hat{q}_n)\dots(m+\hat{q}_1)\bar{\tilde{T}}_T^{\nu_1\dots\nu_m}(p;q;k)\}. \tag{10}$$

A bar indicates bispinor conjugation of the propagator, i.e. Hermitian conjugation and multiplication by the matrix γ^0 on both sides.

Once again we shall use the formulae describing the behaviour of the propagators in the vicinity of the pole so as to obtain the final formula for the transition probability,

$$P(p;q;k) = (2\pi)^4\delta^{(4)}\Big(\sum p_i - \sum q_j + \sum k_l\Big)\prod_i (m^2-p_i^2)$$
$$\times \prod_j (m^2-q_j^2)\prod_l (\mu^2-k_l^2)\tilde{\mathscr{G}}_{\mu_1\nu_1}(k_1)\dots\tilde{\mathscr{G}}_{\mu_m\nu_m}(k_m) \tag{11}$$
$$\times \mathrm{Tr}\{\tilde{G}(p_1)\dots\tilde{G}(p_n)\tilde{T}_T^{\mu_1\dots\mu_m}(p;q;k)\tilde{G}(q_n)\dots\tilde{G}(q_1)\bar{\tilde{T}}_T^{\nu_1\dots\nu_m}(p;q;k)\}.$$

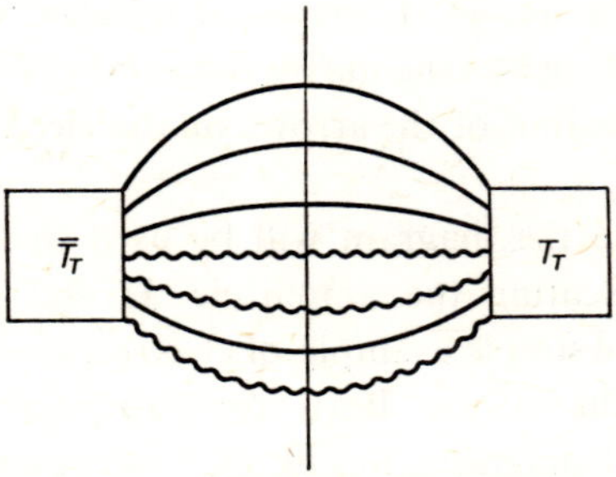

FIG. 25.1. A two-sided Feynman diagram.

The general structure of the expression obtained is illustrated symbolically in Fig. 25.1. The renormalization constants Z_2 and Z_3 do not figure in this formula at all, even though all the propagators in terms of which the transition probability is expressed are unrenormalized. In any event, this formula is invariant under renormalization of propagators,

$$\tilde{G}(p) \to Z_2 \tilde{G}(p),$$

$$\tilde{\mathscr{G}}(k) \to Z_3 \tilde{\mathscr{G}}(k),$$

$$\tilde{T}_T(p; q; k) \to Z_2^{-n} Z_3^{-m/2} \tilde{T}_T(p; q; k).$$

In both cases, whenever unrenormalized or renormalized propagators are inserted when we use formula (11), they should be expressed in terms of the observable parameters m_{obs} and e_{obs}. Consequently, the mass renormalization δm and the charge renormalization δe will appear in the perturbation expansions

$$\delta m = m_{\text{obs}} - m,$$

$$\delta e = e_{\text{obs}} - e.$$

The Feynman diagram technique, which we introduced for propagators and transition amplitudes, can be extended in a natural manner so that it is applicable directly to the calculation of transition probabilities.

For this purpose, we shall introduce *two-sided diagrams* representing products of the amplitudes M and M^*. Diagrams representing the amplitude M will be drawn in the usual fashion, whereas those representing M^* will be drawn as the mirror reflections of the relevant diagrams for M, with the direction of the arrows on the electron lines additionally changed.

A vertical line in the diagram will be used to separate the parts of the diagram representing the amplitude M and the amplitude M^*.

We hope that the simple example of electron scattering in an external electromagnetic field, in the Born approximation, as represented by means of two-sided diagrams in Fig. 25.2 fully explains the above generalization of the diagram technique.

In Fig. 25.3 we give in addition two-sided diagrams representing negaton-negaton scattering in the lowest order of perturbation theory.

Self-energy diagrams representing higher corrections to electron or photon propagators, G or $\mathscr{G}$, can be drawn on either side of the dividing

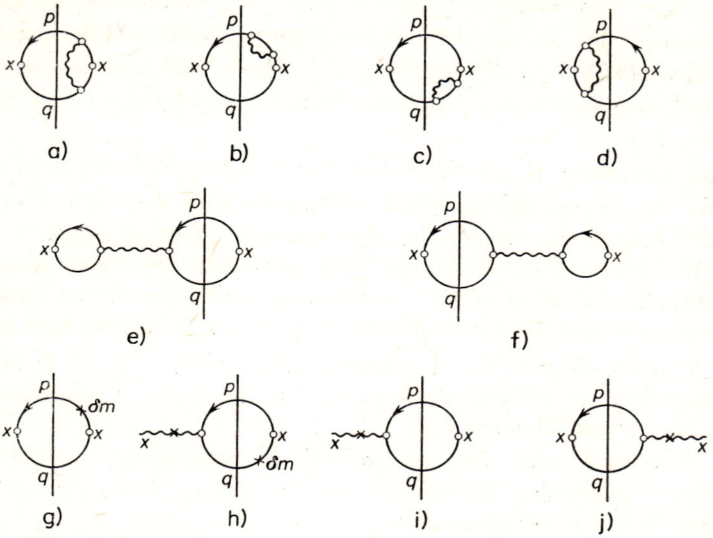

FIG. 25.2. Two-sided diagrams representing electron scattering in an external electromagnetic field in the Born approximation.

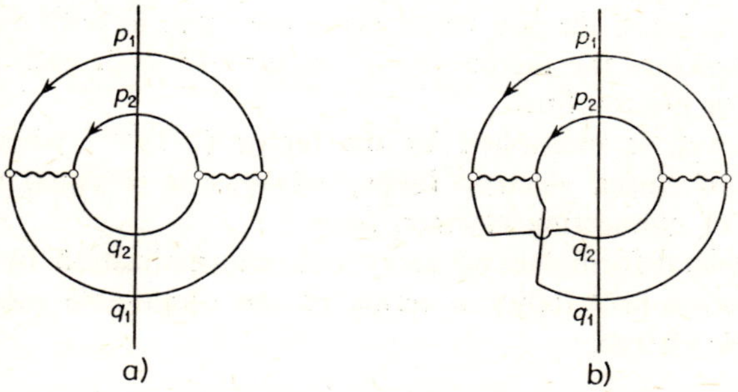

FIG. 25.3. Two-sided diagrams representing negaton-negaton scattering in the lowest order of perturbation theory.

line in two-sided diagrams. This follows from the fact that the self-energy function is self-adjoint in the vicinity of the pole,

$$\tilde{\Sigma}(p) = \bar{\tilde{\Sigma}}(p), \quad \tilde{\Pi}_{\mu\nu}(k) = \tilde{\Pi}^*_{\mu\nu}(k).$$

The reader may verify that this property is possessed by self-energy functions in the second order of perturbation theory.† Thus, instead of diagrams (25.2b) and (25.2c), we could equally well use two diagrams in which the photon line has been shifted to the left-hand side of the dividing line.

Our discussion of two-sided diagrams will be concluded with a list of rules for associating analytical expressions with diagrams. This will be done only in the momentum representation of propagators.

1. We draw all topologically non-equivalent two-sided diagrams with a fixed number of lines connecting both sides of the diagram, one connecting line for each particle in the initial and the final states.
2. To each electron line we assign a propagator, $-i\tilde{S}_F(p)$ or $i\tilde{\tilde{S}}_F(p)$, depending on whether the line lies on the right- or left-hand side of the vertical dividing line.
3. With each wavy line we associate a propagator, $-i\tilde{D}^F_{\mu\nu}(k)$ or $i\tilde{D}^{F*}_{\mu\nu}(k)$, depending on whether the line lies on the right- or left-hand side of the dividing line.
4. Vertices on the right (left) -hand side are represented by $-ie\gamma^\mu(ie\gamma^\mu)$.
5. Self-energy diagrams on the connecting lines are taken into account only once, on one (arbitrary) side of the dividing line. Either propagators or their adjoints may be associated with connecting lines, the end result being the same in both cases.
6. We associate the factor m^2-p^2 or μ^2-k^2 with each connecting electron or photon line.
7. The result is multiplied by the factor $(-1)^{L+n}$, where L is the number of closed electron loops, whereas n is equal to half the number of connecting electron lines.
8. In the final formulae, we carry out renormalization by expressing the transition probability in terms of the observable mass and the observable charge.

Independence of Transition Probabilities from the Compensating Current

The formulae given above for the transition probability also make it possible to prove that transition probabilities do not depend on the form of the compensating current. The full proof of this property has

† The complete proof can be found in the paper by I. Białynicki-Birula (1970).

been given in the paper cited above. Here we shall confine ourselves merely to formulating a theorem to the effect that the transition probability is independent of the choice of current.

Theorem. The transition probability $P(p; q; k)$ does not change under the following changes in the photon propagator:

$$D^{\mathrm{F}}_{\mu\nu}(x-y) \to D^{\mathrm{F}}_{\mu\nu}(x-y)+i\partial_\mu g_\nu(x-y)+i\partial_\nu g_\mu(x-y). \tag{12}$$

The functions $g_\mu(x-y)$ are restricted only by the condition that after the transformation the photon propagator $D^{\mathrm{F}}_{\mu\nu}$ again have the fundamental properties of a Feynman propagator:

$$D^{\mathrm{F}}_{\mu\nu}(x) = \begin{cases} D^{(+)}_{\mu\nu}(x), & x^0 > 0, \\ -D^{(-)}_{\mu\nu}(x), & x^0 < 0, \end{cases}$$

i.e. involve only positive frequencies for positive times and only negative frequencies for negative times.

It follows from this theorem that in the complete quantum electrodynamics, just as in the theory with external current (cf. formula (13.30)), in calculating transition probabilities one may use free photon propagators in the simplest possible form:

$$D^{\mathrm{F}}_{\mu\nu}(x) = -g_{\mu\nu}D_{\mathrm{F}}(x). \tag{13}$$

The transition amplitudes are completely independent of the compensating current; one may even put $a^\mu = 0$, which leads to the photon propagator in the form (13).

The transition probability may be proved to be independent of the compensating current only if the photon mass is assumed not to be zero. For only then transition probabilities describing processes, in which a finite number of photons take place, are different from zero.

The Infrared Catastrophe

Infrared catastrophe, which makes its appearance in the theory with zero photon mass, is caused by the emission of an infinite number of soft photons by accelerated charged particles.

The mathematical apparatus developed in this section for the direct calculation of transition probabilities is also perfectly suited for eliminating the effects of the infrared catastrophe. The conclusions we reached when discussing the infrared catastrophe in Section 15 are also applicable to the complete quantum electrodynamics. We showed there that for zero photon mass, non-zero transition probabilities are obtained by summing the probabilities of the processes for which the final states differ solely as to the number of soft photons they contain. In the notation used now, such an overall probability may be written symbolically as

$$\sum_m \int P(p; q; k; k'_1 \dots k'_m),$$

in which the summation and integration run over the states of the soft photons. This formula is simplified significantly in perturbation theory since only the emission of several soft photons is possible in the lower orders of the theory. In the lowest non-trivial order of perturbation theory it is possible to have either an elastic process without the emission of soft photons or a process with the emission of a single photon. In the simplest case of electron scattering in an external field, the aforementioned overall probability is given by

$$P(p; q) + \int_{|\mathbf{k}| < k_{\min}} d\Gamma P(p; q; k).$$

The divergent expressions of the type $\ln \mu^2/m^2$ in both terms of the sum cancel each other and in the overall probability we may pass to the limit when $\mu \to 0$, obtaining a finite result.

A detailed discussion of the infrared catastrophe may be found in the paper by D. R. Yennie, S. Frautchi and H. Suura (1961). An analysis of the infrared catastrophe with the aid of the propagator theory has been carried out by T. W. B. Kibble (1968).

CHAPTER 8

APPLICATIONS OF QUANTUM ELECTRODYNAMICS

26. TWO-PARTICLE COLLISIONS

IN THIS section we shall use simple examples to illustrate the previously discussed methods of calculating the probabilities of physical processes. Namely, we shall find a general formula for the cross section in collisions of two particles and then apply it to three processes: negaton-negaton scattering, photon-negaton scattering, and photon-photon scattering. The cross sections for these processes will be calculated in the lowest order of perturbation theory, in the case when no external field and no external current are present. The formulae so obtained are relativistic in form and give a good description of both low- and high-energy processes.# In the non-relativistic approximation these formulae go over into the classical formulae of Rutherford and Thomson, respectively. These formulae are utilized to justify the fundamental relationship between the electron charge e_{obs} and the parameter e ($e_{obs} = Z_3^{1/2} e$).

THE GENERAL FORMULAE FOR TWO-PARTICLE PROCESSES

Let us consider an arbitrary process such that only two particles occur in the initial state whereas the final state may contain any number of particles. The transition amplitude for a process of this kind is written as

$$(\Psi_f^{\text{out}}|\Psi_i^{\text{in}}) = \delta_{if} + (\Psi_f^{\text{in}}|\mathrm{T}\Psi_i^{\text{in}}), \tag{1}$$

A detailed analysis of the results obtained is not carried out here. For a more complete analysis of the processes considered, the reader is referred to the book by Akhiezer and Berestetskii and the book by Jauch and Rohrlich.

where

$$\mathsf{T} \equiv \mathsf{S} - 1. \tag{2}$$

The first term in formula (1) corresponds to the transition of particles from the initial to the final state without interacting, whereas the second is the interaction amplitude. The probability of a transition leading from the state Ψ_i to the state Ψ_f is thus given by the formula

$$P_{fi} = |(\Psi_f^{\text{in}}|\mathsf{T}\Psi_i^{\text{in}})|^2 = (\Psi_i^{\text{in}}|\mathsf{T}^\dagger\Psi_f^{\text{in}})(\Psi_f^{\text{in}}|\mathsf{T}\Psi_i^{\text{in}}). \tag{3}$$

Taking the relation between the transition amplitudes and the elements of the S matrix into account, we write the matrix elements of the T operator as

$$(\Psi_f^{\text{in}}|\mathsf{T}\Psi_i^{\text{in}}) = (2\pi)^4 \sum_{s_1, s_2, r_1 \ldots r_k} \int \mathrm{d}\Gamma_{p_1} \ldots \mathrm{d}\Gamma_{p_k} \mathrm{d}\Gamma_{q_1} \mathrm{d}\Gamma_{q_2}$$

$$\times f_{i_1}^*(\mathbf{p}_1, r_1) \ldots f_{i_k}^*(\mathbf{p}_k, r_k)\, \delta^{(4)}\Big(\sum_{i=1}^{k} p_i - q_1 - q_2\Big)$$

$$\times M(\mathbf{p}_1 r_1, \ldots, \mathbf{p}_k r_k; \mathbf{q}_2 s_2, \mathbf{q}_1 s_1) f_{i_1}(\mathbf{q}_2 s_2) f_{i_2}(\mathbf{q}_1, s_1), \tag{4}$$

in which, for the time being, we do not specify what kind of particles are involved in the process. In doing this, we have made use of the fact that the Fourier transform of the connected part of the propagator in the absence of an external field is proportional to the function $\delta^{(4)}$, whose argument is the difference of momenta (cf. formula (25.1)).

With an eye to comparing the predictions of theory with the results of scattering experiments, we are interested in the amplitudes of the transitions from initial states localized in momentum space as precisely as possible. Accordingly, we shall assume that the profiles $f_{i_1}(\mathbf{q}_1, s_1)$ and $f_{i_2}(\mathbf{q}_2, s_2)$ are practically equal to zero for values of the momenta $\mathbf{q}_1$ and $\mathbf{q}_2$ which differ significantly from the average momenta $\mathbf{q}_1^0$ and $\mathbf{q}_2^0$ of the first and second particles. It will furthermore be assumed that $f_{i_1}(\mathbf{q}_1, s_1) = \delta_{s_1 s_1^0} f_{i_1}(\mathbf{q}_1)$ and $f_{i_2}(\mathbf{q}_2, s_2) = \delta_{s_2 s_2^0} f_{i_2}(\mathbf{q}_2)$. If the function M is assumed to be continuous in the variables $\mathbf{q}_2$ and $\mathbf{q}_1$, then for sufficiently narrow profiles f_{i_1} and f_{i_2} we can replace $\mathbf{q}_1$ and $\mathbf{q}_2$ in the function $M(\mathbf{p}_1 r_1 \ldots; \mathbf{q}_2 s_2, \mathbf{q}_1 s_1)$ by the average momenta $\mathbf{q}_1^0$ and $\mathbf{q}_2^0$ and s_1 and s_2 by s_1^0 and s_2^0. In this way, we arrive at the following formula for the probability of an interaction with the transition to an arbitrary state of k final particles:

$$\sum_{\substack{\text{final states of}\\ k \text{ particles}}} P_{f_i} = \sum_{r_1 \dots r_k} \int d\Gamma_{p_1} \dots d\Gamma_{p_k}$$

$$\times |M(\mathbf{p}_1 r_1, \dots, \mathbf{p}_k r_k; \mathbf{q}_2^0 s_2^0, \mathbf{q}_1^0 s_1^0)|^2 \int d\Gamma_{q_1} d\Gamma_{q_2} d\Gamma_{q_1'} d\Gamma_{q_2'}$$

$$\times (2\pi)^4 \delta^{(4)} \Big(\sum_{i=1}^{k} p_i - q_1 - q_2\Big) (2\pi)^4 \delta^{(4)} \Big(\sum_{i=1}^{k} p_i - q_1' - q_2'\Big)$$

$$\times f_{i_1}(\mathbf{q}_1, s_1^0) f_{i_2}(\mathbf{q}_2, s_2^0) f_{i_1}^*(\mathbf{q}_1', s_1^0) f_{i_2}^*(\mathbf{q}_2', s_2^0). \tag{5}$$

This expression may be simplified with the help of the following lemma.

Lemma. The function $Q(p)$,

$$Q(p) = (2\pi)^8 \int d\Gamma_{q_1} d\Gamma_{q_2} d\Gamma_{q_1'} d\Gamma_{q_2'} f_{i_1}(\mathbf{q}_1) f_{i_2}(\mathbf{q}_2) f_{i_1}^*(\mathbf{q}_1') f_{i_2}^*(\mathbf{q}_2')$$
$$\times \delta^{(4)}(p - q_1 - q_2) \delta^{(4)}(p - q_1' - q_2'),$$

when integrated with a slowly varying function (in comparison with the scale of the variation of the wave-packet profiles f) of the variable p, may be replaced by the function $\tilde{Q}(p)$,

$$\tilde{Q}(p) = (2\pi)^4 \delta^{(4)}(p - p_0) \int d^4x |\varphi_{i_1}(x)|^2 |\varphi_{i_2}(x)|^2,$$

where

$$\varphi_i(x) \equiv \int d\Gamma_q e^{-iq\cdot x} f_i(\mathbf{q}),$$

whereas p_0 is the sum of the average four-momenta of the initial particles,

$$p_0 = q_1^0 + q_2^0.$$

To prove this lemma, we integrate the function $Q(p)$ with an arbitrarily chosen function $F(p)$ which satisfies the hypotheses of the lemma and we take the value of the function F at the point p_0 out in front of the integral; this yields

$$\int d^4p F(p) Q(p) = F(p_0) \int d^4p Q(p) = (2\pi)^5 F(p_0) \int d^4x |\varphi_{i_1}(x)|^2 |\varphi_{i_2}(x)|^2.$$

This same result is obtained by integrating the function $\tilde{Q}(p)$ with the function $F(p)$.

We shall employ this lemma to transform the formula for $\sum P_{f_i}$, treating the function $|M|^2$ as a slowly varying function of p. As a result, we arrive at

$$\sum_{\substack{\text{final states of} \\ k \text{ particles}}} P_{f_i} = \int \mathrm{d}^4x |\varphi_{i_1}(x, s_1^0)|^2 |\varphi_{i_2}(x, s_2^0)|^2$$

$$\times (2\pi)^4 \sum_{r_1 \ldots r_k} \int \mathrm{d}\Gamma_{p_1} \ldots \mathrm{d}\Gamma_{p_k} \delta^{(4)}\Big(\sum_{i=1}^{k} p_i - q_1^0 - q_2^0\Big)$$

$$\times |M(\mathbf{p}_1 r_1, \ldots, \mathbf{p}_k r_k; \mathbf{q}_2^0 s_2^0, \mathbf{q}_1^0 s_1^0)|^2. \tag{6}$$

The functions $\varphi_i(x, s)$ describe the space-time structure of wave packets localized in momentum space. For example, the following relation holds for electrons:

$$\chi^{(\pm)}[x|f_i] \simeq u(\mathbf{p}_0, s, \pm 1)\varphi_i(x, s).$$

The expression for the scattering probability thus breaks up into the product of two factors. The first depends only on the initial states of the colliding particles, whereas the second depends only on the dynamical properties of the system under consideration. In formula (6) the space-time integral of the product of the moduli of the functions $\varphi_i(\mathbf{r}, t)$ is a measure of the space-time overlap of the wave packets of the colliding initial particles. This is a measure of the probability of particles meeting. This integral will be put into a form that is easier to interpret; to do this we assume that the spreading of the wave packets φ_{i_1} and φ_{i_2} may be disregarded in the region where the wave functions overlap. With this assumption, the probability "clouds" are displaced without change of shape, that is

$$|\varphi_i(\mathbf{r}, t)|^2 = |\varphi_i(\mathbf{r} - \mathbf{r}_i - \mathbf{v}_i t)|^2. \tag{7}$$

In this way we arrive at

$$\int \mathrm{d}t \, \mathrm{d}^3r |\varphi_{i_1}|^2 |\varphi_{i_2}|^2 = \int \mathrm{d}t \, \mathrm{d}^3r |\varphi_{i_1}(\mathbf{r} - \mathbf{r}_1 - \mathbf{v}_1 t)|^2 |\varphi_{i_2}(\mathbf{r} - \mathbf{r}_2 - \mathbf{v}_2 t)|^2$$

$$= \int \mathrm{d}t \, \mathrm{d}^3r |\varphi_{i_1}(\mathbf{r} - \mathbf{r}_1)|^2 |\varphi_{i_2}(\mathbf{r} - \mathbf{r}_2 + (\mathbf{v}_1 - \mathbf{v}_2)t)|^2.$$

If the z-axis is taken along the direction of the vector $\mathbf{v}_1 - \mathbf{v}_2$, this integral may be written as

$$\int dt\, d^3r |\varphi_{i_1}|^2 |\varphi_{i_2}|^2$$

$$= \frac{1}{|\mathbf{v}_1 - \mathbf{v}_2|^2} \int dx\, dy\, g_1(x-x_1, y-y_1) g_2(x-x_2, y-y_2), \tag{8a}$$

where

$$g_j(x-x_j, y-y_j) = \int dz |\varphi_{i_j}(x-x_j, y-y_j, z)|^2. \tag{8b}$$

For wave packets localized in momentum space, such as we are now considering, the function $|\varphi|^2$ is related to the density of the probability of finding the particle, ϱ, by the formula

$$|\varphi(\mathbf{r}, t)|^2 = \frac{\varrho(\mathbf{r}, t)}{2E_q}, \tag{9}$$

where $E_q = \sqrt{m^2 + \mathbf{q}_0^2}$ is the average energy of the particle. The function $2E_q g_i$ thus is the probability density per unit area perpendicular to the direction $\mathbf{v}_1 - \mathbf{v}_2$.

Expression (8) may be written in the form

$$\int dt\, d^3r |\varphi_{i_1}|^2 |\varphi_{i_2}|^2 = \frac{\eta}{4E_{q_1} E_{q_2} |\mathbf{v}_1 - \mathbf{v}_2|}, \tag{10}$$

where η denotes a measure of the overlap of the probability distributions of the two colliding particles (distributions projected onto a plane perpendicular to $\mathbf{v}_1 - \mathbf{v}_2$). To calculate this expression, it is necessary to have an accurate knowledge of the experimental arrangement used. Most frequently it may be assumed that the probability density projected onto a plane perpendicular to the relative velocity forms a uniform distribution. In this case, the parameter η is equal to the ratio of the area of the overlap, S_{12}, to the product of the areas of the two beams, $S_1 S_2$:

$$\eta = \frac{S_{12}}{S_1 S_2}. \tag{11}$$

Above we made use of the normalization condition for the probabilities:

$$\int d^3r \varrho = \int dx\, dy\, g = 1.$$

As a rule, the participation of many particles is taken into account in scattering experiments. Instead of the probabilities P_{fi} it is then more convenient to operate with the notion of the number of events which we obtain on multiplying P_{fi} by the number of all scattering events. If the colliding beams (or the beam and the target) contain N_1 and N_2 particles, respectively, we multiply the probability by $N_1 N_2$. The factor $N_1 N_2$ characterizes the conditions of the experiment and hence we incorporate it into η. The parameter $\tilde{\eta} = \eta N_1 N_2$ so renormalized is equal to the number of interacting particles per unit area. In the simplest and most common case a beam of particles is directed at a stationary target so that all the particles of the beam strike the target. The factor $\tilde{\eta}$ is then equal to the product of the total flux F of incident particles (the number of particles incident on 1 cm^2), the effective target thickness G (the number of target particles per cm^2 target area), and the beam cross section S,

$$\tilde{\eta} = FGS.$$

In similar fashion $\tilde{\eta}$ can be calculated for intersecting beams. In both cases the parameter $\tilde{\eta}$ is interpreted as the density of collisions per unit area. The dimension of $\tilde{\eta}$ is inverse area.

If we are interested in transitions not to all final states of k particles but only to the regions $\Omega_1, \ldots, \Omega_k$ in the space of momenta $\mathbf{p}_1, \ldots, \mathbf{p}_k$, the integration in formula (6) should be confined to those regions. When the expression then obtained is divided by $\tilde{\eta}$, we obtain the *differential cross section* $\Delta\sigma$:

$$\Delta\sigma = (2\pi)^4 \sum_{r_1 \ldots r_k} \int_{\Omega_1 \ldots \Omega_k} \mathrm{d}\Gamma_{p_1} \ldots \mathrm{d}\Gamma_{p_k} \delta^{(4)}\Big(\sum_{i=1}^{k} p_i - q_1 - q_2\Big) \frac{|M(\mathbf{p}_1 r_1, \ldots, \mathbf{p}_k r_k; \mathbf{q}_1 s_1^0, \mathbf{q}_2 s_2^0)|^2}{4\sqrt{(q_1 q_2)^2 - q_1^2 q_2^2}}. \tag{12}$$

In this formula we have employed the relativistic notation for the expression $E_{q_1} E_{q_2} |\mathbf{v}_1 - \mathbf{v}_2|$ which is valid in all cases when the vectors $\mathbf{v}_1$ and $\mathbf{v}_2$ are parallel or antiparallel.

The transition probability to the final states of particles with particular spins may also be examined. The formula for the cross section will then not involve summation over the indices $r_1, \ldots, r_k$.

NEGATON-NEGATON SCATTERING

Now let us use the foregoing method to determine the differential cross section for negaton-negaton scattering in the second order of perturbation theory. A process of this kind is known as *Møller scattering*.# The function M is determined by using the Feynman diagram technique in the momentum representation. In second-order perturbation theory we consider the two diagrams shown in Fig. 26.1 and consequently obtain:

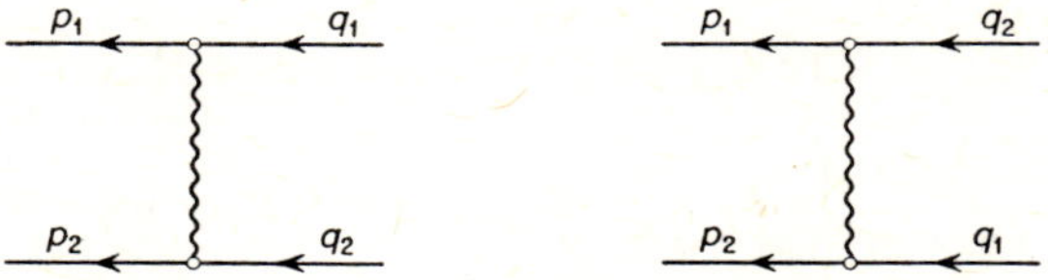

FIG. 26.1. Negaton-negaton scattering in second-order perturbation theory.

$$M(\mathbf{p}_1 r_1, \mathbf{p}_2 r_2; \mathbf{q}_2 s_2, \mathbf{q}_1 s_1)$$

$$= -ie_{\text{obs}}^2 \left[\frac{\{\bar{u}(\mathbf{p}_1, r_1)\gamma^\mu u(\mathbf{q}_1, s_1)\} \cdot \{\bar{u}(\mathbf{p}_2, r_2)\gamma_\mu u(\mathbf{q}_2, s_2)\}}{(p_1 - q_1)^2} - \frac{\{\bar{u}(\mathbf{p}_1, r_1)\gamma^\mu u(\mathbf{q}_2, s_2)\} \cdot \{\bar{u}(\mathbf{p}_2, r_2)\gamma_\mu u(\mathbf{q}_1, s_1)\}}{(p_1 - q_2)^2} \right]. \quad (13)$$

We then calculate the differential cross section for scattering of unpolarized negatons, without measurement of the polarization in the final state. The squared modulus of the function M thus must be averaged with respect to the values of the projections of the negaton spins in the

C. Møller (1932).

initial state and the summation performed over the projections of the negaton spins in the final state. The result we arrive at is:

$$e_{\text{obs}}^2 B \equiv \frac{1}{4} \sum_{r_1, r_2, s_1, s_2} |M(\mathbf{p}_1 r_1, \mathbf{p}_2 r_2; \mathbf{q}_2 s_2, \mathbf{q}_1 s_1)|^2$$

$$= \frac{1}{4} e_{\text{obs}}^2 \Bigg[\frac{1}{(p_1 - q_1)^4} \operatorname{Tr}\{\gamma^\lambda (m + \gamma \cdot p_1) \gamma^\mu (m + \gamma \cdot q_1)\}$$

$$\times \operatorname{Tr}\{\gamma_\lambda (m + \gamma \cdot p_2) \gamma_\mu (m + \gamma \cdot q_2)\}$$

$$+ \frac{1}{(p_1 - q_2)^4} \operatorname{Tr}\{\gamma^\lambda (m + \gamma \cdot p_1) \gamma^\mu (m + \gamma \cdot q_2)\}$$

$$\times \operatorname{Tr}\{\gamma_\lambda (m + \gamma \cdot p_2) \gamma_\mu (m + \gamma \cdot q_1)\}$$

$$- \frac{1}{(p_1 - q_1)^2 (p_1 - q_2)^2} \operatorname{Tr}\{\gamma^\lambda (m + \gamma \cdot p_1) \gamma^\mu (m + \gamma \cdot q_1)$$

$$\times \gamma_\lambda (m + \gamma \cdot p_2) \gamma_\mu (m + \gamma \cdot q_2)\} - \frac{1}{(p_1 - q_1)^2 (p_1 - q_2)^2}$$

$$\times \operatorname{Tr}\{\gamma^\lambda (m + \gamma \cdot p_1) \gamma^\mu (m + \gamma \cdot q_2) \gamma_\lambda (m + \gamma \cdot p_2) \gamma_\mu (m + \gamma \cdot q_1)\} \Bigg]. \tag{14}$$

where m stands for the observable mass m_{obs} and we made use of the relation

$$\sum_r u_a(\mathbf{p}, r) \bar{u}^b(\mathbf{p}, r) = (m + \gamma \cdot p)_a{}^b, \tag{15}$$

which follows from the definition of the bispinors u (cf. Appendix J). The traces of the products of the γ matrices, which figure in formula (14), can be calculated by using formulae given in Appendix J. The final result is written in the form

$$B(s, t, u) = 2 \Bigg[\frac{s^2 + u^2 + 8m^2 t - 8m^4}{t^2}$$

$$+ \frac{s^2 + t^2 + 8m^2 u - 8m^4}{u^2} + \frac{2s^2 - 16m^2 s + 24m^4}{tu} \Bigg], \tag{16}$$

where the Mandelstam variables[#] s, t, and u are defined in the following manner,

$$s \equiv (p_1+p_2)^2 = (q_1+q_2)^2, \tag{17a}$$

$$t \equiv (p_1-q_1)^2 = (p_2-q_2)^2, \tag{17b}$$

$$u \equiv (p_1-q_2)^2 = (p_2-q_1)^2. \tag{17c}$$

The differential cross section (12) in this case (after integration) may be rewritten as

$$\frac{d\sigma}{d\Omega} = \frac{\alpha^2\mathbf{p}_1^2}{|\mathbf{p}_1|\sqrt{m^2+(\mathbf{q}_1+\mathbf{q}_2-\mathbf{p}_1)^2}+(|\mathbf{p}_1|-|\mathbf{q}_1+\mathbf{q}_2|\cos'\theta)\sqrt{m^2+\mathbf{p}_1^2}} \times \frac{B(s,t,u)}{2\sqrt{s(s-4m^2)}}, \tag{18}$$

where $'\theta$ is the angle between the vectors $\mathbf{q}_1+\mathbf{q}_2$ and $\mathbf{p}_1$. Accordingly, in the centre-of-mass system we have

$$\left.\frac{d\sigma}{d\Omega}\right|_{\text{cms}} = \alpha^2\frac{|\mathbf{p}|}{4E_p}\frac{B(s,t,u)}{\sqrt{s(s-4m^2)}}, \tag{19}$$

and in the non-relativistic approximation ($|\mathbf{p}|/m \ll 1$) we obtain the Rutherford formula with exchange (cf. formula (23.27)),

$$\left.\frac{d\sigma}{d\Omega}\right|_{\text{cms}} = \frac{\alpha^2m^2}{16|\mathbf{p}|^4}\left[\frac{1}{\sin^4\left(\frac{1}{2}\theta\right)}+\frac{1}{\cos^4\left(\frac{1}{2}\theta\right)} - \frac{1}{\sin^2\left(\frac{1}{2}\theta\right)\cos^2\left(\frac{1}{2}\theta\right)}\right]. \tag{20}$$

The scattering angle θ and the momentum $\mathbf{p}$ in the centre-of-mass system are related to the Mandelstam variables by the formulae:

$$s = 4(m^2+\mathbf{p}^2), \tag{21a}$$

$$t = -4\mathbf{p}^2\sin^2\left(\frac{1}{2}\theta\right), \tag{21b}$$

$$u = -4\mathbf{p}^2\cos^2\left(\frac{1}{2}\theta\right). \tag{21c}$$

[#] S. Mandelstam (1958).

PHOTON-NEGATON SCATTERING

Let us now calculate, in similar fashion, the cross section for photon-negaton scattering, called *Compton scattering*. In second-order perturbation theory we consider the two diagrams in Fig. 26.2 and get the following expression for the function M:

$$M(\mathbf{p}r, \mathbf{k}_2 t_2; \mathbf{k}_1 t_1, \mathbf{q}s)$$

$$= ie_{\text{obs}}^2 \Big[\bar{u}(\mathbf{p}, r)\gamma\cdot\varepsilon(\mathbf{k}_2, t_2) \frac{1}{m-\gamma\cdot(p+k_2)} \gamma\cdot\varepsilon(\mathbf{k}_1, t_1) u(\mathbf{q}, s)$$

$$+ \bar{u}(\mathbf{p}, r)\gamma\cdot\varepsilon(\mathbf{k}_1, t_1) \frac{1}{m-\gamma\cdot(p-k_1)} \gamma\cdot\varepsilon(\mathbf{k}_2, t_2) u(\mathbf{q}, s)\Big], \qquad (22)$$

where m stands for the observable mass m_{obs}.

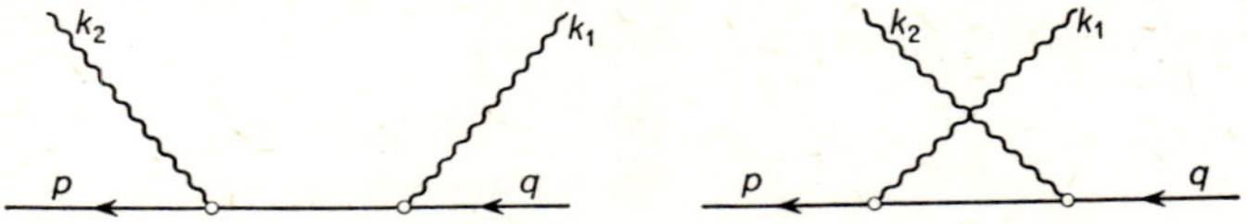

FIG. 26.2. Compton scattering in second-order perturbation theory.

We shall calculate the cross section for scattering on unpolarized negatons in which we do not measure the polarization of the negatons after collision. We take the average of the expression $|M|^2$ over the spin states of the negaton in the initial state and sum over the spin states of the negaton in the final state:

$$\frac{1}{2}\sum_{r,s} |M(\mathbf{p}r, \mathbf{k}_2 t_2; \mathbf{k}_1 t_1, \mathbf{q}s)|^2$$

$$= \frac{1}{2} e_{\text{obs}}^4 \Big[\frac{1}{(m^2-(p+k_2)^2)^2} \operatorname{Tr}\{(m+\gamma\cdot p)\gamma\cdot\varepsilon(\mathbf{k}_2, t_2)$$

$$\times \big(m+\gamma\cdot(p+k_2)\big)\gamma\cdot\varepsilon(\mathbf{k}_1, t_1)(m+\gamma\cdot q)\gamma\cdot\varepsilon^*(\mathbf{k}_1, t_1)$$

$$\times \big(m+\gamma\cdot(p+k_2)\big)\gamma\cdot\varepsilon^*(\mathbf{k}_2, t_2)\}$$

$$+ \frac{1}{(m^2-(p+k_1)^2)^2} \operatorname{Tr}\{(m+\gamma\cdot p)\gamma\cdot\varepsilon(\mathbf{k}_2, t_2)\big(m+\gamma\cdot(p-k_1)\big)$$

$$\times\gamma\cdot\varepsilon(\mathbf{k}_1, t_1)(m+\gamma\cdot q)\gamma\cdot\varepsilon^*(\mathbf{k}_1, t_1)\big(m+\gamma\cdot(p-k_1)\big)\gamma\cdot\varepsilon^*(\mathbf{k}_2, t_2)\}$$

$$+\frac{1}{\big(m^2-(p+k_2)^2\big)\big(m^2-(p-k_1)^2\big)}\operatorname{Tr}\{(m+\gamma\cdot p)\gamma\cdot\varepsilon(\mathbf{k}_2, t_2)$$

$$\times\big(m+\gamma\cdot(p+k_2)\big)\gamma\cdot\varepsilon(\mathbf{k}_1, t_1)(m+\gamma\cdot q)\gamma\cdot\varepsilon^*(\mathbf{k}_2, t_2)$$

$$\times\big(m+\gamma\cdot(p-k_1)\big)\gamma\cdot\varepsilon^*(\mathbf{k}_1, t_1)\}$$

$$+\frac{1}{\big(m^2-(p+k_2)^2\big)\big(m^2-(p-k_1)^2\big)}\operatorname{Tr}\{(m+\gamma\cdot p)\gamma\cdot\varepsilon(\mathbf{k}_2, t_2)$$

$$\times\big(m+\gamma\cdot(p-k_1)\big)\gamma\cdot\varepsilon(\mathbf{k}_1, t_1)(m+\gamma\cdot q)$$

$$\left.\times\gamma\cdot\varepsilon^*(\mathbf{k}_2, t_2)\big(m+\gamma\cdot(p+k_2)\big)\gamma\cdot\varepsilon^*(\mathbf{k}_1, t_1)\}\right]. \tag{23}$$

For simplicity, we assume that we do not measure the polarization of the photons and hence we average over the values of the polarization of the incident photon and sum over the values of the polarization of the scattered photon. To this end we draw upon the following relation which is satisfied by the vectors $\varepsilon_\mu(\mathbf{k}, r)$ in the Coulomb gauge:

$$\sum_{r=\pm 1}\varepsilon_\mu(\mathbf{k}, r)\varepsilon_\nu^*(\mathbf{k}, r) = -g_{\mu\nu}+\frac{k_\mu k_0 g_{0\nu}+k_\nu k_0 g_{0\mu}-k_\mu k_\nu}{|\mathbf{k}|^2}. \tag{24}$$

The second part in this formula contains only terms which are proportional to k_μ or k_ν. Since the function M is invariant under the gauge transformation of the polarization vectors,

$$\varepsilon_\mu \to \varepsilon_\mu + k_\mu \alpha(k),$$

this part makes no contribution to the transition amplitude. On replacing the product of the vectors ε_μ and ε_ν^* by a metric tensor and after computing the traces, we get

$$\frac{1}{4}\sum_{r, s, t_1, t_2}|M(\mathbf{p}r, \mathbf{k}_2 t_2; \mathbf{k}_1 t_1, \mathbf{q}s)|^2$$

$$= 2e_{\text{obs}}^4\left[\frac{4m^4+2m^2(s-m^2)-(s-m^2)(u-m^2)}{(s-m^2)^2}\right.$$

$$\left.+\frac{4m^4+2m^2(u-m^2)-(s-m^2)(u-m^2)}{(u-m^2)^2}+\frac{4m^4+2m^2 s+2m^2 u}{(u-m^2)(s-m^2)}\right], \tag{25}$$

where

$$s = (p+k_2)^2 = (q+k_1)^2, \tag{26a}$$

$$u = (p-k_1)^2 = (q-k_2)^2. \tag{26b}$$

In the laboratory system ($\mathbf{q} = 0$), the variables s and u are expressed in the following manner in terms of the photon energy ω in the initial state and the scattering angle θ:

$$s = m^2+2m\omega, \tag{27a}$$

$$u = m^2 - \frac{2m\omega}{1+(1-\cos\theta)\omega/m}. \tag{27b}$$

Formulae (12), (25), and (27) lead to the following formula for the differential cross section:

$$\left.\frac{d\sigma}{d\Omega}\right|_{\text{lab}} = \frac{r_0^2}{2} \frac{1+\cos^2\theta}{[1+(1-\cos\theta)\omega/m]^2} \times \left[1+\frac{(\omega/m)^2(1-\cos\theta)^2}{(1+\cos^2\theta)[1+(1-\cos\theta)\omega/m]}\right]. \tag{28}$$

This is known as the *Klein–Nishina formula.*# The parameter r_0,

$$r_0 = \frac{e_{\text{obs}}^2}{4\pi mc^2} = 2.8\times 10^{-13}\ \text{cm}, \tag{29}$$

is called the *classical electron radius.* In the non-relativistic approximation ($\omega/m \ll 1$) the Klein–Nishina formula goes over into the classical Thomson formula:

$$\left.\frac{d\sigma}{d\Omega}\right|_{\text{lab}} = \frac{r_0^2}{2}(1+\cos^2\theta). \tag{30}$$

The total cross section in the non-relativistic approximation is obtained by integrating the differential cross section (30) with respect to the angular variables,

$$\sigma = \frac{8\pi}{3} r_0^2. \tag{31}$$

Since this formula is purely classical, it should hold in the limit, as $\omega \to 0$, in all orders of renormalized perturbation theory. Thus, correc-

---

O. Klein and Y. Nishina (1929).

tions of all orders to this formula are equal to zero. If we were to use *unrenormalized* perturbation theory, we would get non-zero constant corrections to formula (31) from the higher orders. On summing these corrections, we have#

$$\sigma = \frac{8\pi}{3}\left(\frac{e^2}{4\pi mc^2}\right)^2 Z_3^2. \tag{32}$$

Thus, if formula (31) together with formula (29) is regarded as the definition of the observable electron charge, the same relation is obtained between the charge e_{obs} and the parameter e as when the Rutherford formulae is used to define the electron charge.

Photon-Photon Scattering

Elastic scattering of a photon by a photon is described by the propagator $T_{\mu\nu\lambda\varrho}(z_1 z_2 z_3 z_4)$. In the lowest (fourth) order of perturbation theory the diagrams corresponding to the strongly connected part

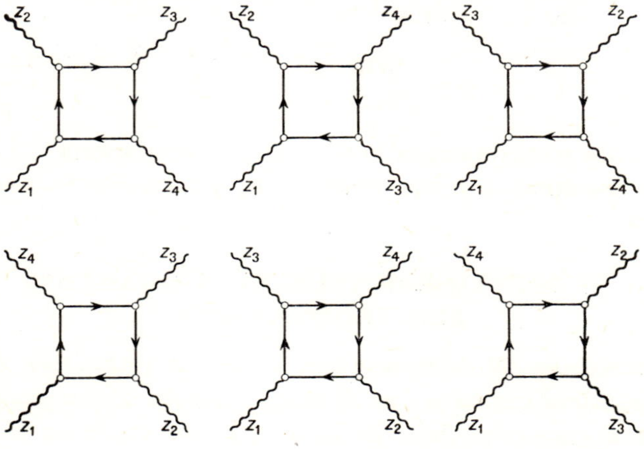

FIG. 26.3. Photon-photon scattering.

$\Delta^{\mu\nu\lambda\varrho}(z_1 z_2 z_3 z_4)$ of this propagator are shown in Fig. 26.3. By virtue of the Furry theorem, identical analytical expressions correspond to diagrams drawn under each other. The transition amplitude and cross

The proof of this assertion was given by W. Thirring (1950).

section for the process considered are given in the form of complicated formulae.# Here we give only the formulae obtained in the low-energy approximation valid when the energy ω of the photons in the centre-of-mass system is much smaller than the electron mass ($\omega \ll m$). The transition amplitude may then be written as

$$(\Psi_f^{\text{out}}|\Psi_i^{\text{in}}) = \left(\Psi_f^{\text{in}}\left|\left[1 - i\int d^4x \mathscr{H}_{\text{int}}^{\text{IN}}(x)\right]\right|\Psi_i^{\text{in}}\right), \tag{33}$$

where

$$\mathscr{H}_{\text{int}}(x) = -\frac{2\alpha^2}{45}\frac{1}{m^4} : [(\mathbf{E}^2 - \mathbf{B}^2)^2 + 7(\mathbf{B}\cdot\mathbf{E})^2]:, \tag{34}$$

and the symbol IN signifies that the operators **B** and **E** are built out of the creation and annihilation operators of IN photons.

The differential and total cross sections for the scattering of two unpolarized photons in the centre-of-mass system are given by

$$\left.\frac{d\sigma}{d\Omega}\right|_{\text{cms}} = \frac{139}{8100}\left(\frac{\alpha}{2\pi}\right)^2 r_0^2 \left(\frac{\omega}{m}\right)^6 (3 + \cos^2\theta)^2, \tag{35}$$

$$\sigma = \frac{1946}{10125}\frac{\alpha^2}{\pi} r_0^2 \left(\frac{\omega}{m}\right)^6, \tag{36}$$

where θ is the scattering angle in the centre-of-mass system.

These quantities are too small to be measurable by experiment at present.

27. NON-LINEAR EFFECTS IN QUANTUM ELECTRODYNAMICS

The scattering of electromagnetic waves on each other described in the preceding section is one of many examples of non-linear effects in quantum electrodynamics. In addition to the scattering of light by light, there are also such phenomena as the scattering of light by an electromagnetic field, e.g. by a nuclear field (Delbrück effect), changes in the propagation of electromagnetic waves in a magnetic field as compared with propagation in the vacuum, the splitting of the photon

These formulae were given in a paper by R. Karplus and M. Neuman in 1951. See also the books of Akhiezer and Berestetskii, and Jauch and Rohrlich.

in external fields, and so on. Such phenomena cannot be described by Maxwellian electrodynamics because that theory is linear. The sum of any two solutions of Maxwell's equations is also a solution.

The situation is different in non-linear theory of the electromagnetic field. Quantum electrodynamics is a non-linear theory, even though it takes its starting point in the linear Maxwell theory. Non-linear effects occur even when there are no real electrons. The self-interaction of electromagnetic fields (photons) arises as a result of the *polarization of the vacuum*,† which is to say as a result of the creation of pairs and their subsequent annihilation. The electromagnetic field creates pairs whose motion in turn is the source of a new field. Since the self-interaction of the electromagnetic field occurs through the motion of particles, it is of a non-local nature. In spite of this, however, these interactions are causal.

Many non-linear effects described exactly by quantum electrodynamics may be described approximately by means of non-linear *classical* electrodynamics, whose general scheme we discussed in Chapter 3, if only the effective Lagrangian of the interaction is chosen suitably. The approximation consists in disregarding the creation and annihilation of real negaton-positon pairs and in disregarding the non-local nature of the interaction of electromagnetic fields. For electromagnetic fields which vary slowly in time (as compared with the parameter $\hbar/mc^2$), the creation and annihilation of real pairs plays a negligible role. For phenomena in which the fields vary in space slowly as compared with the *Compton wavelength of the electron* $\hbar/mc$, the non-local character of the interaction may be neglected. Compared with the macroscopic or even the atomic scale, the phenomena of the creation and annihilation of virtual pairs occur almost at one and the same point.

The applicability of the approximation of the effective local Lagrangian or Hamiltonian may thus be justified by the brief duration and short range of phenomena associated with the propagation of virtual electron pairs. The effective interaction Hamiltonian was given in the preceding section in the discussion on the scattering of low-energy photons in the lowest order of perturbation theory. The complete

† The polarization of the vacuum is discussed in greater length in the next section.

effective Lagrangian is of the form#

$$\mathscr{L}(x) = S(x) - \frac{1}{8\pi^2}\int_0^\infty \frac{d\tau}{\tau^3}\, e^{-m^2\tau}$$

$$\times\left[(e_{\text{obs}}\,\tau)^2 \frac{\operatorname{Re}\cosh(e_{\text{obs}}\,\tau X(x))}{\operatorname{Im}\cosh(e_{\text{obs}}\,\tau X(x))}\, P(x) - 1 + \frac{2}{3}(e_{\text{obs}}\tau)^2 S(x)\right], \tag{1}$$

where

$$X^2 \equiv -2S + 2iP,$$

and m denotes the observable mass of the electron.

Let us write out the first four terms of the asymptotic expansion of the Lagrangian with respect to α,

$$\mathscr{L} \simeq S + \frac{2\alpha^2}{45m^4}(4S^2+7P^2)$$

$$+\frac{32\pi\alpha^3}{315m^8}(8S^3+13SP^2) + \frac{256\pi^2\alpha^4}{945m^{12}}(48S^4+138S^2P^2+19P^4). \tag{2}$$

Replacement of the full expression (1) with the first terms of the asymptotic expansion is justified if the dimensionless parameter of the expansion

$$4\pi\frac{\alpha\hbar^3}{m^4c^5}|F|^2 \simeq \begin{cases} 5\times 10^{-28}(B \text{ in gauss})^2 \\ 6\times 10^{-33}(E \text{ in V/cm})^2 \end{cases} \tag{3}$$

is much smaller than unity. This condition is satisfied veiy vell by all† known macroscopic electromagnetic fields occurring in the universe.

The simplest non-linear effect which lends itself very well to description by using the effective Lagrangian is the influence of a constant and uniform electromagnetic field on the propagation of electromagnetic fields. It turns out## that the electromagnetic field constitutes a birefringent medium for electromagnetic waves. As a result of the bi-

This Lagrangian was given by W. Heisenberg and H. Euler (1936). The form cited here for the effective Lagrangian is due to J. Schwinger (1951 b).

† With the exception, perhaps, of magnetic fields on the surfaces of neutron stars.

The problem of propagation and other non-linear effects has been described in a paper by Z. Białynicka-Birula and I. Białynicki-Birula (1970), to which readers interested in the details are referred.

refringence, waves of different polarizations propagate with different phase and group velocities, $(\mathbf{u}_\pm)$ and $(\mathbf{v}_\pm)$. For fields that are not very strong, these velocities are the following functions of the direction $\mathbf{n}$ of the propagation vector and the field vectors $\mathbf{E}$ and $\mathbf{B}$,

$$\mathbf{u}_\pm = c\mathbf{n}\left[1-\frac{1}{2}\lambda_\pm(\mathbf{n}\times\mathbf{E}+\mathbf{n}\times(\mathbf{n}\times\mathbf{B}))^2\right], \tag{4a}$$

$$\mathbf{v}_\pm = c\mathbf{n}\left[1-\frac{1}{2}\lambda_\pm[(\mathbf{E}^2+\mathbf{B}^2)+(\mathbf{n}\cdot\mathbf{E})^2+(\mathbf{n}\cdot\mathbf{B})^2]\right]$$
$$+c\lambda_\pm[\mathbf{E}(\mathbf{n}\cdot\mathbf{E})+\mathbf{B}(\mathbf{n}\cdot\mathbf{B})-\mathbf{B}\times\mathbf{E}], \tag{4b}$$

where

$$\lambda_+ = 28\frac{\alpha^2\hbar^3}{45m^4c^5},$$

$$\lambda_- = 16\frac{\alpha^2\hbar^3}{45m^4c^5}.$$

28. THE ELECTRON IN A STATIC ELECTROMAGNETIC FIELD

THE EFFECTIVE FIELD AND THE POLARIZATION OF THE VACUUM

Let us first of all consider the simplest problem: the scattering of a negaton in an external electromagnetic field in the absence of an external current in the Born approximation. In this way we obtain a physical interpretation of the function $\mathfrak{A}_{\text{ren}}(x)$. We shall assume that no external field occurs in the remote past and the far future. The amplitude of the negaton scattering in the external field is given by

$$S^{++}[f,g] = (f|g)$$
$$+V[\mathfrak{A}_{\text{ren}}]i\int d^4x\,d^4y\,\bar{\chi}^{(+)}[x|f]D_xG_{\text{ren}}[x,y|\mathfrak{A}_{\text{ren}}]\overleftarrow{D}_y\chi^{(+)}[y|g]. \tag{1}$$

To obtain the Born approximation, we shall expand the propagator $G_{\text{ren}}[x,y|\mathfrak{A}_{\text{ren}}]$ into a series in the function $\mathfrak{A}_{\text{ren}}$ and we retain only the linear term

$$G_{\text{ren}}[x,y|\mathfrak{A}_{\text{ren}}] = G_{\text{ren}}(x-y)-ie_{\text{obs}}\int d^4\xi\,d^4\eta\,d^4\zeta$$
$$\times G_{\text{ren}}(x-\xi)\Gamma^\mu_{\text{ren}}(\xi,\eta,\zeta)\mathfrak{A}_{\mu\text{ren}}(\zeta)G_{\text{ren}}(\eta-y)+\dots. \tag{2}$$

The vacuum-to-vacuum transition amplitude $V[\mathfrak{A}_{\text{ren}}]$ in the case of a weak external field is approximately equal to unity. When we make use of the properties of renormalized propagators, on going over to the momentum representation we finally get the following formula for the S matrix in the momentum representation, which we define in the same way as in the absence of an external field:

$$S_B^{++}(\mathbf{p}, r; \mathbf{q}, s) = \delta_{rs}\,\delta_{\Gamma(m)}(\mathbf{p}, \mathbf{q})$$

$$-ie_{\text{obs}}\overline{u}(\mathbf{p}, r)\tilde{\Gamma}^{\mu}_{\text{ren}}(p, q)\tilde{\mathfrak{A}}_{\mu\text{ren}}(p-q)u(\mathbf{q}, s). \tag{3}$$

The result is to be compared with formula (17.3a) obtained in relativistic quantum mechanics. A correspondence between these formulae exists in the limit when we consider an external field that varies slowly in space and in time. The characteristic length L and the characteristic time T determining the scale of variation of such a field satisfy the inequalities

$$L \ll \hbar/mc = 3.861\times 10^{-11}\ \text{cm}, \tag{4a}$$

$$T \ll h/mc^2 = 1.287\times 10^{-21}\text{sec}, \tag{4b}$$

from which it follows that the Fourier transform $\tilde{\mathfrak{A}}_{\mu\text{ren}}$ of the potential $\mathfrak{A}_{\mu\text{ren}}$ is, practically speaking, equal to zero if $(p-q)^2/m^2$ is of the order of unity. Since on the other hand the characteristic momentum determining the scale or variation of the function $\tilde{\Gamma}^{\mu}_{\text{ren}}$ is mc, in the case of fields characterized by inequalities (4) the vertex function $\tilde{\Gamma}^{\mu}_{\text{ren}}$ may be replaced by its value at the point $p = q$. In turn, the Ward identity (23.23) implies that

$$\tilde{\Gamma}^{\mu}_{\text{ren}}(p, p) = \gamma^{\mu}. \tag{5}$$

In the case of slowly varying fields, formula (3) takes on the form of formula (17.3a) in which the role of the external field $\mathscr{A}_{\mu}$ which appears in quantum mechanics is now played by the field $\mathfrak{A}_{\mu\text{ren}}$. The field $\mathfrak{A}_{\mu\text{ren}}$ is thus interpreted as the *effective field* acting on the electron. This field may be regarded as consisting of the original field $\mathscr{A}_{\mu}$ and the field induced by that original field through the intermediary of the polarization current generated by the motion of pairs. In quantum electrodynamics, just as in classical non-linear electrodynamics, therefore, we are concerned with original fields generated by sources and with effective fields acting on particles.

We shall illustrate this problem with the simple example of a field generated by the static distribution of charge described by the function $\mathscr{I}^0(\mathbf{x})$. The effective potential $\mathfrak{A}_{\mu\mathrm{ren}}$ is a complicated non-linear functional of the function $\mathscr{I}^0(\mathbf{x})$. A relatively simple relationship can, however, be obtained if we confine ourselves to a linear approximation with respect to $\mathscr{I}^0(\mathbf{x})$. First of all, let us consider an arbitrary current $\mathscr{I}^\nu(x)$. In the linear approximation in $\mathscr{I}^\nu(x)$ we have

$$\begin{aligned}\mathfrak{A}_\mu(x) &\simeq \int \mathrm{d}^4x \frac{\delta\mathfrak{A}_\mu(z)}{\delta\mathscr{I}^\nu(x)} \mathscr{I}^\nu(x)\\ &= -\int \mathrm{d}^4x \mathscr{G}_{\mu\nu}(z-x)\mathscr{I}^\nu(x).\end{aligned} \tag{6}$$

To get this formula we have made use of formula (24.7a) for the photon propagator and of the assumption that $\mathscr{A}_\mu$ is absent. The relation between the relevant renormalized quantities can be found from equality (6) if the renormalized current $\mathscr{I}^\nu_{\mathrm{ren}}$ is introduced in addition to $\mathfrak{A}_{\mu\mathrm{ren}}$ and $\mathscr{G}_{\mu\nu\mathrm{ren}}$, which have been defined in Chapter 7. We then obtain the equality

$$\mathfrak{A}_{\mu\mathrm{ren}}(z) = -\int \mathrm{d}^4x \mathscr{G}_{\mu\nu\mathrm{ren}}(z-x)\mathscr{I}^\nu_{\mathrm{ren}}(x). \tag{7}$$

Now let us consider the case of the static charge distribution ($\mathscr{I}^\nu(x) = (\mathscr{I}^0(\mathbf{x}), 0, 0, 0)$ which is of interest to us. We shall use the spectral representation of the photon propagator, from which we can get the representation

$$\begin{aligned}\mathfrak{A}^0_{\mathrm{ren}}(\mathbf{z}) &= \frac{1}{4\pi}\int \mathrm{d}^3x \frac{\mathscr{I}^0_{\mathrm{ren}}(\mathbf{x})}{|\mathbf{z}-\mathbf{x}|}\\ &+ \int_{0+}^{\infty} \frac{\mathrm{d}M^2}{M^2} \varrho_{\mathrm{ren}}(M^2) \frac{1}{4\pi} \int \mathrm{d}^3x \frac{e^{-M|\mathbf{z}-\mathbf{x}|}}{|\mathbf{z}-\mathbf{x}|} \mathscr{I}^0_{\mathrm{ren}}(\mathbf{x}),\end{aligned} \tag{8}$$

where

$$\varrho_{\mathrm{ren}}(M^2) = Z_3^{-1}\varrho(M^2).$$

In order to derive this formula, we must introduce the factor $\exp(-\varepsilon|t|)$ which switches off the charge and, in the final result, perform the transition to the limit when $\varepsilon \to 0$. In the special case of a point charge Ze located at the origin of the coordinate system, we arrive at

$$\mathfrak{A}^0_{\text{ren}}(r) = \frac{1}{4\pi}\frac{Ze}{r} + \frac{Ze}{4\pi}\int\limits_{0+}^{\infty}\frac{\mathrm{d}M^2}{M^2}\,\varrho_{\text{ren}}(M^2)\,\frac{\mathrm{e}^{-Mr}}{r}. \tag{9}$$

For $r \to \infty$ we thus get the usual Coulomb law described by the first term. For values of r which are not too large (compared with $\hbar/mc$) the original field is modified by polarization caused by the introduction of charge. The forces acting in this region are greater (the function ϱ_{ren} is positive) than the purely Coulomb forces. The forces described by the second term in formula (9) are, however, short-range (of the Yukawa type). The function $\varrho_{\text{ren}}(M^2)$ determined in the second order of perturbation theory# is given by

$$\varrho_{\text{ren}}(M^2) = \frac{\alpha}{3\pi}\,\theta(M^2-(2m)^2)\left(1+\frac{2m^2}{M^2}\right)\sqrt{1-\frac{4m^2}{M^2}}\,. \tag{10}$$

This result can be obtained from formula (23.19).

From this we get the following asymptotic form of the effective potential $\mathfrak{A}^0_{\text{ren}}(r)$ for $r \ll \hbar/mc$,

$$\mathfrak{A}^0_{\text{ren}}(r) \simeq \frac{Ze}{4\pi r}\left\{1+\frac{2\alpha}{3\pi}\left[\ln\frac{\hbar}{mcr}-\frac{5}{6}-\ln\gamma\right]\right\}, \tag{11}$$

where $\gamma \simeq 1.781$. This formula expresses a modification of Coulomb's law for short distances. The expression in the braces may be interpreted as a "dielectric constant" which depends on the distance from the centre. The phenomenon described above is known as *polarization of the vacuum*, by analogy with polarization effects in dielectrics. The occurrence of the vacuum polarization effect was confirmed in Lamb's experiments, which are discussed further on in this section.

The potential of the effective electromagnetic field $\mathfrak{A}_{\mu\text{ren}}(z)$ is a real function only in the case of static fields. When fields depend on time, processes of generation of radiation or even pair creation by an external

---

This calculation was carried out by R. Serber (1935), E. A. Uehling (1935), and J. Schwinger (1949 a).

field occur. The function $\mathfrak{A}_{\mu\mathrm{ren}}(z)$ is then complex. When the original external field is static, the equality $\Omega^{\mathrm{out}} = \Omega^{\mathrm{in}}$ holds. The field $\mathfrak{A}_{\mu\mathrm{ren}}(z)$ in that case is a real-valued function. It is interpreted as the average value of the electromagnetic field in the vacuum state.

The Motion of the Electron in a Static Field

Let us now assume that in the remote past and the far future the external electromagnetic field goes over into a static field $\mathscr{A}_\mu(\mathbf{x})$. The same will be true for the complete field $\mathfrak{A}_\mu$:

$$\mathfrak{A}_\mu(\mathbf{x}, t) = \begin{cases} \mathfrak{A}_\mu(\mathbf{x}), & t < T_1, \\ \mathfrak{A}_\mu(\mathbf{x}), & t > T_2. \end{cases} \tag{12}$$

Now let us consider the problem of motion of an electron in such a field in quantum electrodynamics. In our considerations we shall draw upon the results of Section 17 concerning the motion of an electron in a static field, disregarding the influence of the quantum electromagnetic field.

If the field $\mathfrak{A}_\mu(\mathbf{x})$ is one of finite range and if we are interested only in the scattering states, then in accordance with the explanations given in Section 4 use can be made of the same description as employed for electrons without an external field in asymptotic regions. For, in the remote past and the far future wave packets are far from the region in which the external field acts; the motion of the particles then is a free motion.

Frequently, however, we are interested in the motion of electrons which remain permanently within the range of the field in the past and/or in the future. This may be, for instance, the motion of an electron in the field of a nucleus or the motion of an electron in a constant and uniform magnetic field. In such a case, in order to determine the transition amplitude we must first find the electron wave functions $\psi[x|\mathfrak{A}]$ which are solutions of the Dirac equation modified so as to take into account the influence of the quantum electromagnetic field. The photon wave functions should also undergo suitable modification inasmuch as non-linear effects take place. As already mentioned in the preceding section, however, this non-linearity may almost always be disregarded.

The notion of the electron wave function in complete quantum electrodynamics requires some clarification. Up to this point we have not brought in this notion for interacting electrons; hitherto it was sufficient for us to consider only the wave functions of free electrons. Electron wave functions will be introduced in a way so as to preserve the relations between these wave functions and the electron propagators. In Sections 4 and 17 we showed that in the theory of electrons interacting only with the external field, the propagators are constructed bilinearly out of the wave functions. In the special case of the static field the relation between the propagators and the wave functions is particularly simple. In the complete theory we shall write it out in the form

$$G[\mathbf{x}t;\mathbf{y}t'|\mathfrak{A}] = i\theta(t-t')\sum_{E_\alpha>0}\int d\alpha\psi[\mathbf{x},\alpha|\mathfrak{A}]e^{-iE_\alpha(t-t')}\bar{\psi}[\mathbf{y},\alpha|\mathfrak{A}]$$
$$+i\theta(t'-t)\sum_{E_\alpha<0}\int d\alpha\psi[\mathbf{x},\alpha|\mathfrak{A}]e^{-iE_\alpha(t-t')}\bar{\psi}[\mathbf{y},\alpha|\mathfrak{A}]. \quad (13)$$

The integration and summation in the first integral run over all wave functions with positive frequencies, whereas in the second integral they run over functions with negative frequencies. Formula (13) is regarded as an implicit definition of wave functions in the complete quantum electrodynamics. It implies, among other things, that the wave function $\psi[\mathbf{x}t,\alpha|\mathfrak{A}]$ satisfies the same equation as the propagator does, but without the function δ on the right-hand side. For an arbitrarily chosen wave function this equation is of the form#

$$[D_x+e\gamma\cdot\mathfrak{A}(x)]\psi[x|\mathfrak{A}]-\int d^4y\Sigma[x,y|\mathfrak{A}]\psi[y|\mathfrak{A}] = 0. \quad (14)$$

The generalization of the foregoing equation to the case of a time-dependent field does not require any modification; we adopt it in the same form.

Renormalization must be performed in eq. (14) for the electron wave function in the same way as in the equations for the propagators. On renormalizing the self-energy function and expressing all quantities in terms of the observable parameters m_{obs} and e_{obs}, we obtain

$$Z_2[-i\gamma\cdot\partial+m_{\text{obs}}+e_{\text{obs}}\gamma\cdot\mathfrak{A}_{\text{ren}}(x)]\psi[x|\mathfrak{A}_{\text{ren}}]$$
$$-\int d^4y\,\Sigma_{\text{ren}}[x,y|\mathfrak{A}_{\text{ren}}]\psi[y|\mathfrak{A}_{\text{ren}}] = 0. \quad (15)$$

This equation was derived by J. Schwinger (1951 a).

If we knew how to find the solution of eq. (15), the entire problem of motion in a static field would be solved. Unfortunately, as in the case of the equations for propagators, we can merely find approximate solutions by the perturbation method. In what follows we shall discuss a perturbation solution for motion in a static magnetic field, which will enable us to calculate the magnetic moment of the electron, and a perturbation solution in the case of a nuclear field, by means of which the Lamb–Retherford shift is evaluated.

The Magnetic Moment of the Electron

At present we shall use eq. (15) to describe the motion of an electron in a weak electromagnetic field. The function $\sum_{\text{ren}}[x, y|\mathfrak{A}_{\text{ren}}]$ can then be expanded into a series in $\mathfrak{A}_{\text{ren}}$, with only the linear terms being retained:

$$Z_2[D_x+e_{\text{obs}}\,\gamma\cdot\mathfrak{A}_{\text{ren}}(x)]\,\psi[x|\mathfrak{A}_{\text{ren}}]-\int \mathrm{d}^4y\left\{\sum\nolimits_{\text{ren}}(x-y)\right.$$

$$\left.+\int \mathrm{d}^4z\,\frac{\delta\sum_{\text{ren}}[x, y|\mathfrak{A}_{\text{ren}}]}{\delta\mathfrak{A}^{\mu}_{\text{ren}}(z)}\right|_{\mathfrak{A}_{\text{ren}}=0}\left.\mathfrak{A}^{\mu}_{\text{ren}}(z)\right\}\psi[y|\mathfrak{A}_{\text{ren}}]=0. \tag{16}$$

The renormalized functions G_{ren} and Γ_{ren} can be used in order to rewrite this formula as

$$\int \mathrm{d}^4y[G^{-1}_{\text{ren}}(x-y)+e_{\text{obs}}\int \mathrm{d}^4z\Gamma^{\mu}_{\text{ren}}(x, y, z)\,\mathfrak{A}_{\mu\text{ren}}(z)]\,\psi[y|\mathfrak{A}_{\text{ren}}]=0. \tag{17}$$

Going over to the momentum representation, we arrive at

$$\tilde{G}^{-1}_{\text{ren}}(p)\tilde{\psi}(p)+e_{\text{obs}}\int\frac{\mathrm{d}^4q}{(2\pi)^4}\,\tilde{\Gamma}^{\mu}_{\text{ren}}(p, q)\,\tilde{\mathfrak{A}}_{\mu\text{ren}}(p-q)\tilde{\psi}(q)=0. \tag{18}$$

In the absence of an external field, the solution of this equation for negatons is of the form

$$\tilde{\psi}^{(+)}(p)=\delta(p^2-m^2_{\text{obs}})\theta(p_0)\sum_r u(\mathbf{p}, r)f(\mathbf{p}, r), \tag{19}$$

whereas for positons it is

$$\tilde{\psi}^{(-)}(p)=\delta(p^2-m^2_{\text{obs}})\theta(-p_0)\sum_r v(\mathbf{p}, r)f(\mathbf{p}, r). \tag{20}$$

The external electromagnetic field will now be assumed to be static. The field Fourier transform then is of the form

$$\tilde{\mathfrak{A}}_{\mu\text{ren}}(p-q)=2\pi\delta(p_0-q_0)\tilde{\mathfrak{A}}_{\mu\text{ren}}(\mathbf{p}-\mathbf{q}). \tag{21}$$

The wave function $\tilde{\psi}(p_0, \mathbf{p})$ describing a stationary state depends on p_0 only through a δ-function,

$$\tilde{\psi}(p_0, \mathbf{p}) = \delta(p_0 - E)\psi_E(\mathbf{p}). \tag{22}$$

With the help of eqs. (21) and (22), eq. (18) can be transformed into:

$$\tilde{G}^{-1}_{\text{ren}}(E, \mathbf{p})\psi_E(\mathbf{p}) + e_{\text{obs}} \int \frac{d^3 q}{(2\pi)^3} \tilde{\Gamma}^{\mu}_{\text{ren}}(E, \mathbf{p}, E, \mathbf{q}) \tilde{\mathfrak{A}}_{\mu\text{ren}}(\mathbf{p}-\mathbf{q})\psi_E(\mathbf{q}) = 0. \tag{23}$$

The basic difference between eq. (23) and a standard Schrödinger or Dirac equation is that the dependence on the energy E is non-linear.

Since the external field is by assumption weak, we shall use perturbation theory and confine ourselves to the lowest order. From the property (12.5) of the renormalized propagator we can obtain

$$\left.\frac{\partial \tilde{G}^{-1}_{\text{ren}}(E, \mathbf{p})}{\partial E}\right|_{E=E_p} u(\mathbf{p}, r) = -\gamma^0 u(\mathbf{p}, r). \tag{24}$$

Finally we arrive at the following formula for the energy correction ΔE:

$$\Delta E = e_{\text{obs}} \int d\Gamma_p(m_{\text{obs}}) d\Gamma_q(m_{\text{obs}}) \times \sum_{r,s} f^*(\mathbf{p}, r)\bar{u}(\mathbf{p}, r)\tilde{\Gamma}^{\mu}_{\text{ren}}(p, q)\Big|_{p_0=E_p=q_0} u(\mathbf{q}, s)f(\mathbf{q}, s)\tilde{\mathfrak{A}}_{\mu\text{ren}}(\mathbf{p}-\mathbf{q}). \tag{25}$$

The matrix element here for the vertex function $\tilde{\Gamma}^{\mu}_{\text{ren}}$ between two bispinors $\bar{u}$ and u may be expressed in terms of two scalar functions $F_1(k^2)$ and $F_2(k^2)$, defined by

$$\bar{u}(\mathbf{p}, r)\tilde{\Gamma}^{\mu}_{\text{ren}}(p, q)\Big|_{p_0=E_p=q_0} u(\mathbf{q}, s) = \bar{u}(\mathbf{p}, r)\left[\gamma^{\mu}F_1(k^2) - \frac{1}{2m_{\text{obs}}}\gamma^{\mu\nu}k_{\nu}F_2(k^2)\right]u(\mathbf{q}, s), \tag{26}$$

where

$$k_{\mu} = p_{\mu} - q_{\mu}.$$

It follows from the relativistic invariance of the theory that the function $\tilde{\Gamma}^{\mu}_{\text{ren}}(p, q)$ may be written as the sum

$$\tilde{\Gamma}^{\mu}_{\text{ren}}(p, q) = \sum_{\substack{\alpha,\beta,\gamma,\delta=0,1\\ \alpha+\beta+\gamma+\delta=2}} (\gamma\cdot p)^{\alpha}(\gamma\cdot q)^{\beta}\gamma^{\mu}(\gamma\cdot p)^{\gamma}(\gamma\cdot q)^{\delta} C_{\alpha\beta\gamma\delta}(p^2, q^2, p\cdot q). \tag{27}$$

The number of scalar functions $C_{\alpha\beta\gamma\delta}$ may be reduced to two by using the equations satisfied by the bispinors $\bar{u}$ and u and the Ward identity, after the function $\tilde{\Gamma}^{\mu}_{\text{ren}}$ is multiplied by $\bar{u}$ and u.

The scalar functions $F_1(k^2)$ and $F_2(k^2)$ are called *electromagnetic form factors*. It follows from the Ward identity that

$$F_1(0) = 1. \tag{28}$$

The value of the form factor F_2 at zero is related to the *magnetic moment* μ of the electron by

$$\mu = \frac{e_{\text{obs}}\hbar}{2m_{\text{obs}}\,c}\,(1+F_2(0)). \tag{29}$$

To derive this formula, let us calculate, in the non-relativistic approximation, the negaton energy correction (25) in a constant magnetic field **H**. The effective electromagnetic potential in this case has the following Fourier transform:

$$\tilde{\mathfrak{A}}^0_{\text{ren}} = 0, \tag{30a}$$

$$\tilde{\mathfrak{A}}^l_{\text{ren}}(\mathbf{k}) = -(2\pi)^3\,\frac{1}{2}\,(i\nabla_k\times\mathbf{H})^l\,\delta(\mathbf{k}). \tag{30b}$$

In the non-relativistic approximation ($|\mathbf{p}|/m_{\text{obs}} \ll 1$), the formulae for the bispinors given in Appendix J assume the form

$$\bar{u}(\mathbf{p}, r)\gamma^k u(\mathbf{q}, s) \simeq w^\dagger(\mathbf{p}, r)\,[\mathbf{p}+\mathbf{q}+i\boldsymbol{\sigma}\times(\mathbf{p}-\mathbf{q})]_k w(\mathbf{q}, s), \tag{31a}$$

$$\bar{u}(\mathbf{p}, r)i\gamma^{kl}u(\mathbf{q}, s) \simeq 2m_{\text{obs}}\,\varepsilon_{klm}w^\dagger(\mathbf{p}, r)\sigma_m w(\mathbf{q}, s). \tag{31b}$$

Substitution of formulae (26), (30), and (31) into formula (25) for the energy correction in the approximation under consideration yields

$$\Delta E = -\sum_{r,s}\int \mathrm{d}\Gamma_p(m_{\text{obs}})f^*(\mathbf{p}, r)\,w^\dagger(\mathbf{p}, r)\left[\frac{e_{\text{obs}}}{2m_{\text{obs}}}F_1(0)(i\nabla_p\times\mathbf{p}) + \frac{e_{\text{obs}}}{2m_{\text{obs}}}(F_1(0)+F_2(0))\boldsymbol{\sigma}\right]w(\mathbf{p}, s)f(\mathbf{p}, s)\cdot\mathbf{H}. \tag{32}$$

The first term describes the potential energy of the magnetic moment associated with the orbital moment of a negaton in a magnetic field and the second term describes the potential energy of the interaction of the intrinsic magnetic moment of the negaton. The operator of the magnetic moment vector $\boldsymbol{\mu}$ is given by the formula

$$\boldsymbol{\mu} = \frac{e_{\text{obs}}}{2m_{\text{obs}}}\,(F_1(0)+F_2(0))\boldsymbol{\sigma}. \tag{33}$$

On adding the missing powers of $\hbar$ and c, we then obtain formula (29).

The value of the form factor F_2 at zero, which specifies the *anomalous magnetic moment of the electron*, may be calculated by determining the renormalized vertex function. This value has been calculated up to terms of the order α^3. The result obtained is#

$$\mu = \frac{e_{\text{obs}}\hbar}{2m_{\text{obs}}c}\left[1+\frac{\alpha}{2\pi}-0.328479\left(\frac{\alpha}{\pi}\right)^2+(1.49\pm0.25)\left(\frac{\alpha}{\pi}\right)^3\right]. \quad (34)$$

This result is in excellent agreement with experiments. The most precise measurements## have given a value of

$$\mu_{\text{exp}} = \frac{e_{\text{obs}}\hbar}{2m_{\text{obs}}c}\left[1+\frac{\alpha}{2\pi}-0.328479\left(\frac{\alpha}{\pi}\right)^2+(1.68\pm0.33)\left(\frac{\alpha}{\pi}\right)^3\right]. \quad (35)$$

The Lamb–Retherford Shift

To conclude this section we shall employ the equations derived above in order to describe the motion of an electron in the field of a nucleus. The assumption that the field is weak is not justified in this case. Perturbation theory with respect to the parameter α will be used in order to get the equations for ψ. In the lowest order of perturbation theory, eq. (15) is an ordinary Dirac equation in a given electromagnetic field $\mathfrak{A}_u(x)$; the solutions of this equation are known in the case of the Coulomb field. In second-order perturbation theory, eq. (15) becomes

$$\begin{aligned}[D_x+e_{\text{obs}}\gamma\cdot\mathfrak{A}_{\text{ren}}(x)]\psi(x)-\delta m^{(2)}\psi(x)\\ +Z_2^{(2)}[D_x+e_{\text{obs}}\gamma\cdot\mathfrak{A}_{\text{ren}}(x)]\psi(x)\\ +ie_{\text{obs}}^2\int d^4y\gamma^\mu K_F[x,y|\mathfrak{A}_{\text{ren}}]\gamma^\nu D_{\mu\nu}^F(x-y)\psi(y) = 0. \quad (36)\end{aligned}$$

The effective field $\mathfrak{A}_{\text{ren}}$ should be determined in the manner described

---

The α-order correction was determined by J. Schwinger (1948 and 1949b). The α^2-order correction was evaluated by R. Karplus and N. Kroll (1950), but their result contained a computational error. The correct result in this order was obtained by C. Sommerfield (1958) and A. Petermann (1958). Calculations in the order α^3 are the work of many authors and were performed with the aid of computers.

Systematic calculations have been recently performed by M. J. Levine and J. Wright (1917).

J. C. Wesley and A. Rich (1970).

at the beginning of this section. In second-order perturbation theory the effective field of a point nucleus is given by formula (9). Extremely complicated calculations are required to solve eq. (36) and thence to determine the energy level shifts relative to the levels described by the formula

$$E_{nj} = mc^2 \left[1 + \left(\frac{Z\alpha}{n - \left(j+\frac{1}{2}\right) + \sqrt{\left(j+\frac{1}{2}\right)^2 - Z^2\alpha^2}} \right)^2 \right]^{-\frac{1}{2}} \tag{37}$$

which follows from the Dirac equation in relativistic quantum mechanics. Hitherto, the complete solution of this problem is not known even in second-order perturbation theory. The greatest difficulty in this order of perturbation theory is encountered in calculating the function $K_F[x, y|\mathscr{A}]$ for the Coulomb potential; this function has not been determined in closed form. Very good approximate results have, however, been obtained, and not only in the second order of perturbation theory but also in the fourth order. Since a comparison can be made with the results of exceptionally precise experiments, most attention has been devoted to the determination of the separation between the levels $2S_{1/2}$ and $2P_{1/2}$ in the hydrogen atom (and in deuteron and ionized helium). When calculated in relativistic quantum mechanics, these levels coincide since the electron states corresponding to them differ only as to the value of the orbital quantum number l which does not figure in formula (37). This degeneracy is removed by the mutual interaction of electrons and photons. The level splitting determined experimentally for hydrogen is#

$$\Delta E_{\mathrm{H}}^{\mathrm{exp}} = (1057.90 \pm 0.06) \text{ MHz}. \tag{38}$$

The theoretical calculation of the *Lamb–Retherford shift* is the collective work of many authors.## A compilation of the results from these

The first experimental measurement of the energy difference between the $2S_{1/2}$ and $2P_{1/2}$ levels in hydrogen was performed by W. E. Lamb, Jr., and R. C. Retherford (1947). Precise measurements of this quantity were made by S. Triebwasser, E. S. Dayhoff, and W. E. Lamb, Jr. (1953). More recently these results have been confirmed in a paper by R. T. Robiscoe and T. W. Shyn (1970).

The first calculation of the Lamb–Retherford shift was carried out by H. Bethe (1947). A detailed analysis of this problem may be found in the report by S. J. Brodsky (1971).

works yields

$$\Delta E^t_{\mathrm{H}} = (1057.911 \pm 0.012) \text{ MHz}. \tag{39}$$

The theoretical result depends on the value of the fine-structure constant α used in the calculations. The value of α^{-1} used for obtaining the result in (39) was

$$\alpha^{-1} = 137.036. \tag{40}$$

This value is obtained from measurements† of the hyperfine structure of the energy levels of the hydrogen atom and from direct measurements of the quantity $2e_{\mathrm{obs}}/\hbar$ by means of the *Josephson effect* in solid-state physics.‡

29. THE LIMITS OF APPLICABILITY OF QUANTUM ELECTRODYNAMICS

The theory expounded in this book in principle describes a system of mutually interacting electrons and photons isolated from all other particles existing in nature. Since it seems that the interactions of muons with an electromagnetic field are governed by the same laws as the interactions of electrons, the theory might be formulated without difficulty to describe systems of electrons, muons, and photons interacting with each other by electromagnetic forces, weak interactions being disregarded. In reality, electrons and photons do not constitute an isolated system since all other known particles (perhaps with the exception of the neutrino) have an electromagnetic structure,§ and hence interact with photons and through them also with electrons.

† Even though the frequency $\Delta\nu = (1\,420\,405\,751.800 \pm 0.028)$ Hz of the radiation emitted by the hydrogen atom under a change of the mutual orientation of the negaton and proton spin in the ground state is one of the most accurately measured quantities in physics, the value of α determined from $\Delta\nu$ still involves a relatively large theoretical error resulting from the incomplete knowledge of the electromagnetic structure of the proton.

‡ Cf. the survey article by J. Clarke (1970).

§ An electromagnetic structure is possessed not only by charged particles but also by neutral particles such as the neutron, the K^0-meson, or the Λ-hyperon. These particles have a charge distribution and also a distribution of magnetic moment. Even the π^0-meson, although it has no charge distribution, does display a certain electromagnetic structure, as evidenced by its decay into two photons.

Inasmuch as processes of pair creation and annihilation occur, even if other particles could be removed from the immediate vicinity of electrons and photons, the influence of those particles on electromagnetic processes would be exerted through vacuum polarization effects. For example, the possibility of π^+-meson and π^--meson pairs being created in the vacuum modifies electron-electron interactions. For, we must take into account, among other things, the Feynman diagram in Fig. 29.1.

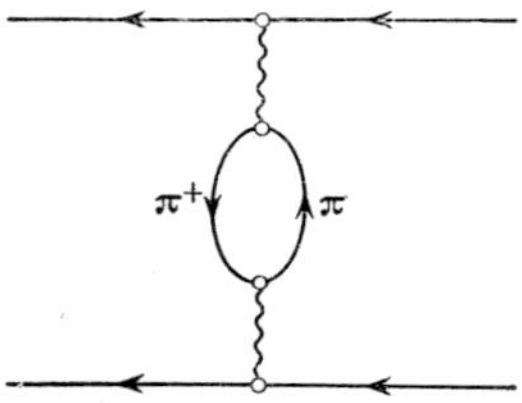

FIG. 29.1. The influence of π^+, π^- pair creation on electron scattering.

Although the system of electrons and photons could never be regarded as a system that is completely isolated from all other particles, it is frequently possible to evaluate the influence of those particles and either to demonstrate that it is negligible or to take approximate account of it in the description of the motion of the electrons and photons. The influence of the creation of pairs of other particles may in general be disregarded since electrons are much lighter (by a factor of at least 200) than all other charged particles. The production of pairs of heavier particles by photons is much less probable than the production of negaton-positon pairs.†

The influence of heavy particles in the vicinity of electrons (this refers chiefly to protons, as well as to neutrons) may be taken into account by introducing a phenomenological external electromagnetic field acting on the electrons and photons. A typical example of such

† The cross section for pair production at low energies is inversely proportional to the squared mass of the particles created. The correction to the cross section when the diagram in Fig. 29.1 is taken into account constitutes about 1/100 000 of the contribution to the cross section from an analogous diagram in which the negaton-positon pair plays the part of the π^+-π^- pair.

procedure is the replacement of the proton by a Coulomb field.† In more precise calculations we also take into account the charge distribution (departure from a point-like distribution) and the magnetic moment of heavy particles.

As the energy of electrons and photons increases, however, it becomes less and less justifiable to replace heavy particles by an external field. In order to be able to make provision for the influence of other particles, it is necessary to know the interaction of photons with those particles. Quantum electrodynamics does not determine these interactions. Quantum electrodynamical methods can be used only to describe "the electron-photon parts" of such processes. A typical example of the procedure employed in such cases is the analysis of the process of electron scattering on protons or on other nuclei.# In second-order perturbation theory with respect to e_{obs} such a process is depicted by the diagram in Fig. 29.2. The results of this experiment cannot be

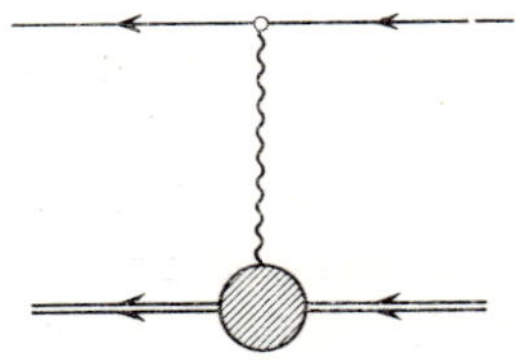

FIG. 29.2. Scattering of an electron on a proton.

predicted only on the basis of the laws of quantum electrodynamics. In addition we must know the mechanism of the photon-proton interaction for which no complete theory has yet been given.

The incomplete knowledge of the mechanism by which photons interact with heavy particles constitutes a major obstacle to the experimental verification of the applicability of the laws of electrodynamics to the description of processes occurring at high energies. For, in de-

† To obtain more accurate results, the kinematic effects (the motion of the proton around the common centre of mass) are also considered.

Experiments consisting of scattering electrons on nuclei were conducted for many years by R. Hofstadter and his co-workers (cf. R. Hofstadter (1963)).

scribing such processes we cannot use the external field approximation. Experiments with high-energy electrons are carried out by using targets made up of heavy particles, or by using two colliding beams travelling in opposite directions.

By comparing the results of various experiments with theoretical calculations, it was found that quantum electrodynamics gives the correct description of the mutual interaction of electrons and photons for energies up to the order of $2-6$ GeV.# The incomplete knowledge of non-electromagnetic interactions which play a part in these experiments does, however, constitute a serious hindrance to analysis of the results obtained. Not until experiments on colliding electron beams are completed will an unambiguous answer be forthcoming to the question: Does quantum electrodynamics "break down" at high energies, and if so, how?

Regardless of what the answer will be, renormalized perturbation theory will remain the fundamental method for calculating the effects of electromagnetic interactions at energies that are not too high.

Comparison of theory with experiment is contained in a review article by B. E. Lautrup, A. Peterman and E. de Rafael (1972). An extensive bibliography on this subject can be found there.

CONCLUDING REMARKS

QUANTUM electrodynamics, as no other physical theory, exists in many, quite different formulations. The reason for this lies both in the distinguished role which quantum electrodynamics has played, and indeed still plays, among quantum field theories (other field theories are patterned after quantum electrodynamics) and in the theoretical complications specific to quantum electrodynamics itself. The ideal agreement of the predictions of this theory with experiment reinforces our conviction that quantum electrodynamics is a successful physical theory. Attempts have been made to eliminate or circumvent the mathematical difficulties it entails by a suitable choice of formulation of the theory.

First and foremost, much effort has gone into eliminating formal infinite expressions which occur at intermediate stages in the calculations in electrodynamics as in other quantum field theories. This means total or partial elimination of renormalization. Such formulations have been devised by such authors as N. N. Bogolyubov and D. Shirkov (cf. their book), K. Nishijima (1960), O. Steinmann (1964), K. Hepp (1966), and P. K. Kuo and D. R. Yennie (1969). Some of these authors have achieved mathematically† more satisfactory formulations of the theory at the expense of its physical content. In our opinion, the renormalization procedure plays an important role in building a bridge between the *primary* quantities figuring in the fundamental equations

† The problem of convergence of perturbation series, however, remains unsolved and there is increasing evidence (see, for instance, a recent paper by B. Simon (1972)) to show that these series are only asymptotic. Recently S. Adler (1973a) has investigated the influence which the order of summation may exert on the sum of the perturbation series.

of the theory and the *secondary* quantities measured in experiments. *Renormalization should be performed in every field theory, regardless of whether the unrenormalized quantities occurring in it are finite or infinite.*

A number of interesting works of recent years have presented an analysis of quantum field equations which could serve as a basis for a theory of interacting fields. In these works (R.A. Brandt (1967), W. Zimmermann (1970), and other papers cited therein) it is shown that the renormalization constants figuring in equations for renormalized field operators can be eliminated by means of a suitable definition of products of operators all taken at the same point. Even though the proof has hitherto been carried out only in perturbation theory, there is some reason to hope that this method will enable results to be obtained without use of expansion in the coupling constant. The aforementioned papers constitute a development and more precise formulation of the method of J. G. Valatin (1954). These investigations were subsequently taken up by K. G. Wilson (1969) and are now being pursued further. The principal idea underlying all of these papers consists in replacing the formal products of the field operators in equations for interacting fields by so-called generalized normal products. These new products are well-defined (at least in perturbation theory) operator-valued distributions. A complete definition of these products cannot be given, however, without introducing many auxiliary distributions and complicated transitions to the limit. A shortcoming of this formulation is that it involves electron field operators which, as demonstrated by the discussion in our book, are objects that have been chosen somewhat unfortunately. For, in the theoretical structure of quantum electrodynamics these operators represent the acts of electron creation and annihilation, which processes are in glaring violation of the principle of charge conservation.

An additional difficulty in formulating quantum electrodynamics is that the photon mass vanishes, which leads to an infinite range of forces and gives rise to the infrared catastrophe. It seems to us that the best method for obviating the difficulties at the intermediate stages is to use the Proca theory. The problem of the gauge invariance of the theory and charge conservation is associated with the vanishing of the photon mass. In this book, we present our own point of view concerning gauge invariance, one based on the concept of a compensating current. Many papers have been devoted to the problem of gauge transformations (L. D. Landau and I. M. Khalatnikov (1955), B. Zumino (1960), I. Białynicki-Birula (1960 and 1962), and S. Mandelstam (1968), where different methods for dealing with gauge invariance were proposed.

In recent years, several papers (M. Baker and K. Johnson (1971a,b), G. Mack and K. Symanzik (1972), S. Adler (1972b)) have drawn attention to the role which transformations of the conformal group may play in quantum electrodynamics. This particular aspect has not been elucidated definitively as yet. First of all, it is not known to what extent the non-vanishing mass of the electron violates the invariance of the complete quantum electrodynamics under conformal transformations.

Appendix A

HILBERT SPACE

Hilbert space $\mathscr{H}$ is the name we give the vector space over the complex number field which is equipped with a scalar product $(\Phi|\Psi)$ and is complete in the norm $||\Psi||$ induced by that scalar product,

$$||\Psi||^2 = (\Psi|\Psi). \tag{1}$$

The *scalar product* is a linear function of the vector Ψ and an antilinear function of the vector Φ, and satisfies the conditions

$$(\Phi|\Psi)^* = (\Psi|\Phi), \tag{2a}$$

$$(\Psi|\Psi) \geqslant 0, \tag{2b}$$

$$(\Psi|\Psi) = 0 \Rightarrow \Psi = 0. \tag{2c}$$

LINEAR OPERATORS

The *domain* $D(A)$ of a linear operator A is the set of vectors on which the action of the operator is defined.

The *range* $R(A)$ of the operator A is the set of vectors resulting from the action of the operator A.

Henceforth we shall concern ourselves solely with linear operators whose domains are dense in $\mathscr{H}$.

The *adjoint operator* $A^\dagger$ to the operator A is defined as follows:

If for a given vector Φ there exists a vector $'\Phi$ such that for each vector Ψ belonging to $D(A)$ the equality

$$(\Phi|A\Psi) = ('\Phi|\Psi)$$

holds, then the vector Φ belongs to the domain of the operator $A^\dagger$ whereas the operation of the operator $A^\dagger$ on Φ is defined by the formula

$$A^\dagger\Phi = '\Phi.$$

The set of all vectors Φ for which corresponding vectors $'\Phi$ exist is the domain of $A^\dagger$.

The operator A is *symmetrical* (*formally self-adjoint*) if the equality

$$(A\Phi|\Psi) = (\Phi|A\Psi) \tag{3}$$

holds for any two vectors belonging to $D(A)$.

An operator A is said to be *self-adjoint* if A is a symmetrical operator and the domains of A and $A^\dagger$ are identical.

An operator B is a *bounded operator* if there exists an upper bound for the expression $||B\Psi||/||\Psi||$. The least upper bound is called the *norm* $||B||$ of the operator B,

$$||B|| = \sup(||B\Psi||/||\Psi||). \tag{4}$$

If an upper bound does not exist, the operator is said to be *unbounded.* A bounded operator can always be extended to all space. The domain of an unbounded operator is a dense set of vectors, different from all space.

A linear operator U, defined in all space, is a *unitary operator* if it satisfies the conditions

$$U^\dagger U = 1, \tag{4a}$$

$$UU^\dagger = 1. \tag{4b}$$

An antilinear operator† T is said to be an *anti-unitary operator* if it satisfies the same conditions (4) as does the unitary operator.

A *projection operator*‡ in Hilbert space is an operator P which satisfies the conditions

$$P^2 = P, \tag{5a}$$

$$P^\dagger = P. \tag{5b}$$

Each projection operator is bounded and its norm (with the exception of the trivial projection operator 0) is equal to unity.

The basic property of self-adjoint operators, on which their application in quantum theories rests in vast measure, is that a *spectral resolution*

† An operator T is said to be anti-linear if it satisfies the condition

$$T(\lambda\Psi+\mu\Phi) = \lambda^* T\Psi+\mu^* T\Phi.$$

‡ It is also called an orthogonal projection operator.

exists for them. This property is described by the following spectral theorem, for which the general proof was given by von Neumann.

Theorem. For each self-adjoint operator A defined in Hilbert space there exists exactly one family of projection operators $E_A(\lambda)$, called the *spectral family*, which meets the following conditions:

1. $E_A(\lambda)$ is a right-continuous function of the real parameter λ;
2. $E_A(\lambda)E_A(\mu) = E_A(\min(\lambda, \mu))$, that is, if $\lambda \leqslant \mu$, then
$$\mathscr{H}_\lambda \subset \mathscr{H}_\mu, \quad \text{where } \mathscr{H}_\lambda = E_A(\lambda)\mathscr{H};$$
3. $\lim\limits_{\lambda\to-\infty} E_A(\lambda) = 0, \quad \lim\limits_{\lambda\to+\infty} E_A(\lambda) = I;$

Both limits are taken to mean strong convergence.†

4. $AE_A(\lambda) = E_A(\lambda)A$;
5. $A = \int\limits_{-\infty}^{+\infty} \lambda \mathrm{d}E_A(\lambda)$.

This integral is taken to mean an operator-valued Stieltjes integral convergent in the norm.

With the help of the spectral theorem, the functions $f(A)$ of the operator A can be defined by the formula

$$f(A) = \int f(\lambda)\mathrm{d}E_A(\lambda). \tag{6}$$

The relation

$$f(A)g(A) = \int f(\lambda)g(\lambda)\mathrm{d}E_A(\lambda) \tag{7}$$

is then satisfied.

If the function $f(\lambda)$ is real-valued, the operator $f(A)$ is self-adjoint. Putting $f(\lambda) = 1$, we obtain the formula

$$I = \int \mathrm{d}E_A(\lambda), \tag{8}$$

which is called the *resolution of unity*.

---

† The sequence of bounded operators A_λ is strongly convergent to the limit $A = \lim\limits_{\lambda\to\infty} A_\lambda$ if for each vector Ψ the equality

$$A\Psi = \lim_{\lambda\to\infty}(A_\lambda \Psi)$$

holds.

The spectral theorem holds also for unitary operators but the spectral family is then a family of operators parametrized by φ varying between the limits $0 \leqslant \varphi < 2\pi$:

$$U = \int_{0}^{2\pi} e^{i\varphi} \mathrm{d}E_U(\varphi). \tag{9}$$

A *spectrum* $\sigma(A)$ can be defined for every linear operator A in Hilbert space. The spectrum of the operator A is defined as the complement of the *set of the resolvent* $\varrho(A)$ of the operator A. The set of the resolvent is a set of complex numbers z such that the operator $A-zI$ has a bounded inverse.

The spectrum of a self-adjoint operator in Hilbert space consists in general of a discrete spectrum and a continuous spectrum, both of which lie on the real axis. The discrete spectrum is the set of eigenvalues. The operator $A-zI$ for z's which are eigenvalues does not have an inverse. The continuous spectrum consists of the numbers z for which the operator $A-zI$ does possess an inverse, but that inverse is an unbounded operator.

APPENDIX B

CHRONOLOGICAL AND NORMAL PRODUCTS

CONSIDER a set of time-dependent operators $A_1, \ldots, A_n$. These may be both field operators and products of field operators taken at the same instant of time. We shall classify the operators A_i into two groups, one comprising fermion operators and the other boson operators. *Fermion operators* will be the name given to the product of an odd number of fermion field operators and an arbitrary number of boson field operators. On the other hand, *boson operators* are products of an even number of fermion fields and an arbitrary number of boson fields.

THE CHRONOLOGICAL PRODUCT

The *chronological product of operators,* $A_1(t_1), \ldots, A_n(t_n)$, is defined as

$$T(A_1(t_1) \ldots A_n(t_n))$$

$$\equiv \sum_{\text{perm}} \varepsilon_F \theta(t_{i_1}-t_{i_2}) \ldots \theta(t_{i_{n-1}}-t_{i_n}) A_{i_1}(t_{i_1}) \ldots A_{i_n}(t_{i_n}), \tag{1}$$

where the coefficient ε_F is equal to $+1$ or -1, depending on whether the permutation $A_{i_1} \ldots A_{i_n}$ is obtained from the permutation $A_1 \ldots A_n$ by an even or odd number of transpositions of fermion operators. Neither the transposition of boson operators nor the transposition of fermion operators with boson operators is taken into account. The definition of the chronological product implies the following property,

$$T(A_1(t_1) \ldots A_n(t_n)) = \varepsilon_F T(A_{i_1}(t_{i_1}) \ldots A_{i_n}(t_{i_n})). \tag{2}$$

Definition (1) will now be used to calculate the derivative of the chronological product of operators with respect to one of the time arguments. To fix our attention, we shall differentiate with respect to the variable t_1. This variable appears several times in each term in the sum (1): as an argument of the operator A_1, and as an argument of the function θ with a plus or minus sign. Differentiation of the function θ gives us the Dirac δ-function with plus or minus sign. Summing all the terms, we get:

$$\frac{\partial}{\partial t_1} T(A_1(t_1) \dots A_n(t_n)) = T\left(\frac{\partial}{\partial t_1} A_1(t_1) \dots A_n(t_n)\right)$$
$$+ \sum_{i=2}^{n} \pm T(A_2(t_2) \dots A_{i-1}(t_{i-1})[A_1(t_1), A_i(t_i)]_{\pm}$$
$$\times \delta(t_1 - t_i) A_{i+1}(t_{i+1}) \dots A_n(t_n)). \tag{3}$$

The minus sign in the sum is taken when both A_1 and an odd number of $A_2 \dots A_{i-1}$ are fermion operators, and the anticommutator when A_1 and A_i are fermion operators.

Let us consider the operator $\mathsf{U}(t, t_0)$ which satisfies the differential equations,

$$i \frac{\partial}{\partial t} \mathsf{U}(t, t_0) = \mathscr{H}(t)\mathsf{U}(t, t_0), \tag{4a}$$

$$-i \frac{\partial}{\partial t_0} \mathsf{U}(t, t_0) = \mathsf{U}(t, t_0)\mathscr{H}(t_0), \tag{4b}$$

and the initial condition,

$$\mathsf{U}(t_0, t_0) = 1. \tag{5}$$

The simultaneous solution of both these equations, satisfying condition (5), is a *chronological exponential operator*,† i.e. the following

† This solution may also be written as

$$\mathsf{U}(t, t_0) = \exp[i\Phi(t, t_0)],$$

where

$$\Phi(t, t_0) = \int_{t_0}^{t} d\tau \mathscr{H}(\tau) + \frac{i}{2} \int_{t_0}^{t} d\tau_2 \int_{t_0}^{t} d\tau_1 \varepsilon(\tau_2 - \tau_1)[\mathscr{H}(\tau_2), \mathscr{H}(\tau_1)] + \dots .$$

The full perturbation series representing the phase operator has been given in a paper by I. Białynicki-Birula, B. Mielnik and J. Plebański (1969).

series of multiple integrals of the chronological products of the operators $\mathscr{H}(\tau_i)$,

$$\mathsf{U}(t, t_0) = T\exp\left(-i\int_{t_0}^{t} d\tau\,\mathscr{H}(\tau)\right)$$

$$\equiv 1+\sum_{n=1}^{\infty}\frac{(-i)^n}{n!}\int_{t_0}^{t} d\tau_1 \ldots \int_{t_0}^{t} d\tau_n\, T(\mathscr{H}(\tau_1)\ldots\mathscr{H}(\tau_n)). \tag{6}$$

Let us verify this statement for the case of equation (4a). For this purpose, we replace this equation and the initial condition by the integral equation:

$$\mathsf{U}(t, t_0) = 1-i\int_{t_0}^{t} d\tau\mathscr{H}(\tau)\mathsf{U}(\tau, t_0).$$

The iterative solution of this equation is of the form

$$\mathsf{U}(t, t_0) = 1+\sum_{n=1}^{\infty}(-i)^n\int_{t_0}^{t} d\tau_n\int_{t_0}^{\tau_n} d\tau_{n-1}\ldots\int_{t_0}^{\tau_2} d\tau_1\mathscr{H}(\tau_n)\ldots\mathscr{H}(\tau_1).$$

The multiple integrals figuring in this formula may be rewritten as

$$\int_{t_0}^{t} d\tau_n\int_{t_0}^{t} d\tau_{n-1}\ldots\int_{t_0}^{t} d\tau_1\theta(\tau_n-\tau_{n-1})\ldots\theta(\tau_2-\tau_1)\mathscr{H}(\tau_n)\ldots\mathscr{H}(\tau_1)$$

$$=\frac{1}{n!}\int_{t_0}^{t} d\tau_n\ldots\int_{t_0}^{t} d\tau_1\sum_{\text{perm}}\theta(\tau_{i_1}-\tau_{i_2})\ldots\theta(\tau_{i_{n-1}}-\tau_{i_n})\mathscr{H}(\tau_{i_1})\ldots\mathscr{H}(\tau_{i_n})$$

$$=\frac{1}{n!}\int_{t_0}^{t} d\tau_n\ldots\int_{t_0}^{t} d\tau_1\, T(\mathscr{H}(\tau_n)\ldots\mathscr{H}(\tau_1)).$$

By virtue of eqs. (4) we have

$$\frac{\partial}{\partial t_1}(\mathsf{U}(t_2, t_1)\mathsf{U}(t_1, t_0)) = 0. \tag{7}$$

It thus follows that the operators U obey the composition law

$$\mathsf{U}(t_2, t_1)\mathsf{U}(t_1, t_0) = \mathsf{U}(t_2, t_0). \tag{8}$$

In similar fashion it may be shown that the operator U is unitary if the operator $\mathscr{H}$ is Hermitian. For, from eq. (4a) and its conjugate we get the equalities:

$$\frac{\partial}{\partial t}(\mathsf{U}^\dagger(t, t_0)\mathsf{U}(t, t_0)) = 0, \tag{9a}$$

$$\frac{\partial}{\partial t}(\mathsf{U}(t, t_0)\mathsf{U}^\dagger(t, t_0)) = 0. \tag{9b}$$

This leads to the unitarity relations:

$$\mathsf{U}^\dagger(t, t_0)\mathsf{U}(t, t_0) = 1, \tag{10a}$$

$$\mathsf{U}(t, t_0)\mathsf{U}^\dagger(t, t_0) = 1. \tag{10b}$$

Comparison of equalities (8) and (10), on the other hand, yields the relations

$$\mathsf{U}^\dagger(t, t_0) = \mathsf{U}(t_0, t), \tag{11a}$$

$$\mathsf{U}^{-1}(t, t_0) = \mathsf{U}(t_0, t). \tag{11b}$$

THE NORMAL PRODUCT

Creation and annihilation operators are said to stand in the *normal order* if all the annihilation operators are to the right of all the creation operators. The normal ordering of an arbitrary product of creation and annihilation operators consists in arranging these operators in the normal order. In doing this, it is convenient to adopt the convention that if an odd number of transpositions are required to put fermion creation and annihilation operators into the normal order, a minus sign is put in front of the result obtained after this normal ordering has been carried out. The operation of arranging creation and annihilation operators in the normal order, along with the aforementioned multiplication by -1, is denoted by colons at the beginning and the end of the expression being ordered. The product of the operators between the colons is called the *normal product* of the operators. The normal product of a sum of operators is, by definition, equal to the sum of the normal products. Some examples of normal products follow:

Fermions:

$$:a^\dagger a: \; = \; -:aa^\dagger: \; = a^\dagger a. \tag{12}$$

Bosons:

$$:c^\dagger c: \, = \, :cc^\dagger: \, = c^\dagger c. \tag{13}$$

Fermions and bosons:

$$:aa^\dagger cc^\dagger: \, = -a^\dagger c^\dagger ac, \tag{14a}$$

$$:a^\dagger ac^\dagger cc: \, = a^\dagger c^\dagger acc. \tag{14b}$$

Appendix C

FUNCTIONAL DIFFERENTIATION

The functional derivative is defined on the basis of the concept of the directional derivative of the functional.

Let $F[\varphi]$ be an arbitrary functional and φ_0 a particular function. If the limit

$$F_{\varphi_0}[\varphi] = \lim_{\lambda \to 0} \frac{F[\varphi + \lambda\varphi_0] - F[\varphi]}{\lambda} \tag{1}$$

exists, then we shall call it the *directional derivative of the functional F* at the point φ in the direction φ_0. The directional derivative is a generalization of the concept of partial derivative. The directional derivative in general depends on φ_0. For example, if the functional $F[\varphi]$ is analytic, i.e. has a convergent series expansion:†

$$F[\varphi] = \sum_{n=0}^{\infty} \frac{1}{n!} \int \mathrm{d}x_1 \dots \int \mathrm{d}x_n f(x_1, \dots, x_n) \varphi(x_1) \dots \varphi(x_n), \tag{2}$$

then

$$F_{\varphi_0}[\varphi] = \sum_{n=0}^{\infty} \frac{1}{n!} \int \mathrm{d}z \int \mathrm{d}x_1 \dots \int \mathrm{d}x_n \times f(z, x_1, \dots, x_n) \varphi_0(z) \varphi(x_1) \dots \varphi(x_n). \tag{3}$$

The directional derivative thus is, in this case, a linear functional of the function φ_0. The distribution determined by this linear functional is called the *functional derivative*. We shall denote it by the symbol

---

† The functions $\varphi(x)$ may be functions of many variables. The coefficients f are, in general, distributions. It will be assumed throughout that they are symmetrical with respect to their arguments.

$\delta F[\varphi]/\delta\varphi(z)$,

$$F_{\varphi_0}[\varphi] = \int \mathrm{d}z \frac{\delta F[\varphi]}{\delta\varphi(z)} \varphi_0(z). \tag{4}$$

The functional derivative could be defined by formula (4) whenever $F_{\varphi_0}[\varphi]$ is a functional that is linear in the function φ_0. If the functional $F[\varphi]$ is analytic, then by comparison of formulae (3) and (4) we get an explicit representation of the functional derivative:

$$\frac{\delta F[\varphi]}{\delta\varphi(x)} = \sum_{n=0}^{\infty} \frac{1}{n!} \int \mathrm{d}x_1 \ldots \int \mathrm{d}x_n f(x, x_1, \ldots, x_n)\varphi(x_1) \ldots \varphi(x_n). \tag{5}$$

A modification of the concept of functional derivative is that of the functional derivative with respect to the hypersurface at the point x. Let $f^\mu(x)$ be a vector field given in space-time. The integral

$$G[\sigma] = \int_\sigma \mathrm{d}\sigma_\mu f^\mu(x) \tag{6}$$

is a functional that is dependent on the function $\sigma(x)$ which determines the shape of the hypersurface σ. The difference of the values of a functional G for two hypersurfaces σ and σ_0, differing in a finite region, can be written in the following form by virtue of Gauss' theorem:

$$G[\sigma] - G[\sigma_0] = \int_\Omega \mathrm{d}^4x \partial_\mu f^\mu(x), \tag{7}$$

where Ω denotes the four-dimensional region contained between the hypersurfaces σ and σ_0. The divergence $\partial_\mu f^\mu(x)$ is called the functional derivative with respect to σ at the point x and is denoted by the symbol $\delta G/\delta\sigma(x)$. Formula (7) then becomes:

$$G[\sigma] - G[\sigma_0] = \int_\Omega \mathrm{d}^4x \frac{\delta G[\sigma]}{\delta\sigma(x)}. \tag{8}$$

Formula (8) is taken as a basis for generalizing the concept of the functional derivative with respect to the hypersurface σ to more complicated cases when the functional $G[\sigma]$ depends in an arbitrary manner on σ (not necessarily in the form (6)). In the general case the functional derivative with respect to σ is defined by the formula

$$\frac{\delta G[\sigma]}{\delta\sigma(x)} \equiv \lim_{\Omega\to x} \frac{G[\sigma]-G[\sigma_0]}{\int\limits_{\Omega} d^4x} \tag{9}$$

where the symbol $\Omega \to x$ means that the region Ω contained between the hypersurfaces σ and σ_0 shrinks to the point x (Fig. C.1).

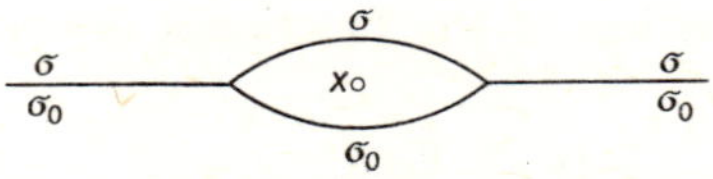

FIG. C.1. The domain of integration in formula (9).

The functional derivative with respect to σ at the point x is well defined only when the right-hand side of definition (9) does not depend on how the limit is evaluated.

APPENDIX D

THE POINCARÉ GROUP

THE set of all events forms a four-dimensional pseudo-euclidean affine space† which we shall call the *Minkowski space.* Points in this space will be denoted by the symbols x, x', etc. Points of Minkowski space are most frequently parametrized with four Cartesian coordinates, each of which runs over all real values. The contravariant and covariant coordinates, x^μ and x_μ $(\mu = 0, 1, 2, 3)$, are related to each other by the metric tensor

$$x_\mu = g_{\mu\nu} x^\nu, \tag{1}$$

$$(g_{\mu\nu}) \equiv \begin{pmatrix} 1 & 0 & 0 & 0 \\ 0 & -1 & 0 & 0 \\ 0 & 0 & -1 & 0 \\ 0 & 0 & 0 & -1 \end{pmatrix}. \tag{2}$$

The pseudo-distance s_{12} between the points x_1 and x_2 (the interval of two events) is defined in Minkowski space:

$$s_{12} \equiv g_{\mu\nu}(x_1^\mu - x_2^\mu)(x_1^\nu - x_2^\nu). \tag{3}$$

Since the matrix $g_{\mu\nu}$ is not positive definite, this pseudo-distance may be positive, negative, or zero.

The expression

$$a^2 \equiv g_{\mu\nu} a^\mu a^\nu \tag{4}$$

is called the length of the vector a^μ. Depending on whether $a^2 > 0$, $a^2 < 0$, or $a^2 = 0$, the vector a^μ is called a *time-like, space-like*, or *nul vector.*

---

† For extended treatment of the material presented in this appendix the reader is referred to the book by Schwartz.

The set of all linear transformations of the Minkowski space into itself, preserving the pseudo-distance (3) between points, forms a group known as the *Poincaré group*. This is a ten-parameter Lie group whose elements are most commonly labelled with a pair consisting of a vector a^μ and a matrix $\Lambda^\mu{}_\nu$. The Minkowski-space transformation corresponding to the pair (a, Λ) associates with the point x of coordinates x^μ a point $'x$ of coordinates $'x^\mu$ which are related to the coordinates x^μ by the formulae

$$'x^\mu = a^\mu + \Lambda^\mu{}_\nu x^\nu. \tag{5}$$

The assumption that the pseudo-distance (3) between the points is invariant implies the following conditions for the components of the matrix $\Lambda^\mu{}_\nu$:

$$\Lambda^\mu{}_\nu \Lambda^\lambda{}_\varrho g_{\mu\lambda} = g_{\nu\varrho}. \tag{6}$$

This equation may also be written in the form of a matrix formula†

$$\hat{\Lambda}^T \hat{g} \hat{\Lambda} = \hat{g}. \tag{7}$$

Equations (6) constitute ten conditions for the sixteen components of the matrix $\hat{\Lambda}$. The remaining six arbitrary parameters, together with the four parameters defining the vector a^μ, form a set of ten parameters of the Poincaré group. It follows from eq. (6) or (7) that the matrix $\Lambda^\mu{}_\nu$ satisfies the conditions

$$|\Lambda^0{}_0| \geqslant 1, \tag{8}$$

$$\det \hat{\Lambda} = \pm 1. \tag{9}$$

The following subgroups may be distinguished in the group of Poincaré transformations:

1. An Abelian *translation group* consisting of all the transformations for which $\hat{\Lambda} = I$, where I is a unit matrix.
2. A homogeneous Poincaré group, known as the *Lorentz group*, consisting of all those transformations for which $a^\mu = 0$.
3. An *orthochronous Poincaré group* (or possibly orthochronous Lorentz group) consisting of the transformations which satisfy the condition

$$\Lambda^0{}_0 \geqslant 1. \tag{10}$$

† To emphasize that a given symbol denotes a matrix we shall put a hat above it.

4. The *proper Poincaré group*† (or possibly the proper Lorentz group) consisting of the transformations which satisfy the conditions

$$\Lambda^0{}_0 \geqslant 1, \tag{11a}$$

$$\det \hat{\Lambda} = 1. \tag{11b}$$

Every transformation of the full Poincaré group may be obtained by the composition of a certain proper Poincaré transformation with one of three inversions: spatial inversion, time inversion, and space-time inversion. The matrices $\hat{\Lambda}$ corresponding to these inversions are of the following form:

$$\textit{spatial inversion} \qquad \hat{\Lambda} = \mathscr{P}, \tag{12a}$$

$$\textit{time inversion} \qquad \hat{\Lambda} = -\mathscr{P}, \tag{12b}$$

$$\textit{space-time inversion} \qquad \hat{\Lambda} = -I, \tag{12c}$$

where

$$\mathscr{P} \equiv \begin{pmatrix} 1 & 0 & 0 & 0 \\ 0 & -1 & 0 & 0 \\ 0 & 0 & -1 & 0 \\ 0 & 0 & 0 & -1 \end{pmatrix}. \tag{13}$$

The inhomogeneous Poincaré transformation (5) is conveniently written in homogeneous form. To this end, we assign to each point x a vector X with five components:

$$X = \begin{vmatrix} x^0 \\ x^1 \\ x^2 \\ x^3 \\ 1 \end{vmatrix}, \tag{14}$$

and to each Poincaré transformation a fifth-order matrix of the form:

$$\hat{\Pi} \equiv \begin{vmatrix} \hat{\Lambda} & \vdots & a \\ \cdots & \cdots & \cdots \\ 0 & \vdots & 1 \end{vmatrix}. \tag{15}$$

† At times, the term orthochronous proper Poincaré group is also used.

The Poincaré transformation (5) may then be rewritten as the formula:

$$'X = \hat{\Pi}X. \tag{16}$$

The composition of two Poincaré transformations corresponds to the matrix multiplication of the matrices $\hat{\Pi}$,

$$\underset{2}{\hat{\Pi}}\underset{1}{\hat{\Pi}} = \left|\begin{array}{c:c} \underset{2}{\hat{\Lambda}}\,\underset{1}{\hat{\Lambda}} & \underset{2}{\hat{\Lambda}}\,\underset{1}{a}+\underset{2}{a} \\ \hdashline 0 & 1 \end{array}\right|. \tag{17}$$

The set of matrices $\hat{\Pi}$ thus forms a representation of the Poincaré group.

Every matrix $\hat{\Pi}$ of the proper Poincaré transformation may be written in exponential form:#

$$\hat{\Pi} = \exp(\alpha^{\mu}\hat{\mathscr{H}}_{\mu}+\frac{1}{2}\,\alpha^{\mu\nu}\hat{\mathscr{H}}_{\mu\nu}), \tag{18}$$

where the ten matrices $\hat{\mathscr{H}}_{\mu}$ and $\hat{\mathscr{H}}_{\mu\nu}$ are defined as follows:

$$\hat{\mathscr{H}}_0 = \begin{vmatrix} 0&0&0&0&1\\ 0&0&0&0&0\\ 0&0&0&0&0\\ 0&0&0&0&0\\ 0&0&0&0&0 \end{vmatrix}, \quad \hat{\mathscr{H}}_1 = \begin{vmatrix} 0&0&0&0&0\\ 0&0&0&0&1\\ 0&0&0&0&0\\ 0&0&0&0&0\\ 0&0&0&0&0 \end{vmatrix},$$

$$\hat{\mathscr{H}}_2 = \begin{vmatrix} 0&0&0&0&0\\ 0&0&0&0&0\\ 0&0&0&0&1\\ 0&0&0&0&0\\ 0&0&0&0&0 \end{vmatrix}, \quad \hat{\mathscr{H}}_3 = \begin{vmatrix} 0&0&0&0&0\\ 0&0&0&0&0\\ 0&0&0&0&0\\ 0&0&0&0&1\\ 0&0&0&0&0 \end{vmatrix}, \tag{19a}$$

Unfortunately, there is no simple proof of this theorem. In the case of proper Lorentz transformations the proof may be found in a paper by J. Ehlers, W. Rindler and I. Robinson (1966). The generalization to the case of proper Poincaré transformations is easily made.

$$\hat{\mathscr{H}}_{01} = \begin{vmatrix} 0 & 1 & 0 & 0 & 0 \\ 1 & 0 & 0 & 0 & 0 \\ 0 & 0 & 0 & 0 & 0 \\ 0 & 0 & 0 & 0 & 0 \\ 0 & 0 & 0 & 0 & 0 \end{vmatrix}, \quad \hat{\mathscr{H}}_{02} = \begin{vmatrix} 0 & 0 & 1 & 0 & 0 \\ 0 & 0 & 0 & 0 & 0 \\ 1 & 0 & 0 & 0 & 0 \\ 0 & 0 & 0 & 0 & 0 \\ 0 & 0 & 0 & 0 & 0 \end{vmatrix},$$

$$\hat{\mathscr{H}}_{03} = \begin{vmatrix} 0 & 0 & 0 & 1 & 0 \\ 0 & 0 & 0 & 0 & 0 \\ 0 & 0 & 0 & 0 & 0 \\ 1 & 0 & 0 & 0 & 0 \\ 0 & 0 & 0 & 0 & 0 \end{vmatrix}, \quad \hat{\mathscr{H}}_{12} = \begin{vmatrix} 0 & 0 & 0 & 0 & 0 \\ 0 & 0 & 1 & 0 & 0 \\ 0 & -1 & 0 & 0 & 0 \\ 0 & 0 & 0 & 0 & 0 \\ 0 & 0 & 0 & 0 & 0 \end{vmatrix}, \tag{19b}$$

$$\hat{\mathscr{H}}_{23} = \begin{vmatrix} 0 & 0 & 0 & 0 & 0 \\ 0 & 0 & 0 & 0 & 0 \\ 0 & 0 & 0 & 1 & 0 \\ 0 & 0 & -1 & 0 & 0 \\ 0 & 0 & 0 & 0 & 0 \end{vmatrix}, \quad \hat{\mathscr{H}}_{31} = \begin{vmatrix} 0 & 0 & 0 & 0 & 0 \\ 0 & 0 & 0 & -1 & 0 \\ 0 & 0 & 0 & 0 & 0 \\ 0 & 1 & 0 & 0 & 0 \\ 0 & 0 & 0 & 0 & 0 \end{vmatrix}.$$

The matrices $\hat{\mathscr{H}}$ satisfy the following commutation rules characteristic of Poincaré transformation generators,

$$[\hat{\mathscr{H}}_\mu, \hat{\mathscr{H}}_\nu] = 0, \tag{20a}$$

$$[\hat{\mathscr{H}}_{\mu\nu}, \hat{\mathscr{H}}_\lambda] = g_{\mu\lambda}\hat{\mathscr{H}}_\nu - g_{\nu\lambda}\hat{\mathscr{H}}_\mu, \tag{20b}$$

$$[\hat{\mathscr{H}}_{\mu\nu}, \hat{\mathscr{H}}_{\lambda\varrho}] = g_{\mu\lambda}\hat{\mathscr{H}}_{\nu\varrho} - g_{\mu\varrho}\hat{\mathscr{H}}_{\nu\lambda} + g_{\nu\varrho}\hat{\mathscr{H}}_{\mu\lambda} - g_{\nu\lambda}\hat{\mathscr{H}}_{\mu\varrho}. \tag{20c}$$

In formulae (20) the signs are opposite to those in the analogous commutation rules for the generators P_μ and $M_{\mu\nu}$, or $\mathscr{T}_\mu$ and $\mathscr{T}_{\mu\nu}$. The difference is due to different composition laws for the transformations. The Poincaré transformation of a field in Minkowski space induces the inverse transformation $\hat{\Pi}^{-1}$ of points of Minkowski space (cf. formula (6.12)). Thus, if transformations of points of Minkowski space into themselves are induced by the Poincaré transformations of fields, what follows from this is the natural composition law (multiplication of matrices from right to left) for the inverse transformations of Minkowski space. The difference in sign results from the fact that in this appendix we have adopted the natural composition law for Poincaré transformations of space, and not for the inverse transformations.

The matrix $\hat{\Pi}$ given by formula (18) may be written as a product of two exponential matrices,

$$\hat{\Pi} = \exp(a^\mu \hat{\mathscr{H}}_\mu)\exp\left(\frac{1}{2}\alpha^{\mu\nu}\hat{\mathscr{H}}_{\mu\nu}\right). \tag{21}$$

The proof of this formula will be carried out by considering a one-parameter family $\hat{\Pi}(\lambda)$ of matrices of Poincaré transformations:

$$\hat{\Pi}(\lambda) \equiv \exp\left(\lambda\alpha^{\mu}\hat{\mathscr{H}}_{\mu}+\frac{1}{2}\lambda\alpha^{\mu\nu}\hat{\mathscr{H}}_{\mu\nu}\right). \tag{22}$$

The matrix $\hat{\Sigma}(\lambda)$,

$$\hat{\Sigma}(\lambda) \equiv \hat{\Pi}(\lambda)\exp\left(-\frac{1}{2}\lambda\alpha^{\mu\nu}\hat{\mathscr{H}}_{\mu\nu}\right), \tag{23}$$

satisfies the following differential equation with respect to the parameter λ,

$$\frac{\mathrm{d}\hat{\Sigma}(\lambda)}{\mathrm{d}\lambda} = \hat{\Sigma}(\lambda)\exp\left(\frac{\lambda}{2}\alpha^{\mu\nu}\hat{\mathscr{H}}_{\mu\nu}\right)\alpha^{\lambda}\hat{\mathscr{H}}_{\lambda}\exp\left(-\frac{\lambda}{2}\alpha^{\varrho\delta}\hat{\mathscr{H}}_{\varrho\delta}\right). \tag{24}$$

Knowing the commutation relations (20), we can use formula (H. 1) to calculate the product of the matrices on the right-hand side of eq. (24). In this way we arrive at the equation:

$$\frac{\mathrm{d}\hat{\Sigma}(\lambda)}{\mathrm{d}\lambda} = \hat{\Sigma}(\lambda)\hat{\mathscr{H}}_{\mu}(\exp(-\lambda\hat{\alpha}))^{\mu}{}_{\nu}\alpha^{\nu}. \tag{25}$$

Since all the matrices $\hat{\mathscr{H}}_{\mu}$ commute with each other, eq. (25) can be solved in the same way as a simple (non-matrix) differential equation. The solution of this equation is

$$\hat{\Sigma}(\lambda) = \exp(\hat{\mathscr{H}}_{\mu}\int_{0}^{\lambda}\mathrm{d}\tau(\exp(-\tau\hat{\alpha}))^{\mu}{}_{\nu}\alpha^{\nu}). \tag{26}$$

On inserting this solution into formula (23) and putting $\lambda = 1$ we obtain the proof of representation (21).

Appendix E

GREEN'S FUNCTIONS

THE Green's function (or fundamental solution) for a differential equation in n variables,

$$\sum_l \sum_{k_1,\ldots,k_l} a^{k_1\ldots k_l}(x_1,\ldots,x_n)\,\partial_{k_1}\ldots\partial_{k_l} f(x_1,\ldots,x_n) = 0, \tag{1}$$

is the name we give to every distribution $G(x_1,\ldots,x_n; y_1,\ldots,y_n)$ dependent on $2n$ variables, which satisfies the equation:

$$\sum_l \sum_{k_1,\ldots,k_l} a^{k_1\ldots k_l}(x_1,\ldots,x_n)\,\partial_{k_1}\ldots\partial_{k_l} G(x_1,\ldots,x_n; y_1,\ldots,y_n) =$$

$$= \delta(x_1-y_1)\ldots\delta(x_n-y_n). \tag{2}$$

The Green's functions for a given differential equation differ from each other as to the boundary conditions they satisfy. In this appendix we shall discuss the properties of the Green's functions for the Schrödinger, Klein-Gordon, d'Alembert, Dirac, and Proca equations. These are equations of type (1) in which the coefficients $a^{k_1\ldots k_l}$ are constant. Equations (1) with constant coefficients have a well-developed mathematical theory of Green's functions, based on distribution theory. The theory of Green's functions is particularly simple in cases when we consider boundary conditions imposed at infinity. The Fourier integral method may then be used. Our discussion will be confined to just such cases.

We shall seek the Green's functions in the form of the following Fourier integral:

$$G(x_1,\ldots,x_n; y_1,\ldots,y_n)$$

$$= \frac{1}{(2\pi)^n}\int \mathrm{d}p_1\ldots\mathrm{d}p_n \exp\left(-i\sum_{k=1}^{n} p_k(x_k-y_k)\right)\tilde{G}(p_1,\ldots,p_n). \tag{3}$$

The Fourier transform of the Green's function $\tilde{G}(p_1, \ldots, p_n)$ satisfies the following equation,

$$\sum_l (-i)^l \sum_{k_1, \ldots, k_l} a^{k_1 \ldots k_l} p_{k_1} \ldots p_{k_l} \tilde{G}(p_1, \ldots, p_n) = 1, \tag{4}$$

by virtue of eq. (2). The solutions of this equation are sought among the distributions. The general solution of this kind is of the form:

$$\tilde{G}(p_1, \ldots, p_n) = \mathrm{P} \frac{1}{\sum_l (-i)^l \sum_{k_1, \ldots, k_l} a^{k_1 \ldots k_l} p_{k_1} \ldots p_{k_l}}$$
$$+ i\pi f(p_1, \ldots, p_n)\, \delta\left(\sum_l (-i)^l \sum_{k_1, \ldots, k_l} a^{k_1 \ldots k_l} p_{k_1} \ldots p_{k_l}\right), \tag{5}$$

where $f(p_1, \ldots, p_n)$ is an arbitrary function† of the variables p_k. The choice of a particular function $f(p_1, \ldots, p_n)$ corresponds to the choice of particular boundary conditions for the Green's functions.

THE SCHRÖDINGER EQUATION

An arbitrary Green's function $\mathscr{K}_I(\mathbf{x}, t)$ for the Schrödinger equation satisfies the equation

$$\left(-i\partial_t - \frac{1}{2m}\Delta\right)\mathscr{K}_I(\mathbf{x}, t) = \delta(\mathbf{x})\,\delta(t). \tag{6}$$

The general solution of this equation may, in accordance with formula (5), be written as:

$$\mathscr{K}_I(\mathbf{x}, t)$$
$$= \int \frac{d\omega}{2\pi} \int \frac{d^3p}{(2\pi)^3} e^{-i\omega t} e^{i\mathbf{p}\cdot\mathbf{x}} \left[\mathrm{P}\frac{1}{\mathbf{p}^2/2m - \omega} + i\pi f(\mathbf{p})\, \delta\left(\frac{1}{2m}\mathbf{p}^2 - \omega\right)\right]. \tag{7}$$

The retarded and advanced Green's functions, $\mathscr{K}_R$ and $\mathscr{K}_A$, are particularly important in regard to applications. In order to obtain these functions we put $f(\mathbf{p}) = 1$ and $f(\mathbf{p}) = -1$, respectively, in formula (7).

† Since the function $f(p_1, \ldots, p_n)$ is a coefficient of the Dirac function, it depends in actual fact on only $n-1$ variables.

The Fourier transform of the functions $\mathscr{K}_{\text{R}}$ and $\mathscr{K}_{\text{A}}$ may be rewritten in the symbolical form:

$$\tilde{\mathscr{K}}_{\text{R}}(\mathbf{p}, \omega) = \frac{1}{\mathbf{p}^2/2m - \omega - i\varepsilon}, \tag{8a}$$

$$\tilde{\mathscr{K}}_{\text{A}}(\mathbf{p}, \omega) = \frac{1}{\mathbf{p}^2/2m - \omega + i\varepsilon}. \tag{8b}$$

This notation means that on integrating with respect to ω and $\mathbf{p}$ in formula (7), we must take the limit when $\varepsilon \to 0$ from the side of the positive values of the parameter ε. Using the residue method, we can perform the integrations in formula (7) for the functions $\mathscr{K}_{\text{R}}$ and $\mathscr{K}_{\text{A}}$,

$$\lim_{\varepsilon \to \text{C}} \int \frac{d\omega}{2\pi} e^{-i\omega t} \frac{1}{\mathbf{p}^2/2m - \omega \mp i\varepsilon}$$

$$= \pm i\theta(\pm t) \exp\left(-i\frac{\mathbf{p}^2}{2m} t\right). \tag{9}$$

We arrive at the same result by calculating separately the integral with the principal value and the integral with the part proportional to the δ-function. Finally, we have the following representations for the functions $\mathscr{K}_{\text{R}}$ and $\mathscr{K}_{\text{A}}$,

$$\mathscr{K}_{\text{R}}(\mathbf{x}, t) = \theta(t)\mathscr{K}(\mathbf{x}, t), \tag{10a}$$

$$\mathscr{K}_{\text{A}}(\mathbf{x}, t) = -\theta(-t)\mathscr{K}(\mathbf{x}, t), \tag{10b}$$

where

$$\mathscr{K}(\mathbf{x}, t) \equiv i \int \frac{d^3 p}{(2\pi)^3} e^{-it\mathbf{p}^2/2m} e^{i\mathbf{p}\cdot\mathbf{x}}. \tag{11}$$

Formulae (10) justify the names of *retarded* and *advanced* Green's functions. The Green's function $\mathscr{K}(\mathbf{x}, t)$ may be expressed in terms of the wave packets $\varphi_n[\mathbf{x}, t | f_n]$ which are solutions of the Schrödinger equation without potential,

$$\mathscr{K}(\mathbf{x}-\mathbf{y}, t-t') = i \sum_n \varphi[\mathbf{x}, t | f_n] \varphi^*[\mathbf{y}, t' | f_n]. \tag{12}$$

THE KLEIN–GORDON AND D'ALEMBERT EQUATIONS

An arbitrary Green's function $\Delta_I(x)$ for the Klein–Gordon equation satisfies the equation

$$(\Box + \mu^2)\Delta_{\text{F}}(x) = \delta^{(4)}(x). \tag{13}$$

The symbol x here denotes a set of four space-time variables. The general solution of this equation is given by

$$\Delta_F(x) = \int \frac{d^4p}{(2\pi)^4} e^{-ip\cdot x}\left[P\frac{1}{\mu^2-p^2} + i\pi f(p)\,\delta(\mu^2-p^2)\right]. \tag{14}$$

The retarded and advanced functions, Δ_R and Δ_A, correspond to the following choice of the functions $f(p) : f(p) = \operatorname{sgn} p_0$, and $f(p) = -\operatorname{sgn} p_0$. Integrating with respect to the variable p_0, we write the functions Δ_R and Δ_A in a form analogous to representation (10) for the functions $\mathscr{K}_R$ and $\mathscr{K}_A$,

$$\Delta_R(x) = \theta(x^0)\Delta(x), \tag{15a}$$

$$\Delta_A(x) = -\theta(-x^0)\Delta(x), \tag{15b}$$

where

$$\Delta(x) \equiv i\int d\Gamma(\mu)(e^{-ip\cdot x} - e^{ip\cdot x})$$

$$= i\int \frac{d^4p}{(2\pi)^3} e^{-ip\cdot x} \operatorname{sgn} p_0\, \delta(\mu^2 - p^2). \tag{16}$$

To get a Feynman and anti-Feynman functions, Δ_F and $\Delta_{\mathcal{F}}$, we must put $f(p) = 1$ and $f(p) = -1$, respectively, in formula (14). The four Green's functions under discussion may be expressed in terms of the Green's function $\bar{\Delta}$, defined by the formula

$$\bar{\Delta}(x) \equiv \int \frac{d^4p}{(2\pi)^4} e^{-ip\cdot x} P\frac{1}{\mu^2-p^2}, \tag{17}$$

and in terms of the functions $\Delta(x)$ and $\Delta^{(1)}(x)$:

$$\Delta^{(1)}(x) \equiv \int d\Gamma(\mu)(e^{-ip\cdot x} + e^{ip\cdot x})$$

$$= \int \frac{d^4p}{(2\pi)^3} e^{-ip\cdot x}\delta(\mu^2 - p^2). \tag{18}$$

The following relations are satisfied:

$$\Delta_R(x) = \bar{\Delta}(x) + \tfrac{1}{2}\Delta(x), \tag{19a}$$

$$\Delta_A(x) = \bar{\Delta}(x) - \tfrac{1}{2}\Delta(x), \tag{19b}$$

$$\Delta_F(x) = \bar{\Delta}(x) + \tfrac{1}{2} i\Delta^{(1)}(x), \tag{19c}$$

$$\Delta_{\text{Ⅎ}}(x) = \bar{\Delta}(x) - \tfrac{1}{2} i\Delta^{(1)}(x). \tag{19d}$$

The functions Δ_F and $\Delta_{\text{Ⅎ}}$ are symmetrical functions of the vector x,

$$\Delta_I(x) = \Delta_I(-x), \quad \text{when} \quad I = \text{F}, \text{Ⅎ},$$

whereas Δ_R and Δ_A are interchanged under the inversion $x \to -x$,

$$\Delta_R(x) = \Delta_A(-x).$$

The function $\bar{\Delta}(x)$ may be expressed in terms of the function $\Delta(x)$ by the formula

$$\bar{\Delta}(x) = \tfrac{1}{2}\varepsilon(x^0)\Delta(x). \tag{20}$$

The difference of the two Green's functions for the Klein–Gordon equation is the solution of that equation. In applications, the functions $\Delta^{(+)}(x)$ and $\Delta^{(-)}(x)$ also frequently appear in addition to the functions $\Delta(x)$ and $\Delta^{(1)}(x)$ already introduced. The former are defined by the formulae:

$$\Delta^{(+)}(x) \equiv i\int d\Gamma(\mu)e^{-ip\cdot x} = \tfrac{1}{2}(\Delta(x) + i\Delta^{(1)}(x)), \tag{21a}$$

$$\Delta^{(-)}(x) \equiv -i\int d\Gamma(\mu)e^{ip\cdot x} = \tfrac{1}{2}(\Delta(x) - i\Delta^{(1)}(x)). \tag{21b}$$

All four functions Δ, $\Delta^{(1)}$, $\Delta^{(+)}$ and $\Delta^{(-)}$ may be written as the differences of Green's functions in the following manner:

$$\Delta(x) = \Delta_R(x) - \Delta_A(x), \tag{22a}$$

$$\Delta^{(1)}(x) = \frac{1}{i}(\Delta_F(x) - \Delta_{\text{Ⅎ}}(x)), \tag{22b}$$

$$\Delta^{(+)}(x) = \Delta_F(x) - \Delta_A(x) = \Delta_R(x) - \Delta_{\text{Ⅎ}}(x), \tag{22c}$$

$$\Delta^{(-)}(x) = \Delta_R(x) - \Delta_F(x) = \Delta_{\text{Ⅎ}}(x) - \Delta_A(x). \tag{22d}$$

From the integral representation (16) of the function $\Delta(x)$ it follows that this function satisfies the initial conditions

$$\Delta(\mathbf{x}, 0) = 0, \tag{23a}$$

$$\frac{\partial}{\partial t}\Delta(\mathbf{x}, t)|_{t=0} = \delta(\mathbf{x}). \tag{23b}$$

When we put $\mu = 0$, the Klein–Gordon equation yields the d'Alembert equation, while the Green's functions (14) give the Green's functions for the d'Alembert equation. The latter functions are denoted by the letter D with appropriate indices. The functions D are related in the same way as are the functions Δ.

If we evaluate the Fourier integrals figuring in the definitions of the various Δ and D functions, we obtain the following expressions:

$$\Delta_F(x) = \frac{1}{4\pi}\delta(x^2) - \frac{\mu}{8\pi\sqrt{x^2}}\theta(x^2)[J_1(\mu\sqrt{x^2}) - iN_1(\mu\sqrt{x^2})]$$
$$+ \frac{\mu i}{4\pi^2\sqrt{-x^2}}\theta(-x^2)K_1(\mu\sqrt{-x^2}), \tag{24a}$$

$$\Delta(x) = \frac{1}{2\pi}\varepsilon(x^0)\delta(x^2) - \frac{\mu}{4\pi\sqrt{x^2}}\varepsilon(x^0)\theta(x^2)J_1(\mu\sqrt{x^2}), \tag{24b}$$

$$\Delta^{(+)}(x) = \frac{1}{4\pi}\varepsilon(x^0)\delta(x^2) + \frac{\mu i}{8\pi\sqrt{x^2}}\theta(x^2)[N_1(\mu\sqrt{x^2})$$
$$+ i\varepsilon(x^0)J_1(\mu\sqrt{x^2})] + \frac{\mu i}{4\pi^2\sqrt{-x^2}}\theta(-x^2)K_1(\mu\sqrt{-x^2}), \tag{24c}$$

$$\Delta^{(-)}(x) = \frac{1}{4\pi}\varepsilon(x^0)\delta(x^2) - \frac{\mu i}{8\pi\sqrt{x^2}}\theta(x^2)[N_1(\mu\sqrt{x^2})$$
$$- i\varepsilon(x^0)J_1(\mu\sqrt{x^2})] - \frac{\mu i}{4\pi^2\sqrt{-x^2}}\theta(-x^2)K_1(\mu\sqrt{-x^2}), \tag{24d}$$

$$D_F(x) = \frac{1}{4\pi}\delta(x^2) - \frac{i}{4\pi^2}\frac{1}{x^2}, \tag{25a}$$

$$D(x) = \frac{1}{2\pi}\varepsilon(x^0)\delta(x^2), \tag{25b}$$

$$D^{(+)}(x) = \frac{1}{4\pi}\varepsilon(x^0)\,\delta(x^2) - \frac{i}{4\pi^2}\frac{1}{x^2}, \tag{25c}$$

$$D^{(-)}(x) = \frac{1}{4\pi}\varepsilon(x^0)\,\delta(x^2) + \frac{i}{4\pi^2}\frac{1}{x^2}. \tag{25d}$$

The other functions may be built out of these four by using the foregoing relations.

THE DIRAC AND PROCA EQUATIONS

The Dirac equation and the Proca equation are, actually speaking sets of differential equations. The definition of Green's function can, however, be generalized in a natural manner so as to be applicable to such problems. The Green's function for the Dirac equation $S_{Ia}{}^{b}(x)$ is a fourth-order matrix, having two bispinor indices, whose components satisfy the set of equations

$$(-i\gamma^{\mu}\partial_{\mu}+m)_{a}{}^{b}\,S_{Ib}{}^{c}(x) = \delta_{a}^{c}\delta^{(4)}(x). \tag{26}$$

Making the substitution

$$S_{Ia}{}^{b}(x) = (i\gamma^{\mu}\partial_{\mu}+m)_{a}{}^{b}\Delta_{I}(x), \tag{27}$$

we can reduce the set (26) to equations for the Green's function of the Klein–Gordon equation. The appropriate Green's functions for the Dirac equations are obtained by the operation $i\gamma^{\mu}\,\partial_{\mu}+m$ on the Green's functions for the Klein–Gordon equation.

The problem of finding the Green's functions for the Proca equation can be likewise reduced to one of solving eq. (13). The set of equations

$$(-g^{\mu\lambda}\Box+\partial^{\mu}\partial^{\lambda}-g^{\mu\lambda}\mu^2)\Delta_{I\lambda\nu}(x) = \delta_{\nu}^{\mu}\delta^{(4)}(x) \tag{28}$$

is reduced to eq. (13) by the substitution

$$\Delta_{I\mu\nu}(x) = -(g_{\mu\nu}+\mu^{-2}\partial_{\mu}\partial_{\nu})\Delta_{I}(x). \tag{29}$$

Functions satisfying the homogeneous Dirac and Proca equations are constructed out of the appropriate functions for the Klein–Gordon equation by means of the operations $(i\gamma^{\mu}\partial_{\mu}+m)$ and $-(g_{\mu\nu}+\mu^{-2}\partial_{\mu}\,\partial_{\nu})$. Among functions of this type an important role is played by the functions $S(x)$ and $\Delta_{\mu\nu}(x)$ which figure in the commutation relations for the operators of the free Dirac field and the free Proca field. These functions satisfy the Dirac and Proca equations, respectively, and the following initial conditions:

$$S_a{}^b(\mathbf{x}, t)|_{t=0} = \gamma_a^{0b}\,\delta(\mathbf{x}), \tag{30a}$$

$$\Delta_{\lambda\varrho}(\mathbf{x}, t)|_{t=0} = \begin{cases} 0 & \lambda = 0,\ \varrho = 0, \\ 0 & \lambda = k,\ \varrho = l, \\ -\mu^{-2}\partial_i\,\delta(\mathbf{x}), & \lambda = i,\ \varrho = 0. \end{cases} \tag{30b}$$

These conditions follow from properties (23) of the function $\Delta(x)$.

Appendix F

THE SYMMETRIC ENERGY-MOMENTUM TENSOR

THE SYMMETRIC energy-momentum tensor is interpreted as a source of the gravitational field and is defined by the formula†

$$\delta_g \int d^4x \mathscr{L} \sqrt{-g} = \frac{1}{2}\int d^4x T_{\mu\nu}\, \delta g^{\mu\nu} \sqrt{-g}$$

$$= -\frac{1}{2}\int d^4x T^{\mu\nu} \delta g_{\mu\nu} \sqrt{-g}, \tag{1}$$

where the field of the metric tensor $g^{\mu\nu}$ is subjected to infinitesimal variation. The symbol g denotes the determinant of the matrix $g_{\mu\nu}$.

To evaluate the variation of the action, we must express the invariants of the electromagnetic field, S and P, in terms of the metric tensor. The field $f_{\mu\nu}$ is regarded as the basic field; the invariants S and P are given by the formulae:

$$S = -\tfrac{1}{4} g^{\mu\lambda} g^{\nu\varrho} f_{\mu\nu} f_{\lambda\varrho}, \tag{2a}$$

$$P = -\tfrac{1}{8}(-g)^{-1/2} \varepsilon^{\mu\nu\lambda\varrho} f_{\mu\nu} f_{\lambda\varrho}. \tag{2b}$$

From this we obtain the following formulae for the variations,

$$\delta_g S = -\tfrac{1}{2} g^{\lambda\varrho} f_{\mu\lambda} f_{\nu\varrho}\, \delta g^{\mu\nu}, \tag{3a}$$

$$\delta_g P = \tfrac{1}{2} P g_{\mu\nu}\, \delta g^{\mu\nu}. \tag{3b}$$

---

† The equivalence of the two definitions follows from the fact that the equality

$$g_{\mu\nu} g^{\nu\lambda} = \delta_\mu^\lambda$$

leads to

$$\delta g_{\mu\nu} = -\delta g^{\lambda\varrho} g_{\lambda\mu} g_{\varrho\nu}.$$

In order to get formula (3b), we have made use of the relation

$$\delta g = -g g_{\mu\nu}\,\delta g^{\mu\nu}. \tag{4}$$

The variation of the action for the electromagnetic field may be rewritten in the form

$$\delta_g \int d^4x \mathscr{L}\sqrt{-g} = \int d^4x \left[\mathscr{L}\,\delta\sqrt{-g} + \frac{\partial\mathscr{L}}{\partial S}\,\delta_g S\sqrt{-g} + \frac{\partial\mathscr{L}}{\partial P}\,\delta_g P\sqrt{-g}\right]$$

$$= \frac{1}{2}\int d^4x \left[\frac{\partial\mathscr{L}}{\partial S} f_{\mu\lambda} f^{\lambda}{}_{\nu} + \frac{\partial\mathscr{L}}{\partial P} P g_{\mu\nu} - g_{\mu\nu}\mathscr{L}\right]\delta g^{\mu\nu}\sqrt{-g}. \tag{5}$$

The formula which results from this for the energy-momentum tensor is as follows:

$$T_{\mu\nu} = \frac{\partial\mathscr{L}}{\partial S} f_{\mu\lambda} f^{\lambda}{}_{\nu} + \frac{\partial\mathscr{L}}{\partial P} P g_{\mu\nu} - g_{\mu\nu}\mathscr{L}. \tag{6}$$

This formula can be put into the form given in Section 6 if we use the identity

$$f_{\mu\lambda}\check{f}^{\lambda\nu} = \delta^{\nu}_{\mu} P. \tag{7}$$

We shall now calculate the variation of the action for the theory of a vector field with mass. The variation of the Lagrangian (14.1) is

$$\delta_g \mathscr{L} = \left(-\tfrac{1}{2} f_{\mu\lambda} f^{\lambda}{}_{\nu} + \tfrac{1}{2}\mu^2 A_\mu A_\nu\right)\delta g^{\mu\nu}. \tag{8}$$

On applying formula (4) to this, we have

$$\begin{aligned}\delta_g \int d^4x \mathscr{L}\sqrt{-g} &= \int d^4x \left(\delta_g \mathscr{L}\sqrt{-g} + \mathscr{L}\,\delta\sqrt{-g}\right)\\ &= \frac{1}{2}\int d^4x \left(f_{\mu\lambda} f^{\lambda}{}_{\nu} + \mu^2 A_\mu A_\nu - g_{\mu\nu}\mathscr{L}\right)\delta g^{\mu\nu}\sqrt{-g}.\end{aligned} \tag{9}$$

The energy-momentum tensor $T^{\mu\nu}$ for the vector field with mass is thus given by formula (14.11).

In the theory of the electromagnetic field interacting with point particles, the energy-momentum tensor $T^{\mu\nu}_{(m)}$ of the particles is obtained when that part of the action (8.4) which contains variables describing the particles is varied with respect to the metric tensor. For the basic variables we take ξ^{μ}_{A}. The expression $-\sum_A e_A \int d\xi^{\mu}_{A} A_\mu(\xi_A)$ thus does not depend on the metric tensor. Since $d\tau_A = \sqrt{g_{\mu\nu} d\xi_A^\mu d\xi_A^\nu}$, the variation of the second term in the expression (8.4) is:

$$\delta_g \Big\{ -\sum_A m_A \int d^4x \int d\tau_A \delta^{(4)}(x-\xi) \Big\}$$

$$= -\frac{1}{2} \sum_A m_A \int d^4x \int \frac{d\xi_A^\mu}{d\tau_A \sqrt{-g}} d\xi_A^\nu \delta^{(4)}(x-\xi)\, \delta g_{\mu\nu} \sqrt{-g}. \quad (10)$$

Whereas, on putting $\sqrt{-g} = 1$, we obtain expression (8.9) for the tensor $T^{\mu\nu}_{(m)}(x)$.

In the theory of the electromagnetic field coupled with a scalar complex field ϕ, the action $W_{em} + W_\phi$ is defined by formulae (8.16) and (8.17). Variation of the action W_{em} with respect to the metric tensor leads, by virtue of formulae (3a) and (4), to the expression:

$$\delta_g W_{em} = \frac{1}{2} \int d^4x (f_{\mu\lambda} f^\lambda{}_\nu - g_{\mu\nu} \mathscr{L}_{em}) \sqrt{-g}\, \delta g^{\mu\nu}. \quad (11)$$

The action W_ϕ will be rewritten in the equivalent form:

$$W_\phi = \int d^4x \sqrt{-g} \Big\{ g^{\mu\nu} \Big[\frac{1}{2} (D_\mu \phi)^* (D_\nu \phi) + \frac{1}{2} (D_\nu \phi)^* (D_\mu \phi) \Big] - m^2 \phi^* \phi \Big\} \quad (12)$$

and we treat $D_\mu \phi$ as the basic variables. We then obtain

$$\delta_g W_\phi = \frac{1}{2} \int d^4x [(D_\mu \phi)^* (D_\nu \phi) + (D_\nu \phi)^* (D_\mu \phi) - g_{\mu\nu} \mathscr{L}_\phi] \sqrt{-g}\, \delta g^{\mu\nu}. \quad (13)$$

When we compare formulae (11) and (13) with formula (1), we arrive at expression (8.20) for the energy-momentum tensor $T^{\mu\nu}(x)$.

APPENDIX G

EVALUATION OF SOME POISSON BRACKETS

WE SHALL evaluate the Poisson brackets of the components of the energy-momentum tensor of the electromagnetic field.

To find the values of the Poisson bracket of the components $T^{00}(\mathbf{x})$ and $T^{00}(\mathbf{y})$, it is sufficient merely to know relations (7.6):

$$\begin{aligned}\{T^{00}(\mathbf{x}), T^{00}(\mathbf{y})\} &= \{\mathscr{H}(\mathbf{x}), \mathscr{H}(\mathbf{y})\} \\ &= -E_k(\mathbf{x})\varepsilon_{klm}\partial_m\delta(\mathbf{x}-\mathbf{y})H_l(\mathbf{y})+H_k(\mathbf{x})\varepsilon_{klm}\partial_m\delta(\mathbf{x}-\mathbf{y})E_l(\mathbf{y}) \\ &= -\varepsilon_{klm}[E_k(\mathbf{y})H_l(\mathbf{y})+E_k(\mathbf{x})H_l(\mathbf{x})]\partial_m\delta(\mathbf{x}-\mathbf{y}) \\ &= -[T^{0k}(\mathbf{x})+T^{0k}(\mathbf{y})]\partial_k\delta(\mathbf{x}-\mathbf{y}). \end{aligned} \tag{1}$$

In evaluations of the Poisson brackets of the components $T^{00}(\mathbf{x})$ and $T^{0k}(\mathbf{y})$, it is convenient to use the equations

$$\{T^{00}(\mathbf{x}), D_k(\mathbf{y})\} = H_m(\mathbf{x})\varepsilon_{mkl}\partial_l\delta(\mathbf{x}-\mathbf{y}), \tag{2a}$$

$$\{T^{00}(\mathbf{x}), B_k(\mathbf{y})\} = -E_m(\mathbf{x})\varepsilon_{mkl}\partial_l\delta(\mathbf{x}-\mathbf{y}). \tag{2b}$$

With the aid of these equations, eqs. (7.3), and the relation

$$F(\mathbf{x})\partial_k\delta(\mathbf{x}-\mathbf{y}) = F(\mathbf{y})\partial_k\delta(\mathbf{x}-\mathbf{y})-\big(\partial_k F(\mathbf{x})\big)\delta(\mathbf{x}-\mathbf{y}), \tag{3}$$

we get

$$\begin{aligned}\{T^{00}(\mathbf{x}), T^{0k}(\mathbf{y})\} &= \{\mathscr{H}(\mathbf{x}), \varepsilon_{klm}D_l(\mathbf{y})B_m(\mathbf{y})\} \\ &= -E_r(\mathbf{x})\varepsilon_{rms}\varepsilon_{klm}\partial_s\delta(\mathbf{x}-\mathbf{y})D_l(\mathbf{y})+H_r(\mathbf{x})\varepsilon_{rls}\varepsilon_{klm}\partial_s\delta(\mathbf{x}-\mathbf{y})B_m(\mathbf{y}) \\ &= -\big[-E_k(\mathbf{x})D_l(\mathbf{x})-H_k(\mathbf{x})B_l(\mathbf{x}) \\ &\quad +\delta_{kl}\big(E_m(\mathbf{x})D_m(\mathbf{x})+H_s(\mathbf{x})B_s(\mathbf{x})-\mathscr{H}(\mathbf{x})\big)\big]\partial_l\delta(\mathbf{x}-\mathbf{y}) \\ &\quad +\mathscr{H}(\mathbf{y})\partial_k\delta(\mathbf{x}-\mathbf{y}) = -[T^{kl}(\mathbf{x})\partial_l+T^{00}(\mathbf{y})\partial_k]\delta(\mathbf{x}-\mathbf{y}). \end{aligned} \tag{4}$$

The Poisson brackets of the components $T^{0k}(\mathbf{x})$ and $T^{0l}(\mathbf{y})$ of the energy–momentum tensor are obtained by algebraic manipulations, wherein use is made of relation (3):

$$\{T^{0k}(\mathbf{x}), T^{0l}(\mathbf{y})\} = \{\varepsilon_{kmn} D_m(\mathbf{x}) B_n(\mathbf{x}), \varepsilon_{lrs} D_r(\mathbf{y}) B_s(\mathbf{y})\}$$
$$= \varepsilon_{kmn}\varepsilon_{lrs}[\varepsilon_{nrp} D_m(\mathbf{x}) B_s(\mathbf{y}) - \varepsilon_{msp} B_n(\mathbf{x}) D_r(\mathbf{y})]\, \partial_p \delta(\mathbf{x}-\mathbf{y})$$
$$= -[\delta_{lp}\varepsilon_{kmn} D_m(\mathbf{y}) B_n(\mathbf{y}) + \delta_{kp}\varepsilon_{lmn} D_m(\mathbf{x}) B_n(\mathbf{x})]\partial_p \delta(\mathbf{x}-\mathbf{y})$$
$$= -[T^{0l}(\mathbf{x})\, \partial_k + T^{0k}(\mathbf{y})\, \partial_l]\, \delta(\mathbf{x}-\mathbf{y}). \tag{5}$$

To find the Poisson brackets (7.17) for the generators of the Poincaré transformations, we integrate equalities (1), (4), and (5) with respect to the variables **x** and **y** and integrate these equalities after multiplying each side of them by x_i and y_i. As a result, we get the following formulae:

$$\begin{aligned}
\{P^0, P^0\} &= 0,\\
\{P^k, P^l\} &= 0,\\
\{P^k, P^0\} &= 0,\\
\{M^k, P^0\} &= 0,\\
\{M^k, P^l\} &= \varepsilon_{klm} P^m,\\
\{M^k, M^l\} &= \varepsilon_{klm} M^m,\\
\{M^k, N^l\} &= \varepsilon_{klm} N^m,\\
\{N^k, P^0\} &= P^k,\\
\{N^k, P^l\} &= \delta_{kl} P^0,\\
\{N^k, N^l\} &= -\varepsilon_{klm} M^m.
\end{aligned} \tag{6}$$

These formulae are the counterparts of formulae (7.17) in three-dimensional notation.

Appendix H

SOME OPERATOR IDENTITIES

In this appendix we shall prove several algebraic† relations between operators which we employ in various sections of the book.

The first relation has the form:

$$e^{A}Be^{-A} = B+\frac{1}{1!}[A, B]+\frac{1}{2!}[A, [A, B]]+ \dots \tag{1}$$

The proof will be carried out by considering a one-parameter family of operators $\exp(\lambda A)$. The operator

$$C(\lambda) = e^{\lambda A}Be^{-\lambda A} \tag{2}$$

satisfies the differential equation

$$\frac{dC(\lambda)}{d\lambda} = [A,C(\lambda)] \tag{3}$$

and the initial condition

$$C(0) = B. \tag{4}$$

The solution of eq. (3) obtained by series expansion in the parameter λ, when $\lambda = 1$, is of the same form as the right-hand side of identity (1).

This same method, based on solving a differential equation, will now be used in order to prove the *Baker–Hausdorff identity*,

$$e^{A+B} = e^{A}e^{B}e^{-1/2[A,B]} \tag{5}$$

which is satisfied by the operators A and B, if

$$[[A, B], A] = 0 = [[A, B], B]. \tag{6}$$

† We shall not consider problems connected with the domains of operators, the convergence of series figuring in the identities, etc.

By virtue of identity (1), the operator

$$C_1(\lambda) \equiv e^{-\lambda B} e^{-\lambda A} e^{\lambda(A+B)} \tag{7}$$

satisfies the differential equation

$$\frac{dC_1(\lambda)}{d\lambda} = -\lambda[A, B] C_1(\lambda) \tag{8}$$

and the initial condition

$$C_1(0) = 1. \tag{9}$$

The solution of eq. (8) which obeys condition (9) is of the form:

$$C_1(\lambda) = e^{-\frac{1}{2}\lambda^2[A,B]}. \tag{10}$$

On substituting this solution in formula (7) and putting $\lambda = 1$, we obtain identity (5).

Two-fold application of formula (5) yields the identity

$$e^A e^B = e^B e^A e^{-[A,B]}, \tag{11}$$

which is satisfied when assumptions (6) are fulfilled.

The proof of the analogous identity for time-dependent operators proceeds in similar fashion. If the operators $A(t)$ and $B(t)$ obey the commutation conditions,

$$[A(t), A(t')] = 0 = [B(t), B(t')], \tag{12a}$$

$$[[A(t), B(t')], A(t'')] = 0 = [[A(t), B(t')], B(t'')], \tag{12b}$$

then the following identity† holds:

$$T\exp\left(\int_{t_0}^{t} d\tau (A(\tau) + B(\tau))\right) = \exp\left(\int_{t_0}^{t} d\tau A(\tau)\right) \exp\left(\int_{t_0}^{t} d\tau B(\tau)\right)$$

$$\times \exp\left(\int_{t_0}^{t} d\tau_1 \int_{t_0}^{t} d\tau_2 \theta(\tau_1 - \tau_2)[B(\tau_1), A(\tau_2)]\right). \tag{13}$$

The proof may be carried out by showing that both sides of equality (13) satisfy the same differential equation in the variable t and the same initial conditions.

---

† This is a particular instance of the application of the so-called *continuous Baker–Hausdorff formula* derived in a paper by I. Białynicki-Birula, B. Mielnik, and J. Plebański (1969).

Identity (13) will be applied to the operators $A(t)$ and $B(t)$ defined as follows:

$$A(t) = i\int d^3x \mathbf{A}^{(-)\mathrm{in}}(\mathbf{x}, t)\cdot \mathscr{J}(\mathbf{x}, t), \tag{14a}$$

$$B(t) = i\int d^3x \mathbf{A}^{(+)\mathrm{in}}(\mathbf{x}, t)\cdot \mathscr{J}(\mathbf{x}, t). \tag{14b}$$

From the commutation relations (13.19) for the operators $\mathbf{A}^{(\pm)\mathrm{in}}(\mathbf{x})$ we obtain:

$$T\exp\left(i\int_{t_0}^{t} dt\int d^3x \mathbf{A}^{\mathrm{in}}(\mathbf{x}, t)\cdot \mathscr{J}(\mathbf{x}, t)\right)$$

$$= \exp\left(i\int_{t_0}^{t} dt\int d^3x \mathbf{A}^{(-)\mathrm{in}}(\mathbf{x}, t)\cdot \mathscr{J}(\mathbf{x}, t)\right)$$

$$\times \exp\left(i\int_{t_0}^{t} dt\int d^3x \mathbf{A}^{(+)\mathrm{in}}(\mathbf{x}, t)\cdot \mathscr{J}(\mathbf{x}, t)\right.$$

$$\times \exp\left(i\int_{t_0}^{t} dt_1\int_{t_0}^{t} dt_2\int d^3x\int d^3y\right.$$

$$\left.\times \mathscr{J}_k(x)\,(\delta_{kl}-\Delta^{-1}\partial_k\partial_l)\,\mathscr{J}_l(y)\theta(t_1-t_2)D^{(+)}(x-y)\right). \tag{15}$$

In the last exponential function on the right-hand side of equality (15), because of the symmetry of the integration region, we can replace the product $\theta(t_1-t_2)D^{(+)}(x-y)$ by its symmetrical part,

$$\theta(t_1-t_2)D^{(+)}(x-y) \to \tfrac{1}{2}[\theta(t_1-t_2)D^{(+)}(x-y)$$
$$+\theta(t_2-t_1)D^{(+)}(y-x)] = \tfrac{1}{2}D_{\mathrm{F}}(x-y). \tag{16}$$

In the limit, when $t\to\infty$ and $t_0\to-\infty$, we get formula (13.24).

In similar manner, we prove the formula:

$$T\exp\int d^4x(\bar{\eta}(x)\psi^{\mathrm{in}}(x)+\bar{\psi}^{\mathrm{in}}(x)\eta(x))$$

$$= :\exp\left[\int d^4x(\bar{\eta}(x)\psi^{\mathrm{in}}(x)+\bar{\psi}^{\mathrm{in}}(x)\eta(x))\right]:$$

$$\times \exp\left(\frac{1}{i}\int d^4x\,d^4y\,\bar{\eta}(x)\,S_{\mathrm{F}}(x-y)\,\eta(y)\right), \tag{17}$$

where $\eta(x)$ and $\bar{\eta}(x)$ are arbitrary auxiliary quantities which anticommute with each other and with the field operators ψ^{in} and $\bar{\psi}^{\text{in}}$.

For the proof we employ formula (13), in which we make the substitutions:

$$A(\tau) = \int d^3x(\bar{\eta}(\mathbf{x}, \tau)\psi^{(+)\text{in}}(\mathbf{x}, \tau) + \overline{\psi^{(-)\text{in}}}(\mathbf{x}, \tau)\eta(\mathbf{x}, \tau)),$$

$$B(\tau) = \int d^3x(\bar{\eta}(\mathbf{x}, \tau)\psi^{(-)\text{in}}(\mathbf{x}, \tau) + \overline{\psi^{(+)\text{in}}}(\mathbf{x}, \tau)\eta(\mathbf{x}, \tau)).$$

Calculation of the expectation value of both sides of equality (17) in the vacuum state Ω^{in} yields

$$\left(\Omega^{\text{in}}|T\exp\int d^4x\,(\bar{\eta}(x)\psi^{\text{in}}(x) + \bar{\psi}^{\text{in}}(x)\eta(x))\,\Omega^{\text{in}}\right)$$
$$= \exp\left(\frac{1}{i}\int d^4x\,d^4y\,\bar{\eta}(x)\,S_F(x-y)\,\eta(y)\right).$$

When we expand both exponential functions in this equality into series and compare the coefficients of the same products of the quantities η and $\bar{\eta}$, we get formula (18.50).

APPENDIX I

SPINORS

IN THIS appendix we shall introduce spinors and present the fundamentals of spinor calculus.# With this in mind let us examine the automorphisms of the ring of Pauli matrices.

The three Pauli matrices σ_i together with the unit matrix I form a basis in the linear space of second-order matrices. These four matrices will be denoted by the symbol σ^μ, where the index μ runs over the values 0, 1, 2, 3,

$$(\sigma^\mu) = (I, \boldsymbol{\sigma}). \tag{1}$$

Each matrix $'\sigma^\mu$ defined by the formula

$$'\sigma^\mu = S^\dagger \sigma^\mu S, \tag{2}$$

where S is an arbitrarily chosen matrix, may be written as a linear combination of the matrices σ^ν with real coefficients $L^\mu{}_\nu$,

$$S^\dagger \sigma^\mu S = L^\mu{}_\nu \sigma^\nu. \tag{3}$$

The coefficients $L^\mu{}_\nu$ satisfy the following relations,

$$L^\mu{}_\lambda L^\nu{}_\varrho g_{\mu\nu} = g_{\lambda\varrho} |\det S|^2. \tag{4}$$

In proving these relations, we may make use of the identity

$$\det(a_\mu \sigma^\mu) = a_\mu a^\mu \tag{5}$$

which holds for every vector a_μ. To do this, we can multiply eq. (3) by an arbitrary vector a_μ, evaluate the determinants on either side by means of identity (5), and then use the arbitrariness in the choice of the vector a_μ.

Spinors were introduced by E. Cartan in 1913. The name *spinor* itself, however, did not make its appearance until electron spin had been discovered. The formulation of spinor calculus given in this supplement is due to B. L. van der Waerden (1932) and L. Infeld and B. L. van der Waerden (1933). For an extension of the information given in this appendix the reader is referred to the book by Corson.

If the unimodularity condition

$$\det S = 1 \tag{6}$$

is imposed on the matrix S, the matrix of the coefficients $L^{\mu}{}_{\nu}$ will be the matrix of the Lorentz transformations $\Lambda^{\mu}{}_{\nu}$. Henceforth we shall assume that the unimodularity condition is always satisfied. Thus, each unimodular matrix S has a corresponding Lorentz transformation,

$$S \rightarrow \Lambda^{\mu}{}_{\nu}(S) \tag{7}$$

It follows from relations (3) that the composition of the corresponding Lorentz transformations is associated with the product of two matrices S; this means that

$$\Lambda^{\mu}{}_{\nu}(S_2\, S_1) = \Lambda^{\mu}{}_{\lambda}(S_2)\Lambda^{\lambda}{}_{\nu}(S_1). \tag{8}$$

In particular, we thus get

$$\Lambda^{\mu}{}_{\nu}(S^{-1}) = \Lambda^{-1\mu}{}_{\nu}(S), \tag{9a}$$

$$\Lambda^{\mu}{}_{\nu}(I) = \delta^{\mu}{}_{\nu}. \tag{9b}$$

To put the aforementioned associations in analytical form, we shall introduce a set of four matrices $\tilde{\sigma}^{\mu}$ defined by the formula

$$(\tilde{\sigma}^{\mu}) = (I, -\boldsymbol{\sigma}). \tag{10}$$

The matrices σ^{μ} and $\tilde{\sigma}^{\mu}$ satisfy the relations

$$\sigma^{\mu}\tilde{\sigma}^{\nu}+\sigma^{\nu}\tilde{\sigma}^{\mu} = 2g^{\mu\nu}I, \tag{11a}$$

$$\tilde{\sigma}^{\mu}\sigma^{\nu}+\tilde{\sigma}^{\nu}\sigma^{\mu} = 2g^{\mu\nu}I, \tag{11b}$$

$$\tfrac{1}{2}\operatorname{Tr}\{\sigma^{\mu}\tilde{\sigma}^{\nu}\} = g^{\mu\nu}. \tag{12}$$

By means of these formulae we obtain the following representation of the coefficients $\Lambda^{\mu}{}_{\nu}$:

$$\Lambda^{\mu}{}_{\nu}(S) = \tfrac{1}{2}\operatorname{Tr}\{S^{\dagger}\sigma^{\mu}S\tilde{\sigma}_{\nu}\}. \tag{13}$$

By virtue of properties (8) and (9), mapping (13) determines the representation of the group† $SL(2, C)$ by the group of Lorentz matrices $\Lambda^{\mu}{}_{\nu}$. This representation is not faithful since the same Lorentz matrix

$$\Lambda^{\mu}{}_{\nu}(S) = \Lambda^{\mu}{}_{\nu}(-S), \tag{14}$$

† This is a group of second-order unimodular matrices with complex coefficients.

corresponds to the matrices S and $-S$. The condition

$$S_1 = \pm S_2 \tag{15}$$

is a necessary as well as a sufficient condition for

$$\Lambda^{\mu}{}_{\nu}(S_1) = \Lambda^{\mu}{}_{\nu}(S_2). \tag{16}$$

This follows from relations (3) and from the fact that the matrices σ^{μ} form an irreducible set.

Formula (13) implies that

$$\Lambda^{0}{}_{0}(S) \geqslant 0. \tag{17}$$

Thus, only the matrices of orthochronous Lorentz transformations can have representation (13). Moreover, it follows from continuity arguments that matrices Λ whose determinant has a value of -1 cannot be represented in the form (13).

We shall prove that each matrix of a proper Lorentz transformation has a representation (13), i.e. that corresponding to each such transformation are exactly two unimodular second-order matrices (S and $-S$) related to Λ by (13).

The proof will consist in constructing a matrix S corresponding to an arbitrary matrix $\Lambda^{\mu}{}_{\nu}$ having a determinant of 1 and satisfying condition (17). For this construction we shall employ six matrices $S^{\mu\nu}$ defined as follows:

$$S^{\mu\nu} \equiv \tfrac{1}{2}(\tilde{\sigma}^{\mu}\sigma^{\nu} - \tilde{\sigma}^{\nu}\sigma^{\mu}), \tag{18}$$

that is,

$$S^{0i} = \sigma_i, \tag{19a}$$

$$S^{ij} = -i\sigma_k \qquad \text{(cyclically)}. \tag{19b}$$

The matrices $S^{\mu\nu}$ satisfy the following algebraic relations with the matrices σ^{μ},

$$\sigma^{\lambda}S^{\mu\nu} + S^{\dagger\mu\nu}\sigma^{\lambda} = 2g^{\mu\lambda}\sigma^{\nu} - 2g^{\nu\lambda}\sigma^{\mu} \tag{20}$$

and commutation relations such as those of the Lorentz transformation generators,

$$[S^{\mu\nu}, S^{\lambda\varrho}] = -g^{\mu\lambda}S^{\nu\varrho} + g^{\mu\varrho}S^{\nu\lambda} - g^{\nu\varrho}S^{\mu\lambda} + g^{\nu\lambda}S^{\mu\varrho}. \tag{21}$$

The matrices $S^{\mu\nu}$ can be used to form two unimodular second-order matrices, differing in sign, out of the coefficients $\alpha_{\mu\nu}$ which determine the matrix of a proper Lorentz transformation (cf. Appendix D):

$$S(\Lambda) = \pm\exp(-\tfrac{1}{4}\alpha_{\mu\nu}S^{\mu\nu}). \tag{22}$$

Each of these matrices satisfies the condition

$$S^{\dagger}(\Lambda)\sigma^{\mu}S(\Lambda) = (\exp(-\hat{\alpha}))^{\mu}{}_{\nu}\sigma^{\nu}. \tag{23}$$

The proof of this relation will be carried out by introducing a one-parameter family of transformations $S_{\lambda}(\Lambda)$,

$$S_{\lambda}(\Lambda) \equiv \pm\exp(-\tfrac{1}{4}\lambda\alpha_{\mu\nu}S^{\mu\nu}). \tag{24}$$

The matrices $\sigma^{\mu}_{(\lambda)}$, defined by the formula

$$\sigma_{(\lambda)} \equiv S^{\dagger}_{\lambda}(\Lambda)\sigma^{\mu}S_{\lambda}(\Lambda), \tag{25}$$

satisfy the differential equation

$$\frac{d\sigma^{\mu}_{(\lambda)}}{d\lambda} = -\alpha_{\nu}{}^{\mu}\sigma^{\nu}_{(\lambda)} \tag{26}$$

by virtue of relations (20). The solution of this equation is of the form

$$\sigma^{\mu}_{(\lambda)} = (\exp(-\lambda\hat{\alpha}))^{\mu}{}_{\nu}\sigma^{\nu}. \tag{27}$$

Putting $\lambda = 1$, we get formula (23) from this and from definition (25).

The matrix $S(\Lambda)$ may also be determined by solving the set of eq. (13) for the elements of this matrix. In this way, we obtain

$$S(\Lambda) = \pm[4-\mathrm{Tr}\,\Lambda^2+i\varepsilon^{\mu\nu\lambda\varrho}\Lambda_{\mu\nu}\Lambda_{\lambda\varrho}]^{-1/2}$$
$$\times[\Lambda^{\mu}{}_{\mu}+(\Lambda^{i}{}_{0}+\Lambda^{0}{}_{i}+i\varepsilon_{ijk}\Lambda^{j}{}_{k})\sigma_{i}].$$

In this way we have associated two matrices $S(\Lambda)$, defined by formula (22), with each proper Lorentz transformation Λ. As follows from formula (8), this association satisfies the conditions

$$S(\Lambda_2\Lambda_1) = \pm S(\Lambda_2)S(\Lambda_1) \tag{28}$$

and is known as the *two-valued representation*[#] of the proper Lorentz group. Two-valued representations appear here since the Lorentz group, just as the group of three-dimensional rotations, is a doubly-

---

A more extensive treatment of this topic may be found in the book by Hamermesh.

connected space. In order to avoid using multi-valued representations, covering groups are introduced into the theory of group representations. The *covering group* $\tilde{G}$ for an m-connected group G is a simply-connected group with the group G as its homomorphic image (the kernel of this homomorphism contains m elements). Because of its simple connectedness, the covering group has only single-valued representations. The covering group for a proper Lorentz group is the group $SL(2, C)$ whereas for a proper Poincaré group it is the non-homogeneous† $SL(2, C)$ group.

In addition to representation (22), the proper Lorentz group has one more non-equivalent two-valued representation in terms of unimodular second-order matrices. This representation is defined as follows:

$$\tilde{S}(\Lambda) = \pm\exp\left(\tfrac{1}{4}\alpha_{\mu\nu}S^{\dagger\mu\nu}\right). \tag{29}$$

The matrices $\tilde{S}_{\mu\nu} \equiv -S^{\dagger}{}_{\mu\nu}$ also satisfy the commutation relations for the generators of Lorentz transformations. These matrices cannot, however, be obtained from the matrices $S^{\mu\nu}$ by a similarity transformation. The matrices $\tilde{S}(\Lambda)$ satisfy the relations

$$\tilde{S}^{\dagger}(\Lambda)\tilde{\sigma}^{\mu}\tilde{S}(\Lambda) = \left(\exp(-\hat{\alpha})\right)^{\mu}{}_{\nu}\tilde{\sigma}^{\nu}. \tag{30}$$

The proof of this formula is based on the algebraic relations

$$\tilde{S}^{\dagger\mu\nu}\tilde{\sigma}^{\lambda}+\tilde{\sigma}^{\lambda}\tilde{S}^{\mu\nu} = 2g^{\mu\lambda}\tilde{\sigma}^{\nu}-2g^{\nu\lambda}\tilde{\sigma}^{\mu} \tag{31}$$

and is analogous to that for formula (23).

Now let us introduce two two-dimensional vector spaces over the field of complex numbers. Vectors belonging to these spaces will be denoted by the symbols ψ, φ, etc. and $\tilde{\psi}$, $\tilde{\varphi}$, etc.

Each proper Lorentz transformation Λ may have associated with it the transformation of each of these spaces into itself as defined by the formulae

$$\psi \to {}'\psi = S(\Lambda)\psi, \tag{32a}$$

$$\tilde{\varphi} \to {}'\tilde{\varphi} = \tilde{S}(\Lambda)\quad . \tag{32b}$$

† This is a group of elements (a, S) with the group operation $(a_2, S_2)(a_1, S_1) = (a_2+\Lambda(S_2)a_1, S_2S_1)$.

Two-dimensional vectors transforming† under Lorentz transformations according to the transformation laws (32a) and (32b) are called *spinors of the first kind* and *spinors of the second kind.*

Spinors of the first and second kind may depend on the point in Minkowski space. We then speak of spinor fields in space. Such a field is denoted by the symbols $\psi(x)$, $\tilde{\varphi}(x)$, etc.

An electron in an external electromagnetic field $\mathscr{A}_\mu(x)$ is described by two spinor fields $\psi(x)$ and $\tilde{\varphi}(x)$ which satisfy a set of differential equations of the form

$$\sigma^\mu\big(i\partial_\mu - e\mathscr{A}_\mu(x)\big)\psi(x) = m\tilde{\varphi}(x), \tag{33a}$$

$$\tilde{\sigma}^\mu\big(i\partial_\mu - e\mathscr{A}_\mu(x)\big)\tilde{\varphi}(x) = m\psi(x). \tag{33b}$$

The invariance of the forms of these equations with respect to proper Poincaré transformations follows from relations (23) and (30) and from the following transformation laws for spinor fields:

$$^{\mathrm{IL}}\psi(x) = S(\Lambda)\psi\big(\Lambda^{-1}(x-a)\big), \tag{34a}$$

$$^{\mathrm{IL}}\tilde{\varphi}(x) = \tilde{S}(\Lambda)\tilde{\varphi}\big(\Lambda^{-1}(x-a)\big). \tag{34b}$$

To ensure the invariance of eqs. (33) with respect to the full Poincaré group, it is necessary to introduce the transformations of spinors from one space to the other. Under spatial inversion spinors of the first kind go over into spinors of the second kind, and conversely. Under spatial and time inversions, the transformation laws of the pair $(\psi, \tilde{\varphi})$ are

$$^{\mathrm{P}}\psi(x) = \tilde{\varphi}(\mathscr{P}x), \tag{35a}$$

$$^{\mathrm{P}}\tilde{\varphi}(x) = \psi(\mathscr{P}x), \tag{35b}$$

$$^{\mathrm{T}}\psi(x) = \big(i\hat{\varepsilon}\psi(-\mathscr{P}x)\big)^*, \tag{36a}$$

$$^{\mathrm{T}}\tilde{\varphi}(x) = \big(i\hat{\varepsilon}\tilde{\varphi}(-\mathscr{P}x)\big)^*, \tag{36b}$$

where

$$\hat{\varepsilon} = \begin{pmatrix} 0 & -1 \\ 1 & 0 \end{pmatrix}. \tag{37}$$

---

† Since the representations $S(\Lambda)$ and $\tilde{S}(\Lambda)$ of the proper Lorentz group are two-valued, the transformation laws of bispinors are always defined only up to the sign. This arbitrariness does not, however, lead to any ambiguity in observables since spinor transformation laws are obeyed only by the wave functions and field operators of fermions, whose phases are in any event not measurable.

Similarly, under charge conjugation a spinor of the first kind $\psi(x)$ and of the second kind $\tilde{\varphi}(x)$ change places with each other,

$$^{C}\psi(x) = -\hat{\varepsilon}\tilde{\varphi}^{*}(x), \tag{38a}$$

$$^{C}\tilde{\varphi}(x) = \hat{\varepsilon}\psi^{*}(x). \tag{38b}$$

Four-dimensional objects consisting of a spinor of the first kind ψ and a spinor of the second kind $\tilde{\varphi}$ are called *bispinors*. Under proper Lorentz transformations, bispinors transform according to the reducible representation of the group $S \oplus \tilde{S}$. Bispinors, however, constitute an irreducible representation of an orthochronous Lorentz group.

To rewrite eq. (33) in a form most often used in quantum mechanics, four matrices defined as follows are introduced:

$$\gamma^{\mu} \equiv \begin{pmatrix} 0 & \tilde{\sigma}^{\mu} \\ \sigma^{\mu} & 0 \end{pmatrix}. \tag{39}$$

With their help, we write eqs. (33) as

$$\left(-i\gamma^{\mu}\partial_{\mu} + m + e\gamma^{\mu}\mathscr{A}_{\mu}(x)\right)\psi(x) = 0, \tag{40}$$

where

$$\psi(x) \equiv \begin{pmatrix} \psi(x) \\ \tilde{\varphi}(x) \end{pmatrix}. \tag{41}$$

Spinors provide a convenient and universal mathematical apparatus for investigating various geometrical objects which occur in the special, as well as the general, theory of relativity. *Spinor indices* are introduced in order to facilitate the notation of involved formulae containing spinors. These indices are denoted by the capital Roman letters A, B, etc., for spinors of the first kind and by capital Roman letters with dots $\dot{A}$, $\dot{B}$, etc., for spinors of the second kind. All of these indices take on the values 1 and 2. We shall assume that the spinor indices are also subject to the summation convention. It is convenient to place the undotted indices of spinors of the first kind as subscripts of the symbol ψ, and the dotted indices of spinors of the second kind as superscripts of the symbol $\tilde{\varphi}$. Transformation laws (32) in this notation are of the form

$$'\psi_{A} = S_{A}{}^{B}\psi_{B}, \tag{42a}$$

$$'\tilde{\varphi}^{\dot{A}} = \tilde{S}^{\dot{A}}{}_{\dot{B}}\tilde{\varphi}^{\dot{B}}. \tag{42b}$$

The condition for the unimodularity of the matrices $S_A{}^B$ and $S^{\dot{A}}{}_{\dot{B}}$ may be rewritten as

$$S_A{}^C S_B{}^D \varepsilon^{AB} = \varepsilon^{CD}, \tag{43a}$$

$$\varepsilon^{\dot{C}\dot{D}} \tilde{S}^{\dot{A}}{}_{\dot{C}} \tilde{S}^{\dot{B}}{}_{\dot{D}} = \varepsilon^{\dot{A}\dot{B}} \tag{43b}$$

where

$$(\varepsilon^{AB}) = \begin{pmatrix} 0 & -1 \\ 1 & 0 \end{pmatrix} = (\varepsilon^{\dot{A}\dot{B}}). \tag{44}$$

The Dirac equation (40) in this notation is of the form

$$\sigma^{\mu\dot{A}B}\big(i\partial_\mu - e\mathscr{A}_\mu(x)\big)\psi_B(x) = m\tilde{\varphi}^{\dot{A}}(x), \tag{45a}$$

$$\tilde{\sigma}^{\mu}_{A\dot{B}}\big(i\partial_\mu - e\mathscr{A}_\mu(x)\big)\tilde{\varphi}^{\dot{B}}(x) = m\psi_A(x). \tag{45b}$$

APPENDIX J

THE PROPERTIES OF SOLUTIONS OF THE DIRAC EQUATION

THE bispinors $u(\mathbf{p}, r)$ and $v(\mathbf{p}, r)$ are defined as solutions of the algebraic equations

$$(\gamma \cdot p - m)u(\mathbf{p}, r) = 0, \tag{1a}$$

$$(\gamma \cdot p + m)v(\mathbf{p}, r) = 0, \tag{1b}$$

which satisfy the following *orthogonality* and *normalization conditions*:

$$\bar{u}(\mathbf{p}, r)u(\mathbf{p}, s) = 2m\,\delta_{rs}, \tag{2a}$$

$$\bar{v}(\mathbf{p}, r)v(\mathbf{p}, s) = -2m\,\delta_{rs}. \tag{2b}$$

The bispinors u and v also satisfy the *completeness relations*

$$\sum_r u_a(\mathbf{p}, r)\bar{u}^b(\mathbf{p}, r) = (\gamma \cdot p + m)_a{}^b, \tag{3a}$$

$$\sum_r v_a(\mathbf{p}, r)\bar{v}^b(\mathbf{p}, r) = (\gamma \cdot p - m)_a{}^b. \tag{3b}$$

Relations (3) may be derived directly from the explicit representation (16.52) of the bispinors u and v, or from the following general considerations. The matrix $\sum_r u_a(\mathbf{p}, r)\bar{u}^b(\mathbf{p}, r)$ may be written as a linear combination of sixteen Dirac matrices since these matrices constitute a basis in the linear space of fourth-order matrices. Because of the transformation properties of the matrix in question under Lorentz transformations, only the coefficients of the scalar and vector parts can be non-zero. Hence

$$\sum_r u_a(\mathbf{p}, r)\bar{u}^b(\mathbf{p}, r) = (Am + B\gamma \cdot p)_a{}^b. \tag{4}$$

It follows from the normalization condition (2a) that the square of matrix (4) is equal to that matrix multiplied by $4m$. Thus, we finally arrive at $A = 1$ and $B = 1$.

The summation relations (16.59) are obtained from formula (3) by multiplying the bispinors $u(\mathbf{p}, r)$ and $v(\mathbf{p}, r)$, respectively, by $e^{-ip \cdot x}$

and $e^{ip\cdot x}$, and the conjugate bispinors by $e^{ip\cdot y}$ and $e^{-ip\cdot y}$, and integrating with respect to p. In doing this we make use of the definition (16.60) of the function $S^{(\pm)}$ and the definition (E.21) of the function $\Delta^{(\pm)}$.

In the Dirac representation of γ matrices, the general solution of eqs. (1) may be written as:

$$u(\mathbf{p}, r) = \sqrt{E_p+m}\begin{pmatrix} w(\mathbf{p}, r) \\ \dfrac{\boldsymbol{\sigma}\cdot\mathbf{p}}{E_p+m}\, w(\mathbf{p}, r)\end{pmatrix}, \tag{5a}$$

$$v(\mathbf{p}, r) = \sqrt{E_p+m}\begin{pmatrix} \dfrac{\boldsymbol{\sigma}\cdot\mathbf{p}}{E_p+m}\, \tilde{w}(\mathbf{p}, r) \\ \tilde{w}(\mathbf{p}, r)\end{pmatrix}. \tag{5b}$$

where $w(\mathbf{p}, r)$ and $\tilde{w}(\mathbf{p}, r)$ form two sets of linearly independent bispinors.

Γ_A	C_A	D_A
I	$1-\mathbf{k}_1\cdot\mathbf{k}_2-i\boldsymbol{\sigma}\cdot(\mathbf{k}_1\times\mathbf{k}_2)$	$-\boldsymbol{\sigma}\cdot(\mathbf{k}_1-\mathbf{k}_2)$
γ^0	$1+\mathbf{k}_1\cdot\mathbf{k}_2+i\boldsymbol{\sigma}\cdot(\mathbf{k}_1\times\mathbf{k}_2)$	$\boldsymbol{\sigma}\cdot(\mathbf{k}_1+\mathbf{k}_2)$
$\boldsymbol{\gamma}$	$\mathbf{k}_1+\mathbf{k}_2+i\boldsymbol{\sigma}\times(\mathbf{k}_1-\mathbf{k}_2)$	$(1-\mathbf{k}_1\cdot\mathbf{k}_2)\boldsymbol{\sigma}-i(\mathbf{k}_1\times\mathbf{k}_2)$ $+(\boldsymbol{\sigma}\cdot\mathbf{k}_1)\mathbf{k}_2+(\boldsymbol{\sigma}\cdot\mathbf{k}_2)\mathbf{k}_1$
$\boldsymbol{\mu}$	$(1+\mathbf{k}_1\cdot\mathbf{k}_2)\boldsymbol{\sigma}+i(\mathbf{k}_1\times\mathbf{k}_2)$ $-(\boldsymbol{\sigma}\cdot\mathbf{k}_1)\mathbf{k}_2-(\boldsymbol{\sigma}\cdot\mathbf{k}_2)\mathbf{k}_1$	$-(\mathbf{k}_1-\mathbf{k}_2)-i\boldsymbol{\sigma}\times(\mathbf{k}_1+\mathbf{k}_2)$
$\boldsymbol{\pi}$	$-i(\mathbf{k}_1-\mathbf{k}_2)+\boldsymbol{\sigma}\times(\mathbf{k}_1+\mathbf{k}_2)$	$(1+\mathbf{k}_1\cdot\mathbf{k}_2)\boldsymbol{\sigma}-(\mathbf{k}_1\times\mathbf{k}_2)$ $-(\boldsymbol{\sigma}\cdot\mathbf{k}_1)\mathbf{k}_2-(\boldsymbol{\sigma}\cdot\mathbf{k}_2)\mathbf{k}_1$
$\gamma_5\gamma_0$	$-\boldsymbol{\sigma}\cdot(\mathbf{k}_1+\mathbf{k}_2)$	$-(1+\mathbf{k}_1\cdot\mathbf{k}_2)-i\boldsymbol{\sigma}\cdot(\mathbf{k}_1\times\mathbf{k}_2)$
$\gamma_5\boldsymbol{\gamma}$	$-(1-\mathbf{k}_1\cdot\mathbf{k}_2)\boldsymbol{\sigma}-i(\mathbf{k}_1\times\mathbf{k}_2)$ $-(\boldsymbol{\sigma}\cdot\mathbf{k}_1)\mathbf{k}_2-(\boldsymbol{\sigma}\cdot\mathbf{k}_2)\mathbf{k}_1$	$-(\mathbf{k}_1+\mathbf{k}_2)-i\boldsymbol{\sigma}\times(\mathbf{k}_1-\mathbf{k}_2)$
γ_5	$-\boldsymbol{\sigma}\cdot(\mathbf{k}_1-\mathbf{k}_2)$	$(1+\mathbf{k}_1\cdot\mathbf{k}_2)-i\boldsymbol{\sigma}\cdot(\mathbf{k}_1\times\mathbf{k}_2)$

The table above gives the form of sixteen C_A matrices and sixteen D_A matrices which are related in the following manner to the matrix elements of the sixteen linearly independent matrices Γ_A:

$$\sqrt{N_1 N_2}\,\bar{u}(\mathbf{p}_1, r)\Gamma_A u(\mathbf{p}_2, s) = w^\dagger(\mathbf{p}_1, r) C_A w(\mathbf{p}_2, s),$$

$$\sqrt{N_1 N_2}\,\bar{u}(\mathbf{p}_1, r)\Gamma_A v(\mathbf{p}_2, s) = w^\dagger(\mathbf{p}_1, r) D_A \tilde{w}(\mathbf{p}_2, s),$$

$$\sqrt{N_1 N_2}\,\bar{v}(\mathbf{p}_1, r)\Gamma_A v(\mathbf{p}_2, s) = \varepsilon_A \tilde{w}^\dagger(\mathbf{p}_1, r) C_A \tilde{w}(\mathbf{p}_2, s),$$

$$\sqrt{N_1 N_2}\,\bar{v}(\mathbf{p}_1, r)\Gamma_A u(\mathbf{p}_2, s) = \varepsilon_A \tilde{w}^\dagger(\mathbf{p}_1, r) D_A w(\mathbf{p}_2, s),$$

where

$$\varepsilon_A = \begin{cases} +1 \text{ for } \gamma^\mu, \gamma_5\gamma^\mu, \\ -1 \text{ for } I, i\gamma^{\mu\nu}, \gamma_5 . \end{cases}$$

In the foregoing we have used the notation,

$$\mathbf{k}_1 = N_1 \mathbf{p}_1, \ \mathbf{k}_2 = N_2 \mathbf{p}_2,$$
$$N_1 = (E_{p_1}+m)^{-1}, N_2 = (E_{p_2}+m)^{-1},$$
$$\mu_i = \tfrac{1}{2} i\varepsilon_{ijk}\gamma^{jk}, \quad \pi_i = \gamma^{0i}.$$

In conclusion we shall give, without proof, a set of useful algebraic formulae containing various combinations of the matrices γ.

$$\gamma^\mu\gamma_\mu = 4,$$
$$\gamma^\mu\gamma_\lambda\gamma_\mu = -2\gamma_\lambda,$$
$$\gamma^\mu\gamma_\lambda\gamma_\varrho\gamma_\mu = 4g_{\lambda\varrho},$$
$$\gamma^\mu\gamma_{\alpha_1} \ldots \gamma_{\alpha_{2k+1}}\gamma_\mu = -2\gamma_{\alpha_{2k+1}} \ldots \gamma_{\alpha_1},$$
$$\gamma^\mu\gamma_{\alpha_1} \ldots \gamma_{\alpha_{2k}}\gamma_\mu = 2\gamma_{\alpha_{2k}}\gamma_{\alpha_1} \ldots \gamma_{\alpha_{2k-1}} + 2\gamma_{\alpha_{2k-1}} \ldots \gamma_{\alpha_1}\gamma_{\alpha_{2k}},$$
$$\tfrac{1}{4}\operatorname{Tr}\{\gamma_\mu\gamma_\nu\} = g_{\mu\nu},$$
$$\tfrac{1}{4}\operatorname{Tr}\{\gamma_\mu\gamma_\nu\gamma_\lambda\gamma_\varrho\} = g_{\mu\nu}g_{\lambda\varrho} - g_{\mu\lambda}g_{\nu\varrho} + g_{\mu\varrho}g_{\nu\lambda},$$
$$\tfrac{1}{4}\operatorname{Tr}\{\gamma_{\mu_1} \ldots \gamma_{\mu_n}\} = \begin{cases} 0, & n \text{ odd}, \\ \sum \varepsilon_p g_{\mu_{i_1}\mu_{i_2}} \ldots g_{\mu_{i_{n-1}}\mu_{i_n}}, & n \text{ even} \end{cases} \tag{7}$$

The sum in the last formula contains only those terms to which there corresponds the partition of the indices $\mu_1, \ldots, \mu_n$ into pairs $(\mu_{i_1}, \mu_{i_2}) \ldots (\mu_{i_{n-1}}, \mu_{i_n})$ such that for each pair $(\mu_{i_k}, \mu_{i_{k+1}})$ it satisfies the condition $i_k < i_{k+1}$.

Appendix K

REGULARIZATION

REGULARIZATION is the name we give every procedure by means of which we introduce a family of regular functions $\Delta_F^{\text{Reg}}(x;\beta)$, $S_F^{\text{Reg}}(x;\beta)$, etc., possessing the following properties:

1. For non-zero values of the parameter β regularized propagators are ordinary functions of the variable x, and not distributions.

2. In the limit, when $\beta \to 0$, regularized propagators go over into the corresponding distributions.

Many methods of regularization are to be found in the literature on quantum field theory. One of those used most frequently is the *Pauli–Villars regularization.*# This method consists in introducing regularized propagators in the form †

$$\Delta_F^{\text{Reg}}(x) = \int \frac{d^4p}{(2\pi)^4}\left(\frac{1}{m^2-p^2-i\varepsilon}+\sum_{i=1}^{N}\frac{C_i}{M_i^2-p^2-i\varepsilon}\right)e^{-ip\cdot x} \tag{1}$$

where the coefficients C_i and the auxiliary masses are chosen so that the Fourier transform of the propagator tend to zero as $(p^2)^{-N-1}$. This requirement imposes the following conditions upon the coefficients C_i:

$$1+\sum_{i=1}^{N} C_i = 0,$$

$$m^2+\sum_{i=1}^{N} C_i M_i^2 = 0,$$

W. Pauli and F. Villars (1949).

† The propagators S_F^{Reg} are obtained from the propagators Δ_F^{Reg} by the operation $(i\gamma^\mu\partial_\mu + m)$.

$$m^4+\sum_{i=1}^{N} C_i M_i^4 = 0,$$

$$\vdots$$

$$m^{2N-2}+\sum_{i=1}^{N} C_i M_i^{2N-2} = 0.$$

The greater the value of N, the more regular the function Δ_F^{Reg}. The parameters M_i are related to the regularization parameter so that $M_i \to \infty$, when $\beta \to 0$. In the simplest case, when $N = 1$, we still get a singular function but the singularity is only logarithmic. (Singularities of the type $\delta(x^2)$ appear in the propagator Δ_F). We then have

$$\Delta_F^{\text{Reg}}(x) = \Delta_F(x, m)-\Delta_F(x, M)$$

$$= -\frac{\theta(x^2)}{8\pi\sqrt{x^2}}[mH_1^{(2)}(m\sqrt{x^2})-MH_1^{(2)}(M\sqrt{x^2})]$$

$$+\frac{i\theta(-x^2)}{4\pi^2\sqrt{-x^2}}[mK_1(m\sqrt{-x^2})-MK_1(M\sqrt{-x^2})].$$

This function tends to the propagator $\Delta_F(x, m)$ in the limit when $\beta \to 0$ $(M \to \infty)$. On putting $N = 2$, we obtain a regularized propagator as a function without singularities. The principal advantage of the Pauli–Villars regularization is that the additional terms by which the function Δ_F^{Reg} differs from the propagator Δ_F are also Feynman propagators (but for fictitious particles with masses M_i), to which standard computational techniques may be applied.

In quantum electrodynamics one often uses a modification of the Pauli–Villars regularization consisting in the regularization of the products of the propagators S_F which correspond to the chains of electron lines:

$$S_F(x_1) \dots S_F(x_n) \to [S_F(x_1) \dots S_F(x_n)]^{\text{Reg}}$$

$$= S_F(x_1, m) \dots S_F(x, m) + \sum_{i=1}^{N} C_i S_F(x_1, M_i) \dots S_F(x_n, M_i). \qquad (2)$$

This modification ensures the gauge invariance of the theory.

A different method of regularization is employed in the calculations in this book. In this method four-dimensional space-time is replaced by N-dimensional space-time. In what follows we shall assume that one dimension has the nature of time, whereas $N-1$ dimensions have the character of space. The parameter β is related to the number of dimensions N by

$$\beta = 2-\frac{N}{2}$$

and will be treated formally as a continuous parameter. For a parameter β in the interval $1 > \beta > 0$ the integral representing the function $\Delta_F^{\text{Reg}}(x)$,

$$\Delta_F^{\text{Reg}}(x) = \int \frac{d^N p}{(2\pi)^N} \frac{e^{-ip\cdot x}}{m^2-p^2-i\varepsilon}, \tag{3}$$

is still divergent when $x = 0$, it is true, but if only $\beta > 0$, then all divergent integrals over momentum space in the expressions for propagators in quantum electrodynamics (these integrals are always at most logarithmically divergent) become convergent. This regularization is used in Chapter 7.

Appendix L

METHODS OF CALCULATING INTEGRALS OVER MOMENTUM SPACE

All expressions for propagators in perturbation theory obtained by applying the diagram technique in momentum representation are in the form of the integrals

$$I = \int d^4 l_1 \dots d^4 l_m \frac{P(l_1 \dots l_m)}{(m_1^2 - q_1^2 - i\varepsilon) \dots (m_n^2 - q_n^2 - i\varepsilon)}, \tag{1}$$

where $P(l_1 \dots l_m)$ is a polynomial in the coordinates of the vectors $l_1, \dots, l_m$; the integration runs over all independent vector variables which remain after the δ functions have been used, whereas the momenta $q_1, \dots, q_n$ are four-momenta of the internal lines of the diagram. They are constructed out of the external momenta p_i and the momenta l_i. Integration with respect to $l_1, \dots, l_m$ is made possible by the identity

$$\frac{1}{a_1 \dots a_n} = (n-1)! \int_0^1 d\alpha_1 \dots \int_0^1 d\alpha_n \frac{\delta(\sum \alpha_i - 1)}{(\alpha_1 a_1 + \alpha_2 a_2 + \dots + \alpha_n a_n)^n}. \tag{2}$$

By means of this identity, we reduce integral (1) to the form

$$I = \int_0^1 d\alpha_1 \dots \int_0^1 d\alpha_n \int d^4 l_1 \dots d^4 l_m$$

$$\times \frac{P(l_1 \dots l_m)\, \delta(\sum \alpha_i - 1)}{[\alpha_1(m_1^2 - q_1^2) + \dots + \alpha_n(m_n^2 - q_n^2) - i\varepsilon]^n}. \tag{3}$$

Since the denominator contains a quadratic form in the variables $l_1, \dots, l_m$, integration with respect to $l_1, \dots, l_m$ can be performed.

Integration with respect to one of the variables α_i in formula (2) can be carried out. On introducing new variables, we obtain a different, convenient integral representation of the quotient $1/(a_1 \ldots a_n)$,

$$\frac{1}{a_1 \ldots a_n} = (n-1)! \int_0^1 u_1^{n-2} \mathrm{d}u_1 \int_0^1 u_2^{n-3} \mathrm{d}u_2 \ldots \int_0^1 \mathrm{d}u_{n-1}$$

$$\times [a_1 u_1 \ldots u_{n-1} + a_2 u_1 \ldots u_{n-2}(1-u_{n-1}) + \ldots + a_n(1-u_1)]^{-n}. \quad (4)$$

For $n = 2$ we obtain the formula# which we used in Section 23,

$$\frac{1}{a_1 a_2} = \int_0^1 \frac{\mathrm{d}u}{[a_1 u + a_2(1-u)]^2}. \quad (5)$$

Another method of combining denominators utilizes the following integral representation of the Fourier transforms of the propagator:

$$\frac{1}{m^2 - p^2 - i\varepsilon} = -i \int_0^\infty \mathrm{d}\lambda \, \mathrm{e}^{i\lambda(m^2 - p^2 - i\varepsilon)}. \quad (6)$$

The product of the Fourier transforms of free propagators occurring in the expressions for the propagators may be written as

$$\prod_{i=1}^{n} \frac{1}{m_i^2 - p_i^2 - i\varepsilon}$$

$$= (-i)^n \int_0^\infty \mathrm{d}\lambda_1 \ldots \int_0^\infty \mathrm{d}\lambda_n \mathrm{e}^{-i \sum_{i=1}^{n} \lambda_i (m_i^2 - p_i^2 - i\varepsilon)}. \quad (7)$$

Thanks to this representation, integration with respect to the variables l_i may also be performed without difficulty since the exponent contains a quadratic form of the coordinates of the vectors l_i.

This formula was used by R. P. Feynman (1949b) and J. Schwinger (1949b) and this is why we have referred to the aforementioned method of calculating integrals as the *Feynman–Schwinger method of combining denominators.*

Formula (7) may also be used to prove formula (2). For this purpose it is sufficient to substitute unity under the integral sign in the form

$$1 = \int_0^\infty \frac{d\lambda}{\lambda}\,\delta\left(1 - \frac{\sum \lambda_j}{\lambda}\right),$$

then change the variables,

$$\lambda_i \to \lambda\alpha_i,$$

and perform the elementary integration with respect to λ.

Appendix M

REPRESENTATION OF THE *S* MATRIX AS A DOUBLE LIMIT OF THE PROPAGATOR

In this appendix we shall show that the S matrix $S(\mathbf{p}, \mathbf{q})$ can be obtained by two limiting transitions in the propagator $\mathcal{K}_R[\mathbf{x}t; \mathbf{y}t'|U]$. These will be the transitions when the points $(\mathbf{x}, t)$ and $(\mathbf{y}, t')$ tend to infinity along straight lines determined by the velocities $\mathbf{v}$ and $\mathbf{v}'$. Let us, therefore, substitute $\mathbf{v}t$ for $\mathbf{x}$ and $\mathbf{v}'t'$ for $\mathbf{y}$ and take the limits when $t \to \infty$ and $t' \to -\infty$. These two limiting transitions can be viewed as uniform motions of the particle in the remote past with the velocity $\mathbf{v}$ and in the far future with the velocity $\mathbf{v}'$

To derive final formulae we will need an explicit form of the free propagator $\mathcal{K}_R$,

$$\mathcal{K}_R(\mathbf{x}-\mathbf{y}, t-t') = i\left[\frac{m}{2\pi i\hbar(t-t')}\right]^{\frac{3}{2}} \exp\left[\frac{im(\mathbf{x}-\mathbf{y})^2}{2\hbar(t-t')}\right], \tag{1}$$

which can be obtained by integration from eq. (E.11). The following asymptotic formulae can be obtained from eq. (1),

$$\mathcal{K}_R(\mathbf{v}t-\mathbf{y}, t-t') \underset{t\to\infty}{\to} i\left(\frac{m}{2\pi i\hbar t}\right)^{\frac{3}{2}}$$

$$\times \exp\left(\frac{im\mathbf{v}^2 t}{2\hbar}\right) \exp\left(\frac{im\mathbf{v}^2 t'}{2\hbar} - \frac{im\mathbf{v}\cdot\mathbf{y}}{\hbar}\right),$$

$$\mathcal{K}_R(\mathbf{x}-\mathbf{v}'t', t-t') \underset{t'\to-\infty}{\to} i\left(\frac{m}{2\pi i\hbar(-t')}\right)^{\frac{3}{2}}$$

$$\times \exp\left(-\frac{im\mathbf{v}'^2 t'}{2\hbar}\right) \exp\left(-\frac{im\mathbf{v}'^2 t}{2\hbar} + \frac{im\mathbf{v}'\cdot\mathbf{x}}{\hbar}\right).$$

With the help of these asymptotic formulae and the perturbative expansion (4.14) of the propagator $\mathcal{K}_R[\mathbf{x}t; \mathbf{y}t'|U]$ one can transform the right-hand side of eq. (4.28) to the form:

$$S(\mathbf{p}, \mathbf{q}) = -\left(\frac{2\pi i\hbar}{m}\right)^3 \lim_{t\to\infty} \lim_{t'\to-\infty} t^{3/2}|t'|^{3/2}$$

$$\times \exp\left(-\frac{im\mathbf{v}^2 t}{2\hbar}\right) \exp\left(\frac{im\mathbf{v}'^2 t'}{2\hbar}\right) \mathscr{K}_{\mathrm{R}}[\mathbf{v}t, t; \mathbf{v}'t', t'|U]. \qquad (2)$$

The Dirac function $\delta(\mathbf{p}-\mathbf{q})$ appearing in eq. (4.28) may cause some difficulties in the proof of eq. (2), but one can overcome them by first integrating with the arbitrary trial function $f(\mathbf{q})$ and then taking the limits.

We believe that eq. (2) has a more transparent physical interpretation than the equivalent equation (4.27) since now, as we have already pointed out at the beginning, the trajectory of a particle with initial and final velocities makes its appearance.

The method described here can also be extended, with only slight changes, to the relativistic case when the electron motion is described by the Dirac equation.

REFERENCES

ARTICLES

ADLER, S., Short-distance behaviour of quantum electrodynamics and an eigenvalue condition for α, *Phys. Rev.* **D5**, 3021 (1972a).

ADLER, S., Massless, Euclidean quantum electrodynamics on the 5-dimensional uni hypersphere, *Phys. Rev.* **D6**, 3445 (1972b).

BAKER, M. and JOHNSON, K., Asymptotic form of the electron propagator and the self-mass of the electron, *Phys. Rev.* D **3**, 2516 (1971a).

BAKER, M. and JOHNSON, K., Simplified equation for the bare charge in renormalized quantum electrodynamics, *Phys. Rev.* D **3**, 2541 (1971b).

BARGMANN, V., Note on Wigner's theorem on symmetry operations, *J. Math. Phys.* **5**, 862 (1964).

BATEMAN, H., *Proc. London Math. Soc.* (2) **7** (1909). This is cited after the book: H. BATEMAN, *The Mathematical Analysis of Electrical and Optical Wave Motion*, Dover Publications, Inc., New York 1955.

BETHE, H., The electromagnetic shift of energy levels, *Phys. Rev.* **72**, 241 (1947).

BIAŁYNICKI-BIRULA, I., On the gauge properties of Green's functions, *Nuovo Cimento* **17**, 951 (1960).

BIAŁYNICKI-BIRULA, I., On the gauge covariance of quantum electrodynamics, *J. Math. Phys.* **3**, 1094 (1962).

BIAŁYNICKI-BIRULA, I., On the electron propagator in the two-component formulation of quantum electrodynamics, *Bull. Acad. Polon. Sci. Cl. III* **12**, 231 (1964).

BIAŁYNICKI-BIRULA, I., Commutation relations for energy-momentum tensor, *Nuovo Cimento* **35**, 697 (1965a).

BIAŁYNICKI-BIRULA, I., Simplified renormalization theory in quantum electrodynamics, *Bull. Acad. Polon. Sci. Cl. III* **13**, 499 (1965b).

BIAŁYNICKI-BIRULA, I., MIELNIK, B. and PLEBAŃSKI, J., Explicit solution of the continuous Baker–Campbell–Hausdorff problem and a new expression for the phase operator, *Ann. Phys.* **51**, 187 (1969).

BIAŁYNICKA-BIRULA, Z. and BIAŁYNICKI-BIRULA, I., Nonlinear effects in quantum electrodynamics. Photon propagation and photon splitting in an external field. *Phys. Rev.* D **2**, 2341 (1970).

BIAŁYNICKI-BIRULA, I., Renormalization. diagrams and gauge invariance, *Phys Rev.* D **2**, 2877 (1970).

BIAŁYNICKI-BIRULA, I., Classical electrodynamics in two dimensions: exact solution, *Phys. Rev.* D **3**, 864 (1971).

BIAŁYNICKI-BIRULA, I. and BIAŁYNICKA-BIRULA, Z., Magnetic monopoles in the hydrodynamic formulation of quantum mechanics, *Phys. Rev.* D **3**, 2410 (1971).

BIAŁYNICKI-BIRULA, I., and BIAŁYNICKA-BIRULA, Z., Quantum electrodynamics of intense photon beams, *Phys. Rev.* **A8**, 3146 (1973).

BLEULER, K., Eine neue Methode zur Behandlung der longitudinalen und skalaren Photonen, *Helv. Phys. Acta* **23**, 567 (1950).

BLOCH, F. and NORDSIECK, A., Note on the radiation field of the electron, *Phys. Rev.* **52**, 54 (1936).

BOHR, N. and ROSENFELD, L., Zur Frage der Messbarkeit der Elektromagnetischen Feldgrössen, *Det. Kgl. Dan. Vid. Selskab. Mat.-fys. Med.* **12**, 8 (1933).

BORN, M., On the quantum theory of the electromagnetic field, *Proc. Roy. Soc.* A **143**, 410 (1934).

BORN, M. and INFELD, L., Foundations of the new field theory, *Proc. Roy. Soc.* A **144**, 425 (1934a).

BORN, M. and INFELD, L., On the quantization of the new field equations, I., *Proc. Roy. Soc.* A. **147**, 522 (1934b).

BORN, M. and INFELD, L., On the quantization of the new field theory, II, *Proc. Roy. Soc.* A **150**, 141 (1935).

BRANDT, R. A., Derivation of renormalized relativistic perturbation theory from finite local field equations, *Ann. Phys.* (*USA*) **44**, 221 (1967).

BRODSKY, S. J., Radiative problems and quantum electrodynamics, in: *Proceedings 1971 International Symposium on Electron and Photon Interactions at High Energies*, Cornell University, Ithaca N. Y. 1971.

BROWN, L. M., Two-component fermion theory, *Phys. Rev.* **111**, 957 (1958).

CLARKE, J., The Josephson effect and e/h, *Am. J. of Phys.* **38**, 1071 (1970).

DIRAC, P. A. M., The quantum theory of the emission and absorption of radiation, *Proc. Roy. Soc.* A **114**, 243 (1927).

DIRAC, P. A. M., Electrons and protons, *Proc. Roy. Soc.* A **126**, 360 (1930).

DIRAC, P. A. M., Quantized singularities in the electromagnetic field, *Proc. Roy. Soc.* **A 133**, 60 (1931).

DIRAC, P. A. M., Classical theory of radiating electrons, *Proc. Roy. Soc.* A **167**, 148 (1938).

DYSON, F. J., The S matrix in quantum electrodynamics, *Phys. Rev.* **75**, 486 (1949).

EHLERS, J., RINDLER, W. and ROBINSON, I., Quaternions, bivectors and the Lorentz group, in: *Perspectives in Geometry and Relativity*, Indiana University Press, Bloomington, Ind. 1966.

FERMI, E., Quantum theory of radiation, *Rev. Mod. Phys.* **4**, 87 (1932).

FEYNMAN, R. P., The theory of positrons, *Phys. Rev.* **76**, 749 (1949a).

FEYNMAN, R. P., Space-time approach to quantum electrodynamics, *Phys. Rev.* **76**, 769 (1949b).

FEYNMAN, R. P., Mathematical formulation of the quantum theory of electromagnetic interaction, *Phys. Rev.* **80**, 440 (1950).

FOCK, V., Konfigurationsraum und zweite Quantelung, *Zeits. Phys.* **75**, 622 (1932).

FRADKIN, E. S., On some general relations in quantum electrodynamics, *ZhÉTF* **29**, 258 (1955a) (in Russian).

FRADKIN, E. S., On the theory of quantized fields, *ZhÉTF* **29**, 121 (1955b) (in Russian).

FRADKIN, E. S.,The Green's function method in quantum field theory and in quantum statistics, *Trudy Fizicheskogo Instituta im. P. N. Lebedeva* **29**, 7 (1965) (in Russian).

FREESE, E., Many-point correlation-functions in quantum field theory, *Nuovo Cimento* **2**, 50 (1955).

FURRY, W. H., A symmetry theorem in the positron theory, *Phys. Rev.* **51**, 125 (1937).

GLAUBER, R. J., The quantum theory of optical coherence, *Phys. Rev.* **130**, 2529 (1963a).

GLAUBER, R. J., Coherent and incoherent states of the radiation field, *Phys. Rev.* **131**, 2766 (1963b).

GOLDHABER, A. S. and NIETO, M. M., Terrestrial and extraterrestrial limits on the photon mass, *Rev. Mod. Phys.* **43**, 277 (1971).

GREEN, H. S., A Pre-renormalized quantum electrodynamics, *Proc. Phys. Soc.* A **66**, 873 (1953).

GUPTA, S. N., Theory of longitudinal photons in quantum electrodynamics, *Proc. Phys. Soc.* A **63**, 681 (1950).

HEISENBERG, W., Die "beobachtbaren Grössen" in der Theorie der Elementarteilchen, *Z. Phys.* **120**, 513 (1943).

HEISENBERG, W. and EULER, H., Folgerungen aus der Diracschen Theorie des Positrons, *Z. Physik* **98**, 714 (1936).

HEPP, K., Proof of the Bogoliubov–Parasiuk theorem on renormalization, *Comm. Math. Phys.* **2**, 301 (1966).

HOFSTADTER, R., *Nuclear and Nucleon Structure*, W. A. Benjamin, Inc., New York 1963.

INFELD, L. and VAN DER WAERDEN, B. L., Die Wellengleichung des Electrons in der allgemeinen Relativitätstheorie, *Sitz. Ber. der Preuss. Akad.* **9**, 380 (1933).

JORDAN, P. and WIGNER, E., Über das Paulische Äquivalenzverbot, *Z. Phys.* **47**, 631 (1928).

KÄLLÉN, G., On the definition of the renormalization constants in quantum electrodynamics, *Helv. Phys. Acta* **25**, 417 (1952).

KARPLUS, R. and KROLL, N., Fourth order corrections in quantum electrodynamics and the magnetic moment of the electron, *Phys. Rev.* **77**, 536 (1950).

KARPLUS, R. and NEUMAN, M., The scattering of light by light, *Phys. Rev.* **83**, 776 (1951).

KIBBLE, T. W. B., Coherent soft-photon states and infrared divergences, *I. J. Math.*

Phys. **9**, 315 (1968); *II*, *Phys. Rev.* **173**, 1527 (1968); *III*, *Phys. Rev.* **174**, 1882 (1968); *IV Phys. Rev.* **175**, 1624 (1968).

KLEIN, O. and NISHINA, Y., Über die Streuung von Strahlung durch freie Elektronen nach der neuen relativistischen Quantendynamik von Dirac, *Z. Phys.* **52**, 853 (1929).

KUO, P. K. and YENNIE, D. R., Renormalization theory, *Ann. Phys.* **51**, 496 (1969).

LAMB, W. E., JR. and RETHERFORD, R. C., Fine structure of the hydrogen atom by a microwave method, *Phys. Rev.* **72**, 241 (1947).

LANCZOS, C., Die tensoranalytischen Beziehunger der Diracschen Gleichung, *Z. Phys.* **57**, 447 (1929).

LANDAU, L. D. and KHALATNIKOV, I. M., Gauge transformations of the Green's functions of charged particles, *ZhÉTF* **29**, 89 (1955) (in Russian).

LANGERHOLC, J. and SCHROER, B., Can current operators determine a complete theory?, *Commun. Math. Phys.* **4**, 123 (1967).

LAUTRUP, B. E., PETERMAN, A., DE RAFAEL, E., Recent Developments in the Comparison between Theory and Experiments in Quantum Electrodynamics, *Physics Reports* 3C, 193 (1972).

LEHMANN, J., Über Eigenschaften von Ausbreitungsfunktionen und Renormierungkonstanten quantisierter Felder, *Nuovo Cimento* **11**, 342 (1954).

LEVINE, M. J., Sixth-order magnetic moment of the electron, *Phys. Rev. Letters* **26** 1351 (1971).

MACK, G. and SYMANZIK, K., Currents, stress tensor and generalized unitarity in conformal invariant quantum field theory, *Comm. Math. Phys.* **27**, 247 (1972).

MANDELSTAM, S., Determination of the pion-nucleon scattering amplitude from dispersion relations and unitarity. General Theory, *Phys. Rev.* **112**, 1344 (1958).

MANDELSTAM, S., Quantum electrodynamics without potentials, *Ann. Phys.* (*USA*) **19**, 1 (1962).

MANDELSTAM, S., Feynman rules for electromagnetic and Yang–Mills fields from the gauge-independent field-theoretical formalism, *Phys. Rev.* **175**, 1580 (1968).

MØLLER, C., Zur Theorie des Durchgangs schneller Elektronen durch Materie, *Ann. Phys.* (*Germany*) **14**, 568 (1932).

NISHIJIMA, K., Asymptotic conditions and perturbation theory, *Phys. Rev.* **119**, 485 (1960).

PAULI, W., Contributions mathématiques à la théorie des matrices de Dirac, *Ann. Inst. Poincaré* **6**, 109 (1936).

PAULI, W., The connection between spin and statistics, *Phys. Rev.* **58**, 716 (1940).

PAULI, W. and VILLARS, F., On the invariant regularization in relativistic quantum theory, *Rev. Mod. Phys.* **21**, 434 (1949).

PETERMAN, A., Fourth order magnetic moment of the electron, *Nuclear Phys.* **5**, 677 (1958).

PROCA, A., Sur la théorie ondulatoire des électrons positifs et négatifs, *J. Phys. Radium* **7**, 347 (1936).

ROBISCOE, R. T. and SHYN, T. W., Kinematic corrections to atomic beam experiments, *Phys. Rev. Letters* **24**, 559 A (1970).

SALAM, A., Overlapping divergences and the *S*-matrix, *Phys. Rev.* **82**, 217 (1951a).

SALAM, A., Divergent integrals in renormalizable field theories, *Phys. Rev.* **84**, 426 (1951b).

SCHWINGER, J., On quantum electrodynamics and the magnetic moment of the electron, *Phys. Rev.* **73**, 416 (1948).

SCHWINGER, J., Quantum electrodynamics, vacuum polarization and self-energy, *Phys. Rev.* **75**, 651 (1949a).

SCHWINGER, J., Quantum electrodynamics, III: the electromagnetic properties of the electron-radiative corrections to scattering, *Phys. Rev.* **76**, 790 (1949b)

SCHWINGER, J., On the Green's functions of quantized fields, *Proc. Nat. Acad. Sci. (USA)* **37**, 452 (1951a).

SCHWINGER, J., On gauge invariance and vacuum polarization, *Phys. Rev.* **82**, 664 (1951b).

SCHWINGER, J., The theory of quantized fields, III, *Phys. Rev.* **91**, 728 (1953a).

SCHWINGER, J., The theory of quantized fields, IV, *Phys. Rev.* **92**, 1283 (1953b).

SCHWINGER, J., The theory of quantized fields, V, *Phys. Rev.* **93**, 615 (1954a).

SCHWINGER, J., The theory of quantized fields, VI, *Phys. Rev.* **94**, 1362 (1954b).

SCHWINGER, J., Non-Abelian gauge fields. Relativistic invariance, *Phys. Rev.* **127**, 324 (1962).

SCHWINGER, J., Energy momentum density in field theory, *Phys. Rev.* **130**, 800 (1963).

SERBER, R., Linear modifications in the Maxwell equations, *Phys. Rev.* **48**, 49 (1935).

SIMON, B., Summability methods, the strong asymptotic condition and unitarity: *QFT, Phys. Rev. Letters* **28**, 1145 (1972).

SOMMERFIELD, C., The magnetic moment of the electron, *Ann. Phys. (USA)* **5**, 26 (1958).

STARUSZKIEWICZ, A., On affine properties of the light cone and their application in the quantum electrodynamics, *Acta Phys. Polon.* **B 4**, 57 (1973).

STEINMANN, O., Perturbation theory in the LSZ-formalism, *Ann. Phys. (USA)* **29**, 76 (1964).

TAKAHASHI, Y., On the generalized Ward identity, *Nuovo Cimento* **6**, 371 (1957).

THIRRING, W., Radiative corrections in the non-relativistic limit, *Phil. Mag.* **41**, 1193 (1950).

TRIEBWASSER, S., DAYHOFF, E. S. and LAMB, W. E., Jr., Fine structure of the hydrogen atom, V, *Phys. Rev.* **89**, 98 (1953).

UEHLING, E. A., Polarization effects in the positron theory, *Phys. Rev.* **48**, 55 (1935).

VALATIN, J. G., On the definition of finite operator quantities in quantum electrodynamics, *Proc. Roy. Soc.* **226**, 254 (1954).

VAN DER WAERDEN, B. L., *Die Gruppentheoretische Methode in der Quantenmechanik* Springer, Berlin 1932.

WARD, J. C., An identity in quantum electrodynamics, *Phys. Rev.* **78**, 182 (1950).

WHEELER, J. A., On the mathematical description of light nuclei by the method of resonating group structure, *Phys. Rev.* **52**, 1107, 1937.

WEINBERG, S., High-energy behavior in quantum field theory, *Phys. Rev.* **118**, 838 (1960).

WESLEY, J. C. and RICH, A., Preliminary results of a new electron g-2 measurement, *Phys. Rev. Letters* **24**, 1320 (1970).

WICHMANN, E. H. and CRICHTON, J. H., Cluster decomposition properties of the S matrix, *Phys. Rev.* **132**, 2788 (1963).

WIGHTMAN, A. S., Quantum field theory in terms of vacuum expectation values, *Phys. Rev.* **101**, 860 (1956).

WILLIAMS, E. R., FALLER, J. E. and HILL, H. A., New experimental test of Coulomb's law: A laboratory upper limit on the photon rest mass, *Phys. Rev. Letters* **26**, 721 (1971).

WILSON, K. G., Non-Lagrangian models of current algebra, *Phys. Rev.* **179**, 1499 (1969).

YENNIE, D. R., FRAUTCHI, S. and SUURA, H., The infrared divergence phenomena and high-energy processes, *Ann. Phys.* (*USA*) **13**, 379 (1961).

ZIMMERMANN, W., Local operator products and renormalization in quantum field theory, in: *Lectures on Elementary Particles and Quantum Field Theory*, Brandeis University Summer Institute in Theoretical Physics, Waltham, Mass. (1970).

ZUMINO, B., Gauge properties of propagators in quantum electrodynamics, *J. Math. Phys.* **1**, 1 (1960).

TEXTBOOKS AND MONOGRAPHS

AKHIEZER, A. I. and BERESTETSKII, V. B., *Quantum Electrodynamics*, John Wiley and Sons Inc., New York 1965.

BJORKEN, J. D. and DRELL, S. D., *Relativistic Quantum Fields*, McGraw-Hill Book Company, New York 1965.

BOGOLYUBOV, N. N. and SHIRKOV, D. V., *Introduction to the Theory of Quantized Fields*, John Wiley and Sons Inc., New York 1959.

BOGOLYUBOV, N. N., LOGUNOV, A. A. and TODOROV, I. T., *The Foundations of the Axiomatic Approach to Quantum Field Theory*, Nauka, Moscow 1969 (in Russian).

CORSON, E. M., *Introduction to Tensors, Spinors and Relativistic Wave Equations*, Blackie and Son, London 1953.

DIRAC, P. A. M., *The Principles of Quantum Mechanics*, Clarendon Press, Oxford 1958.

FRIED, H, M., *Functional Methods and Models in Quantum Field Theory*. M. I. T. Press, Cambridge, Mass. 1972.

HAMERMESH, M., *Group Theory*, Addison–Wesley Publishing Co., Reading, Mass. 1962.

HAMILTON, J., *The Theory of Elementary Particles*, Clarendon Press, Oxford 1961.

HEITLER W., *The Quantum Theory of Radiation*, Clarendon Press, Oxford 1954.

JACKSON, J. D., *Classical Electrodynamics*, John Wiley and Sons Inc., New York 1962.

JAUCH, J. M., *Foundations of Quantum Mechanics*, Addison–Wesley Publishing Co., Reading, Mass. 1968.

JAUCH, J. M. and ROHRLICH, F., *The Theory of Photons and Electrons*, Addison–Wesley Publishing Co., Reading, Mass. 1959.

KLAUDER, J. R. and SUDARSHAN, E. C. G., *Fundamentals of Quantum Optics*, W. A. Benjamin Inc., New York 1968.

KRAMERS, H. A., *Quantum Mechanics*, North-Holland Publishing Co., Amsterdam 1957.

LANDAU, L. D. and LIFSHITZ, E. M., *The Classical Theory of Fields*, Pergamon Press, Oxford 1962.

MERZBACHER, E., *Quantum Mechanics*, 2nd ed., John Wiley and Sons Inc., New York 1970.

NEWTON, R. G., *Scattering Theory of Waves and Particles*, McGraw-Hill Book Company, New York 1966.

PAULI, W., *Theory of Relativity*, Pergamon Press, Oxford 1958.

RZEWUSKI, J., *Field Theory*: Vol. I, *Classical Theory* 1958, Vol. II *Functional Formulation of S Matrix Theory* 1969, PWN (Polish Scientific Publishers) Warsaw and Illiffe Books Ltd., London.

SCHWARTZ, H. M., *Introduction to Special Relativity*, McGraw-Hill Book Company New York 1968.

SCHWINGER, J., *Particles, Sources, and Fields*, Addison–Wesley Publishing Co. Reading, Mass. 1970.

STREATER, R. F. and WIGHTMAN, A. S., *PCT, Spin and Statistics and All That*, W. A. Benjamin Inc., New York 1964.

INDEX OF SYMBOLS

a_k, $a_k^\dagger$ **(5.14), (5.15), (18.1)**

$a(\mathbf{p}, \lambda)$, $a^\dagger(\mathbf{p}, \lambda)$ **(5.45a, b)**, (18.5), (18.6)

$a[f]$, $a^\dagger[f]$ **(5.42), (18.5)**

$a^\mu(z, x)$ (22.8), (22.9)

A_{fi} **(4.15)**

$A(n_1 \ldots n_k \ldots; n'_1 \ldots n'_k \ldots)$, $A(\{n_i\}, \{n'_j\})$ **(12.7)**

$A(\{g_1\}, \{g_2\})$ **(5.47)**

$A_\mu(x)$ **(6.3)**, (14.4)

$\mathbf{A}$ **(16.10a)**

$\mathscr{A}_\mu(x)$ (13.44), (16.1)

$\mathscr{A}_\mu[z|a]$ **(22.12)**

b_k, $b_k^\dagger$ **(18.1)**, (18.4)

$b(\mathbf{p}, r)$, $b^\dagger(\mathbf{p}, r)$ **(18.5)**

$b[g]$, $b^\dagger[g]$ (18.5)

$\mathbf{B}$ **(6.2b)**

B **(16.15)**

$\mathscr{B}(x)$ **(11.49b)**

$\mathscr{B}[x|f_i]$ **(11.50b)**

$\mathscr{B}(z)$ (22.3)

c_i, $c_i^\dagger$, $c[f_i]$, $c^\dagger[f_i]$ **(11.10)**

$c(\mathbf{k}, \lambda)$, $c^\dagger(\mathbf{k}, \lambda)$ (11.7), (11.8)

$c_{\mathbf{k},\pm}$, $c^\dagger_{\mathbf{k},\pm}$ (11.25), (11.27)

$c(\mathbf{k}, r)$, $c^\dagger(\mathbf{k}, r)$ (14.24)

$c_\mu(\mathbf{k})$ **(14.25)**

$C[\mathscr{A}]$ (17.58), **(17.81)**

$C[\mathscr{A}, \mathscr{I}]$ **(18.58)**, (20.2)

$C[x, y|\mathscr{A}]$ (17.72)

$C[x_1 \ldots x_n, y_n \ldots y_1|\mathscr{A}]$ **(18.47)**

C **(16.10b)**

$\mathscr{C}[\mathscr{A}]$ **(13.47)**

$\mathbf{D}$ **(6.9a)**

D_μ **(8.14)**

D_j **(9.30)**

D_x **(16.42)**

$D[f]$ **(13.48)**

$D(x)$ **(9.9)**, (E.25b)

$D^{(1)}(x)$ **(12.12)**

$D^{(+)}(x)$ **(13.16)**, (E.25c)

$D^{(-)}(x)$ **(13.17)**, (E.25d)

$D_{\mathrm{F}}(x)$ **(13.25)**, (E.25a)

$D_{\mathrm{R,A}}(x)$ (9.51)

$D[\sigma]$ **(9.46)**

$D^{\mathrm{F}}_{\mu\nu}(x)$ **(13.30)**, (22.16), (22.43), (25.13)

$D^{\mathrm{F}}_{\mu\nu}[w, w'|a]$ **(22.11)**

$\mathsf{D}(\alpha)$ **(11.32)**

$\mathsf{D}(\{\alpha\})$ **(11.36)**

$\mathscr{D}^{\mathrm{F}}_{\mu\nu}(x, n)$ **(13.29)**

$e(\mathbf{k})$ **(9.23)**, (9.24)

$e_{\mu\nu}(\mathbf{k})$ (9.19)

e_{obs} **(23.25)**

$\mathbf{E}$ **(6.2a)**

$E[x_1 x_2, y_2 y_1 | \mathscr{A}, \mathscr{I}]$ **(22.28)**

$\mathscr{E}(x)$ **(11.49a)**

$\mathscr{E}[x|f_i]$ **(11.50a)**

$f(\mathbf{k}, \lambda)$ (9.28), (9.33), (9.34), (9.61)

$f_{\mu\nu}(x)$ (6.3), (6.4), **(14.2)**, (9.54), (9.56), **(11.13)**, (12.3)

$f_{\mathbf{k},\pm}$ (9.40)

$f_{\mu}(\mathbf{k})$ **(13.11)**, **(14.6)**

$f(\mathbf{p}_1 r_1, \ldots, \mathbf{p}_n r_n; \mathbf{q}_1 s_1, \ldots, \mathbf{q}_m s_m)$ **(18.7)**

$\mathbf{F}(x)$ **(9.7)**

$F_{\mu\nu}(x)$ **(9.3)**, (9.60)

$g^{(n)}(\mathbf{p}_1 \sigma_1, \ldots, \mathbf{p}_n \sigma_n)$ (5.41)

$G[x, y | \mathscr{A}, \mathscr{I}]$ **(22.17)**

$G[x_1 \ldots x_n, y_n \ldots y_1 | \mathscr{A}, \mathscr{I}]$ **(22.19)**

$\tilde{G}^{(2)}(p)$ (23.1)

$G^{(n,n)}_{\mathbf{E,B}}(x_1 l_1, \ldots, x_n l_n; x_{n+1} l_{n+1}, \ldots x_{2n} l_{2n})$ **(11.48)**

$\mathscr{G}_{\mu\nu}[z_1 z_2 | \mathscr{A}, \mathscr{I}]$ **(20.12)**

$\tilde{\mathscr{G}}^{(2)}_{\mu\nu}(k)$ (23.13)

$\tilde{\mathscr{G}}(k^2)$ (21.34)

$\mathscr{G}^{(n,m)}(i_1 \ldots i_n, i_{n+1} \ldots i_{n+m})$ **(11.46)**

$h^{\mu\nu}(x)$ **(6.6)**

H, $H(t)$ **(1.8)**, (1.23), (4.6), (5.38), (5.54), (9.33a), (11.20a), (14.14a)

H_D **(17.51)**

$\mathbf{H}$ **(6.9b)**

$\mathscr{H}^{\mathrm{in}}_{\mathrm{int}}(x)$ **(26.34)**

$\mathscr{I}(\mathbf{x}, t)$ (9.49)

$\mathscr{I}^{\mathrm{tr}}(\mathbf{x}, \mathrm{t})$ **(13.3)**

$\mathscr{I}^{\mu}(x)$ (9.50), (12.1a), (15.7)

$\mathscr{I}^{\mu}[z | \mathscr{A}, \mathscr{I}]$ **(22.39)**

$\mathscr{I}^{\mu}_{c}(z)$ (22.9)

$j^{\mu}(x)$ **(6.7)**, **(8.1a)**, **(18.19)**, (18.22)

$\mathscr{J}^{\lambda}[z | \mathscr{A}, \mathscr{I}]$ **(24.13)**

$K[x_1 x_2, y_2 y_1]$ **(24.53)**

$K^{\lambda}[\sigma]$ **(9.47)**

$K^{\mu\nu}$ **(14.19)**

$K_{\mathrm{F}}[x, y | \mathscr{A}]$ **(17.18)**

$K_{\mathrm{F}}[x_1 \ldots x_n, y_1 \ldots y_n | \mathscr{A}]$ **(18.52)**

$\mathscr{K}(\mathbf{x}, \mathrm{t})$ (5.46}), **(E.11)**

$\mathscr{K}_{\mathrm{I}}(\mathbf{x}, t)$ **(E.6)**

$\mathscr{K}_{\mathrm{R,A}}(\mathbf{x}, t)$ **(E.10)**

$\mathscr{K}[x, y, z_1 z_2]$ **(24.43)**

$\mathscr{K}[\mathbf{x}, t; \mathbf{y}, t_0 | U]$ **(4.7)**

$\mathscr{K}_{\mathrm{R,A}}[\mathbf{x}, t; \mathbf{y}, t_0 | U]$, $\mathscr{K}_{\mathrm{R,A}}[x, y | U]$ **(4.9)**

$\mathscr{K}_{\mathrm{R}}[x_1 \ldots x_n; y_n \ldots y_1 | U]$ **(5.72)**

$L^{\mu}{}_{\nu}$ **(I.3)**

m_{obs} **(23.5)**

$\hat{\mathbf{M}}$, $\mathbf{M}$ **(5.33d)**, **(6.25c)**, **(9.33c)**, (11.20c), **(14.14c)**, (14.15c)

$M(p; q; k)$ **(25.2)**, (26.4)

$M^{\mu\nu}$ **(6.20b)**, (9.59), (10.6)

$M^{\mu\nu\lambda}(x)$ **(6.18)**

$M^{\mu\nu}_{z}$ **(21.13)**

$\mathscr{M}(p, q)$ **(17.70)**

n_i **(5.12)**

N **(5.13)**

N_i **(5.12)**

$\mathbf{N}$ **(6.25d)**, (9.33d), (11.20d)

P **(6.1b)**

$\hat{\mathbf{P}}, \mathbf{P}$ (5.33c), (9.33b), (11.20b), (14.14b), (14.15b)

$\mathbf{P}$ **(16.23b)**, (16.26a)

$P(p; q; k)$ **(25.7)**, (25.11)

P_{fi} **(26.3)**

P^μ **(6.20a)**, (9.57), (10.6)

$P_{\{n_i\}}$ (11.56)

$\mathscr{P}^\mu{}_\nu$ (6.13), (6.14), (16.21), (16.22)

r_0 **(26.29)**

$R[x_1 x_2, y_2 y_1]$ **(24.54)**

$\mathscr{R}[x, y, z_1 z_2]$ **(24.44)**

s **(26.17a)**

S **(6.1a)**

$S(\Lambda)$ **(I.22)**

S **(1.27)**, (5.48), (5.55), (13.31), (14.38), (21.20), (22.34), (22.37)

$\mathsf{S}[U]$ (5.67)

$\mathsf{S}[\mathscr{A}]$ **(13.45)**, **(18.28)**, (18.30), (18.56)

$\mathsf{S}[\mathscr{A}, \mathscr{I}]$ (22.34), (22.37)

$S[f, g]$ (4.22a)

$S(\mathbf{p}; \mathbf{q})$ **(4.26)**

$S_{\mathrm{B}}^{\pm\pm}[f, g]$ **(17.1)**

$S_{\mathrm{Q}}^{\pm\pm}[f, g]$ (17.8), (17.11), (22.32)

$S^{\pm\pm}[f, g]$ (17.61)

$S^{\pm\pm}(\mathbf{p}, r; \mathbf{q}, s)$, $S(\mathbf{p}, r, \lambda; \mathbf{q}, s, \lambda')$ **(17.2)**, (17.63)

$S^{(n)}(\mathbf{p}_1\sigma_1, \ldots, \mathbf{p}_n\sigma_n; \mathbf{p}'_n\sigma'_n, \ldots, \mathbf{p}'_1\sigma'_1)$ **(5.50)**

$S[f_1 \ldots f_n, g_1 \ldots g_n]$ (5.7)

$S^{\mu\nu}$ **(I.18)**

$\mathbf{S}_{\mathrm{L}}$ **(16.23a)**, (16.24)

$S(x)$ **(18.12)**

$S^{(\pm)}(x)$ **(16.59)**

$S_{\mathrm{F}}(x)$ **(17.7)**, (22.43)

$S_{\mathrm{J}}(x)$ **(17.17)**

$S_{\mathrm{F}}(x_1 \ldots x_n, y_1 \ldots y_n)$ **(17.75)**

t **(26.17b)**

$\hat{T}$ (5.33a)

T **(26.2)**, (26.4)

$\mathbf{T}$ **(16.23c)**

$T^{\mu\nu}(x)$ (6.15), (8.1b), (8.9), (8.20), (9.32), (14.11), (18.23), (F.1)

$T(x_1 \ldots x_n; y_n \ldots y_1)$ **(5.65)**

$T[x_1 \ldots x_n; y_n \ldots y_1 | U]$ **(5.69)**

$T_0[x_1 \ldots x_n; y_n \ldots y_1 | U]$ **(5.77)**

$T_{\mu_1 \ldots \mu_k}[z_1 \ldots z_k | \mathscr{A}, \mathscr{I}]$ **(13.53)**, **(20.1)**

$T[x_1 \ldots x_n, y_n \ldots y_1 | \mathscr{A}, \mathscr{I}_c + \mathscr{I}]$ **(22.5)**

$T[x_1 \ldots x_n, y_n \ldots y_1, z_1 \ldots z_k | \mathscr{A}, \mathscr{I}_c + \mathscr{I}]$ **(22.6)**

$\mathscr{T}[x_1 \ldots x_n, y_n \ldots y_1, z_1 \ldots z_k | \mathscr{A}, \mathscr{I}_c + \mathscr{I}]$ **(22.7)**

u **(26.17c)**

$u(\mathbf{p}, r)$ **(16.44a)**

$u(\mathbf{p}, s, \lambda)$ **(22.45a)**

$\hat{U}$ (5.33b)

$\mathsf{U}(a, \Lambda)$ (10.2), (13.36)

$\mathsf{U}(t, t_1)$ (1.7), (1.9)

$\mathsf{U}_S(t, t_0)$ (1.19)

$U(\mathbf{x}, t)$ (4.6)

$\tilde{U}(\mathbf{k}, \omega)$ (4.28)

$v(\mathbf{p}, r)$ **(16.44b)**
$\hat{V}$ (5.33e)
$V(\mathbf{x}_i - \mathbf{x}_j)$ (5.2)
$V[\mathscr{A}, \mathscr{I}]$ **(20.8)**, (22.37)
$\mathsf{V}(t, t_0)$ **(1.25)**

Z_3 (21.4), (21.34), (22.34), (22.36), (24.62)
Z_2 (22.34), (22.36), (24.62)

α (28.40)

γ^μ **(16.2)**, (16.6)
γ_5 **(16.5)**
$\gamma^{\mu\nu}$ **(16.4)**
$\gamma(z)$ (17.74)
$\gamma^\mu(x, y, z)$ (24.32)
$\Gamma^\mu[x, y, z | \mathscr{A}, \mathscr{I}]$ **(22.25)**, (24.52), (24.59)
$\Gamma^{\mu_1 \dots \mu_k}[x, y, z_1 \dots z_k | \mathscr{A}, \mathscr{I}]$ **(22.27)**
$\tilde{\Gamma}^{\mu(2)}(p, q)$ (23.20)

$\Delta(x)$ **(E.16)**
$\overline{\Delta}(x)$ **(E.17)**
$\Delta^{(\pm)}(x)$ **(E.21)**
$\Delta_{\mathrm{I}}(x)$ **(E.13)**
$\Delta_{\mathrm{F}}(x)$ **(14.41)**, **(17.22a)**, (E.19c)
$\Delta_{\mathrm{R,A}}(x)$ **(17.22)**, (E.15)
$\Delta_{\lrcorner}(x)$ **(17.22b)**, (E.19d)
$\Delta(x, \mu)$ **(14.28a)**
$\Delta^{(\pm)}(x, \mu)$ **(14.28b)**
$\Delta_{\mu\nu}(x)$ **(14.27)**
$\Delta_{\mu\nu}{}^{\mathrm{F}}(x)$ **(14.40)**
$\Delta^{\mathrm{F}}{}_{\mu\nu}[w, w|a]$ **(22.33)**

$\Delta^\nu[z]$ (20.17)
$\Delta^{\mu\nu\lambda}[z_1 z_2 z_3]$ (20.19)

$\hat{\varepsilon}$ **(I.37)**
$\varepsilon_\mu(\mathbf{k})$ **(9.62)**, (9.64)
$\varepsilon_\mu(\mathbf{k}, r)$ **(14.7)**
$\varepsilon_\mu(\mathbf{k}, r, \varkappa)$ **(22.45b)**

$\iota(\mathbf{k}, \lambda)$ **(9.67)**
$\iota_{(e,m)}(JM\omega)$ (9.80)
$\iota_{\mathrm{QED}}(\mathbf{k}, \lambda)$ **(21.37)**

η (26.10)

$\Lambda(p, s)$ **(25.8a)**
$\Lambda_\mu{}^\nu$ (10.2), **(D.5)**, (I.13)
$\Lambda_{\mathrm{R}}^{\mu(2)}(p, q)$ (23.21)

μ **(28.29)**, (28.34)
$\boldsymbol{\mu}$ **(28.33)**

$\nu(n_1 \dots n_i)$ **(5.16)**
$\nu(n_1 \dots n_k \dots)$ **(18.2)**

$\Pi^{\lambda\varrho}[z_1 z_2 | \mathscr{A}, \mathscr{I}]$ (20.18), (24.39), (24.58)
$\tilde{\Pi}^{\lambda\varrho(2)}(k)$ **(23.14)**
$\pi(x)$ **(14.12)**

ϱ (1.2), (11.59)
$\varrho_S(t, t_0)$ (1.15)
$\varrho_D(t)$ (1.21)

ϱ_L **(12.60)**
ϱ_T **(12.61)**
$\varrho(M^2)$ (21.34)
$\varrho^{in}(\mathbf{x}, t)$ (5.66)
$\varrho_{\mu\nu}(k, r)$ **(25.8b)**

$\Sigma[x, y|\mathscr{A}, \mathscr{I}]$ (22.43), **(24.38)**, **(24.51)**
$\tilde{\Sigma}^{(2)}(p)$ **(23.2)**

$\varphi[\mathbf{x}, t|f]$ **(4.21)**
$\varphi^{\pm}(\mathbf{x}, \mathbf{p})$ **(4.19)**
$\varphi_\mu[x|f]$ **(13.10)**
$\varphi_{\mu\nu}[x|f]$ **(11.16)**
$\tilde{\varphi}(x)$ **(I.32b)**, (I.33)
$\varphi[\mathbf{x}_1\sigma_1, \ldots, \mathbf{x}_n\sigma_n, t|g]$ (5.40)
$\varphi_{n_1\ldots n_k}(x_1\sigma_1, \ldots, \mathbf{x}_n\sigma_n)$ **(5.9)**
$\boldsymbol{\varphi}$ (14.14)
$\phi(x)$ (8.13)
Φ_n (4.17)
$\Phi^{(\pm)}(\mathbf{p})$ (4.20)
$\Phi^{(n)}(\mathbf{x}_1\sigma_1, \ldots, \mathbf{x}_n\sigma_n, t)$ (5.21)
$\Phi_{n_1\ldots n_k\ldots}, \Phi_{\{n_k\}}$ (5.11), (5.20)
Φ_α **(11.30)**
$\Phi\{\alpha_k\}, \Phi(\alpha_1, \ldots, \alpha_k, \ldots)$ **(11.28)**, (11.29)
$\Phi^{\mu\nu}[z|f], \Phi^{*\mu\nu}[z|f]$ **(21.9)**

$\chi^{(\pm)}[x|g]$ **(16.56)**
$\chi^{(\pm)}(x; p, r)$ **(16.55)**

$\psi(\mathbf{x}, t)$ **(4.1)**
$\psi(x)$ (16.1)
$\psi_F^{\pm}[x|g]$ **(17.10)**
$\psi_F[x|g]$ **(17.13)**
$\psi_\text{⅃}(x)$ **(17.21)**
$\psi_{R,A}[\mathbf{x}, t|g]$ (4.22)
$\psi(x_1\sigma_1, \ldots, x_n\sigma_n, t)$ (5.1), (5.2)
$\psi_{R,A}[\mathbf{x}_1 \ldots \mathbf{x}_n, t|g_1 \ldots g_n]$ (5.6)
$\psi(n_1, \ldots, n_k, \ldots, t)$ **(5.10)**
$\psi(x)$ **(I.32)**, (I.33)
$\boldsymbol{\psi}(\mathbf{x}\sigma t), \boldsymbol{\psi}^\dagger(\mathbf{x}\sigma t)$ **(5.22)**
$\boldsymbol{\psi}_t[h], \boldsymbol{\psi}_t^\dagger[h]$ **(5.23)**, (5.24)
$\boldsymbol{\psi}(x)$ **(18.10)**, (18.11)
$\boldsymbol{\psi}_{N,P}(x), \boldsymbol{\psi}_{N^+,P^+}(x)$ **(18.8)**, **(18.9)**
$\Psi(t)$ (4.1)
$\Psi_B(t)$ (4.17)
$\Psi_{SC}(t)$ (4.20)
$\Psi\{g\}$ (5.44)
$\Psi^{(n)}(\mathbf{x}_1\sigma_1, \ldots, \mathbf{x}_n\sigma_n, t)$ (5.21)
$\Psi^{in}_{out\,\{n_k, m_k\}}, \Psi^{in}_{out\; n_1\ldots n_k\ldots, m_1\ldots m_k\ldots}$ (18.24)

Ω (5.20), **(11.19)**, (12.6)
$\boldsymbol{\Omega}_\pm$ **(1.28)**

SUBSCRIPTS AND SUPERSCRIPTS

O^{in} — *incoming* — refers to the field in the remote past

O^{out} — *outgoing* — refers to the field in the far future

O^{rad} — *radiation* — refers to the radiation field

${}^{IL}O$ — object O transformed under an *inhomogeneous Lorentz transformation*

${}^{I}O$ — object O transformed under *Lorentz transformation*

${}^{Tr}O$ — object O *translated*

^{P}O — *parity transformation* — object O transformed under space inversion

^{T}O — object O transformed under *time reversal*

^{C}O — object O transformed under *charge conjugation*

^{G}O — object O transformed under *gauge transformation*

O_{ren} — object O *renormalized*

T^{C} — *connected part* of the propagator

T^{L} — *linked part* of the propagator

T_{S} — *strongly connected part* of the propagator

T_{T} — *truncated* part of the propagator

BRACKETS

$\{\ ,\ \}$ — Poisson bracket

$[\ ,\ ]_{-}$ — commutator

$[\ ,\ ]_{+}$ — anticommutator

$(\ |\ )$ — scalar product

$:\quad:$ — normal product

$T(\ )$ — chronological product

SUBJECT INDEX

Acceleration as conformal transformations **135**
advanced Green's function 266, **493**
advanced potential **139**
advanced propagator **28**
d'Alembert equation **124**
  Feynman function for 202
  Green's functions for **139, 496**
amplitude
  probability **8**
  quasi **254**
  scattering **33**, 41, **62**
  transition **188**, 301
  vacuum-to-vacuum transition **189**, 258, 292, **329**
angular momentum **54**
  density tensor **92**
  of electromagnetic field **95**
  tensor **93**
annihilation operator
  in non-relativistic theory **43**
  negaton and positon **294**
  photon **164**
  (see also pair annihilation)
anomalous magnetic moment of the electron **464**
asymptotic conditions (for)
  Feynman wave function **260**
  field operators in non-relativistic theory **63**
  field operators in quantum electrodynamics **341, 372**
  Green's functions for Dirac equation **268**
anti-Feynman Green's function (for)
  Dirac equation **267**
  Klein–Gordon equation **266**, 495
anti-Feynman solution of Dirac equation **262**

Baker–Hausdorff identity **504**
Baker–Hausdorff continous **505**
bispinor **230, 514**
black-body radiation 185
Bogolyubov causality condition **19, 67**, 205, 303
Born approximation **35**
Born–Infeld theory **113**
bosons **39**
bound states **30**

canonical quantization **23**
  of the electromagnetic field **156**
canonical transformations 24, 101
causality condition **19, 67**, 205, 303
charge **94**
  conjugation **239**
  conservation of **121**
  observable **399**
  renormalization **399**
chronological product **65, 477**
chronological exponential operator **478**
classical electron radius **450**
coherence of electromagnetic radiation 178

full **179**
first-order **179**
M-th order **180**
coherent state **173**
combinatorial factor of the diagram **325**
commutation relations (for)
boson and fermion creation and annihilation operators **46**
canonical quantization **24**
electromagnetic field operators **156, 161**
electron field operators **297**
field operators **50, 52**
generators of Poincaré transformations **159**
photon creation and annihilation operators **164**
potential operators **200**
complete orthonormal set (of)
one-particle functions **42**
wave-packet profiles **163**
completeness condition **200,** 216
completeness relations **245**
compensating current **353**
composition law (for)
evolution operators **10**
exponential chronological operators **65**
Compton scattering 448
Compton wavelength of the electron **2, 453**
conformal transformations **134**
connected diagram **326**
connected part (of)
diagram **326**
propagator **330**
(see also strongly connected part)
conservation laws **94**
conservation of charge **121**
continuity equation (for)
angular momentum density tensor **92**
current 89, 298, 313
energy-momentum tensor **92,** 313
correlation functions **178, 179**
Coulomb gauge 197, 358
Coulomb interaction 370
energy 203
modification by vacuum polarization **458**
creation operator
for negaton and positon **294**
for photon **164**
in non-relativistic theory **45**
cross section 439
differential **444**
crossing symmetry **286**
current
compensating **353**
complete **407**
continuity equation **89,** 298, 313
external 186
four-vector **89**
operator in quantum electrodynamics 313
operator in relativistic quantum mechanics **298**
renormalized **352**
soft (hard) part of **227**
transformation property 90

Delbrück effect 452
density operator **6**
diagonal representation **183**
for black-body radiation **185**
for ideal single-mode laser **184**
diagonal representation of density operator **183**
diagram
electron self-energy **362**
Feynman in non-relativistic theory **82**
Feynman in momentum representation **377**
Feynman in quantum electrodynamics **325**

in non-relativistic quantum mechanics **79**
in relativistic quantum mechanics **251**
irreducible **419**
photon self-energy **335**
reducible **420**
skeleton **421**
strongly (weakly) connected **335**
truncated **337, 365**
two-sided **434**
dielectric permeability **90**
differential cross section **444**
dilatations as conformal transformations **135**
Dirac equation **230**
Feynman (anti-Feynman) Green's function for **267**
Green's functions for **265, 497**
solutions with negative (positive) frequency 241
symmetries of **236**
without potential **241**
Dirac Hamiltonian **273**
Dirac picture **14**
Dirac representation of γ matrices **231**
displacement operator **175**
dual rotations **117**

effective field
in classical theory 108, 114
in quantum electrodynamics **456**
effective Lagrangian **454**
electric displacement **89**
electric field intensity **88**
electromagnetic form factors **463**
electron
Compton wavelength 2, **453**
field operator **297**
line **320**
observable charge **399**
observable mass **385**
propagator **360**
n-electron propagator **355**
scattering in electromagnetic field 247
scattering in static electromagnetic field 273
self-energy diagram **362**
self-energy function **363**
wave function **230**, 260
electron-photon process **352**
electron-photon propagator **355**
energy density **95**
energy operator **11**
(see also Hamiltonian)
energy-momentum four-vector **93**
energy-momentum tensor **499**
for charged fluid **107**
for point particles **114**
in Proca theory **217**
of electromagnetic field **91**
operator **299**
evolution equation **11**, 65
evolution operator **10**
exponential chronological operator **64**
external sources **138**, 186

fermions **39**
Feynman diagram (in)
momentum representation **377**
non-relativistic theory **82**
quantum electrodynamics **325**
(see also diagram)
Feynman Green's function (for)
d'Alembert equation **202, 496**
Dirac equation **267**
Klein–Gordon equation **222, 494**
Proca equation **222**
Feynman–Schwinger method 387, **523**
Feynman propagator **254**
Feynman wave function **255**
Fock basis **47**
Fock parametrization of the Hilbert space **49**, 58
Fock space **164**

form factors **463**
fundamental amplitude **301**
fundamental dynamical postulate **314**
functional Ward identity **411**
function
  first (second) kind **419**
  self-energy for electron **363**
  self-energy for photon **336**
  (see also wave functions and Green's function)
fully coherent state **179**
Furry's theorem **375**

gauge
  conditions for the potential (propagator) **358**
  Landau **358**
  Lorentz **358**
  of the potential **119**
  radiation (Coulomb) **143**, 197, 358
gauge transformation (of)
  bispinor field **240**
  charged field **119**
  electron field **313**
  electron wave function **240**
  Green's function for Dirac equation **271**
  vectors ε^{μ} **143**
generalized state vector **59**
generalized Poisson bracket **98**
generalized Ward identity **413**
generating functionals **101**
generators (of)
  canonical transformations **101**
  conformal transformations **137**
  Poincaré transformations **104, 159**
Green's function **491**
  advanced (retarded) 493, 494
  d'Alembert equation **139, 496**
  Dirac equation **267, 497**
  Klein–Gordon equation **266, 493**
  Proca equation **222, 497**
  Schrödinger equations 28, **492**
  (see also Feynman Green's function)

Hamiltonian
  Dirac **273**
  effective 453
  generator for evolution operator **11**
  in classical theory of the electromagnetic field **97**
  in Schrödinger picture 12
  interaction (in Dirac picture) **15**
  single-particle non-relativistic **27**
helicity **170**
hard part of the current **227**
Heisenberg picture **11**

ideal single-mode laser state **184**
infrared catastrophe **224**, 437
inhomogeneous Dirac equation **267**
inhomogeneous Schrödinger equation **28**
irreducible diagram **415**
insert
  propagator **420**
  vertex **420**
inversion
  space-time **487**
  spatial **90**, 237, 271, 273, **487**
  time 91, 237, 272, 273, **487**

Josephson effect 466

Källén–Lehmann representation **350**
Klein–Gordon equation **216**
  Green's functions for **266, 493**
Klein–Nishina formula **450**

Lamb–Retherford shift **465**
linked part of the diagram (propagator) **326**

locality **299, 313**
long-wavelength part of the radiation **153**
Lorentz condition **196**, 216
Lorentz gauge **358**
Lorentz group **486**

magnetic field intensity **89**
magnetic induction **88**
magnetic moment of the electron **463**
  anomalous **464**
magnetic monopoles **118**
magnetic permeability **90**
Majorana representation of γ matrices **231**
Mandelstam variables **447**
many-electron propagator **307**
many-particle correlation **374**
many-photon propagator **213, 317**
mass
  observable **385**
  of the photon 224
  shell **284**
  renormalization **384**
Maxwell's equations **123**
Maxwell's theory **123**
minimal coupling method **119**, 241
mixed state **8**
møment of energy **95**
Møller scattering **445**
multipole expansion **144**

negative frequencies **127**
negatons **250**
normal order **480**
normal product **67, 480**

occupation number **43**
  operator **44**
  representation **42**
  wave function **43**

operator
  annihilation (creation) **45, 164, 294**
  components of energy-momentum tensor **299**
  density **6**
  electric current density **298**
  energy **11**
  evolution **10**
  generators of Poincaré group 159
  occupation number **43**
  (see also field operator, S operator)
orthochronous Poincaré (Lorentz) group **486**

pair creation (annihilation) **250**
Pauli exclusion principle 40, 43
Pauli–Villars regularization **519**
perturbation
  expansion of propagator in non-relativistic theory **78**
  series for scattering amplitude in relativistic mechanics **252**
  solution for non-relativistic scattering amplitude **34**
  theory in quantum electrodynamics 316
  renormalized **426**
photon 165
  annihilation (creation) operator **164**
  counting experiment **181**
  helicity **170**
  line **320**
  number operator 173
  propagator **332**
  process **339**
  self-energy diagram **335**
  self-energy function **336**
  soft **227**
photon-photon scattering **451**
physical field **312**
Poincaré group **486**
  as a symmetry group of amplitudes 205

as a symmetry group of classical electrodynamics **90**
as a symmetry group of quantum electrodynamics **157, 312**
generators **104, 159**
orthochronous **486**
proper **487**
Poisson bracket **23**, 218
generalized **98**
Poisson distribution 183, 191
polarization
photon **170**
vacuum 453, **458**
positive frequencies **127**
positons **250**
P representation **183**
potential
advanced (retarded) **139**
commutation relations for **200**
conditions for **358**
Coulomb (radiation) gauge 197
equations for **197**
four-vector **88**
for free field 198
gauge transformation of **119**
Lorentz condition **196**
operator **195**
principle of conservation of charge **121**
principle of universality of electromagnetic interactions 211
Proca equations **215**
Green's functions for **222, 497**
Proca theory **215**
propagator
electron **360**
electron-photon **355**
Feynman **264**
insert **420**
Källén–Lehmann representation of photon **350**
many-electron **307**
many-particle **71**
many-photon **213, 317**
n-electron **355**
photon **332**
Källén–Lehmann representation of **350**
proper **329**
renormalized **385, 394, 423**
retarded (advanced) **28**
pure state **8**

quasi-amplitude **254**

radiation field **139**
radiation gauge **143**, 197, 358
reducible diagram **420**
reduction formulae **69, 309, 371**
regularization 386, **519**
Pauli–Villars **519**
relativistic invariance **167, 312**
renormalization **402**
charge **399**
constant Z_2 **368**
constant Z_3 of the photon propagator **341**
mass **384**
of an external current **351**
theory 382
renormalized current **352**
renormalized electron-photon propagator **403**
renormalized functions **423**
renormalized perturbation theory **426**
renormalized photon propagator **394**
renormalized propagator 385
renormalized transition probabilities 430
renormalized vertex function **397**
retardation condition for propagators **76**
retarded Green's function 493
retarded potential **139**
retarded propagator **28**
Rutherford formula **399**

scattering
  amplitude **33**, 41, 62
  Compton **448**
  Møller **445**
  photon-photon **451**
  state **30**
Schrödinger coordinate representation **26**
Schrödinger equation 13, **39**
  inhomogeneous **28**
  Green's functions for 28, **492**
Schrödinger picture **12**
second-quantization 38
self-energy function (of)
  electron **363**
  photon **336**
skeleton diagram **421**
S matrix **35**, **63**, 380
soft photons **227**
soft part of the current **227**
S operator **15**
  in non-relativistic theory 36, 62
  in quantum electrodynamics **372**
  in quantum theory of electromagnetic field **201**
  in Proca theory **222**
  in relativistic quantum mechanics **303**
  in the subspace of photon states **346**
space (spatial) inversion **90**, **487**
space-time inversion **487**
spectral condition **312**
spinor 508
  first (second) kind **513**
spinor representation of γ matrices **232**
spontaneous emission **194**
stimulated emission **194**
strongly connected diagram (part of) **335**
structure constants **23**
sum rule **350**
symmetry 19
  crossing **286**
  group **21**
  of amplitudes **205**
  of Dirac equation **236**
  transformations **21**
  (see also Poincaré group as a symmetry group)

tensor wave function of the photon **166**
theorem on the connection between spin and statistics **39**
Thomson formula **450**
time inversion **487**
time reversal **91**
topologically non-equivalent diagrams **322**
transformation laws (for)
  bispinor field **237**
  electromagnetic field and current **90**
  electromagnetic potential **236**
  function $f(k, \lambda)$ **133**
  scattering amplitudes **282**
  wave-packet profiles **246**
translation group **486**
transtition amplitude
  induced in intense photon beam 213
  in quantum electrodynamics 368
  in quantum theory of electromagnetic field **188**
  in relativistic quantum mechanics **301**
  simple representation of **207**
  vacuum-to-vacuum **189**, 258, 292, **329**
truncated diagram (part of) **337**, **365**
two-sided diagram **434**

unitarity conditions **37**
universality of electromagnetic interactions 211, 316

vacuum part of the diagram **326**
vacuum state vector **47**, 294, 312

vacuum-to-vacuum transition amplitude **189**, 258, 292, **329**
vector wave function of the photon **199**
vertex
  function **364**
  insert **420**
  of diagram **323**

Ward's identity **398**
  functional **411**
  generalized **413**
wave function
  anti-Feynman 262
  electron **230**
  Feynman **255**, 260
  many-particle 38
  non-relativistic single-particle 26
  occupation number representation **43**
  photon tensor **166**
  photon vector **199**
wave-packet profile **32**, **244**
weakly connected diagram **335**
Wick's rotation **388**
Wigner factor **46**

Other Titles in the Series in Natural Philosophy

Vol. 1. DAVYDOV—Quantum Mechanics
Vol. 2. FOKKER—Time and Space, Weight and Inertia
Vol. 3. KAPLAN—Interstellar Gas Dynamics
Vol. 4. ABRIKOSOV, GOR'KOV and DZYALOSHINSKII—Quantum Field Theoretical Methods in Statistical Physics
Vol. 5. OKUN'—Weak Interaction of Elementary Particles
Vol. 6. SHKLOVSKII—Physics of the Solar Corona
Vol. 7. AKHIEZER *et al.*—Collective Oscillations in a Plasma
Vol. 8. KIRZHNITS—Field Theoretical Methods in Many-body Systems
Vol. 9. KLIMONTOVICH—The Statistical Theory of Non-equilibrium Processes in a Plasma
Vol. 10. KURTH—Introduction to Stellar Statistics
Vol. 11. CHALMERS—Atmospheric Electricity (2nd Edition)
Vol. 12. RENNER—Current Algebras and their Applications
Vol. 13. FAIN and KHANIN—Quantum Electronics, Volume 1—Basic Theory
Vol. 14. FAIN and KHANIN—Quantum Electronics, Volume 2—Maser Amplifiers and Oscillators
Vol. 15. MARCH—Liquid Metals
Vol. 16. HORI—Spectral Properties of Disordered Chains and Lattices
Vol. 17. SAINT JAMES, THOMAS and SARMA—Type II Superconductivity
Vol. 18. MARGENAU and KESTNER—Theory of Intermolecular Forces (2nd Edition)
Vol. 19. JANCEL—Foundations of Classical and Quantum Statistical Mechanics
Vol. 20. TAKAHASHI—An Introduction to Field Quantization
Vol. 21. YVON—Correlations and Entropy in Classical Statistical Mechanics
Vol. 22. PENROSE—Foundations of Statistical Mechanics
Vol. 23. VISCONTI—Quantum Field Theory, Volume 1
Vol. 24. FURTH—Fundamental Principles of Modern Theoretical Physics
Vol. 25. ZHELEZNYAKOV—Radioemission of the Sun and Planets
Vol. 26. GRINDLAY—An Introduction to the Phenomenological Theory of Ferroelectricity
Vol. 27. UNGER—Introduction to Quantum Electronics
Vol. 28. KOGA—Introduction to Kinetic Theory: Stochastic Processes in Gaseous Systems

Vol. 29. GALASIEWICZ—Superconductivity and Quantum Fluids
Vol. 30. CONSTANTINESCU and MAGYARI—Problems in Quantum Mechanics
Vol. 31. KOTKIN and SERBO—Collection of Problems in Classical Mechanics
Vol. 32. PANCHEV—Random Functions and Turbulence
Vol. 33. TALPE—Theory of Experiments in Paramagnetic Resonance
Vol. 34. TER HAAR—Elements of Hamiltonian Mechanics (2nd Edition)
Vol. 35. CLARKE and GRAINGER—Polarized Light and Optical Measurement
Vol. 36. HAUG—Theoretical Solid State Physics, Volume I
Vol. 37. JORDAN and BEER—The Expanding Earth
Vol. 38. TODOROV—Analytical Properties of Feynman Diagrams in Quantum Field Theory
Vol. 39. SITENKO—Lectures in Scattering Theory
Vol. 40. SOBEL'MAN—Introduction to the Theory of Atomic Spectra
Vol. 41. ARMSTRONG and NICHOLLS—Emission, Absorption and Transfer of Radiation in Heated Atmospheres
Vol. 42. BRUSH—Kinetic Theory, Volume 3
Vol. 43. BOGOLYUBOV—A Method for Studying Model Hamiltonians
Vol. 44. TSYTOVICH—An Introduction to the Theory of Plasma Turbulence
Vol. 45. PATHRIA—Statistical Mechanics
Vol. 46. HAUG—Theoretical Solid State Physics, Volume 2
Vol. 47. NIETO—The Titius-Bode Law of Planetary Distances: Its History and Theory
Vol. 48. WAGNER—Introduction to the Theory of Magnetism
Vol. 49. IRVINE—Nuclear Structure Theory
Vol. 50. STROHMEIER—Variable Stars
Vol. 51. BATTEN—Binary and Multiple Systems of Stars
Vol. 52. ROUSSEAU and MATHIEU—Problems in Optics
Vol. 53. BOWLER—Nuclear Physics
Vol. 54. POMRANING—The Equations of Radiation Hydrodynamics
Vol. 55. BELINFANTE—A Survey of Hidden-Variables Theories
Vol. 56. SCHEIBE—The Logical Analysis of Quantum Mechanics
Vol. 57. ROBINSON—Macroscopic Electromagnetism
Vol. 58. GOMBÁS and KISDI—Wave Mechanics and Its Applications
Vol. 59. KAPLAN and TSYTOVICH—Plasma Astrophysics
Vol. 60. KOVÁCS and ZSOLDOS—Dislocations and Plastic Deformation
Vol. 61. AUVRAY and FOURRIER—Problems in Electronics
Vol. 62. MATHIEU—Optics, Parts 1 and 2
Vol. 63. ATWATER—Introduction to General Relativity
Vol. 64. MULLER—Quantum Mechanics: A Physical World Picture
Vol. 65. BILENKY—Introduction to Feyman
Vol. 66. VODAR and ROMAND—Some Aspects of Vacuum Ultraviolet Radiation Physics
Vol. 67. WILLET—Gas Lasers: Population Inversion Mechanisms
Vol. 68. AKHIEZER—Plasma Electrodynamics, Volume 1
Vol. 69. GLASBY—The Nebular Variables